$f(t)$	$\mathscr{L}\{f(t)\} = F(s)$
35. $\sin kt \cosh kt$	$\dfrac{k(s^2 + 2k^2)}{s^4 + 4k^4}$
36. $\cos kt \sinh kt$	$\dfrac{k(s^2 - 2k^2)}{s^4 + 4k^4}$
37. $\cos kt \cosh kt$	$\dfrac{s^3}{s^4 + 4k^4}$
38. $J_0(kt)$	$\dfrac{1}{\sqrt{s^2 + k^2}}$
39. $\dfrac{e^{bt} - e^{at}}{t}$	$\ln \dfrac{s - a}{s - b}$
40. $\dfrac{2(1 - \cos kt)}{t}$	$\ln \dfrac{s^2 + k^2}{s^2}$
41. $\dfrac{2(1 - \cosh kt)}{t}$	$\ln \dfrac{s^2 - k^2}{s^2}$
42. $\dfrac{\sin at}{t}$	$\arctan\left(\dfrac{a}{s}\right)$
43. $\dfrac{\sin at \cos bt}{t}$	$\dfrac{1}{2}\arctan \dfrac{a + b}{s} + \dfrac{1}{2}\arctan \dfrac{a - b}{s}$
44. $\delta(t)$	1
45. $\delta(t - t_0)$	e^{-st_0}
46. $e^{at}f(t)$	$F(s - a)$
47. $f(t - a)\mathscr{U}(t - a)$	$e^{-as}F(s)$
48. $\mathscr{U}(t - a)$	$\dfrac{e^{-as}}{s}$
49. $f^{(n)}(t)$	$s^nF(s) - s^{(n-1)}f(0) - \cdots - f^{(n-1)}(0)$
50. $t^nf(t)$	$(-1)^n \dfrac{d^n}{ds^n} F(s)$
51. $\displaystyle\int_0^t f(\tau)g(t - \tau)\, d\tau$	$F(s)G(s)$

DIFFERENTIAL EQUATIONS WITH COMPUTER LAB EXPERIMENTS

Dennis G. Zill
Loyola Marymount University

with the assistance of
Warren S. Wright
Loyola Marymount University

PWS Publishing Company • Boston

I(T)P **An International Thomson Publishing Company**

New York • London • Bonn • Boston • Detroit • Madrid • Melbourne
Mexico City • Paris • Singapore • Tokyo • Toronto • Washington
Albany NY • Belmont CA • Cincinnati OH

PWS PUBLISHING COMPANY
20 Park Plaza, Boston, MA 02116-4324

I(T)P™

International Thomson Publishing
The trademark ITP is used under license.

This book is printed on recycled, acid-free paper.

Library of Congress Cataloging-in-Publication Data

Zill, Dennis G.
Differential equations with computer lab experiments / by Dennis G. Zill
p. cm.
Includes index.
ISBN 0-534-93785-3
1. Differential equations—Data processing. I. Title.
QA371.5.D37Z55 1994
515′.35—dc20 94-21602
CIP

Sponsoring Editor: Steve Quigley
Editorial Assistant: John Ward
Developmental Editor: Barbara Lovenvirth
Production Editor: Patricia Adams
Interior Designer: Monique Calello
Interior Illustrator: Network Graphics
Cover Designer: Patricia Adams
Cover Photo: © Joel Gordon 1983
Marketing Manager: Marianne Rutter
Manufacturing Coordinator: Lisa Flanagan
Cover Printer: Henry N. Sawyer Company
Text Printer and Binder: R.R. Donnelley & Sons/Crawfordsville

Printed and bound in the United States of America.
94 95 96 97 98 99 — 10 9 8 7 6 5 4 3 2 1

CONTENTS

PREFACE

Differential Equations with Computer Lab Experiments was written to provide an alternative to more traditional differential equations texts. The text is intended for those instructors who wish to integrate the computer in the teaching of the theory and applications of the subject. It reflects, in part, how my own course in differential equations is evolving along with our departmental computer laboratory.

It is fair to state that the emphasis of the other two texts in this series, *A First Course in Differential Equations, Fifth Edition,* and *Differential Equations with Boundary-Value Problems, Third Edition,* is on applications and methods of solutions of ordinary differential equations. Although there are an abundance of applications throughout this text, the primary goals are to encourage the use of the growing number of computer aids in the study of differential equations, to introduce qualitative and numerical aspects of differential equations early in the course, and to motivate students toward independent thinking and outside reading.

CONTENT

The standard first section, on terminology, is followed by a discussion of direction fields, ODE solvers, and phase portraits of first- and second-order differential equations. The concepts of phase portraits, phase line and phase plane analysis, and stability are presented in a spiral fashion, beginning at a fairly intuitive level, first in Section 1.2, and then repeated in Chapters 2, 3, and 4. The traditional manner of presenting these concepts, namely, gathered together in a chapter on systems on differential equations, tends to be an overwhelming accumulation of concepts, theory, and terminology. Since we will discuss these subjects at an introductory level, commensurate with the abilities of a typical post-calculus sophomore, we have purposely chosen not to include certain aspects of stability theory. This material may be embellished by individual instructors.

Students are introduced to systems of differential equations early and often throughout the text. Reduction of a second-order differential equation to a system of first-order equations is first illustrated in Section 1.2. Numerical methods are first presented in Chapter 2 and the use of ODE Solvers are stressed throughout Chapters 2, 3, and 4.

Chapter 4, ''Systems of First-Order Differential Equations,'' assumes a familiarity with some aspects of matrix algebra: addition and multiplication of matrices, inverse of a square matrix, and solving systems of algebraic equations by elementary row operations on an augmented matrix. As most students will not have been introduced to these concepts, an introduction to some core material is given in Appendix I. If time permits, this material can be inserted into the flow of Chapter 4 and covered in the traditional lecture manner.

The text contains Project Problems and Computer Lab Experiments. To provide time for students to work on these new types of problems, some of the more traditional material has been condensed. Project problems are included in most sections, and consist of two types: discovery and writing. They are intended to either expand on material found in the section, or introduce new concepts not covered.

The project problems can be used for classroom discussion, or as assignments to single individuals or groups either for purposes of grading or class presentation. The problems are often open-ended and some are challenging, requiring the student to do additional reading and writing. *All the project problems, however, are intended to be directed by the instructor.*

As stated, a major emphasis of this text is to involve the use computers in the teaching of differential equations. *Computer Lab Experiments,* specifically designed for this purpose are provided in a separate manual packaged with this text. These experiments are designed to complement the various topics and are keyed to appropriate sections of the text.

The lab experiments encompass a wide range of problems emphasizing such diverse areas as engineering, physics, economics, science, and many more. Each requires the use of a computer software package: an ODE solver, a graphing program, or a computer algebra system.

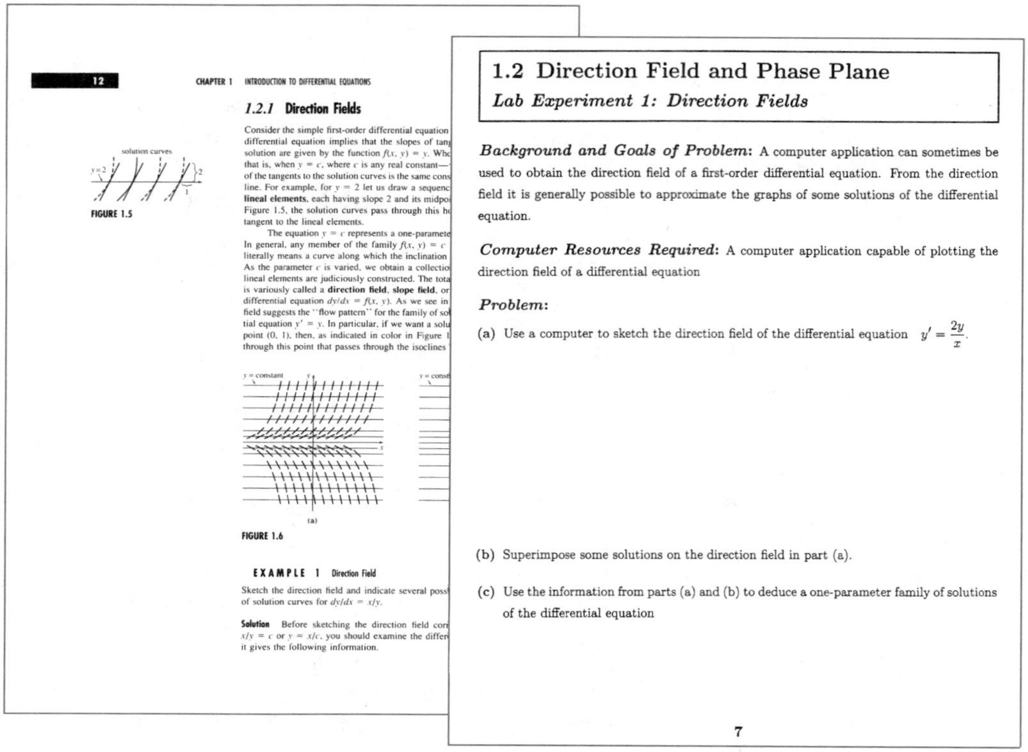

12 CHAPTER 1 INTRODUCTION TO DIFFERENTIAL EQUATIONS

1.2.1 Direction Fields

Consider the simple first-order differential equation
differential equation implies that the slopes of tan
solution are given by the function $f(x, y) = y$. Wh
that is, when $y = c$, where c is any real constant—
of the tangents to the solution curves is the same cons
line. For example, for $y = 2$ let us draw a sequenc
lineal elements, each having slope 2 and its midpo
Figure 1.5, the solution curves pass through this h
tangent to the lineal elements.

The equation $y = c$ represents a one-paramete
In general, any member of the family $f(x, y) = c$
literally means a curve along which the inclination
As the parameter c is varied, we obtain a collectio
lineal elements are judiciously constructed. The tota
is variously called a **direction field**, **slope field**, or
differential equation $dy/dx = f(x, y)$. As we see in
field suggests the "flow pattern" for the family of so
tial equation $y' = y$. In particular, if we want a solu
point (0, 1), then, as indicated in color in Figure 1
through this point that passes through the isoclines

FIGURE 1.5

FIGURE 1.6

EXAMPLE 1 Direction Field

Sketch the direction field and indicate several poss
of solution curves for $dy/dx = x/y$.

Solution Before sketching the direction field cor
$x/y = c$ or $y = x/c$, you should examine the differ
it gives the following information.

1.2 Direction Field and Phase Plane

Lab Experiment 1: Direction Fields

Background and Goals of Problem: A computer application can sometimes be used to obtain the direction field of a first-order differential equation. From the direction field it is generally possible to approximate the graphs of some solutions of the differential equation.

Computer Resources Required: A computer application capable of plotting the direction field of a differential equation

Problem:

(a) Use a computer to sketch the direction field of the differential equation $y' = \dfrac{2y}{x}$.

(b) Superimpose some solutions on the direction field in part (a).

(c) Use the information from parts (a) and (b) to deduce a one-parameter family of solutions of the differential equation

7

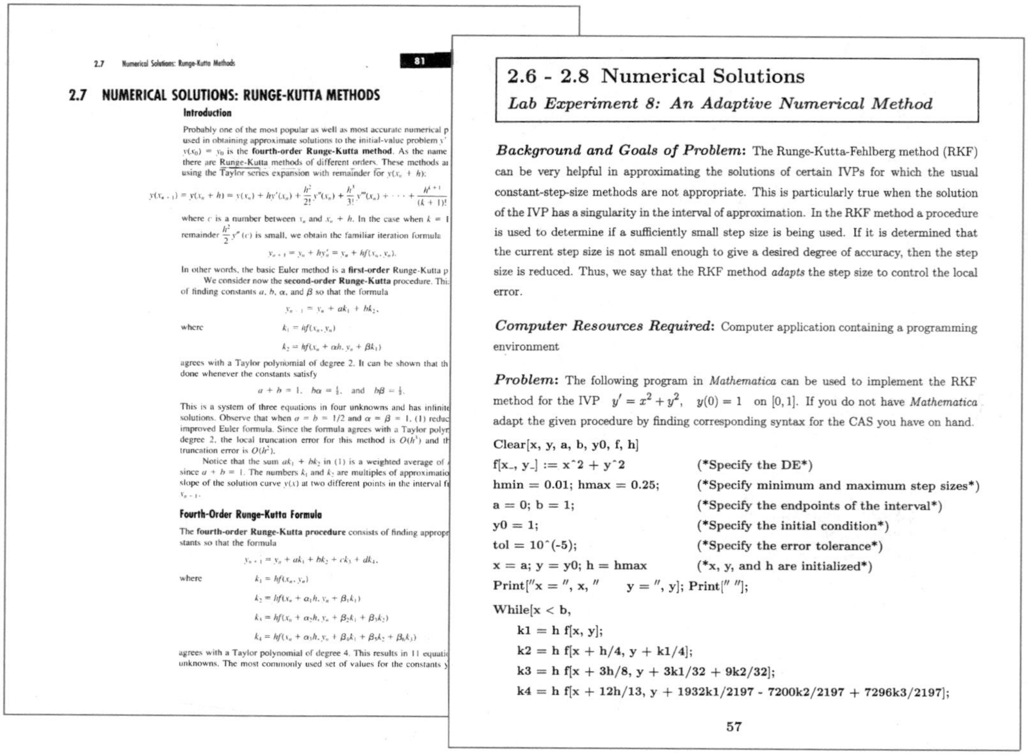

2.7 Numerical Solutions: Runge-Kutta Methods 81

2.7 NUMERICAL SOLUTIONS: RUNGE-KUTTA METHODS

Introduction

Probably one of the most popular as well as most accurate numerical p
used in obtaining approximate solutions to the initial-value problem y'
$y(x_0) = y_0$ is the **fourth-order Runge-Kutta method**. As the name
there are Runge-Kutta methods of different orders. These methods a
using the Taylor series expansion with remainder for $y(x_n + h)$:

$$y(x_{n+1}) = y(x_n + h) = y(x_n) + hy'(x_n) + \frac{h^2}{2!}y''(x_n) + \frac{h^3}{3!}y'''(x_n) + \cdots + \frac{h^{k+1}}{(k+1)!}$$

where c is a number between x_n and $x_n + h$. In the case when $k = 1$
remainder $\frac{h^2}{2}y''(c)$ is small, we obtain the familiar iteration formula

$$y_{n+1} = y_n + hy'_n = y_n + hf(x_n, y_n).$$

In other words, the basic Euler method is a **first-order** Runge-Kutta p
We consider now the **second-order Runge-Kutta** procedure. Thi
of finding constants a, b, α, and β so that the formula

$$y_{n+1} = y_n + ak_1 + bk_2,$$

where

$$k_1 = hf(x_n, y_n)$$
$$k_2 = hf(x_n + \alpha h, y_n + \beta k_1)$$

agrees with a Taylor polynomial of degree 2. It can be shown that th
done whenever the constants satisfy

$$a + b = 1, \quad b\alpha = \tfrac{1}{2}, \quad \text{and} \quad b\beta = \tfrac{1}{2}.$$

This is a system of three equations in four unknowns and has infinite
solutions. Observe that when $a = b = 1/2$ and $\alpha = \beta = 1$, (1) reduc
improved Euler formula. Since the formula agrees with a Taylor poly
degree 2, the local truncation error for this method is $O(h^3)$ and th
truncation error is $O(h^2)$.

Notice that the sum $ak_1 + bk_2$ in (1) is a weighted average of
since $a + b = 1$. The numbers k_1 and k_2 are multiples of approximatio
slope of the solution curve $y(x)$ at two different points in the interval fr
x_{n+1}.

Fourth-Order Runge-Kutta Formula

The **fourth-order Runge-Kutta procedure** consists of finding appropr
stants so that the formula

$$y_{n+1} = y_n + ak_1 + bk_2 + ck_3 + dk_4,$$

where

$$k_1 = hf(x_n, y_n)$$
$$k_2 = hf(x_n + \alpha_1 h, y_n + \beta_1 k_1)$$
$$k_3 = hf(x_n + \alpha_2 h, y_n + \beta_2 k_1 + \beta_3 k_2)$$
$$k_4 = hf(x_n + \alpha_3 h, y_n + \beta_4 k_1 + \beta_5 k_2 + \beta_6 k_3)$$

agrees with a Taylor polynomial of degree 4. This results in 11 equati
unknowns. The most commonly used set of values for the constants y

2.6 - 2.8 Numerical Solutions

Lab Experiment 8: An Adaptive Numerical Method

Background and Goals of Problem: The Runge-Kutta-Fehlberg method (RKF) can be very helpful in approximating the solutions of certain IVPs for which the usual constant-step-size methods are not appropriate. This is particularly true when the solution of the IVP has a singularity in the interval of approximation. In the RKF method a procedure is used to determine if a sufficiently small step size is being used. If it is determined that the current step size is not small enough to give a desired degree of accuracy, then the step size is reduced. Thus, we say that the RKF method *adapts* the step size to control the local error.

Computer Resources Required: Computer application containing a programming environment

Problem: The following program in *Mathematica* can be used to implement the RKF method for the IVP $y' = x^2 + y^2$, $y(0) = 1$ on $[0, 1]$. If you do not have *Mathematica* adapt the given procedure by finding corresponding syntax for the CAS you have on hand.

```
Clear[x, y, a, b, y0, f, h]
f[x_, y_] := x^2 + y^2                  (*Specify the DE*)
hmin = 0.01; hmax = 0.25;               (*Specify minimum and maximum step sizes*)
a = 0; b = 1;                           (*Specify the endpoints of the interval*)
y0 = 1;                                 (*Specify the initial condition*)
tol = 10^(-5);                          (*Specify the error tolerance*)
x = a; y = y0; h = hmax                 (*x, y, and h are initialized*)
Print["x = ", x, "      y = ", y]; Print[" "];
While[x < b,
    k1 = h f[x, y];
    k2 = h f[x + h/4, y + k1/4];
    k3 = h f[x + 3h/8, y + 3k1/32 + 9k2/32];
    k4 = h f[x + 12h/13, y + 1932k1/2197 - 7200k2/2197 + 7296k3/2197];
```

57

3.7 Applications: Initial-Value Probl

Lab Experiment 2: The Aging Spring

Background and Goals of Problem: In the first set of remark
indicated that the differential equation

$$mx'' + ke^{-\alpha t}x = 0, \quad \alpha > 0, \; k > 0$$

serves as one model for the motion of a mass on an aging spring. In this e
asked to investigate the motion of a mass predicted by this model.

Note to the Instructor: You might have the students write up their res
below and collect them before proceeding with the rest of the experimen

Computer Resources Required: ODE solver

Problem:

(a) Using only the differential equation, describe the motion of the mass
time. Then predict the motion of the mass as $t \to \infty$.

89

3.9 - 3.10 Systems of Differential Equations; Numerical Solutions

Lab Experiment 6: Coupled Moons of Saturn

Background and Goals of Problem: In November of 1980, the Voyager I probe confirmed the existence of two moons occupying nearly the same orbit around the planet Saturn. The mean orbital radii of the moons Janus and Epimetheus are approximately 50 kilometers apart. The natural question is: Why don't the moons collide? The answer is that they don't collide because they "switch" orbits. To understand the process remember that the smaller the orbit the faster a moon travels. So when Epimetheus is in the inner orbit, it is travelling faster than the moon Janus and therefore Epimetheus will "catch up". But as the moons get closer together the gravitational attraction between them increases. This has the effect of pulling Epimetheus toward Janus and into the outer orbit. At the same time Janus is pulled back towards Epimetheus and into the inner orbit. Now with Janus in the inner orbit it is travelling faster than Epimetheus and pulls away.

Computer Resources Required: CAS and a computer implementation of the fourth-order Runge-Kutta method for a system of two equations

Problem: If the position of the first moon is (x_1, y_1) and the position of the second moon is (x_2, y_2), then the equations of motion of the moons are given by

$$\begin{aligned} x_1'' &= k(x_2 - x_1)/H^3 - x_1/r_1^3 \\ x_2'' &= k(x_1 - x_2)/H^3 - x_2/r_2^3 \end{aligned} \quad \text{and} \quad \begin{aligned} y_1'' &= k(y_2 - y_1)/H^3 - y_1/r_1^3 \\ y_2'' &= k(y_1 - y_2)/H^3 - y_2/r_2^3 \end{aligned}$$

where $H(t) = \sqrt{(x_1 - x_2)^2 - (y_1 - y_2)^2}$ is the distance between the two moons and $r_1 = \sqrt{x_1^2 + y_1^2}$ and $r_2 = \sqrt{x_2^2 + y_2^2}$ are the distances between each moon and the center of Saturn, whose coordinates are assumed to be $(0, 0)$. A dimensionless constant, $k = 2 \times 10^{-8}$.

113

ANCILLARY MATERIAL

- *Complete Solutions Manual* for instructors provides complete, worked-out solutions to all problems, project problems (excepting the writing projects) and computer lab assignments.
- *Student Solutions Manual* (Warren S. Wright) provides complete solutions to every fourth problem in each exercise set, with the exception of the project problems.
- *ODE Solver: Numerical Procedures for Ordinary Differential Equations* (Thomas Kiffe/William Rundell) is a guide to most of the standard numerical routines used to solve initial- and boundary-value problems for ordinary differential equations. Using this software, the user is able to analyze and compare the output of multiple numerical calculations quickly and conveniently without any need for programming. For both IBM and Compatibles and Macintosh.
- *Grapher Version 2.2* (Steve Scarborough) is a flexible graphing software program for the Macintosh. This versatile collection of graphing utilities can plot rectangular curves, polar curves, and interpolating polynomials, as well as graph parametric equations, systems of two first-order differential equations, series, direction fields, and more.

- *The PWS Notebook Series* are electronic problem sets keyed to the text including practical problem-solving suggestions. Notebooks are compatible for use with both *Maple* (Mac and Windows) and *Mathematica* (Mac and Windows).
- *Computer Programs for BASIC, FORTRAN and Pascal* (C.J. Knickerbocker) contain a listing of computer programs of many of the numerical methods considered in the text. This package is available in either IBM or Macintosh.

ACKNOWLEDGMENTS

The idea for this text was inspired by an NSF workshop on teaching differential equations with computer experiments conducted at Harvey Mudd College in the summer of 1992. I am indebted to Robert Borelli and Courtney Coleman.

I would like to express my appreciation to the following individuals for their influence in shaping this text:

Philip Crooke
Vanderbilt University

John H. Ellison
Grove City College

Juan A. Gatica
University of Iowa

Terry L. Herdman
Virginia Polytechnic Institute and State University

J. Myron Hood
California Polytechnic State University

Roger Marty
Cleveland State University

Barbara J. Shabell
California State Polytechnic University

Daniel C. Sloughter
Furman University

I'd also like to thank Michael Cullen, my colleague at Loyola Marymount University, for reading the manuscript and making suggestions for some of the project problems, C.J. Knickerbocker, St. Lawrence University, for his many good suggestions for computer lab assignments, and the staff at PWS Publishing Co., namely, Patty Adams, production editor, for another job well done, Elise Kaiser, Monique Calello, and last, but never least, Barbara Lovenvirth, developmental editor, for adroitly putting all the pieces of the puzzle together.

Dennis G. Zill
Los Angeles

SOFTWARE BIBLIOGRAPHY

The following is a noncomprehensive list of the some software that is useful in working in calculus and differential equations. Of course, not all programs do the same things. In addition, the programs range from the very user friendly *Grapher, ODE Solver,* and *MacMath* to the very powerful and very sophisticated computer algebra systems (CAS) such as *Mathematica, Maple, and Macsyma.*

Axiom, NAG, Inc.

DE Graph, copyright Henry C. Pinkham, Mathematics Department, Columbia University

DERIVE, Soft Warehouse, Inc.

Differential Systems, copyright Drexel University

Grapher, PWS Publishing Co.

MacMath, Springer-Verlag, Inc.

Macsyma, Symbolics, Inc.

Maple, Brooks-Cole Publishing Co.

MathCad, Math Soft, Inc.

Mathematica, Wolfram Research, Inc.

MATLAB, The Math Works, Inc.

MATHLIB, Innosoft International, Inc.

ODE Solver, PWS Publishing Co.

PhasePlane, Brooks-Cole Publishing Co.

Phaser, comes with the text *Differential and Difference Equations Through Computer Experiments,* H. Koçak, Springer-Verlag, 1990

Theorist, (Student Edition) PWS Publishing Co.

DEQSolve, Innosoft International, Inc.

ODE Toolkit, copyright Harvey Mudd College

MDEP, US Naval Academy

CHAPTER 1

INTRODUCTION TO DIFFERENTIAL EQUATIONS

1.1 DEFINITIONS AND TERMINOLOGY

Introduction

The words *differential* and *equations* certainly suggest solving some kind of equation that contains derivatives. In fact, the preceding sentence tells the complete story about the course that you are about to begin. But before you start solving anything, you must learn some of the basic definitions and terminology of the subject.

A Differential Equation

In calculus you learned that the derivative dy/dx of a function $y = f(x)$ is itself another function of x found by some appropriate rule. For example, if $y = e^{x^2}$ then $dy/dx = 2xe^{x^2}$. Replacing e^{x^2} by the symbol y then gives

$$\frac{dy}{dx} = 2xy. \tag{1}$$

The problem we face in this course is not: Given a function $y = f(x)$, find its derivative. Rather, our problem is: If we are given a differential equation such as (1), is there some way or method by which we can find the unknown function $y = f(x)$?

DEFINITION 1.1 Differential Equation

An equation containing the derivatives of one or more dependent variables, with respect to one or more independent variables, is said to be a **differential equation** (DE).

Differential equations are classified by **type**, **order**, and **linearity**.

Classification by Type

If an equation contains only ordinary derivatives of one or more dependent variables with respect to a single independent variable, it is said to be an **ordinary differential equation** (ODE). For example,

$$\frac{dy}{dx} + 10y = e^x \quad \text{and} \quad \frac{d^2y}{dx^2} - \frac{dy}{dx} + 6y = 0$$

are ordinary differential equations. An equation involving the partial derivatives of one or more dependent variables of two or more independent variables is called a **partial differential equation** (PDE). For example,

$$\frac{\partial u}{\partial y} = -\frac{\partial v}{\partial x} \quad \text{and} \quad \frac{\partial^2 u}{\partial x^2} = \frac{\partial^2 u}{\partial t^2} - 2\frac{\partial u}{\partial t}$$

are partial differential equations. We shall consider only the theory and applications of ordinary differential equations in this text.

Classification by Order

The **order of a differential equation** (ODE or PDE) is the order of the highest derivative in the equation. For example,

$$\underset{\substack{\text{second-order}\\ \downarrow}}{\frac{d^2y}{dx^2}} + 5\left(\underset{\substack{\text{first-order}\\ \downarrow}}{\frac{dy}{dx}}\right)^3 - 4y = e^x$$

is a second-order differential equation. Since the differential equation $(y - x)\,dx + 4x\,dy = 0$ can be put into the form

$$4x\frac{dy}{dx} + y = x$$

by dividing by the differential dx, it is an example of a first-order ordinary differential equation.

A general nth-order, ordinary differential equation is often represented by the symbolism

$$F(x, y, y', \ldots, y^{(n)}) = 0, \tag{2}$$

where $y^{(n)} = d^ny/dx^n$. The following is a special case of (2).

Classification as Linear or Nonlinear

A differential equation is said to be **linear** if it can be written in the form

$$a_n(x)\frac{d^ny}{dx^n} + a_{n-1}(x)\frac{d^{n-1}y}{dx^{n-1}} + \cdots + a_1(x)\frac{dy}{dx} + a_0(x)y = g(x).$$

It should be observed that linear differential equations are characterized by two properties:

(i) The dependent variable y and all its derivatives are of the first degree; that is, the power of each term involving y is 1.

(ii) Each coefficient depends on only the independent variable x.

Functions of y or its derivatives such as $e^{y'}$ or $\sin y$ cannot appear in a linear equation. A differential equation that is not linear is said to be **nonlinear**. The equations

$$xdy + ydx = 0, \quad y'' - 2y' + y = 0, \quad x^3\frac{d^3y}{dx^3} + 3x\frac{d^2y}{dx^2} - 5y = e^x$$

are linear first-, second-, and third-order ordinary differential equations, respectively. On the other hand,

$$\underset{\substack{\text{coefficient}\\ \text{depends on } y}}{(1+y)y' + 2y = e^x}, \qquad \underset{\substack{\text{nonlinear}\\ \text{function of } y}}{\frac{d^2y}{dx^2} + \sin y = 0}, \qquad \underset{\text{power not 1}}{\frac{d^4y}{dx^4} + y^2 = 0}$$

are nonlinear first-, second-, and fourth-order ordinary differential equations, respectively.

Solutions

As mentioned before, our goal in this course is to solve, or find **solutions** of, differential equations.

DEFINITION 1.2 Solution of a Differential Equation

Any function f defined on some interval I, which when substituted into a differential equation reduces the equation to an identity, is said to be a **solution** of the equation on the interval.

In other words, a solution of an ordinary differential equation (2) is a function f that possesses at least n derivatives and

$$F(x, f(x), f'(x), \ldots, f^{(n)}(x)) = 0 \quad \text{for all } x \text{ in } I.$$

We say that $y = f(x)$ *satisfies* the differential equation. The interval I could be an open interval (a, b), a closed interval $[a, b]$, an infinite interval (a, ∞), and so on.

EXAMPLE 1 Verification of a Solution

Verify that $y = x^4/16$ is a solution of the nonlinear equation

$$\frac{dy}{dx} = xy^{1/2}$$

on the interval $(-\infty, \infty)$.

Solution One way of verifying that the given function is a solution is to write the differential equation as $dy/dx - xy^{1/2} = 0$ and then see, after substituting, whether the sum $dy/dx - xy^{1/2}$ is zero for every x in the interval. Using

$$\frac{dy}{dx} = 4\frac{x^3}{16} = \frac{x^3}{4} \quad \text{and} \quad y^{1/2} = \left(\frac{x^4}{16}\right)^{1/2} = \frac{x^2}{4},$$

we see that

$$\frac{dy}{dx} - xy^{1/2} = \frac{x^3}{4} - x\left(\frac{x^4}{16}\right)^{1/2} = \frac{x^3}{4} - \frac{x^3}{4} = 0$$

for every real number. ●

EXAMPLE 2 Verification of a Solution

The function $y = xe^x$ is a solution of the linear equation

$$y'' - 2y' + y = 0$$

on $(-\infty, \infty)$. To see this, we compute

$$y' = xe^x + e^x \quad \text{and} \quad y'' = xe^x + 2e^x.$$

Observe $\quad y'' - 2y' + y = (xe^x + 2e^x) - 2(xe^x + e^x) + xe^x = 0$

for every real number. ●

Not every differential equation that we write necessarily has a solution. You are encouraged to think about this and to read, perhaps even attempt to solve, Problem 44 in Exercises 1.1.

Explicit and Implicit Solutions

You should be familiar with the notions of explicit and implicit functions from your study of calculus. Similarly, solutions of differential equations can be further distinguished as either explicit solutions or implicit solutions. A solution of an ordinary differential equation (2) that can be written in the form $y = f(x)$ is said to be an **explicit solution**. We have already seen in our initial discussion that $y = e^{x^2}$ is an explicit solution of $dy/dx = 2xy$. In Examples 1 and 2, $y = x^4/16$ and $y = xe^x$ are explicit solutions of $dy/dx = xy^{1/2}$ and $y'' - 2y' + y = 0$, respectively. Note, too, that in Examples 1 and 2 each differential equation possesses the constant solution $y = 0$, $-\infty < x < \infty$. An explicit solution of a differential equation that is identically zero on an interval I is referred to as a **trivial solution**. A relation $G(x, y) = 0$ is said to be an **implicit solution** of an ordinary differential equation (2) on an interval I, provided it defines one or more explicit solutions on I.

EXAMPLE 3 Verification of an Implicit Solution

The relation $x^2 + y^2 - 4 = 0$ is an implicit solution of the differential equation

$$\frac{dy}{dx} = -\frac{x}{y} \tag{3}$$

on the interval $-2 < x < 2$. By implicit differentiation we obtain

$$\frac{d}{dx}x^2 + \frac{d}{dx}y^2 - \frac{d}{dx}4 = 0 \quad \text{or} \quad 2x + 2y\frac{dy}{dx} = 0.$$

Solving the last equation for the symbol dy/dx gives (3). ●

The relation $x^2 + y^2 - 4 = 0$ in Example 3 defines two explicit differentiable functions on the interval $(-2, 2)$: $y = \sqrt{4 - x^2}$ and $y = -\sqrt{4 - x^2}$. Also any relation of the form $x^2 + y^2 - c = 0$ *formally* satisfies (3) for any constant c. However, it is understood that the relation should always make sense in the real number system; thus, for example, we cannot say that $x^2 + y^2 + 1 = 0$ determines a solution of the differential equation.

Because the distinction between an explicit and implicit solution should be intuitively clear, we will not belabor the issue by always saying "here is an explicit (or implicit) solution."

More Terminology

The study of differential equations is similar to that of integral calculus. A solution is sometimes referred to as an **integral** of the equation, and its graph is called an **integral curve** or a **solution curve**. When evaluating an antiderivative or indefinite integral in calculus, we use a single constant c of integration. Analogously, when solving a first-order differential equation $F(x, y, y') = 0$ we usually obtain a solution containing a single arbitrary constant or parameter c. A solution containing an arbitrary constant represents a set $G(x, y, c) = 0$ of solutions, called a **one-parameter family of solutions**. When solving an nth-order differential equation $F(x, y, y', \ldots, y^{(n)}) = 0$ we seek an ***n*-parameter family of solutions** $G(x, y, c_1, c_2, \ldots, c_n) = 0$. This simply means that a single differential equation can possess an infinite number of solutions corresponding to the unlimited choices for the parameter(s). A solution of a differential equation that is free of arbitrary parameters is called a **particular solution**. For example, by direct substitution we can show that any function in the one-parameter family $y = ce^{x^2}$, where c is an arbitrary constant, also satisfies equation (1). The original solution $y = e^{x^2}$ corresponds to $c = 1$ and thus is a particular solution of the equation. Figure 1.1 shows some of the integral curves in this family. The trivial solution $y = 0$, corresponding to $c = 0$, is also a particular solution of (1).

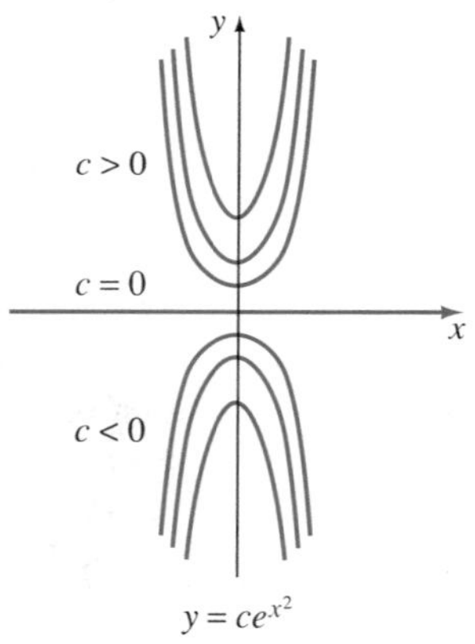

FIGURE 1.1

EXAMPLE 4 Particular Solutions

The function $y = c_1e^x + c_2e^{-x}$ is a two-parameter family of solutions of the linear second-order equation $y'' - y = 0$. Some particular solutions are $y = 0$ $(c_1 = c_2 = 0)$, $y = e^x$ $(c_1 = 1, c_2 = 0)$, and $y = 5e^x - 2e^{-x}$ $(c_1 = 5, c_2 = -2)$. ●

In all the preceding examples we have used x and y to denote the independent and dependent variables, respectively. But in practice these two variables are represented by many different symbols. For example, we could denote the independent variable by t and the dependent variable by x.

EXAMPLE 5 Using Different Symbols

The functions $x = c_1 \cos 4t$ and $x = c_2 \sin 4t$, where c_1 and c_2 are arbitrary constants, are solutions of the differential equation

$$x'' + 16x = 0.$$

For $x = c_1 \cos 4t$, the first two derivatives with respect to t are $x' = -4c_1 \sin 4t$ and $x'' = -16c_1 \cos 4t$. Substituting x'' and x then gives

$$x'' + 16x = -16c_1 \cos 4t + 16(c_1 \cos 4t) = 0.$$

In like manner, for $x = c_2 \sin 4t$ we have $x'' = -16c_2 \sin 4t$ and so

$$x'' + 16x = -16c_2 \sin 4t + 16(c_2 \sin 4t) = 0.$$

Finally, it is easy to verify that the linear combination of solutions, or two-parameter family, $x = c_1 \cos 4t + c_2 \sin 4t$ is a solution of the given equation. ●

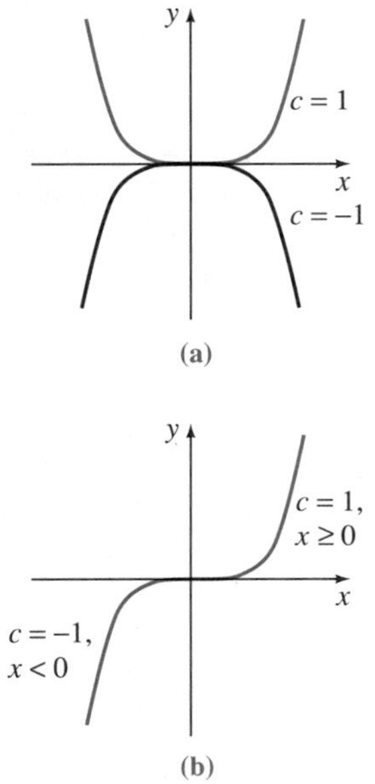

FIGURE 1.2

The next example shows that a solution of a differential equation can be a piecewise defined function.

EXAMPLE 6 A Piecewise Defined Solution

You should verify that any function in the one-parameter family $y = cx^4$ is a solution of the differential equation $xy' - 4y = 0$ on the interval $(-\infty, \infty)$. See Figure 1.2(a). The piecewise defined differentiable function

$$y = \begin{cases} -x^4 & x < 0 \\ x^4 & x \geq 0 \end{cases}$$

is a particular solution of the equation but cannot be obtained from the family $y = cx^4$ by a single choice of c. See Figure 1.2(b). ●

Sometimes a differential equation possesses a solution that cannot be obtained by specializing the parameters in a family of solutions. Such a solution is called a **singular solution**.

EXAMPLE 7 A Singular Solution

In Section 2.2 we will prove that a one-parameter family of solutions of $y' = xy^{1/2}$ is given by $y = (x^2/4 + c)^2$. When $c = 0$, the resulting particular solution is $y = x^4/16$. In this case the trivial solution $y = 0$ is a singular solution of the equation, because it cannot be obtained from the family for any choice of the parameter c. ●

If *every* solution of an nth-order equation $F(x, y, y', \ldots, y^{(n)}) = 0$ on an interval I can be obtained from an n-parameter family $G(x, y, c_1, c_2, \ldots, c_n) = 0$ by appropriate choices of the parameters c_i, $i = 1, 2, \ldots, n$, we then say that the family is the **general solution** of the differential equation. In solving linear differential equations we shall impose relatively simple restrictions on the coefficients of the equation; with these restrictions we can always be assured not only that a solution does exist on an interval but also that a family of solutions does indeed yield all possible solutions. Nonlinear equations, with the exception of some first-order equations, are usually difficult or impossible to solve in terms of familiar elementary functions (finite combinations of integer powers of x, roots, exponential and logarithmic functions, trigonometric and inverse trigonometric functions). Furthermore, if we happen to have a family of solutions for a nonlinear equation, it is not obvious when this family constitutes a general solution. On a practical level then, the designation "general solution" is applied only to linear differential equations.

Systems of Differential Equations

Up to this point we have discussed single differential equations containing one unknown function. But often, in theory as well as in many applications, we must deal with *systems* of differential equations. A **system of ordinary differential equations** is two or more equations involving the derivatives of two or more unknown functions of a single independent variable. For example, if x and y denote dependent variables and t the independent variable, the following is a system of two first-order differential equations:

$$\begin{aligned} \frac{dx}{dt} &= 3x - 4y \\ \frac{dy}{dt} &= x + y \end{aligned} \qquad \textbf{(4)}$$

A solution of a system such as (4) is a pair of differentiable functions $x = f(t)$, $y = g(t)$ that satisfy each equation of the system on some interval.

ODE Solvers

It is sometimes possible to obtain an *approximate* graphic representation of a solution of a differential equation or a system of differential equations without actually obtaining an explicit or implicit solution. This graphic representation requires a computer application with a feature known as an **ODE solver**. Each software supplement to this text has an ODE solver for first- and second-order differential equations as well as for systems of two differential equations. In the case of, say, the first-order differential equation $dy/dx = f(x, y)$, we need only

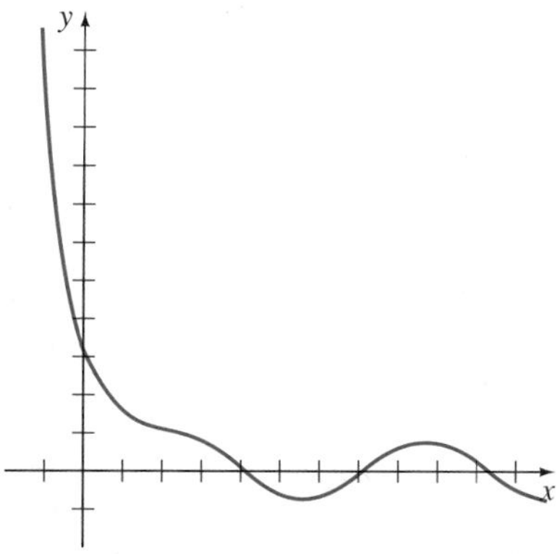

FIGURE 1.3

supply $f(x, y)$ and specify a value $y(x_0) = y_0$, where x_0 and y_0 are arbitrary constants. The solver then gives us the (approximate) solution curve that passes through the point (x_0, y_0). The problem of finding a solution of $dy/dx = f(x, y)$ that additionally satisfies $y(x_0) = y_0$ is called an **initial-value problem**. The prescribed value $y(x_0) = y_0$ is called an **initial condition**. For example, with the aid of an ODE solver we get the graph of the solution of the initial-value problem

$$\frac{dy}{dx} = -y + \sin x, \quad y(0) = 3,$$

shown in Figure 1.3. Observe that the solution curve passes through (0, 3).

More will be said about initial-value problems in Chapters 2 and 3. The theory behind ODE solvers—namely, approximation methods for differential equations and systems—will be discussed in Sections 2.6–2.8.

EXERCISES 1.1

Answers to odd-numbered problems begin on page AN-1.

In Problems 1–10 state whether the given differential equation is linear or nonlinear. Give the order of the equation.

1. $(1 - x)y'' - 4xy' + 5y = \cos x$
2. $x\dfrac{d^3y}{dx^3} - 2\left(\dfrac{dy}{dx}\right)^4 + y = 0$
3. $yy' + 2y = 1 + x^2$
4. $x^2\,dy + (y - xy - xe^x)\,dx = 0$
5. $x^3y^{(4)} - x^2y'' + 4xy' - 3y = 0$
6. $\dfrac{d^2y}{dx^2} + \dfrac{dy}{dx} + y = \cos(x + y)$
7. $\dfrac{dy}{dx} = \sqrt{1 + \left(\dfrac{d^2y}{dx^2}\right)^2}$
8. $\dfrac{d^2r}{dt^2} = -\dfrac{k}{r^2}$
9. $(\sin x)y''' - (\cos x)y' = 2$
10. $(1 - y^2)\,dx + x\,dy = 0$

In Problems 11–32 verify that the indicated function is a solution of the given differential equation. Where appropriate, c_1 and c_2 denote constants.

11. $2y' + y = 0; \quad y = e^{-x/2}$
12. $y' + 4y = 32; \quad y = 8$
13. $\dfrac{dy}{dx} - 2y = e^{3x}; \quad y = e^{3x} + 10e^{2x}$
14. $\dfrac{dy}{dt} + 20y = 24; \quad y = \dfrac{6}{5} - \dfrac{6}{5}e^{-20t}$
15. $y' = 25 + y^2; \quad y = 5\tan 5x$
16. $\dfrac{dy}{dx} = \sqrt{\dfrac{y}{x}}; \quad y = (\sqrt{x} + c_1)^2, x > 0, c_1 > 0$
17. $y' + y = \sin x; \quad y = \dfrac{1}{2}\sin x - \dfrac{1}{2}\cos x + 10e^{-x}$
18. $2xy\,dx + (x^2 + 2y)\,dy = 0; \quad x^2y + y^2 = c_1$
19. $\dfrac{dX}{dt} = (2 - X)(1 - X); \quad \ln\dfrac{2 - X}{1 - X} = t$
20. $y' + 2xy = 1; \quad y = e^{-x^2}\int_0^x e^{t^2}\,dt + c_1e^{-x^2}$
21. $(x^2 + y^2)\,dx + (x^2 - xy)\,dy = 0; \quad c_1(x + y)^2 = xe^{y/x}$
22. $y'' + y' - 12y = 0; \quad y = c_1e^{3x} + c_2e^{-4x}$
23. $y'' - 6y' + 13y = 0; \quad y = e^{3x}\cos 2x$
24. $\dfrac{d^2y}{dx^2} - 4\dfrac{dy}{dx} + 4y = 0; \quad y = e^{2x} + xe^{2x}$
25. $y'' = y; \quad y = \cosh x + \sinh x$
26. $y'' + y = \tan x; \quad y = -(\cos x)\ln(\sec x + \tan x)$
27. $x\dfrac{d^2y}{dx^2} + 2\dfrac{dy}{dx} = 0; \quad y = c_1 + c_2x^{-1}$
28. $x^2y'' - xy' + 2y = 0; \quad y = x\cos(\ln x), x > 0$
29. $x^2y'' - 3xy' + 4y = 0; \quad y = x^2 + x^2\ln x, x > 0$
30. $y''' - y'' + 9y' - 9y = 0;$ $y = c_1\sin 3x + c_2\cos 3x + 4e^x$
31. $y''' - 3y'' + 3y' - y = 0; \quad y = x^2e^x$
32. $x^3\dfrac{d^3y}{dx^3} + 2x^2\dfrac{d^2y}{dx^2} - x\dfrac{dy}{dx} + y = 12x^2;$ $y = c_1x + c_2x\ln x + 4x^2, x > 0$

In Problems 33 and 34 verify that the indicated piecewise defined function is a solution of the given differential equation.

33. $xy' - 2y = 0;\quad y = \begin{cases} -x^2, & x < 0 \\ x^2, & x \geq 0 \end{cases}$

34. $(y')^2 = 9xy;\quad y = \begin{cases} 0, & x < 0 \\ x^3, & x \geq 0 \end{cases}$

35. A one-parameter family of solutions for

$$y' = y^2 - 1 \quad \text{is} \quad y = \frac{1 + ce^{2x}}{1 - ce^{2x}}.$$

By inspection, determine a singular solution of the differential equation.

36. On page 6 we saw that $y = \sqrt{4 - x^2}$ and $y = -\sqrt{4 - x^2}$ are solutions of $dy/dx = -x/y$ on the interval $(-2, 2)$. Explain why

$$y = \begin{cases} \sqrt{4 - x^2}, & -2 < x < 0 \\ -\sqrt{4 - x^2}, & 0 \leq x < 2 \end{cases}$$

is not a solution of the differential equation on the interval.

In Problems 37 and 38 find values of m so that $y = e^{mx}$ is a solution of each differential equation.

37. $y'' - 5y' + 6y = 0$ **38.** $y'' + 10y' + 25y = 0$

In Problems 39 and 40 find values of m so that $y = x^m$ is a solution of each differential equation.

39. $x^2y'' - y = 0$ **40.** $x^2y'' + 6xy' + 4y = 0$

In Problems 41 and 42 verify that the given pair of functions is a solution of the given system of differential equations.

41. $\dfrac{dx}{dt} = x + 3y,$

$\dfrac{dy}{dt} = 5x + 3y;$

$x = e^{-2t} + 3e^{6t},\ y = -e^{-2t} + 5e^{6t}$

42. $\dfrac{d^2x}{dt^2} = 4y + e^t,$

$\dfrac{d^2y}{dt^2} = 4x - e^t;$

$x = \cos 2t + \sin 2t + \frac{1}{5}e^t,\ y = -\cos 2t - \sin 2t - \frac{1}{5}e^t$

43. ***WRITING PROJECT*** Write a short report on the solution of **Clairaut's differential equation**. Show how a Clairaut equation possesses a singular solution that is defined parametrically. Use a specific example and a figure to illustrate the connection between the family of solutions and the singular solution. If possible, introduce the concept of an **envelope** of a family of curves.

44. ***PROJECT PROBLEM***

(a) Make up two differential equations that do not possess any real solutions.

(b) In this section we discussed the fact that we normally try to find an n-parameter family of solutions $G(x, y, c_1, c_2, \ldots, c_n) = 0$ for a given nth-order differential equation $F(x, y, y', \ldots, y^{(n)}) = 0$. You should not get the impression that an nth-order differential equation *must* possess an n-parameter family of solutions. Make up a differential equation that possesses a solution but not a family of solutions. Do not think profound thoughts.

45. ***PROJECT PROBLEM*** Information about the nature of a solution of a differential equation can often be obtained from the equation itself.

(a) Review your calculus text about the geometric significance of the derivatives dy/dx and d^2y/dx^2.

(b) If a solution $y(x)$ of the differential equation $dy/dx = 2y - 8y^2$ has maximum and minimum values, explain why it is at least plausible that $y_{\min} = 0$ and $y_{\max} = \frac{1}{4}$.

(c) In Example 5 we saw that differential equation $d^2x/dt^2 = -16x$ has periodic solutions (sines and cosines). Explain why the graph in Figure 1.4 cannot be the graph of a solution.

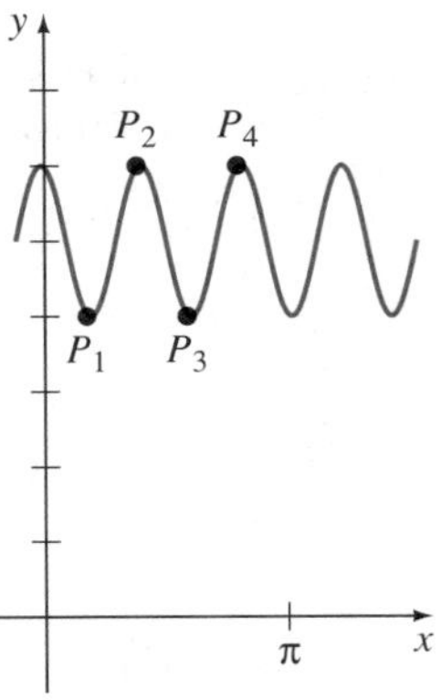

FIGURE 1.4

(d) Give a sound mathematical argument, based only on the differential equation, why it is unlikely that $d^2y/dx^2 = ay$, $a > 0$ possesses periodic solutions. Sketch several plausible graphs of solutions of the equation. Explain your reasoning.

46. *PROJECT PROBLEM* Many of the functions that we work with in applicable mathematics are defined by means of integrals. Before starting the two parts of this problem, look up **Leibniz's rule** for differentiation with respect to a parameter of a definite integral under the integral sign. Be sure to consider two cases of this rule: constant limits of integration and variable limits of integration.

(a) Verify that the indicated function is a solution of the given differential equation.

$$\frac{d^2y}{dx^2} + \omega^2 y = f(x); \qquad y = \frac{1}{\omega}\int_0^x f(t)\sin\omega(x-t)\,dt$$

$$xy'' + y' + xy = 0; \qquad y = \frac{1}{\pi}\int_0^\pi \cos(x\sin t)\,dt$$

$$\frac{dy}{dx} = -\frac{1}{1+x^2}; \qquad y = \int_0^\infty e^{-xt}\frac{\sin t}{t}\,dt, \quad x > 0$$

(b) Find a linear second-order differential equation for which

$$y = \int_0^\infty \frac{e^{-t}}{1+xt}\,dt, x \geq 0$$

is a solution.

47. *PROJECT PROBLEM* Occasionally, instead of seeking an n-parameter family of solutions (curves) for an nth-order differential equation, we have to turn the problem around: Starting with an n-parameter family of curves, can we find an associated nth-order differential equation that describes the family? In parts (a)–(d), first find a Cartesian equation of the family of curves and then find the differential equation of the family whose order matches the number of parameters in the Cartesian equation. Make sure your differential equation does not contain any arbitrary parameters.

(a) Family of straight lines passing through (2, 5)
(b) Family of parabolas with vertex on the x-axis
(c) Family of circles of radius 1 centered on the x-axis
(d) Family of circles of radius 1
(e) Inspect the differential equations in parts (a)–(c) for and note any singular solutions. Consult your calculus text and supply an interpretation for your answer in part (d).

1.2 SOME QUALITATIVE CONSIDERATIONS

Introduction

In Section 1.1 you may have gotten the impression that a solution of a differential equation is some function or expression that we can obtain and then examine. We have seen in that section that a differential equation does not have to possess a solution, but it is important to be aware of the fact that even if a solution is proven to exist there may be no method or technique for finding and displaying it. Thus it is often important to analyze a differential equation *qualitatively*. In addition to seeking answers to fundamental questions about solutions—

Does a solution exist? Is a solution unique?

qualitative analysis consists of discerning information about the behavior of solutions—

How does a solution behave near a certain point? What is its behavior for large values of the independent variable?

from the differential equation itself.

In the discussion that follows we consider several means by which solutions can be analyzed without actually solving the differential equation.

1.2.1 Direction Fields

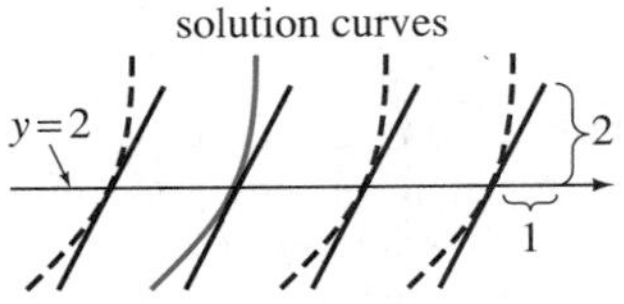

FIGURE 1.5

Consider the simple first-order differential equation $dy/dx = y$. Specifically, the differential equation implies that the slopes of tangent lines to the graph of a solution are given by the function $f(x, y) = y$. When $f(x, y)$ is held constant—that is, when $y = c$, where c is any real constant—we are stating that the slope of the tangents to the solution curves is the same constant value along a horizontal line. For example, for $y = 2$ let us draw a sequence of short line segments, or **lineal elements**, each having slope 2 and its midpoint on the line. As shown in Figure 1.5, the solution curves pass through this horizontal line at every point tangent to the lineal elements.

The equation $y = c$ represents a one-parameter family of horizontal lines. In general, any member of the family $f(x, y) = c$ is called an **isocline**, which literally means a curve along which the inclination of the tangents is the same. As the parameter c is varied, we obtain a collection of isoclines on which the lineal elements are judiciously constructed. The totality of these lineal elements is variously called a **direction field**, **slope field**, or **lineal element field** of the differential equation $dy/dx = f(x, y)$. As we see in Figure 1.6(a), the direction field suggests the "flow pattern" for the family of solution curves of the differential equation $y' = y$. In particular, if we want a solution that passes through the point (0, 1), then, as indicated in color in Figure 1.6(b), we construct a curve through this point that passes through the isoclines with the appropriate slopes.

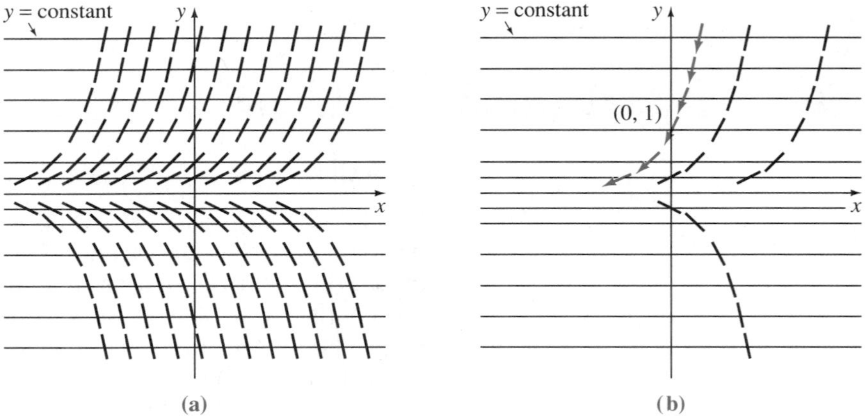

FIGURE 1.6

EXAMPLE 1 Direction Field

Sketch the direction field and indicate several possible members of the family of solution curves for $dy/dx = x/y$.

Solution Before sketching the direction field corresponding to the isoclines $x/y = c$ or $y = x/c$, you should examine the differential equation and see that it gives the following information.

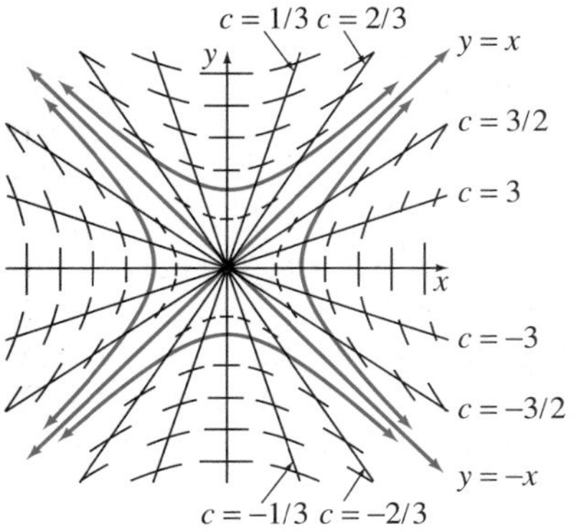

FIGURE 1.7

(a) If a solution curve crosses the x-axis ($y = 0$), it does so tangent to a vertical lineal element at every point, except possibly (0, 0).

(b) If a solution curve crosses the y-axis ($x = 0$), it does so tangent to a horizontal lineal element at every point, except possibly (0, 0).

(c) The lineal elements corresponding to the isoclines $c = 1$ and $c = -1$ are collinear with the lines $y = x$ and $y = -x$, respectively. Indeed, $y = x$ and $y = -x$ are particular solutions of the given differential equation (verify this). Note that, *in general*, isoclines are not solutions of a differential equation.

Figure 1.7 shows the direction field and several possible solution curves in color. Remember, on any isocline all the lineal elements are parallel. Also, the lineal elements may be drawn in such a manner as to suggest the flow of a particular curve. In other words, imagine choosing the isoclines so close together that if the lineal elements were connected, we would have a polygonal curve suggestive of the shape of a smooth solution curve.

EXAMPLE 2 Approximate Solution

The differential equation $dy/dx = x^2 + y^2$ cannot be solved in terms of elementary functions. Use a direction field to locate an approximate solution satisfying $y(0) = 1$.

Solution The isoclines are concentric circles defined by $x^2 + y^2 = c$, $c > 0$. For $c = \frac{1}{4}$, $c = 1$, $c = \frac{9}{4}$, and $c = 4$, we obtain circles with radii $\frac{1}{2}$, 1, $\frac{3}{2}$, and 2, as shown in Figure 1.8(a). The lineal elements constructed on each circle have a slope corresponding to the chosen value of c. It seems plausible from inspection of Figure 1.8(a) that an approximate solution curve passing through the point (0, 1) might have the shape given in Figure 1.8(b).

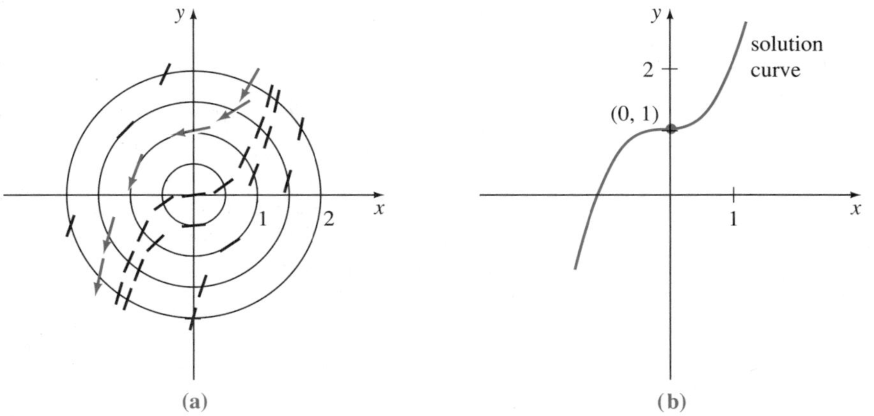

FIGURE 1.8

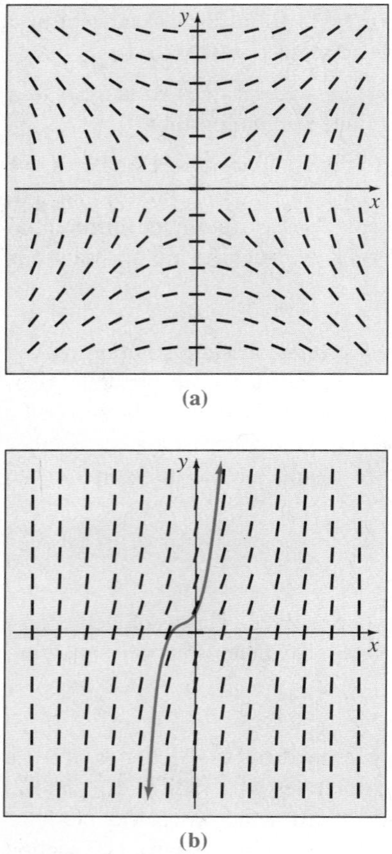

FIGURE 1.9

Use of Computers

Sketching a direction field is straightforward but time consuming; it is probably one of those tasks about which an argument can be made for doing it by hand once or twice in a lifetime, but is overall most efficiently carried out by means of computer software. Using $dy/dx = x/y$ and the direction field feature in the software supplement to this text we obtain Figure 1.9(a). Observe that in this computer version of Figure 1.7 the lineal elements are drawn uniformly spaced on their isoclines (which themselves are not drawn). The resulting direction field suggests the flow of the solution curves even more strongly. In Figure 1.9(b), using an ODE solver, we have superimposed the approximate solution curve for the differential equation in Example 2 that passes through (0, 1) on top of its computer-generated direction field.

1.2.2 Phase Portraits

Although x and y are the symbols we usually use when discussing a differential equation, in practice the symbols can be anything and are often suggestive of the context in which we are working. In the discussion that follows, it is convenient to denote t and x as the independent and dependent variables, respectively.

First-Order Differential Equations

A first-order equation $F(x, x') = 0$—that is, a differential equation in which the independent variable t does not appear explicitly—is said to be an **autonomous equation**. By solving $F(x, x') = 0$ for the derivative*, an autonomous first-order equation can be expressed as

$$\frac{dx}{dt} = f(x), \tag{1}$$

where f and its derivative f' are assumed to be continuous functions of x on some interval I. The differential equations

$$\overbrace{\frac{dx}{dt} = x - x^3}^{f(x)\ \downarrow} \qquad \text{and} \qquad \overbrace{\frac{dx}{dt} = 2tx}^{f(t,\,x)\ \downarrow}$$

are autonomous and nonautonomous, respectively. Many differential equations encountered in applications are autonomous and of the form (1).

We say that x_1 is a **critical point** of the autonomous differential equation (1) if $f(x_1) = 0$. Critical points are also called **equilibrium points** and **stationary points**. Since substitution of $x(t) = x_1$ makes both sides of the differential equation (1) zero we see that

A critical point is a constant solution of the autonomous equation.

Such constant solutions of (1) are often referred to as **equilibrium solutions**.

---

* In this text we assume that a differential equation $F(x, y, y', \ldots, y^{(n)}) = 0$ can always be solved explicitly for the highest-order derivative $y^{(n)}$. There are exceptions.

Constant solutions of (1) are horizontal lines in the tx-plane. If we can determine the algebraic sign of the derivative dx/dt by examining $f(x)$, then we can tell when a nonconstant solution $x(t)$ is increasing or decreasing. Recall from calculus that if the derivative of a differentiable function is positive (negative) on an interval, then the function is increasing (decreasing) on the interval.

EXAMPLE 3 Autonomous First-Order DE

Inspection of the right-hand member of the autonomous equation

$$\frac{dx}{dt} = x(a - bx) \quad a > 0, b > 0$$

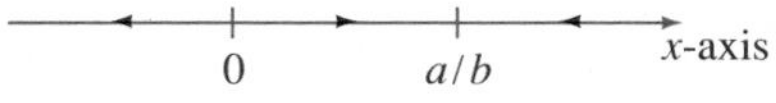

FIGURE 1.10

shows that 0 and a/b are critical points. By putting the points 0 and a/b on a horizontal line, we divide the line into three intervals: $-\infty < x < 0, 0 < x < a/b$, and $a/b < x < \infty$. The arrows on the line shown in Figure 1.10 indicate the algebraic sign of dx/dt—that is, the sign of $f(x) = x(a - bx)$ on these intervals—and whether $x(t)$ is increasing or decreasing. The following table explains the figure.

Interval	Sign of $f(x)$	$x(t)$	Arrow
$(-\infty, 0)$	minus	decreasing	points to left
$(0, a/b)$	plus	increasing	points to right
$(a/b, \infty)$	minus	decreasing	points to left

Figure 1.10 is called the **phase portrait** of the differential equation $dx/dt = x(a - bx)$. The horizontal line is called the **phase line**. A phase portrait such as this can also be interpreted in terms of the motion of a moving particle. If we imagine that $x(t)$ denotes position of a particle at time t on a horizontal line whose positive x-direction is to the right, then dx/dt is the velocity of the particle. Positive velocity indicates motion to the right and negative velocity indicates that the particle is moving to the left. If a particle is placed at a critical point, then it must remain there for all time, whence the origin of the alternative name *stationary point*.

We shall elaborate further on the concept of phase portraits of autonomous first-order equations in Section 2.1.

Second-Order Differential Equations

A second-order differential equation is said to be **autonomous** if it has no explicit dependence on the independent variable t. Thus an autonomous second-order equation has either the form $F(x, x', x'') = 0$ or, analogous to equation (1),

$$\frac{d^2x}{dt^2} = g(x, x'), \tag{2}$$

where g is assumed to be continuously differentiable in some region in the xx' plane. For example, $x'' + 2x' + x = 0$ (or $x'' = -2x' - x$) is an autonomous equation, but $x'' + 2x' + x = \sin t$ and $x'' = t^2x$ are nonautonomous. In the next example we consider an autonomous equation for which we already know a two-parameter family of solutions. The focus of the example is not on the integral curves of the equation but rather on curves that are determined by the variables x and dx/dt.

EXAMPLE 4 Autonomous Second-Order DE

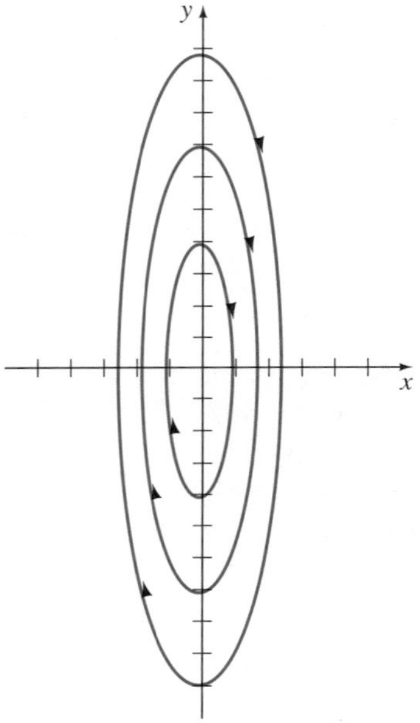

FIGURE 1.11

In Example 5 of Section 1.1 we saw that

$$x = c_1 \cos 4t + c_2 \sin 4t \tag{3}$$

was a two-parameter family of solutions of the differential equation $x'' + 16x = 0$. (With the methods in Chapter 3 you will be able to verify that (3) is actually the general solution of the equation.) Let us denote the derivative dx/dt of (3) by the symbol y:

$$y = -4c_1 \sin 4t + 4c_2 \cos 4t. \tag{4}$$

For each choice of c_1 and c_2, equations (3) and (4) constitute a set of parametric equations of a curve in the xy-plane. With the aid of a computer graphing program, we get the family of curves shown in Figure 1.11. The apparent fact that the curves are ellipses is confirmed by eliminating the parameter t from (3) and (4). Squaring these expressions and adding gives

$$x^2 + \frac{y^2}{4^2} = c_1^2 + c_2^2. \tag{5}$$

The arrowheads indicate the orientation induced on the curves by taking increasing values of t from $-\infty < t < \infty$. ●

In general, the plane of the variables x and y—that is, x and dx/dt—is called the **phase plane** of the differential equation $F(x, x', x'') = 0$. If $x = f(t)$ is a real solution of $F(x, x', x'') = 0$ on some interval and $y = dx/dt = g(t)$, then curves in the phase plane defined parametrically by the equations $x = f(t)$, $y = g(t)$, are called **trajectories**. The direction corresponding to increasing values of the parameter t is said to be the **positive direction** on the trajectory. The graphs of the family of trajectories of a differential equation $F(x, x', x'') = 0$ is called its **phase portrait** or **phase-plane diagram**.

You should not think that there is a one-to-one correspondence between solutions of a differential equation and trajectories in the phase plane. In Example 4, we know that $x = c_2 \cos 4t + c_2 \sin 4t$ is a solution of the equation $x'' + 16x = 0$ for any choice of c_1 and c_2. For the choices $c_1 = 1$, $c_2 = 0$, and $c_1 = 0$, $c_2 = 1$ we obtain, in turn, the particular solutions $x = \cos 4t$ and $x = \sin 4t$. But (5) shows that these different choices for c_1 and c_2 yield the same ellipse, $x^2 + y^2/4^2 = 1$. In other words, many different solutions of the original differential equation $F(x, x', x'') = 0$ could have the same trajectory in the phase plane.

While we cannot do justice to the importance of the phase plane at this point, it is possible to give some meaning to these concepts. Let's once again consider the differential equation $x'' + 16x = 0$ in Example 4, but this time let us ignore the fact that we possess an explicit two-parameter family of solutions and suppose instead that we have on hand only the phase portrait of the differential equation (this is possible, as we shall see in the next example). What information does Figure 1.11 give?

The simple closed curves in Figure 1.11 indicate that an object returns to a given point after a certain time and keeps returning to that point subsequently. In other words, the closed trajectory indicates that the unknown solutions corresponding to that trajectory are *periodic*. Note, too, that the common center (0, 0) of all the closed trajectories also corresponds to a constant ($x = 0$) solution of the equation $x'' + 16x = 0$.

To further interpret the phase-portrait in Figure 1.11, it is helpful to think in dynamic terms, as we did in the discussion following Example 3, by letting the variable t represent time. A trajectory then is a set of points (x, y) where the first-coordinate $x(t)$ gives the position of a moving object and the second coordinate $y(t) = dx/dt$ is the velocity of the object. A trajectory is not a representation of the path taken by the object (that would be a graph of x versus t), but rather a trajectory describes a *state* of the object, in other words, where it is and how fast it is moving. Moreover, by substituting $y = dx/dt$ in the original differential equation we find $dy/dt = -16x$, which is an equation that gives the acceleration of the object. Thus, in our dynamic interpretation, the center (0, 0) could be called an **equilibrium** or **stationary point**, because the object is at rest (velocity $y = 0$) and is free of forces (acceleration $dy/dt = 0$).

Use of Computers

In addition to plotting direction fields for first-order equations, many of the programs listed in the preface can produce phase portraits for second-order equations. To do this we need to express the second-order differential equation as a system of two first-order equations. The system can be obtained by simply substituting $y = dx/dt$ in $F(x, x', x'') = 0$ and solving the resulting first-order equation $F(x, y, y') = 0$ for the symbol y', say, $dy/dt = g(x, y)$. When we input the system of two first-order differential equations

$$\begin{aligned} \frac{dx}{dt} &= y \\ \frac{dy}{dt} &= g(x, y), \end{aligned} \tag{6}$$

computer software enables us to plot a trajectory (that is, a graph of y versus x) that passes through a specified point (x_0, y_0).

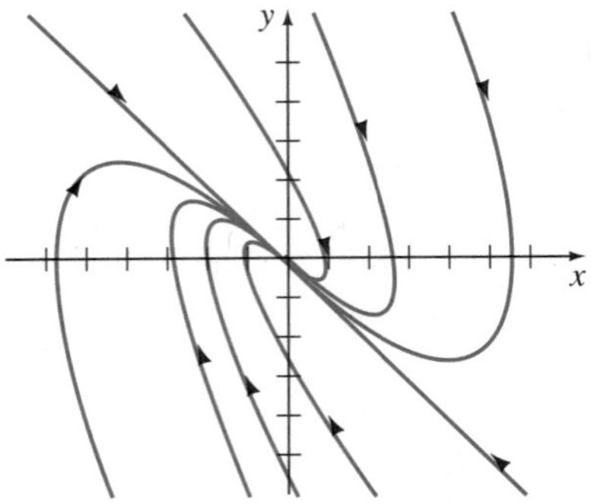

FIGURE 1.12

EXAMPLE 5 Phase Portrait

Find the phase portrait of the differential equation $x'' + 2x' + x = 0$.

Solution If $y = dx/dt$ then, after solving the given equation for $x'' = dy/dt$, we get $dy/dt = -x - 2y$. The phase portrait of the equation, or equivalently, the phase portrait of the system

$$\frac{dx}{dt} = y$$

$$\frac{dy}{dt} = -x - 2y$$

in Figure 1.12 was obtained, using software, for $-10 \le t \le 10$. The figure shows nine trajectories chosen to pass through the points (5, 4), (2, 3), (1, 0), (−4, 4), (−5, 2), (−2, −2), (−1, −3), (1, −4), and (5, −5). The arrowheads mark these points as well as indicate the positive direction on each trajectory. Note that, because the trajectories are not closed curves, the corresponding solutions of the original differential equation are not periodic. If x is a solution of the equation corresponding to any of these trajectories, observe that $x \to 0$ as $t \to \infty$. ●

EXERCISES 1.2

Answers to odd-numbered problems begin on page AN-2.

1.2.1

In Problems 1–4 use the given computer-generated direction field to sketch several possible solution curves for the indicated differential equation.

1. $y' = xy$

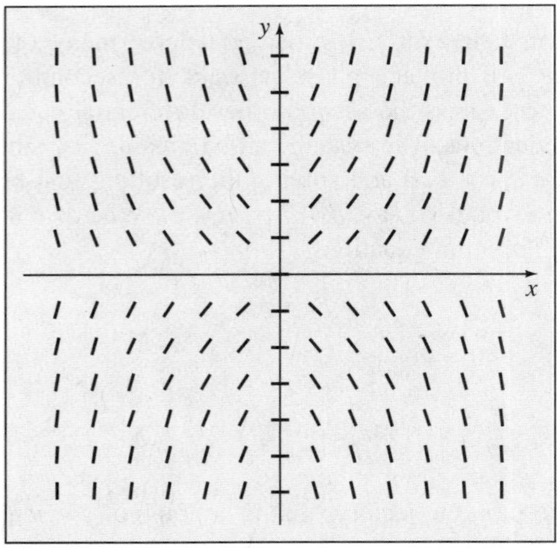

FIGURE 1.13

2. $y' = 1 - xy$

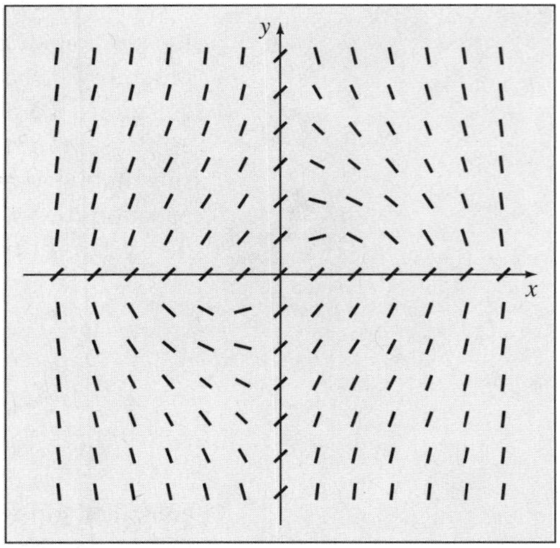

FIGURE 1.14

3. $y' = y - x$

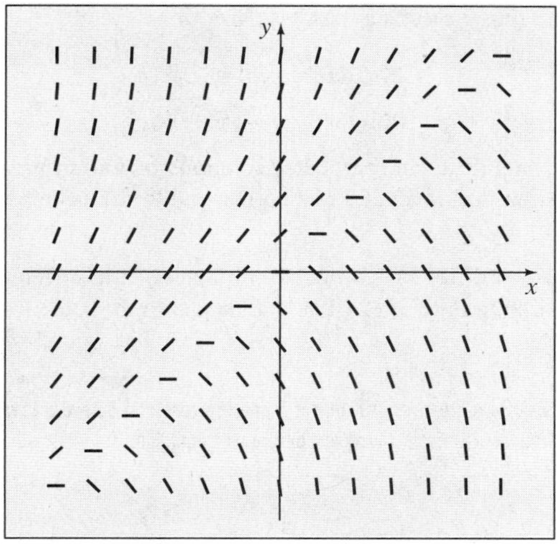

FIGURE 1.15

4. $y' = \cos x/\sin y$

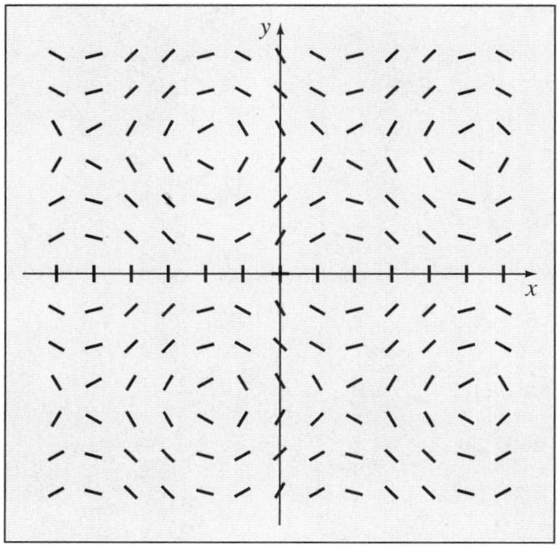

FIGURE 1.16

In Problems 5–12 sketch, or use a computer to obtain, the direction field for the given differential equation. Indicate several possible solution curves.

5. $y' = x$

6. $y' = x + y$

7. $y\dfrac{dy}{dx} = -x$

8. $\dfrac{dy}{dx} = \dfrac{1}{y}$

9. $\dfrac{dy}{dx} = 0.2x^2 + y$

10. $\dfrac{dy}{dx} = xe^y$

11. $y' = y - \cos\dfrac{\pi}{2}x$

12. $y' = 1 - \dfrac{y}{x}$

1.2.2

In Problems 13–20 find the critical points and phase portrait of the given autonomous first-order differential equation.

13. $\dfrac{dx}{dt} = x^2 - 3x$

14. $\dfrac{dx}{dt} = x^2 - x^3$

15. $\dfrac{dx}{dt} = (x - 2)^2$

16. $\dfrac{dx}{dt} = 10 + 3x - x^2$

17. $\dfrac{dx}{dt} = x^2(4 - x^2)$

18. $\dfrac{dx}{dt} = x(2 - x)(4 - x)$

19. $\dfrac{dx}{dt} = x\ln(x + 2)$

20. $\dfrac{dx}{dt} = \dfrac{xe^x - 9x}{e^x}$

In Problems 21 and 22 the given curve is a trajectory in the phase plane for an autonomous second-order differential equation $F(x, x', x'') = 0$, where $dx/dt = y$. Assume $t_i > 0$, $i = 0, 1, 2, 3, 4$. Use the data from the figure to graph a corresponding solution curve.

21.

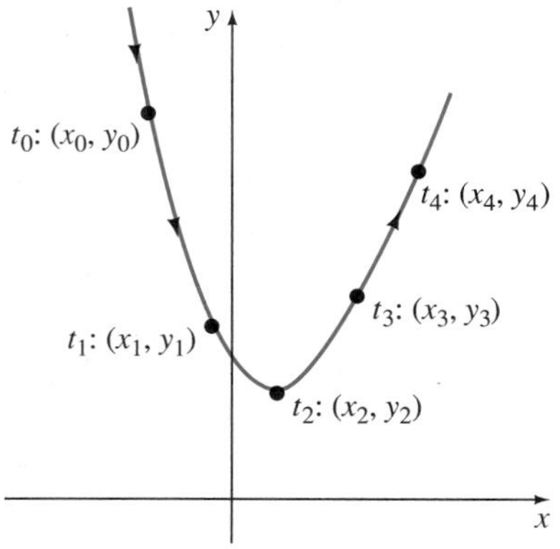

FIGURE 1.17

22.

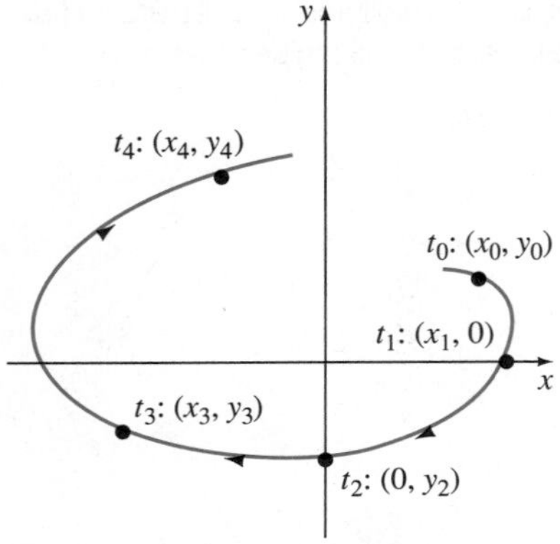

FIGURE 1.18

In Problems 23 and 24, $x(t)$ is a two-parameter family of solutions for the given differential equation. Proceed as in Example 4 and use a computer graphing program to obtain a phase portrait of the given differential equation. Use arrows to indicate the positive direction on each trajectory.

23. $x'' + x' - 2x = 0, \quad x(t) = c_1e^{t} + c_2e^{-2t}$

24. $x'' - 4x' + 4x = 0, \quad x(t) = c_1e^{2t} + c_2te^{2t}$

25. Find two different, particular solutions of the differential equation in Problem 23 that correspond to the same trajectory.

26. Find two different, particular solutions of the differential equation in Problem 24 that correspond to the same trajectory.

In Problems 27–30 proceed as in Example 5 and use computer software to obtain a phase portrait directly from the given differential equation.

27. $x'' + 2x' + 2x = 0$

28. $x'' + x' = 0$

29. $x'' + x^3 = 0$

30. $x'' + x' + x^2 = 0$

CHAPTER 2

FIRST-ORDER DIFFERENTIAL EQUATIONS

2.1 PRELIMINARY THEORY

Initial-Value Problem

We are often interested in solving a first-order differential equation $dy/dx = f(x, y)$ subject to a prescribed side condition $y(x_0) = y_0$, where x_0 and y_0 are arbitrarily specified real numbers. The problem

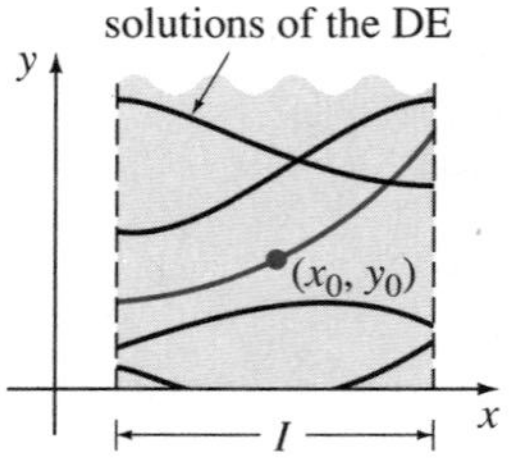

FIGURE 2.1

$$\begin{aligned} &\textit{Solve:} && \frac{dy}{dx} = f(x, y) \\ &\textit{Subject to:} && y(x_0) = y_0 \end{aligned} \quad \textbf{(1)}$$

is called an **initial-value problem** (IVP). The side condition $y(x_0) = y_0$ is called an **initial condition**. In geometric terms we are seeking at least one solution of the differential equation on an interval I containing x_0 so that a solution curve passes through the point (x_0, y_0). See Figure 2.1.

EXAMPLE 1 **Solutions of Different IVPs**

It is readily verified that $y = ce^x$ is a one-parameter family of solutions of the simple first-order equation $y' = y$ on the interval $(-\infty, \infty)$. If we specify an initial condition, say, $y(0) = 3$, then substituting $x = 0$, $y = 3$ in the family determines the constant $3 = ce^0 = c$. Thus the function $y = 3e^x$ is a solution of the initial-value problem

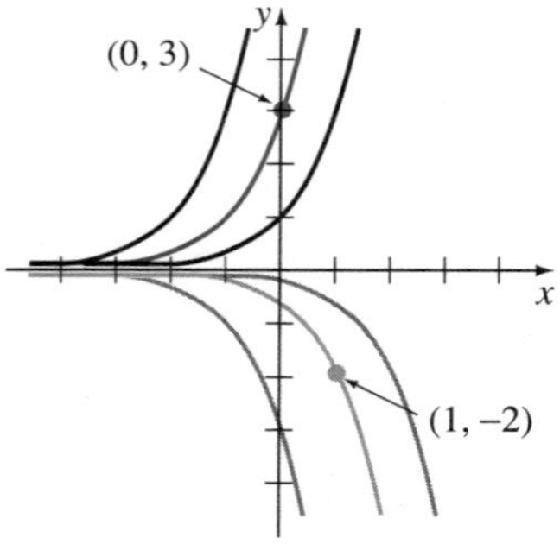

FIGURE 2.2

$$y' = y, \quad y(0) = 3.$$

If we now demand that a solution of the differential equation pass through the point $(1, -2)$ rather than $(0, 3)$, then $y(1) = -2$ would yield $-2 = ce$ or $c = -2e^{-1}$. The function $y = -2e^{x-1}$ is a solution of the initial-value problem

$$y' = y, \quad y(1) = -2.$$

The graphs of these two functions are shown in color in Figure 2.2. ●

Two fundamental questions arise in considering an initial-value problem such as (1):

Does a solution of the problem exist? If a solution exists, is it unique?

In other words, does the differential equation $dy/dx = f(x, y)$ possess a solution whose graph passes through (x_0, y_0) and, if it does, is there precisely *one* such solution?

As the next example shows, the answer to the second question is sometimes no.

EXAMPLE 2 An IVP Can Have Several Solutions

You should verify that each of the functions $y = 0$ and $y = x^4/16$ satisfies the differential equation and initial condition in the problem

$$\frac{dy}{dx} = xy^{1/2}, \quad y(0) = 0.$$

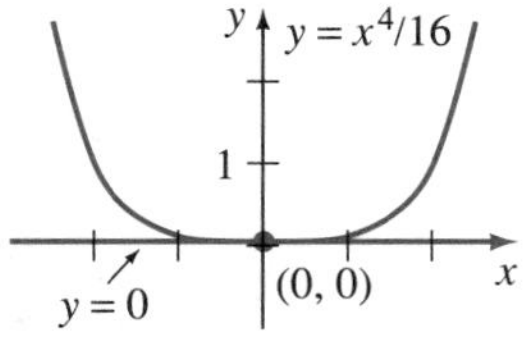

FIGURE 2.3

As illustrated in Figure 2.3, the graphs of both functions pass through the same point (0, 0). ●

It is often desirable to know before tackling an initial-value problem whether a solution exists and, when it does, whether it is the only solution of the problem. The following theorem gives conditions that are sufficient to guarantee that a solution exists and that it is also a unique solution.

THEOREM 2.1 Existence of a Unique Solution

Let R be a rectangular region in the xy-plane defined by $a \le x \le b$, $c \le y \le d$ that contains the point (x_0, y_0) in its interior. If $f(x, y)$ and $\partial f/\partial y$ are continuous on R, then there exists an interval I centered at x_0 and a unique function $y(x)$ defined on I satisfying the initial-value problem (1).

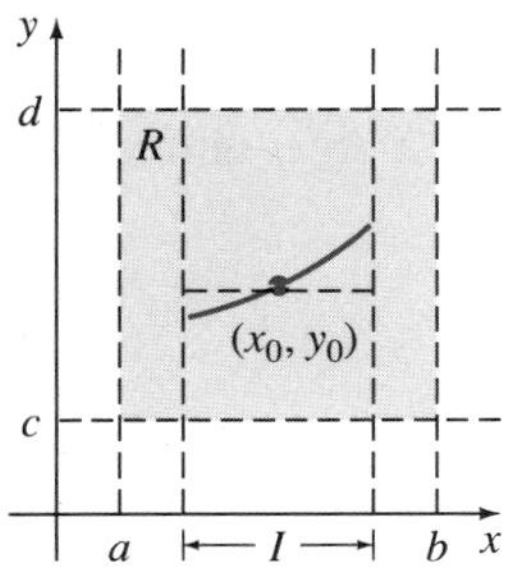

FIGURE 2.4

The foregoing is one of the most popular existence and uniqueness theorems for first-order differential equations, because the criteria of continuity of $f(x, y)$ and $\partial f/\partial y$ are relatively easy to check. In general, it is not always possible to find a specific interval I on which a solution is defined without actually solving the differential equation (see Problem 16 in Exercises 2.1). The geometry of Theorem 2.1 is illustrated in Figure 2.4.

EXAMPLE 3 Example 2 Revisited

We saw in Example 2 that the differential equation $dy/dx = xy^{1/2}$ possesses at least two solutions whose graphs pass through (0, 0). The functions

$$f(x, y) = xy^{1/2} \quad \text{and} \quad \frac{\partial f}{\partial y} = \frac{x}{2y^{1/2}}$$

are continuous in the upper half-plane defined by $y > 0$. We conclude from Theorem 2.1 that through any point (x_0, y_0), $y_0 > 0$ (say, for example, (0, 1)), there is some interval around x_0 on which the given differential equation has a unique solution. ●

EXAMPLE 4 Existence/Uniqueness

Theorem 2.1 guarantees that there exists an interval about $x = 0$ on which $y = 3e^x$ is the only solution of the initial-value problem of Example 1:

$$y' = y, \quad y(0) = 3.$$

This follows from the fact that $f(x, y) = y$ and $\partial f/\partial y = 1$ are continuous throughout the entire xy-plane. It can be further shown that this interval is $(-\infty, \infty)$. ●

EXAMPLE 5 Interval of Existence

For the equation $dy/dx = x^2 + y^2$, we observe that both $f(x, y) = x^2 + y^2$ and $\partial f/\partial y = 2y$ are continuous throughout the entire xy-plane. Therefore, through *any* given point (x_0, y_0) there passes one and only one solution curve. Note, however, that this does *not* mean the interval I containing x_0 guaranteed by Theorem 2.1 is necessarily $(-\infty, \infty)$. See *Computer Lab Experiment* 2 for Section 2.1. ●

Autonomous Equations and Stability

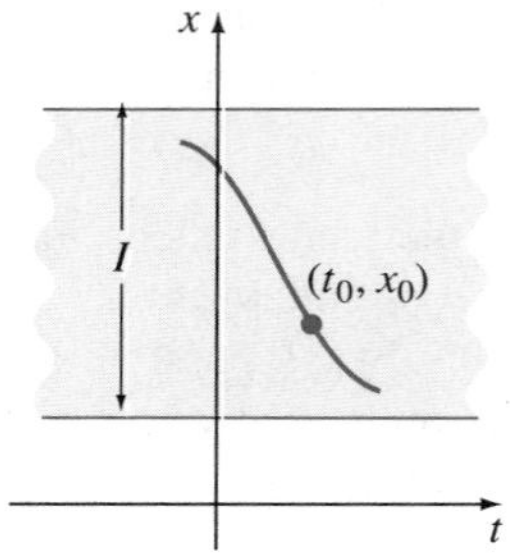

FIGURE 2.5

In Section 1.2 we introduced the notion of an autonomous first-order differential equation

$$\frac{dx}{dt} = f(x), \tag{2}$$

where f and f' are continuous on some interval I. Since the function f is independent of the variable t we may consider it defined for $-\infty < t < \infty$ (or $0 \le t \le \infty$). Since the fundamental results of Theorem 2.1 hold in the strip shown in Figure 2.5, only one solution curve of (2) passes through any point (t_0, x_0) in this strip. Moreover, it is important to realize that the graphs of solutions of (2) cannot intersect. See Problem 29 in Exercises 2.1.

Recall that x_1 is a critical point of (2) if $f(x_1) = 0$. Critical points are further classified as either stable or unstable. In terms of the motion of a particle whose position is given by a solution $x(t)$, a critical point x_1 is said to be **stable** provided that, whenever its initial position x_0 is sufficiently close to x_1, the particle remains close to x_1 for all time $t > t_0$. More precisely,

> *A critical point x_1 is stable if, for every arbitrarily specified neighborhood N_ε of radius ε centered at x_1—in other words, N_ε is the interval $(x_1 - \varepsilon, x_1 + \varepsilon)$, $\varepsilon > 0$—there exists some neighborhood N_δ of x_1 defined by $(x_1 - \delta, x_1 + \delta)$, $\delta > 0$, such that whenever the particle starts at a point $x(t_0) = x_0$ within N_δ, then it will remain in N_ε subsequently.*

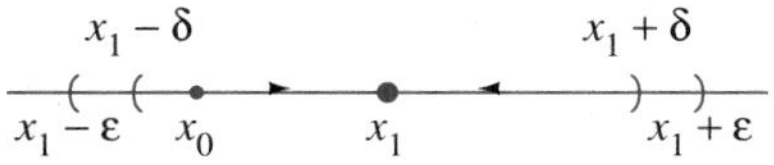

(a) asymptotically stable critical point

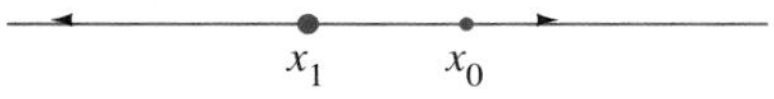

(b) unstable critical point

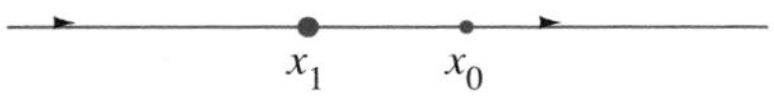

(c) unstable critical point

FIGURE 2.6

Some stable critical points are also asymptotically stable. Using our particle analogy, we say that x_1 is an **asymptotically stable** critical point if the particle not only stays close to x_1 whenever it starts sufficiently close to x_1 but additionally approaches x_1 after a long period of time. That is,

A critical point x_1 is asymptotically stable if it is stable and $\lim_{t\to\infty} x(t) = x_1$.

Geometrically this means that the graph of a solution $x(t)$ satisfying an initial condition that is sufficiently close to an asymptotically stable critical point x_1 has a horizontal asymptote $x(t) = x_1$. A critical point x_1 is said to be **unstable** if it is not stable. In terms of our moving particle analogy, a critical point is unstable if there are slight changes in the position of the particle from x_1 that result in the particle moving away, or receding, from x_1. You might recognize Figure 2.6 as the basic idea of the phase portrait of an autonomous first-order differential equation. Note that in Figure 2.6(c) the arrowheads point in the same direction; nonetheless, we classify x_1 as unstable because at least *some* of the solutions, say starting at the x_0 indicated, move away from x_1.

It can be proved from Theorem 2.1 that a solution of (2) is either constant or strictly monotonic (increasing or decreasing). This means that the differential equation can possess no oscillatory solutions nor nonconstant solutions that have a relative extremum (maximum or minimum).

In Example 3 of Section 1.2 we obtained the phase portrait of the differential equation

$$\frac{dx}{dt} = x(a - bx), \quad a > 0, b > 0. \tag{3}$$

This nonlinear equation is quite famous and is called the **logistic equation**. We shall solve (3) subject to the initial condition $x(0) = x_0$ in Section 2.5. But even without solving the equation we can sketch solution curves based solely on information obtained from its phase portrait in Figure 1.10, the classification of its critical points, and from Theorem 2.1.

EXAMPLE 6 A Phase Portrait Revisited

Reexamining the phase portrait in Figure 1.10, we see immediately that the critical points 0 and a/b of (3) are unstable and asymptotically stable, respectively. The behavior of the graph of a solution $x(t)$ of (3) depends on the interval $-\infty < x < 0$, $0 < x < a/b$, or $a/b < x < \infty$, in which the initial point $x(0) = x_0$ is situated. For some solutions, the lines defined by the constant solutions $x = 0$ and $x = a/b$ are horizontal asymptotes. In view of Theorem 2.1, a solution curve can never cross one of these asymptotes.

(i) If $x_0 > a/b$, then $x(t)$ is a decreasing function. As $t \to \infty$, the graph of $x(t)$ approaches the horizontal asymptote $x = a/b$ from above.

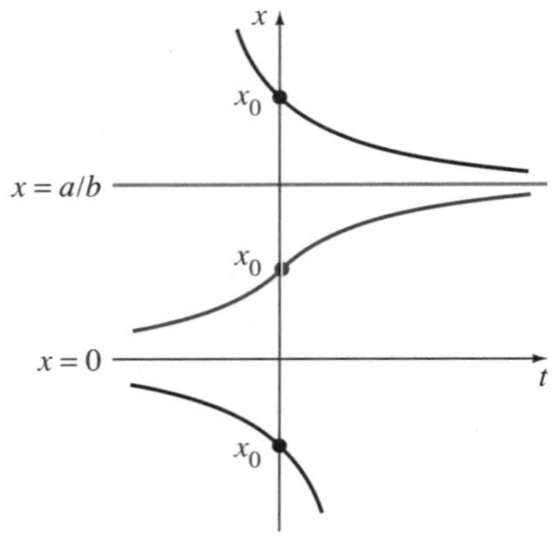

FIGURE 2.7

(ii) If $0 < x_0 < a/b$, then $x(t)$ is an increasing bounded function. As $t \to \infty$, the graph of $x(t)$ approaches the horizontal asymptote $x = a/b$ from below.

(iii) If $x_0 < 0$, then $x(t)$ decreases without bound.

Graphs of $x(t)$ illustrating these three cases are given in Figure 2.7. ●

Note In Example 6, 0 is an unstable critical point and so, as illustrated in Figure 2.7, solutions that are initially close to 0 must move away from 0. In particular, if $x_0 < 0$ we see that $x(t)$ decreases without bound. You should *not* interpret this statement to mean that the values of $x(t)$ become unbounded as $t \to \infty$. In a situation such as this, a solution may become unbounded or infinite in *finite time*, in other words, $x(t) \to -\infty$ as $t \to t_1$. See Problem 27 in Exercises 2.1.

EXERCISES 2.1

Answers to odd-numbered problems begin on page AN-3.

In Problems 1–10 determine a region of the *xy*-plane for which the given differential equation would have a unique solution through a point (x_0, y_0) in the region.

1. $\dfrac{dy}{dx} = y^{2/3}$ **2.** $\dfrac{dy}{dx} = \sqrt{xy}$

3. $x\dfrac{dy}{dx} = y$ **4.** $\dfrac{dy}{dx} - y = x$

5. $(4 - y^2)y' = x^2$ **6.** $(1 + y^3)y' = x^2$

7. $(x^2 + y^2)y' = y^2$ **8.** $(y - x)y' = y + x$

9. $\dfrac{dy}{dx} = x^3 \cos y$ **10.** $\dfrac{dy}{dx} = (x - 1)e^{y/(x-1)}$

In Problems 11 and 12 determine by inspection at least two solutions of the given initial-value problem.

11. $y' = 3y^{2/3}, \quad y(0) = 0$ **12.** $x\dfrac{dy}{dx} = 2y, \quad y(0) = 0$

13. By inspection determine a solution of the nonlinear differential equation $y' = y^3$ satisfying $y(0) = 0$. Is the solution unique?

14. By inspection find a solution of the initial-value problem

$$y' = |y - 1|, \quad y(0) = 1.$$

State why the conditions of Theorem 2.1 do not hold for this differential equation. Although we shall not prove it, the solution to this initial-value problem is unique.

15. Verify that $y = cx$ is a solution of the differential equation $xy' = y$ for every value of the parameter c. Find at least two solutions of the initial-value problem $xy' = y$, $y(0) = 0$. Observe that the piecewise defined function

$$y = \begin{cases} 0, & x < 0 \\ x, & x \geq 0 \end{cases}$$

satisfies the condition $y(0) = 0$. Is it a solution of the initial-value problem?

16. **(a)** Consider the differential equation $y' = 1 + y^2$. Determine a region in the *xy*-plane for which the equation has a unique solution whose graph passes through a point (x_0, y_0) in the region.

(b) Formally show that $y = \tan x$ satisfies the differential equation and the condition $y(0) = 0$.

(c) Explain why $y = \tan x$ is not a solution of the initial-value problem $y' = 1 + y^2$, $y(0) = 0$ on the interval $(-2, 2)$.

(d) Explain why $y = \tan x$ is a solution of the initial-value problem in part (c) on the interval $(-1, 1)$.

In Problems 17–20 determine whether Theorem 2.1 guarantees that the differential equation $y' = \sqrt{y^2 - 9}$ possesses a unique solution through the given point.

17. $(1, 4)$ **18.** $(5, 3)$ **19.** $(2, -3)$ **20.** $(-1, 1)$

21. Consider the autonomous first-order differential equation $dx/dt = x - x^3$ and the initial condition $x(0) = x_0$. Sketch the graph of a typical solution when x_0 has the given values.

(a) $x_0 > 1$ (b) $0 < x_0 < 1$
(c) $-1 < x_0 < 0$ (d) $x_0 < -1$

22. Consider the autonomous first-order differential equation $dx/dt = x^2 - x^4$ and the initial condition $x(0) = x_0$. Sketch the graph of a typical solution when x_0 has the given values.

(a) $x_0 > 1$ (b) $0 < x_0 < 1$
(c) $-1 < x_0 < 0$ (d) $x_0 < -1$

23. The autonomous differential equation

$$m\frac{dv}{dt} = mg - kv,$$

where m, g, and k are positive constants, governs the velocity of a body falling under the influence of gravity. Because the term $-kv$ represents air resistance, the velocity of the body does not increase indefinitely. Show that the limiting or terminal velocity of the body is mg/k.

24. The autonomous differential equation

$$m\frac{dv}{dt} = mg - kv^2,$$

where m, g, and k are positive constants, governs the velocity of a body falling at great speed under the influence of gravity. In this case the air resistance term is $-kv^2$. Show that the limiting or terminal velocity of the body is $\sqrt{mg/k}$.

25. The rate at which a drug disseminates into the bloodstream is governed by the differential equation

$$\frac{dX}{dt} = A - BX,$$

where A and B are positive constants. The function $X(t)$ describes the concentration of the drug in the bloodstream at any time t. Sketch the graph of the solution of the differential equation satisfying $X(0) = 0$. What is the limiting value of X as $t \to \infty$?

26. When forgetfulness is taken into account, the rate of memorization of a subject is given by

$$\frac{dA}{dt} = k_1(M - A) - k_2A,$$

where $k_1 > 0$, $k_2 > 0$, $A(t)$ is the amount of material memorized in time t, M is the total amount to be memorized, and $M - A$ is the amount remaining to be memorized. Find the limiting value of A as $t \to \infty$ and interpret the result.

27. When two chemicals are combined, the rate at which a new compound is formed is governed by the differential equation

$$\frac{dX}{dt} = k(\alpha - X)(\beta - X),$$

where $k > 0$ is a constant of proportionality and $\beta > \alpha > 0$. Here $X(t)$ denotes the number of grams of the new compound formed in time t. A reaction described by this kind of differential equation is said to be **second-order**.

(a) Discuss the behavior of X as $t \to \infty$.
(b) Consider the case when $\alpha = \beta$. What is the behavior of X as $t \to \infty$ if $X(0) < \alpha$? From the phase portrait of the differential equation can you predict the behavior of X as $t \to \infty$ if $X(0) > \alpha$?
(c) The explicit solution of the differential equation in the case when $k = 1$ and $\alpha = \beta$ is $X(t) = \alpha - 1/(t + c)$. Find a solution satisfying $X(0) = \alpha/2$. Find a solution satisfying $X(0) = 2\alpha$. Graph these two solutions. Does the behavior of the solutions as $t \to \infty$ agree with your answers to part (b)?

28. The following **derivative test** enables us to distinguish between asymptotically stable and unstable critical points without the necessity of constructing a phase portrait of the differential equation.

Let x_1 be a critical point of the autonomous first-order differential equation (2). If $f'(x_1) < 0$ then x_1 is asymptotically stable, whereas if $f'(x_1) > 0$ then x_1 is unstable.

Use this derivative test to classify the critical points of the following differential equations.

(a) $dx/dt = \sin x$
(b) $dx/dt = \sin 2x - \sin x$
(c) $dx/dt = x^2 e^{2x} - x$

29. *WRITING PROJECT* Write a short report explaining why the family of curves shown in Figure 2.8 cannot be a family of solution curves of the differential equation $dy/dx = f(x, y)$ where f is a polynomial function in the variables x and y.

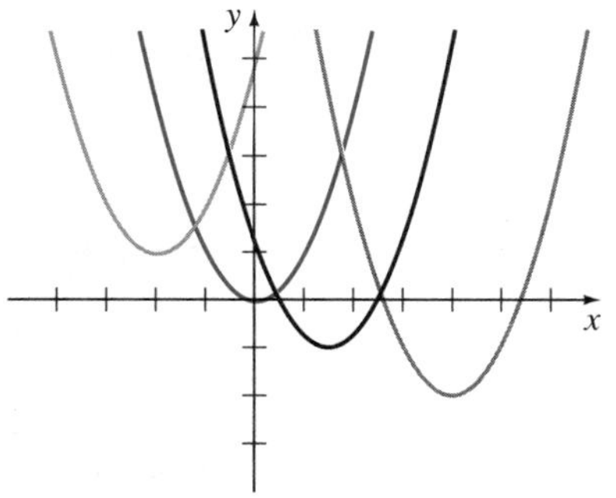

FIGURE 2.8

30. *PROJECT PROBLEM* We have seen that qualitative information about solutions of differential equations can often be obtained directly from the equation; in other words, we can discern certain properties that solutions must possess without actually solving the equation itself. Suppose for the purposes of this problem that we do not know any explicit solutions of the equation $y' - y = 0$. From Theorem 2.1 we know that the initial-value problem $y' - y = 0$, $y(0) = 1$ has a unique solution—call it $E(x)$—on some interval I. Let us assume that it has been shown that I is the interval $(-\infty, \infty)$.

(a) Explain why $E(x)$ possesses derivatives of all orders on the interval.

(b) What is $E^{(n)}(x)$?

(c) Verify that $y = cE(x)$ is a one-parameter family of solutions of the differential equation on $(-\infty, \infty)$.

(d) Verify that $y = E(x_1 + x)$, where x_1 is any real number, is a solution of the initial-value problem $y' - y = 0$, $y(0) = E(x_1)$. Use this fact to prove that $E(x_1 + x) = E(x_1)E(x)$.

(e) Show that $E(-x) = 1/E(x)$.

(f) Give a plausibility argument to justify the conclusions that, on the interval $(0, \infty)$, the function $E(x)$ is strictly increasing and that its graph is concave upward. What is $\lim_{x\to\infty} E(x)$?

(g) Explain why $0 < E(x) < 1$ on $(-\infty, 0)$. What is $\lim_{x\to-\infty} E(x)$?

(h) Obtain an approximation for the values of $E(\frac{1}{2})$ and $E(-\frac{1}{2})$. Sketch a graph of $E(x)$.

2.2 SEPARABLE VARIABLES

To the Student In solving differential equations you will often have to integrate, and this, in turn, may require some special technique of integration. It is worth a few minutes of your time to review your calculus text or, if a CAS is available, to review the command syntax for carrying out basic integration, integration by parts, and partial fractions.

We begin our study of the methodology of solving first-order equations with the simplest of all differential equations.

If $g(x)$ is a given continuous function, then the first-order equation

$$\frac{dy}{dx} = g(x) \tag{1}$$

can be solved by integration. The solution of (1) is

$$y = \int g(x)\,dx + c.$$

For example, if

$$\frac{dy}{dx} = 1 + e^{2x} \quad \text{then} \quad y = \int (1 + e^{2x})\,dx = x + \frac{1}{2}e^{2x} + c.$$

Equation (1), as well as its method of solution, is just a special case of the following.

DEFINITION 2.1 Separable Equation

A differential equation of the form

$$\frac{dy}{dx} = \frac{g(x)}{h(y)}$$

is said to be **separable** or to have **separable variables**.

Observe that a separable equation can be written as

$$h(y)\frac{dy}{dx} = g(x). \tag{2}$$

We can see immediately that (2) reduces to (1) when $h(y) = 1$.

Now if $y = f(x)$ denotes a solution of (2), we must have

$$h(f(x))f'(x) = g(x),$$

and therefore

$$\int h(f(x))f'(x)\,dx = \int g(x)\,dx + c. \tag{3}$$

But $dy = f'(x)\,dx$, so (3) is the same as

$$\int h(y)\,dy = \int g(x)\,dx + c. \tag{4}$$

Method of Solution

Equation (4) indicates the procedure for solving separable differential equations. A one-parameter family of solutions, usually given implicitly, is obtained by integrating both sides of $h(y)\,dy = g(x)\,dx$.

Note There is no need to use two constants in the integration of a separable equation, because if we write $\int h(y)\,dy + c_1 = \int g(x)\,dx + c_2$ the difference $c_2 - c_1$ can be denoted by a single constant, as in (4). In many instances throughout the following chapters, we relabel constants in a manner convenient to a given equation. For example, multiples of constants or combinations of constants can sometimes be replaced by one constant.

EXAMPLE 1 Solving a Separable DE

Solve $$(1 + x)\,dy - y\,dx = 0.$$

Solution Dividing by $(1 + x)y$, we can write $dy/y = dx/(1 + x)$, from which it follows that

$$\int \frac{dy}{y} = \int \frac{dx}{1 + x}.$$

$$\ln|y| = \ln|1 + x| + c_1$$
$$y = e^{\ln|1+x|+c_1}$$
$$= e^{\ln|1+x|} \cdot e^{c_1}$$
$$= |1 + x|e^{c_1}$$
$$= \pm e^{c_1}(1 + x) \quad \leftarrow \begin{cases} |1 + x| = 1 + x, x \geq -1 \\ |1 + x| = -(1 + x), x < -1 \end{cases}$$

Relabeling $\pm e^{c_1}$ as c then gives $y = c(1 + x)$.

Alternative Solution Since each integral results in a logarithm, a judicious choice for the constant of integration is $\ln|c|$ rather than c:

$$\ln|y| = \ln|1 + x| + \ln|c| \quad \text{or} \quad \ln|y| = \ln|c(1 + x)|$$

so that $$y = c(1 + x).$$

Even if the indefinite integrals are not *all* logarithms, it may still be advantageous to use $\ln|c|$. However, no firm rule can be given. ●

EXAMPLE 2 An Initial-Value Problem

Solve the initial-value problem

$$\frac{dy}{dx} = -\frac{x}{y}, \quad y(4) = 3.$$

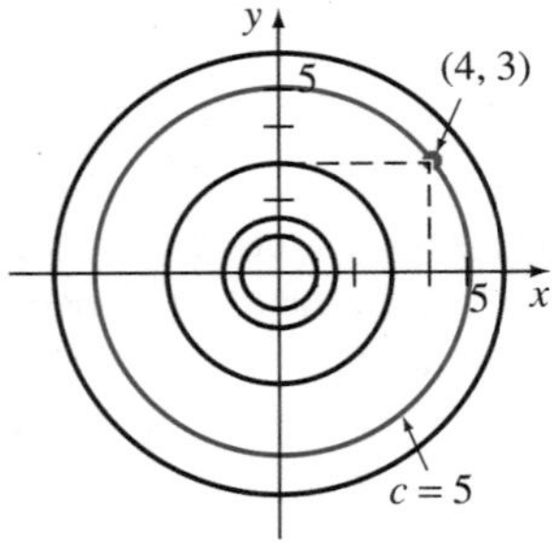

FIGURE 2.9

Solution From $y\,dy = -x\,dx$ we get

$$\int y\,dy = -\int x\,dx \quad \text{and} \quad \frac{y^2}{2} = -\frac{x^2}{2} + c_1.$$

We can write this solution as $x^2 + y^2 = c^2$ by replacing the constant $2c_1$ by c^2. The solution represents a family of concentric circles.

Now when $x = 4$, $y = 3$ so that $16 + 9 = 25 = c^2$. Thus the initial-value problem determines $x^2 + y^2 = 25$. In view of Theorem 2.1, we can conclude that it is the only circle of the family passing through the point (4, 3). See Figure 2.9. ●

EXAMPLE 3 Losing a Solution

Solve $$xy^4\,dx + (y^2 + 2)e^{-3x}\,dy = 0. \tag{5}$$

Solution By multiplying the given equation by e^{3x} and dividing by y^4, we obtain

$$xe^{3x}\,dx + \frac{y^2 + 2}{y^4}\,dy = 0 \quad \text{or} \quad xe^{3x}\,dx + (y^{-2} + 2y^{-4})\,dy = 0. \tag{6}$$

Using integration by parts on the first term yields

$$\frac{1}{3}xe^{3x} - \frac{1}{9}e^{3x} - y^{-1} - \frac{2}{3}y^{-3} = c_1.$$

The one-parameter family of solutions can also be written as

$$e^{3x}(3x - 1) = \frac{9}{y} + \frac{6}{y^3} + c, \tag{7}$$

where the constant $9c_1$ is rewritten as c. ●

Two points are worth mentioning at this time. First, unless it is important or convenient, there is no need to try to solve an expression representing a family of solutions for y explicitly in terms of x. Equation (7) shows that this task may present more problems than just the drudgery of symbol pushing. As a consequence, it is often the case that the interval over which a solution is valid is not apparent. Second, some care should be exercised when separating variables to make certain that divisors are not zero. A constant solution may sometimes get lost in the shuffle of solving the problem. In Example 3 observe that $y = 0$ is a perfectly good solution of (5) but is not a member of the set of solutions defined by (7).

EXAMPLE 4 An Initial-Value Problem

Solve the initial-value problem

$$\frac{dy}{dx} = y^2 - 4, \quad y(0) = -2.$$

Solution We put the equation into the form

$$\frac{dy}{y^2 - 4} = dx \tag{8}$$

and use partial fractions on the left side. We have

$$\left[\frac{-1/4}{y+2} + \frac{1/4}{y-2}\right] dy = dx \tag{9}$$

so that

$$-\frac{1}{4}\ln|y + 2| + \frac{1}{4}\ln|y - 2| = x + c_1. \tag{10}$$

Thus

$$\ln\left|\frac{y-2}{y+2}\right| = 4x + c_2 \quad \text{and} \quad \frac{y-2}{y+2} = ce^{4x},$$

where we have replaced $4c_1$ by c_2 and e^{c_2} by c. Finally, solving the last equation for y, we get

$$y = 2\frac{1 + ce^{4x}}{1 - ce^{4x}}. \tag{11}$$

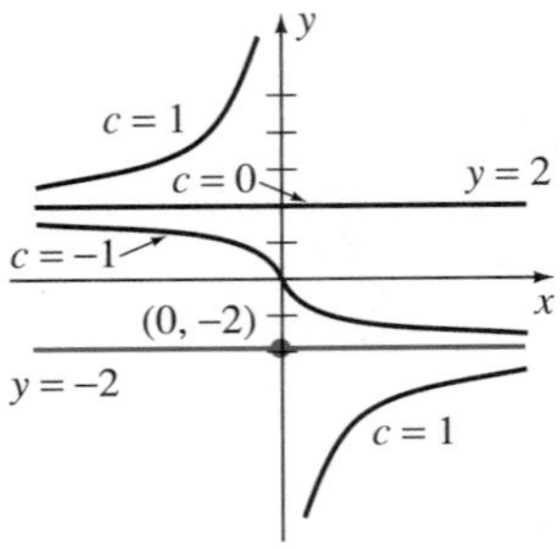

FIGURE 2.10

Substituting $x = 0$, $y = -2$ leads to the mathematical dilemma

$$-2 = 2\frac{1+c}{1-c} \quad \text{or} \quad -1 + c = 1 + c \quad \text{or} \quad -1 = 1.$$

The last equality prompts us to examine the differential equation a little more carefully. The fact is, the equation

$$\frac{dy}{dx} = (y+2)(y-2)$$

is satisfied by two constant functions, namely, $y = -2$ and $y = 2$. Inspection of equations (8), (9), and (10) clearly indicates we must preclude $y = -2$ and $y = 2$ at those steps in our solution. But it is interesting to observe that we can subsequently recover the solution $y = 2$ by setting $c = 0$ in equation (11). However, there is no finite value of c that will ever yield the solution $y = -2$. This latter constant function is the only solution to the original initial-value problem. See Figure 2.10. ●

If, in Example 4, we had used $\ln|c|$ for the constant of integration, then the form of the one-parameter family of solutions would be

$$y = 2\frac{c + e^{4x}}{c - e^{4x}}. \tag{12}$$

Note that (12) reduces to $y = -2$ when $c = 0$, but now there is no finite value of c that will give the constant solution $y = 2$.

If an initial condition leads to a particular solution by finding a specific value of the parameter c in a family of solutions for a first-order differential equation, it is a natural inclination of most students (and instructors) to relax and be content. In Section 2.1, however, we saw that a solution of an initial-value problem may not be unique. For example, the problem

$$\frac{dy}{dx} = xy^{1/2}, \quad y(0) = 0 \tag{13}$$

has at least two solutions, namely $y = 0$ and $y = x^4/16$. We are now in a position to solve the equation. Separating variables

$$y^{-1/2}\,dy = x\,dx$$

and integrating give

$$2y^{1/2} = \frac{x^2}{2} + c_1 \quad \text{or} \quad y = \left(\frac{x^2}{4} + c\right)^2$$

when $x = 0$, $y = 0$, so necessarily $c = 0$. Therefore $y = x^4/16$. The solution $y = 0$ was lost by dividing by $y^{1/2}$. In addition, the initial-value problem (13) possesses infinitely more solutions, since for any choice of the parameter $a \geq 0$ the piecewise defined function

$$y = \begin{cases} 0, & x < a \\ \dfrac{(x^2 - a^2)^2}{16}, & x \geq a \end{cases}$$

satisfies both the differential equation and initial condition. See Figure 2.11.

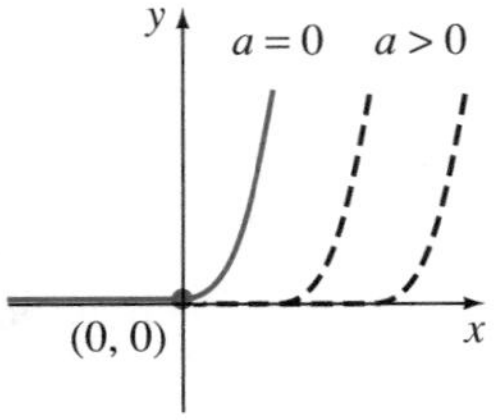

FIGURE 2.11

Substitutions

We solve a differential equation by recognizing it as a certain kind of equation, say, separable, and then carrying out a procedure consisting of equation-specific mathematical steps that yields a sufficiently differentiable function that satisfies the equation. Often the first step in solving a given differential equation consists of transforming it into another differential equation by means of a **substitution.** For example, suppose we wish to transform the first-order equation $dy/dx = f(x, y)$ by the substitution $y = \phi(x, u)$, where u is regarded as a function of the variable x. If ϕ possesses first-partial derivatives, then the Chain Rule gives

$$\frac{dy}{dx} = \phi_x(x, u) + \phi_u(x, u)\frac{du}{dx}.$$

By replacing dy/dx by $f(x, y)$ and y by $\phi(x, u)$ in the foregoing derivative we get the new first-order differential equation

$$f(x, \phi(x, u)) = \phi_x(x, u) + \phi_u(x, u)\frac{du}{dx},$$

which, after solving for du/dx, has the form $du/dx = F(x, u)$. If we can determine a solution $u = g(x)$ of this second equation, then a solution of the original differential equation is $y = \phi(x, g(x))$.

Use of Substitutions: Homogeneous Equations

If a function f possesses the property

$$f(tx, ty) = t^{\alpha}f(x, y)$$

for some real number α, then f is said to be a **homogeneous function** of degree α. For example, $f(x, y) = x^3 + y^3$ is homogeneous of degree 3 because

$$f(tx, ty) = (tx)^3 + (ty)^3 = t^3(x^3 + y^3) = t^3f(x, y),$$

whereas $f(x, y) = x^3 + y^3 + 1$ is not homogeneous.

A differential equation of the form $M(x, y)\,dx + N(x, y)\,dy = 0$ is said to be **homogeneous** if both coefficients M and N are homogeneous functions of the *same* degree. Such an equation can be reduced to separable variables by *either* the substitution $y = ux$ or $x = vy$, where u and v are new dependent variables.

EXAMPLE 5 Solving a Homogeneous Differential Equation

Solve $$(x^2 + y^2)\,dx + (x^2 - xy)\,dy = 0.$$

Solution Inspection of $M(x, y) = x^2 + y^2$ and $N(x, y) = x^2 - xy$ shows that these coefficients are homogeneous functions of degree 2. If we let $y = ux$, then

$$dy = u\,dx + x\,du$$

so that, after substituting, the given equation becomes

$$(x^2 + u^2x^2)\,dx + (x^2 - ux^2)[u\,dx + x\,du] = 0$$

$$x^2(1 + u)\,dx + x^3(1 - u)\,du = 0$$

$$\frac{1 - u}{1 + u}\,du + \frac{dx}{x} = 0$$

$$\left[-1 + \frac{2}{1 + u}\right]du + \frac{dx}{x} = 0 \quad \leftarrow \text{long division}$$

After integration the last line gives

$$-u + 2\ln|1 + u| + \ln|x| = \ln|c|$$

$$-\frac{y}{x} + 2\ln\left|1 + \frac{y}{x}\right| + \ln|x| = \ln|c|.$$

Using the properties of logarithms, we can write the preceding solution as

$$\ln\left|\frac{(x + y)^2}{cx}\right| = \frac{y}{x}.$$

The definition of a logarithm then yields $(x + y)^2 = cxe^{y/x}$. ●

Although either of the indicated substitutions can be used for every homogeneous differential equation, in practice we try $x = vy$ whenever the function $M(x, y)$ is simpler than $N(x, y)$. Also it could happen that after using one substitution, we may encounter integrals that are difficult or impossible to evaluate in closed form; switching substitutions may result in an easier problem.

Use of Substitutions: Second-Order Equations

Some second-order differential equations $F(x, y, y', y'') = 0$ can be solved by reducing the equation to a first-order equation by means of the substitution $u = y'$. The next example illustrates this technique. Note that since we are solving a second-order equation, its solution will contain two arbitrary constants.

EXAMPLE 6 Solving a Second-Order Differential Equation

Solve
$$y'' = 2x(y')^2.$$

Solution If we let $u = y'$, then $du/dx = y''$. After substituting, the equation reduces to separable form:

$$\frac{du}{dx} = 2xu^2 \quad \text{or} \quad \frac{du}{u^2} = 2x\,dx$$

$$\int u^{-2}\,du = \int 2x\,dx$$

$$-u^{-1} = x^2 + c_1^2.$$

The reason we write the constant of integration as c_1^2 should be obvious in the next few steps. Since $u^{-1} = 1/y'$, it follows that

$$\frac{dy}{dx} = -\frac{1}{x^2 + c_1^2} \quad \text{or} \quad dy = -\frac{dx}{x^2 + c_1^2}$$

$$\int dy = -\int \frac{dx}{x^2 + c_1^2}$$

$$y = -\frac{1}{c_1}\tan^{-1}\frac{x}{c_1} + c_2.$$

●

Trajectories and the Phase Plane

In Section 1.2 we introduced the concept of the phase plane. The next example illustrates how to obtain a Cartesian equation for a family of trajectories of an autonomous differential equation $F(x, x', x'') = 0$.

EXAMPLE 7 Cartesian Equation of Trajectories

We considered the differential equation $x'' + 16x = 0$ in Example 4 of Section 1.2. Proceeding as in Example 5 of that section, the substitution $y = dx/dt$ enables us to write the equation as an equivalent system of first-order equations

$$\frac{dx}{dt} = y$$

$$\frac{dy}{dt} = -16x.$$

By dividing the second equation of this system by the first equation and using the fact that

$$\frac{dy}{dx} = \frac{dy/dt}{dx/dt}$$

we obtain

$$\frac{dy}{dx} = -\frac{16x}{y}. \qquad \textbf{(14)}$$

A solution of (14) is a trajectory of $x'' + 16x = 0$. Equation (14) is separable, so integration of $16x\,dx + y\,dy = 0$ yields

$$16x^2 + y^2 = c^2. \qquad \textbf{(15)}$$

The last equation, representing a family of ellipses with center (0, 0), is the same family defined by the parametric equations (3) and (4) of Section 1.2, except that (15) does not assign any direction to a trajectory. ●

Remarks

(i) In some of the preceding examples we saw that the constant in the one-parameter family of solutions for a first-order differential equation can be relabeled when convenient. Also, it can easily happen that two individuals solving the same equation arrive at dissimilar answers. For example, by separation of variables we can show that one-parameter families of solutions for $(1 + y^2)\,dx + (1 + x^2)\,dy = 0$ are

$$\arctan x + \arctan y = c \quad \text{or} \quad \arctan x + \arctan y = \arctan c$$

$$\text{or} \quad \frac{x+y}{1-xy} = c.$$

As you work your way through the next several sections, bear in mind that families of solutions may be *equivalent* in the sense that one family may be obtained from another by either relabeling the constant or applying algebra and trigonometry.

(ii) Definition 2.1 indicates that *every* autonomous differential equation $dx/dt = f(x)$, or, using the symbols x and y, $dy/dx = f(y)$, is separable Observe in Example 4 that -2 and 2 are critical points of the autonomous equation $dy/dx = y^2 - 4$. It is evident in Figure 2.10 that these two critical points are, in turn, asymptotically stable and unstable.

EXERCISES 2.2

Answers to odd-numbered problems begin on page AN-5.

In Problems 1–34 solve the given differential equation by separation of variables.

1. $\dfrac{dy}{dx} = \sin 5x$
2. $\dfrac{dy}{dx} = (x+1)^2$
3. $dx + e^{3x}\,dy = 0$
4. $dx - x^2\,dy = 0$
5. $(x+1)\dfrac{dy}{dx} = x + 6$
6. $e^x\dfrac{dy}{dx} = 2x$
7. $xy' = 4y$
8. $\dfrac{dy}{dx} + 2xy = 0$
9. $\dfrac{dy}{dx} = \dfrac{y^3}{x^2}$
10. $\dfrac{dy}{dx} = \dfrac{y+1}{x}$
11. $\dfrac{dx}{dy} = \dfrac{x^2y^2}{1+x}$
12. $\dfrac{dx}{dy} = \dfrac{1+2y^2}{y\sin x}$
13. $\dfrac{dy}{dx} = e^{3x+2y}$
14. $e^xy\dfrac{dy}{dx} = e^{-y} + e^{-2x-y}$
15. $(4y + yx^2)\,dy - (2x + xy^2)\,dx = 0$
16. $(1 + x^2 + y^2 + x^2y^2)\,dy = y^2\,dx$
17. $2y(x+1)\,dy = x\,dx$
18. $x^2y^2\,dy = (y+1)\,dx$
19. $y\ln x\dfrac{dx}{dy} = \left(\dfrac{y+1}{x}\right)^2$
20. $\dfrac{dy}{dx} = \left(\dfrac{2y+3}{4x+5}\right)^2$
21. $\dfrac{dS}{dr} = kS$
22. $\dfrac{dQ}{dt} = k(Q - 70)$
23. $\dfrac{dP}{dt} = P - P^2$
24. $\dfrac{dN}{dt} + N = Nte^{t+2}$
25. $\sec^2 x\,dy + \csc y\,dx = 0$
26. $\sin 3x\,dx + 2y\cos^3 3x\,dy = 0$
27. $e^y \sin 2x\,dx + \cos x(e^{2y} - y)\,dy = 0$
28. $\sec x\,dy = x\cot y\,dx$

29. $(e^y + 1)^2 e^{-y}\,dx + (e^x + 1)^3 e^{-x}\,dy = 0$

30. $\dfrac{y\,dy}{x\,dx} = (1 + x^2)^{-1/2}(1 + y^2)^{1/2}$

31. $(y - yx^2)\dfrac{dy}{dx} = (y + 1)^2$ 32. $2\dfrac{dy}{dx} - \dfrac{1}{y} = \dfrac{2x}{y}$

33. $\dfrac{dy}{dx} = \dfrac{xy + 3x - y - 3}{xy - 2x + 4y - 8}$ 34. $\dfrac{dy}{dx} = \dfrac{xy + 2y - x - 2}{xy - 3y + x - 3}$

In Problems 35–38 solve the given differential equation subject to the indicated initial condition.

35. $\dfrac{dx}{dy} = 4(x^2 + 1), \quad x\left(\dfrac{\pi}{4}\right) = 1$

36. $\dfrac{dy}{dx} = \dfrac{y^2 - 1}{x^2 - 1}, \quad y(2) = 2$

37. $x^2y' = y - xy, \quad y(-1) = -1$

38. $y' + 2y = 1, \quad y(0) = \frac{5}{2}$

In Problems 39 and 40 find a solution of the given differential equation that passes through the indicated points.

39. $\dfrac{dy}{dx} - y^2 = -9$ (a) $(0, 0)$ (b) $(0, 3)$ (c) $\left(\dfrac{1}{3}, 1\right)$

40. $x\dfrac{dy}{dx} = y^2 - y$ (a) $(0, 1)$ (b) $(0, 0)$ (c) $\left(\dfrac{1}{2}, \dfrac{1}{2}\right)$

In Problems 41–56 solve the given differential equation by using an appropriate substitution.

41. $(x - y)\,dx + x\,dy = 0$ 42. $(x + y)\,dx + x\,dy = 0$

43. $x\,dx + (y - 2x)\,dy = 0$ 44. $y\,dx = 2(x + y)\,dy$

45. $(y^2 + yx)\,dx - x^2\,dy = 0$

46. $(y^2 + yx)\,dx + x^2\,dy = 0$

47. $\dfrac{dy}{dx} = \dfrac{y - x}{y + x}$ 48. $\dfrac{dy}{dx} = \dfrac{x + 3y}{3x + y}$

49. $-y\,dx + (x + \sqrt{xy})\,dy = 0$ 50. $x\dfrac{dy}{dx} - y = \sqrt{x^2 + y^2}$

51. $y'' + (y')^2 + 1 = 0$ 52. $y'' = 1 + (y')^2$

53. $xy'' - y' = 0$ 54. $y'' + (\tan x)y' = 0$

55. $y'' + 2y(y')^3 = 0$ 56. $y^2y'' = y'$

$\left[\textit{Hint: } \text{Let } u = y' \text{ so that } y'' = \dfrac{du}{dx} = \dfrac{du}{dy}\dfrac{dy}{dx} = \dfrac{du}{dy}u.\right]$

57. Find the most general function whose square plus the square of its derivative is equal to 1.

58. ***PROJECT PROBLEM*** Prove that there are more than 1,650,000 digits in the y-coordinate of the point of intersection of the solution curves for the initial-value problems

$$\frac{dy}{dx} = y, \quad y(0) = 1 \qquad \text{and} \qquad \frac{dy}{dx} = y + \frac{y}{x \ln x}, \quad y(e) = 1.$$

In Problems 59–62 proceed as in Example 7 to find a Cartesian equation for the family of trajectories for the given autonomous differential equation.

59. $x'' - 9x = 0$ 60. $x'' + x' = 0$

61. $x'' + x - x^3 = 0$ 62. $x'' - 2x' + x = 0$

63. ***WRITING PROJECT*** The "snowplow problem" is a classic and appears in many differential equations texts but is due originally to Ralph Palmer Agnew:

> *One day it started snowing at a heavy and steady rate. A snowplow started out at noon, going 2 miles the first hour and 1 mile the second hour. What time did it start snowing?*

It is a good first problem for you to study that illustrates the part differential equations play in the modeling of certain physical situations. The differential equation that arises is first-order and separable. Write a report on the content and solution of the "snowplow problem." Try to find the text: *Differential Equations*, Ralph Palmer Agnew, McGraw-Hill Book Co.

2.3 EXACT EQUATIONS

Although the simple equation $ydx + xdy = 0$ is separable, it is also equivalent to the differential of the product of x and y; that is,

$$y\,dx + x\,dy = d(xy) = 0.$$

By integrating we immediately obtain the implicit solution $xy = c$.

You might remember from calculus that if $z = f(x, y)$ is a function having continuous first partial derivatives in a region R of the xy-plane, then its differential (also called the total differential) is

$$dz = \frac{\partial f}{\partial x}\,dx + \frac{\partial f}{\partial y}\,dy. \tag{1}$$

Now if $f(x, y) = c$, it follows from (1) that

$$\frac{\partial f}{\partial x}\,dx + \frac{\partial f}{\partial y}\,dy = 0. \tag{2}$$

In other words, given a family of curves $f(x, y) = c$, we can generate a first-order differential equation by computing the total differential. For example, if $x^2 - 5xy + y^3 = c$, then (2) gives

$$(2x - 5y)\,dx + (-5x + 3y^2)\,dy = 0 \quad \text{or} \quad \frac{dy}{dx} = \frac{5y - 2x}{-5x + 3y^2}.$$

For our purposes, it is more important to turn the problem around; namely, given an equation such as

$$\frac{dy}{dx} = \frac{5y - 2x}{-5x + 3y^2}, \tag{3}$$

can we identify the equation as being equivalent to the statement

$$d(x^2 - 5xy + y^3) = 0?$$

DEFINITION 2.4 Exact Equation

A differential expression $M(x, y)\,dx + N(x, y)\,dy$ is an **exact differential** in a region R of the xy-plane if it corresponds to the differential of some function $f(x, y)$. A differential equation

$$M(x, y)\,dx + N(x, y)\,dy = 0$$

is said to be an **exact equation** if the expression on the left side is an exact differential.

EXAMPLE 1 An Exact DE

The equation $x^2y^3\,dx + x^3y^2\,dy = 0$ is exact, because it is recognized that

$$d\left(\frac{1}{3}x^3y^3\right) = x^2y^3\,dx + x^3y^2\,dy.$$

Notice in Example 1, if $M(x, y) = x^2y^3$ and $N(x, y) = x^3y^2$ that $\partial M/\partial y = 3x^2y^2 = \partial N/\partial x$. The next theorem shows that this equality of partial derivatives is no coincidence.

THEOREM 2.2 Criterion for an Exact Differential

Let $M(x, y)$ and $N(x, y)$ be continuous and have continuous first partial derivatives in a rectangular region R defined by $a < x < b$, $c < y < d$. Then a necessary and sufficient condition that $M(x, y)\,dx + N(x, y)\,dy$ be an exact differential is

$$\frac{\partial M}{\partial y} = \frac{\partial N}{\partial x}. \tag{4}$$

Proof of the Necessity For simplicity let us assume that $M(x, y)$ and $N(x, y)$ have continuous first partial derivatives for all (x, y). Now if the expression $M(x, y)\,dx + N(x, y)\,dy$ is exact, there exists some function f such that for all x in R,

$$M(x, y)\,dx + N(x, y)\,dy = \frac{\partial f}{\partial x}\,dx + \frac{\partial f}{\partial y}\,dy.$$

Therefore,
$$M(x, y) = \frac{\partial f}{\partial x}, \qquad N(x, y) = \frac{\partial f}{\partial y},$$

and
$$\frac{\partial M}{\partial y} = \frac{\partial}{\partial y}\left(\frac{\partial f}{\partial x}\right) = \frac{\partial^2 f}{\partial y\,\partial x} = \frac{\partial}{\partial x}\left(\frac{\partial f}{\partial y}\right) = \frac{\partial N}{\partial x}.$$

The equality of the mixed partials is a consequence of the continuity of the first partial derivatives of $M(x, y)$ and $N(x, y)$. ○

The sufficiency part of Theorem 2.2 consists of showing that there exists a function f for which $\partial f/\partial x = M(x, y)$ and $\partial f/\partial y = N(x, y)$ whenever (4) holds. The construction of the function f actually reflects a basic procedure for solving exact equations.

Method of Solution

Given an equation of the form $M(x, y)\,dx + N(x, y)\,dy = 0$, determine whether the equality in (4) holds. If it does, then there exists a function f for which

$$\frac{\partial f}{\partial x} = M(x, y).$$

We can find f by integrating $M(x, y)$ with respect to x, while holding y constant:

$$f(x, y) = \int M(x, y)\,dx + g(y), \tag{5}$$

where the arbitrary function $g(y)$ is the "constant" of integration. Now differentiate (5) with respect to y and assume $\partial f/\partial y = N(x, y)$:

$$\frac{\partial f}{\partial y} = \frac{\partial}{\partial y}\int M(x, y)\,dx + g'(y) = N(x, y).$$

This gives
$$g'(y) = N(x, y) - \frac{\partial}{\partial y}\int M(x, y)\,dx. \tag{6}$$

Finally, integrate (6) with respect to y and substitute the result in (5). The solution of the equation is $f(x, y) = c$.

Note Some observations are in order. First, it is important to realize that the expression $N(x, y) - (\partial/\partial y)\int M(x, y)\,dx$ in (6) is independent of x, because

$$\frac{\partial}{\partial x}\left[N(x, y) - \frac{\partial}{\partial y}\int M(x, y)\,dx\right] = \frac{\partial N}{\partial x} - \frac{\partial}{\partial y}\left(\frac{\partial}{\partial x}\int M(x, y)\,dx\right) = \frac{\partial N}{\partial x} - \frac{\partial M}{\partial y} = 0.$$

Second, we could just as well start the foregoing procedure with the assumption that $\partial f/\partial y = N(x, y)$. After integrating N with respect to y and then differentiating that result, we find the analogues of (5) and (6) would be, respectively,

$$f(x, y) = \int N(x, y)\,dy + h(x) \quad \text{and} \quad h'(x) = M(x, y) - \frac{\partial}{\partial x}\int N(x, y)\,dy.$$

In either case *none of these formulas should be memorized.*

EXAMPLE 2 Solving an Exact DE

Solve
$$2xy\,dx + (x^2 - 1)\,dy = 0.$$

Solution With $M(x, y) = 2xy$ and $N(x, y) = x^2 - 1$ we have

$$\frac{\partial M}{\partial y} = 2x = \frac{\partial N}{\partial x}.$$

Thus the equation is exact and so, by Theorem 2.2, there exists a function $f(x, y)$ such that

$$\frac{\partial f}{\partial x} = 2xy \quad \text{and} \quad \frac{\partial f}{\partial y} = x^2 - 1.$$

From the first of these equations we obtain, after integrating,

$$f(x, y) = x^2y + g(y).$$

Taking the partial derivative of the last expression with respect to y and setting the result equal to $N(x, y)$ give

$$\frac{\partial f}{\partial y} = x^2 + g'(y) = x^2 - 1. \quad \leftarrow N(x, y)$$

It follows that
$$g'(y) = -1 \quad \text{and} \quad g(y) = -y.$$

The constant of integration need not be included in the preceding line because the solution is $f(x, y) = c$. Some of the family of curves $x^2y - y = c$ are given in Figure 2.12.

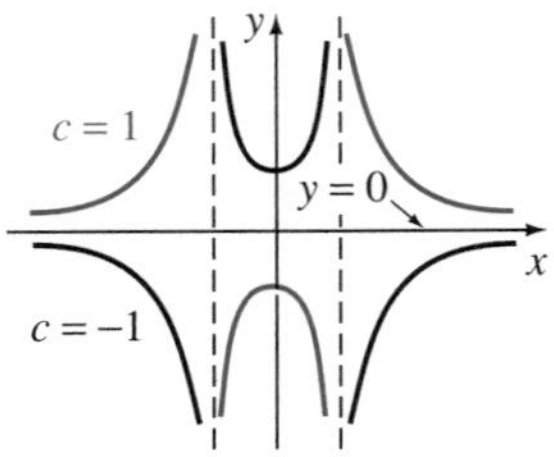

FIGURE 2.12

Note The solution of the equation is *not* $f(x, y) = x^2y - y$. Rather it is $f(x, y) = c$ or $f(x, y) = 0$ if a constant is used in the integration of $g'(y)$. Observe that the equation could also be solved by separation of variables.

EXAMPLE 3 Solving an Exact DE

Solve $(e^{2y} - y\cos xy)\,dx + (2xe^{2y} - x\cos xy + 2y)\,dy = 0.$

Solution The equation is exact because

$$\frac{\partial M}{\partial y} = 2e^{2y} + xy\sin xy - \cos xy = \frac{\partial N}{\partial x}.$$

Hence a function $f(x, y)$ exists for which

$$M(x, y) = \frac{\partial f}{\partial x} \quad \text{and} \quad N(x, y) = \frac{\partial f}{\partial y}.$$

Now for variety we shall start with the assumption that $\partial f/\partial y = N(x, y)$;

that is,
$$\frac{\partial f}{\partial y} = 2xe^{2y} - x\cos xy + 2y$$

$$f(x, y) = 2x\int e^{2y}\,dy - x\int \cos xy\,dy + 2\int y\,dy.$$

Remember, the reason x can come out in front of the symbol $\int$ is that in the integration with respect to y, x is treated as an ordinary constant. It follows that

$$f(x, y) = xe^{2y} - \sin xy + y^2 + h(x)$$

$$\frac{\partial f}{\partial x} = e^{2y} - y\cos xy + h'(x) = e^{2y} - y\cos xy, \quad \leftarrow M(x, y)$$

and so $h'(x) = 0$ or $h(x) = c$. Hence a family of solutions is

$$xe^{2y} - \sin xy + y^2 + c = 0.$$

●

EXAMPLE 4 An Initial-Value Problem

Solve the initial-value problem

$$(\cos x\sin x - xy^2)\,dx + y(1 - x^2)\,dy = 0, \qquad y(0) = 2.$$

Solution The equation is exact because

$$\frac{\partial M}{\partial y} = -2xy = \frac{\partial N}{\partial x}.$$

Now
$$\frac{\partial f}{\partial y} = y(1 - x^2)$$

$$f(x, y) = \frac{y^2}{2}(1 - x^2) + h(x)$$

$$\frac{\partial f}{\partial x} = -xy^2 + h'(x) = \cos x\sin x - xy^2.$$

The last equation implies that $h'(x) = \cos x \sin x$. Integrating gives

$$h(x) = -\int (\cos x)(-\sin x\, dx) = -\frac{1}{2}\cos^2 x.$$

Thus $\quad \dfrac{y^2}{2}(1 - x^2) - \dfrac{1}{2}\cos^2 x = c_1 \quad$ or $\quad y^2(1 - x^2) - \cos^2 x = c,$

where $2c_1$ has been replaced by c. The initial condition $y = 2$ when $x = 0$ demands that $4(1) - \cos^2(0) = c$ or that $c = 3$. Thus a solution of the problem is

$$y^2(1 - x^2) - \cos^2 x = 3.$$

Remark When testing an equation for exactness, make sure it is of the precise form $M(x, y)\, dx + N(x, y)\, dy = 0$. Sometimes a differential equation is written $G(x, y)\, dx = H(x, y)\, dy$. In this case, rewrite it as $G(x, y)\, dx - H(x, y)\, dy = 0$ and then identify $M(x, y) = G(x, y)$ and $N(x, y) = -H(x, y)$ before using (4).

EXERCISES 2.3

Answers to odd-numbered problems begin on page AN-5.

In Problems 1–24 determine whether the given equation is exact. If it is exact, solve.

1. $(2x - 1)\, dx + (3y + 7)\, dy = 0$
2. $(2x + y)\, dx - (x + 6y)\, dy = 0$
3. $(5x + 4y)\, dx + (4x - 8y^3)\, dy = 0$
4. $(\sin y - y \sin x)\, dx + (\cos x + x \cos y - y)\, dy = 0$
5. $(2y^2x - 3)\, dx + (2yx^2 + 4)\, dy = 0$
6. $\left(2y - \dfrac{1}{x} + \cos 3x\right)\dfrac{dy}{dx} + \dfrac{y}{x^2} - 4x^3 + 3y \sin 3x = 0$
7. $(x + y)(x - y)\, dx + x(x - 2y)\, dy = 0$
8. $\left(1 + \ln x + \dfrac{y}{x}\right) dx = (1 - \ln x)\, dy$
9. $(y^3 - y^2 \sin x - x)\, dx + (3xy^2 + 2y \cos x)\, dy = 0$
10. $(x^3 + y^3)\, dx + 3xy^2\, dy = 0$
11. $(y \ln y - e^{-xy})\, dx + \left(\dfrac{1}{y} + x \ln y\right) dy = 0$
12. $\dfrac{2x}{y}\, dx - \dfrac{x^2}{y^2}\, dy = 0$
13. $x\dfrac{dy}{dx} = 2xe^x - y + 6x^2$
14. $(3x^2y + e^y)\, dx + (x^3 + xe^y - 2y)\, dy = 0$
15. $\left(1 - \dfrac{3}{x} + y\right) dx + \left(1 - \dfrac{3}{y} + x\right) dy = 0$
16. $(e^y + 2xy \cosh x)y' + xy^2 \sinh x + y^2 \cosh x = 0$
17. $\left(x^2y^3 - \dfrac{1}{1 + 9x^2}\right)\dfrac{dx}{dy} + x^3y^2 = 0$
18. $(5y - 2x)y' - 2y = 0$
19. $(\tan x - \sin x \sin y)\, dx + \cos x \cos y\, dy = 0$
20. $(3x \cos 3x + \sin 3x - 3)\, dx + (2y + 5)\, dy = 0$
21. $(1 - 2x^2 - 2y)\dfrac{dy}{dx} = 4x^3 + 4xy$
22. $(2y \sin x \cos x - y + 2y^2e^{xy^2})\, dx = (x - \sin^2 x - 4xye^{xy^2})\, dy$
23. $(4x^3y - 15x^2 - y)\, dx + (x^4 + 3y^2 - x)\, dy = 0$
24. $\left(\dfrac{1}{x} + \dfrac{1}{x^2} - \dfrac{y}{x^2 + y^2}\right) dx + \left(ye^y + \dfrac{x}{x^2 + y^2}\right) dy = 0$

In Problems 25–30 solve the given differential equation subject to the indicated initial condition.

25. $(x + y)^2\, dx + (2xy + x^2 - 1)\, dy = 0, \quad y(1) = 1$
26. $(e^x + y)\, dx + (2 + x + ye^y)\, dy = 0, \quad y(0) = 1$

27. $(4y + 2x - 5)\,dx + (6y + 4x - 1)\,dy = 0, \quad y(-1) = 2$

28. $\left(\dfrac{3y^2 - x^2}{y^5}\right)\dfrac{dy}{dx} + \dfrac{x}{2y^4} = 0, \quad y(1) = 1$

29. $(y^2 \cos x - 3x^2y - 2x)\,dx + (2y \sin x - x^3 + \ln y)\,dy = 0,$
$y(0) = e$

30. $\left(\dfrac{1}{1 + y^2} + \cos x - 2xy\right)\dfrac{dy}{dx} = y(y + \sin x), \quad y(0) = 1$

In Problems 31–34 find the value of k so that the given differential equation is exact.

31. $(y^3 + kxy^4 - 2x)\,dx + (3xy^2 + 20x^2y^3)\,dy = 0$

32. $(2x - y \sin xy + ky^4)\,dx - (20xy^3 + x \sin xy)\,dy = 0$

33. $(2xy^2 + ye^x)\,dx + (2x^2y + ke^x - 1)\,dy = 0$

34. $(6xy^3 + \cos y)\,dx + (kx^2y^2 - x \sin y)\,dy = 0$

35. Determine a function $M(x, y)$ so that the following differential equation is exact:

$$M(x, y)\,dx + \left(xe^{xy} + 2xy + \frac{1}{x}\right)dy = 0.$$

36. Determine a function $N(x, y)$ so that the following differential equation is exact:

$$\left(y^{1/2}x^{-1/2} + \frac{x}{x^2 + y}\right)dx + N(x, y)\,dy = 0.$$

37. *WRITING PROJECT* It is sometimes possible to convert a nonexact differential equation $M(x, y)\,dx + N(x, y)\,dy = 0$ into an exact equation by multiplying it by a function $\mu(x, y)$, called an **integrating factor**. Write a report on the concept of integrating factors. In this report illustrate how to solve a nonexact differential equation by this method. Address the question whether $M(x, y)\,dx + N(x, y)\,dy = 0$ and $\mu(x, y)M(x, y)\,dx + \mu(x, y)N(x, y)\,dy = 0$ are equivalent equations in the sense that a solution of one is also a solution of the other. Finally, determine whether there is a way of finding integrating factors.

38. *PROJECT PROBLEM* This problem is more of theoretical interest than of practical importance.

(a) Review the definition of a homogeneous function in Section 2.2. Show that if f is homogeneous of degree α, then f satisfies the partial differential equation

$$x\frac{\partial f}{\partial x} + y\frac{\partial f}{\partial y} = \alpha f.$$

(b) If the differential equation $M(x, y)\,dx + N(x, y)\,dy = 0$ is both homogeneous and exact, show that a one-parameter family of solutions is given by $x\,M(x, y) + y\,N(x, y) = c$. Illustrate this result by means of several examples.

2.4 LINEAR EQUATIONS

In Chapter 1 we defined the general form of a linear differential equation of order n to be

$$a_n(x)\frac{d^ny}{dx^n} + a_{n-1}(x)\frac{d^{n-1}y}{dx^{n-1}} + \cdots + a_1(x)\frac{dy}{dx} + a_0(x)y = g(x).$$

Remember that linearity means that all coefficients are functions of x only, and that y and all its derivatives are raised to the first power. Now when $n = 1$, we obtain a linear first-order equation.

DEFINITION 2.5 Linear Equation

A differential equation of the form

$$a_1(x)\frac{dy}{dx} + a_0(x)y = g(x) \tag{1}$$

is said to be a **linear equation**.

Dividing both sides of (1) by the lead coefficient $a_1(x)$ yields a more useful form, the **standard form**, of a linear equation:

$$\frac{dy}{dx} + P(x)y = f(x). \tag{2}$$

We seek a solution of (2) on an interval I for which both functions P and f are continuous.

In the discussion that follows we illustrate a property and a procedure and end up with a formula representing a solution of (2). But more than the formula, the property and the procedure are important, because these two concepts carry over to linear equations of higher order.

The Property

If y_c is a solution of

$$\frac{dy}{dx} + P(x)y = 0 \tag{3}$$

and y_p is a particular solution of (2) you should be able to verify by direct substitution that the **sum** of the two solutions, $y = y_c + y_p$, is also a solution of (2). We can find y_c by separation of variables. Writing (3) as

$$\frac{dy}{y} + P(x)\,dx = 0,$$

integrating, and solving for y gives $y_c = c_1 e^{-\int P(x)\,dx}$. For convenience let us write $y_c = c_1 y_1(x)$, where $y_1 = e^{-\int P(x)\,dx}$. The fact that $dy_1/dx + P(x)y_1 = 0$ will be used immediately in finding y_p.

The Procedure

We can now find a particular solution of equation (2) by a procedure known as **variation of parameters**. The basic idea here is to find a function u_1 so that $y_p = u_1(x)y_1(x)$, where y_1 is defined in the preceding paragraph, is a solution of (2). In other words, our assumption for y_p is the same as $y_c = c_1 y_1(x)$, except that c_1 is replaced by the "variable parameter" u_1.

Now substituting $y_p = u_1 y_1$ into (2) gives

$$\frac{d}{dx}[u_1 y_1] + P(x)u_1 y_1 = f(x)$$

$$u_1 \frac{dy_1}{dx} + y_1 \frac{du_1}{dx} + P(x)u_1 y_1 = f(x)$$

$$u_1 \underbrace{\left[\frac{dy_1}{dx} + P(x)y_1\right]}_{\text{zero}} + y_1 \frac{du_1}{dx} = f(x)$$

so that $$y_1 \frac{du_1}{dx} = f(x).$$

By separating variables, we find

$$du_1 = \frac{f(x)}{y_1(x)}\,dx \quad \text{and} \quad u_1 = \int \frac{f(x)}{y_1(x)}\,dx.$$

From the definition of y_1 we see that

$$y_p = u_1 y_1 = e^{-\int P(x)\,dx} \int e^{\int P(x)\,dx} f(x)\,dx.$$

Thus

$$y = y_c + y_p = c_1 e^{-\int P(x)\,dx} + e^{-\int P(x)\,dx} \int e^{\int P(x)\,dx} f(x)\,dx. \tag{4}$$

Thus if (2) has a solution, it must be of form (4). Conversely, it is a straightforward exercise in differentiation to verify that (4) constitutes a one-parameter family of solutions of equation (2).

Do not memorize the formula given in (4). There is an equivalent but easier way of solving (2). If (4) is multiplied by

$$e^{\int P(x)\,dx} \tag{5}$$

and then $$e^{\int P(x)\,dx} y = c_1 + \int e^{\int P(x)\,dx} f(x)\,dx \tag{6}$$

is differentiated $$\frac{d}{dx}\left[e^{\int P(x)\,dx} y\right] = e^{\int P(x)\,dx} f(x), \tag{7}$$

we get

$$e^{\int P(x)\,dx} \frac{dy}{dx} + P(x)\, e^{\int P(x)\,dx} y = e^{\int P(x)\,dx} f(x). \tag{8}$$

By dividing this last result by $e^{\int P(x)\,dx}$ we obtain (2).

Method of Solution

The recommended method of solving (2) actually consists of (6)–(8) worked in reverse order. In other words, if (2) is multiplied by (5) we get (8). The left side of (8) is recognized as the derivative of the product of $e^{\int P(x)\,dx}$ and y. This gets us to (7). We then integrate both sides of (7) to get the solution (6). Because we can solve (2) by integration after multiplication by $e^{\int P(x)\,dx}$, we call this function an **integrating factor** for the differential equation. For convenience we can summarize these results as follows.

Solving a Linear First-Order Equation

(i) To solve a linear first-order equation, first put it into the form (2); that is, make the coefficient of dy/dx unity.

(ii) Identify $P(x)$ and find the integrating factor $e^{\int P(x)\,dx}$.

(iii) Multiply the equation obtained in step (i) by the integrating factor:

$$e^{\int P(x)\,dx}\frac{dy}{dx} + P(x)e^{\int P(x)\,dx}y = e^{\int P(x)\,dx}f(x).$$

(iv) The left side of the equation in step (iii) is the derivative of the product of the integrating factor and the dependent variable y; that is,

$$\frac{d}{dx}[e^{\int P(x)\,dx}y] = e^{\int P(x)\,dx}f(x).$$

(v) Integrate both sides of the equation found in step (iv).

EXAMPLE 1 **Solving a Linear DE**

Solve
$$x\frac{dy}{dx} - 4y = x^6e^x.$$

Solution By dividing by x we get the standard form

$$\frac{dy}{dx} - \frac{4}{x}y = x^5e^x. \tag{9}$$

From this form we identify $P(x) = -4/x$, and so the integrating factor is

$$e^{-4\int dx/x} = e^{-4\ln|x|} = e^{\ln x^{-4}} = x^{-4}.$$

Here we have used the basic identity $b^{\log_b N} = N$, $N > 0$. Now we multiply (9) by this term

$$x^{-4}\frac{dy}{dx} - 4x^{-5}y = xe^x \tag{10}$$

and obtain
$$\frac{d}{dx}[x^{-4}y] = xe^x.* \tag{11}$$

It follows from integration by parts that

$$x^{-4}y = xe^x - e^x + c \quad \text{or} \quad y = x^5e^x - x^4e^x + cx^4.$$ ●

* You should perform the indicated differentiations a few times in order to be convinced that all equations, such as (9), (10), and (11), are formally equivalent.

EXAMPLE 2 Solving a Linear DE

Solve $$\frac{dy}{dx} - 3y = 0.$$

Solution This linear equation can be solved by separation of variables. Alternatively, since the equation is already in the standard form (2), we see that the integrating factor is $e^{\int(-3)\,dx} = e^{-3x}$. Multiplying the given equation by this factor gives $e^{-3x}dy/dx - 3e^{-3x}y = 0$. This last equation is the same as

$$\frac{d}{dx}[e^{-3x}y] = 0.$$

Integrating then gives $e^{-3x}y = c$ and therefore $y = ce^{3x}$. ●

General Solution

If it is assumed that $P(x)$ and $f(x)$ are continuous on an interval I and that x_0 is any point in the interval, then it follows from Theorem 2.1 that there exists only one solution of the initial-value problem

$$\frac{dy}{dx} + P(x)y = f(x), \quad y(x_0) = y_0. \tag{12}$$

But we saw earlier that (2) possesses a family of solutions and that every solution of the equation on the interval I is of form (4). Thus, obtaining the solution of (12) is a simple matter of finding an appropriate value of c, in (4). Consequently, we are justified in calling (4) the **general solution** of the differential equation. In retrospect, you should recall that in several instances we found singular solutions of nonlinear equations. This cannot happen in the case of a linear equation when proper attention is paid to solving the equation over a common interval on which $P(x)$ and $f(x)$ are continuous.

EXAMPLE 3 A General Solution

Find the general solution of $$(x^2 + 9)\frac{dy}{dx} + xy = 0.$$

Solution We write $$\frac{dy}{dx} + \frac{x}{x^2 + 9}y = 0.$$

The function $P(x) = x/(x^2 + 9)$ is continuous on $(-\infty, \infty)$. Now the integrating factor for the equation is

$$e^{\int x\,dx/(x^2+9)} = e^{\frac{1}{2}\int 2x\,dx/(x^2+9)} = e^{\frac{1}{2}\ln(x^2+9)} = \sqrt{x^2+9}$$

so that $$\sqrt{x^2+9}\,\frac{dy}{dx} + \frac{x}{\sqrt{x^2+9}}y = 0.$$

Integrating $\quad \dfrac{d}{dx}[\sqrt{x^2+9}\,y] = 0 \quad$ gives $\quad \sqrt{x^2+9}\,y = c.$

Hence the general solution on the interval is

$$y = \frac{c}{\sqrt{x^2+9}}.$$

Except in the case when the lead coefficient is 1, the recasting of equation (1) into the standard form (2) requires division by $a_1(x)$. If $a_1(x)$ is not a constant, then close attention should be paid to the points where $a_1(x) = 0$. Specifically, in (2) the points at which $P(x)$ (formed by dividing $a_0(x)$ by $a_1(x)$) is discontinuous are potentially troublesome. See *Computer Lab Experiment* 2 for this section.

EXAMPLE 4 An Initial-Value Problem

Solve the initial-value problem $x\dfrac{dy}{dx} + y = 2x, \quad y(1) = 0.$

Solution Write the given equation as

$$\frac{dy}{dx} + \frac{1}{x}y = 2$$

and observe that $P(x) = 1/x$ is continuous on any interval not containing the origin. In view of the initial conditions, we solve the problem on the interval $(0, \infty)$.

The integrating factor is $e^{\int dx/x} = e^{\ln x} = x$, and so

$$\frac{d}{dx}[xy] = 2x$$

gives $xy = x^2 + c$. Solving for y yields the general solution

$$y = x + \frac{c}{x}. \tag{13}$$

But $y(1) = 0$ implies $c = -1$. Hence we obtain

$$y = x - \frac{1}{x}, \quad 0 < x < \infty. \tag{14}$$

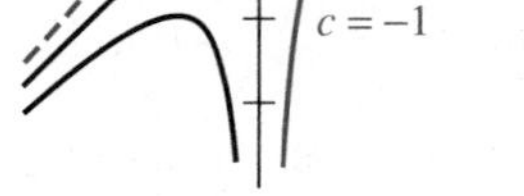

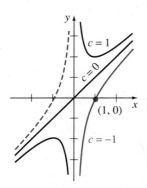

FIGURE 2.13

Considered as a one-parameter family of curves, the graph of (13) is given in Figure 2.13. The solution (14) of the initial-value problem is indicated by the solid colored portion of the graph.

The next example illustrates how to solve (2) when f has a jump discontinuity.

EXAMPLE 5 A Discontinuous $f(x)$

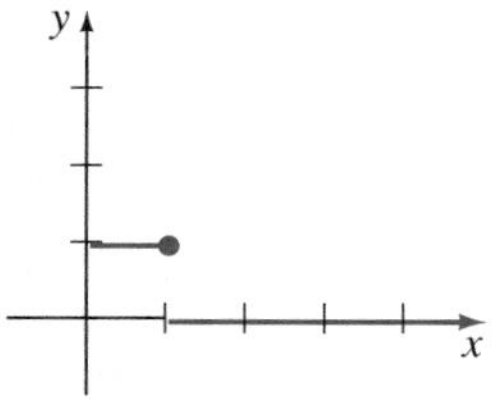

FIGURE 2.14

Find a continuous solution satisfying

$$\frac{dy}{dx} + y = f(x), \quad \text{where} \quad f(x) = \begin{cases} 1, & 0 \le x \le 1 \\ 0, & x > 1 \end{cases}$$

and the initial condition $y(0) = 0$.

Solution From Figure 2.14 we see that f is piecewise continuous with a discontinuity at $x = 1$. Consequently, we solve the problem in two parts corresponding to the two intervals over which f is defined. For $0 \le x \le 1$ we have

$$\frac{dy}{dx} + y = 1 \quad \text{or, equivalently,} \quad \frac{d}{dx}[e^x y] = e^x.$$

Integrating the last equation and solving for y gives $y = 1 + c_1 e^{-x}$. Since $y(0) = 0$ we must have $c_1 = -1$ and therefore

$$y = 1 - e^{-x}, \quad 0 \le x \le 1.$$

For $x > 1$,

$$\frac{dy}{dx} + y = 0,$$

leads to $y = c_2 e^{-x}$. Hence we can write

$$y = \begin{cases} 1 - e^{-x}, & 0 \le x \le 1 \\ c_2 e^{-x}, & x > 1. \end{cases}$$

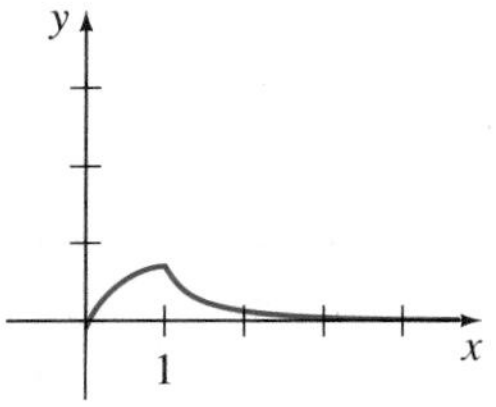

FIGURE 2.15

Now, in order that y be a continuous function, we certainly want $\lim_{x \to 1^+} y(x) = y(1)$. This latter requirement is equivalent to $c_2 e^{-1} = 1 - e^{-1}$, or $c_2 = e - 1$. As Figure 2.15 shows, the function

$$y = \begin{cases} 1 - e^{-x}, & 0 \le x \le 1 \\ (e - 1)e^{-x}, & x > 1 \end{cases}$$

is continuous but not differentiable at $x = 1$. ●

Use of Substitutions: Bernoulli's Equation

The differential equation

$$\frac{dy}{dx} + P(x)y = f(x)y^n, \tag{15}$$

where n is any real number, is called **Bernoulli's equation**. Note that for $n = 0$ and $n = 1$, equation (15) is linear. For $n \neq 0$ and $n \neq 1$, the substitution $u = y^{1-n}$ reduces any equation of form (15) to a linear equation.

EXAMPLE 6 Solving a Bernoulli DE

Solve
$$x\frac{dy}{dx} + y = x^2y^2.$$

Solution We first rewrite the equation as

$$\frac{dy}{dx} + \frac{1}{x}y = xy^2$$

by dividing by x. With $n = 2$, we next substitute

$$y = u^{-1} \quad \text{and} \quad \frac{dy}{dx} = -u^{-2}\frac{du}{dx} \quad \leftarrow \text{chain rule}$$

into the given equation and simplify. The result is

$$\frac{du}{dx} - \frac{1}{x}u = -x.$$

The integrating factor for this linear equation on, say, $(0, \infty)$ is

$$e^{-\int dx/x} = e^{-\ln x} = e^{\ln x^{-1}} = x^{-1}.$$

Integrating
$$\frac{d}{dx}[x^{-1}u] = -1$$

gives
$$x^{-1}u = -x + c \quad \text{or} \quad u = -x^2 + cx.$$

Since $u = y^{-1}$ we have $y = 1/u$ and so a solution of the equation is

$$y = \frac{1}{-x^2 + cx}.$$

●

Note that we have not obtained the general solution of the original nonlinear differential equation in Example 6 because $y = 0$ is a singular solution of the equation.

Remark Occasionally a first-order differential equation is not linear in one variable but is linear in the other variable. For example, the differential equation

$$\frac{dy}{dx} = \frac{1}{x + y^2}$$

is not linear in the variable y. But its reciprocal

$$\frac{dx}{dy} = x + y^2 \quad \text{or} \quad \frac{dx}{dy} - x = y^2$$

is linear in the variable x. You should verify that the integrating factor $e^{\int(-1)\,dy} = e^{-y}$ and integration by parts yield $x = -y^2 - 2y - 2 + ce^y$.

EXERCISES 2.4

Answers to odd-numbered problems begin on page AN-6.

In Problems 1–40 find the general solution of the given differential equation. State an interval on which the general solution is defined.

1. $\dfrac{dy}{dx} = 5y$
2. $\dfrac{dy}{dx} + 2y = 0$
3. $3\dfrac{dy}{dx} + 12y = 4$
4. $x\dfrac{dy}{dx} + 2y = 3$
5. $\dfrac{dy}{dx} + y = e^{3x}$
6. $\dfrac{dy}{dx} = y + e^x$
7. $y' + 3x^2y = x^2$
8. $y' + 2xy = x^3$
9. $x^2y' + xy = 1$
10. $y' = 2y + x^2 + 5$
11. $(x + 4y^2)\,dy + 2y\,dx = 0$
12. $\dfrac{dx}{dy} = x + y$
13. $x\,dy = (x \sin x - y)\,dx$
14. $(1 + x^2)\,dy + (xy + x^3 + x)\,dx = 0$
15. $(1 + e^x)\dfrac{dy}{dx} + e^xy = 0$
16. $(1 - x^3)\dfrac{dy}{dx} = 3x^2y$
17. $\cos x\dfrac{dy}{dx} + y \sin x = 1$
18. $\dfrac{dy}{dx} + y \cot x = 2 \cos x$
19. $x\dfrac{dy}{dx} + 4y = x^3 - x$
20. $(1 + x)y' - xy = x + x^2$
21. $x^2y' + x(x + 2)y = e^x$
22. $xy' + (1 + x)y = e^{-x} \sin 2x$
23. $\cos^2 x \sin x\,dy + (y \cos^3 x - 1)\,dx = 0$
24. $(1 - \cos x)\,dy + (2y \sin x - \tan x)\,dx = 0$
25. $y\,dx + (xy + 2x - ye^y)\,dy = 0$
26. $(x^2 + x)\,dy = (x^5 + 3xy + 3y)\,dx$
27. $x\dfrac{dy}{dx} + (3x + 1)y = e^{-3x}$
28. $(x + 1)\dfrac{dy}{dx} + (x + 2)y = 2xe^{-x}$
29. $y\,dx - 4(x + y^6)\,dy = 0$
30. $xy' + 2y = e^x + \ln x$
31. $\dfrac{dy}{dx} + y = \dfrac{1 - e^{-2x}}{e^x + e^{-x}}$
32. $\dfrac{dy}{dx} - y = \sinh x$
33. $y\,dx + (x + 2xy^2 - 2y)\,dy = 0$
34. $y\,dx = (ye^y - 2x)\,dy$
35. $\dfrac{dr}{d\theta} + r \sec \theta = \cos \theta$
36. $\dfrac{dP}{dt} + 2tP = P + 4t - 2$
37. $(x + 2)^2\dfrac{dy}{dx} = 5 - 8y - 4xy$
38. $(x^2 - 1)\dfrac{dy}{dx} + 2y = (x + 1)^2$
39. $y' = (10 - y) \cosh x$
40. $dx = (3e^y - 2x)\,dy$

In Problems 41–50 solve the given differential equation subject to the indicated initial condition.

41. $\dfrac{dy}{dx} + 5y = 20, \quad y(0) = 2$
42. $y' = 2y + x(e^{3x} - e^{2x}), \quad y(0) = 2$
43. $L\dfrac{di}{dt} + Ri = E; \quad L, R, \text{ and } E \text{ constants}, i(0) = i_0$
44. $y\dfrac{dx}{dy} - x = 2y^2, \quad y(1) = 5$
45. $y' + (\tan x)y = \cos^2 x, \quad y(0) = -1$
46. $\dfrac{dQ}{dx} = 5x^4Q, \quad Q(0) = -7$
47. $\dfrac{dT}{dt} = k(T - 50); \quad k \text{ a constant}, T(0) = 200$
48. $x\,dy + (xy + 2y - 2e^{-x})\,dx = 0, \quad y(1) = 0$
49. $(x + 1)\dfrac{dy}{dx} + y = \ln x, \quad y(1) = 10$
50. $xy' + y = e^x, \quad y(1) = 2$

In Problems 51–54 find a continuous solution satisfying the given differential equation and the indicated initial condition.

51. $\dfrac{dy}{dx} + 2y = f(x), \quad f(x) = \begin{cases} 1, & 0 \le x \le 3 \\ 0, & x > 3 \end{cases}, \quad y(0) = 0$

52. $\frac{dy}{dx} + y = f(x), \quad f(x) = \begin{cases} 1, & 0 \le x \le 1 \\ -1, & x > 1 \end{cases}, \quad y(0) = 1$

53. $\frac{dy}{dx} + 2xy = f(x), \quad f(x) = \begin{cases} x, & 0 \le x < 1 \\ 0, & x \ge 1 \end{cases}, \quad y(0) = 2$

54. $(1 + x^2)\frac{dy}{dx} + 2xy = f(x),$

$$f(x) = \begin{cases} x, & 0 \le x < 1 \\ -x, & x \ge 1 \end{cases}, \quad y(0) = 0$$

In Problems 55–58 solve the given Bernoulli equation.

55. $x\frac{dy}{dx} + y = \frac{1}{y^2}$

56. $\frac{dy}{dx} - y = e^x y^2$

57. $\frac{dy}{dx} = y(xy^3 - 1)$

58. $x\frac{dy}{dx} - (1 + x)y = xy^2$

In Problems 59–60 solve the given Bernoulli equation subject to the indicated initial condition.

59. $x^2\frac{dy}{dx} - 2xy = 3y^4, \quad y(1) = \frac{1}{2}$

60. $y^{1/2}\frac{dy}{dx} + y^{3/2} = 1, \quad y(0) = 4$

61. The following system of differential equations is encountered in the study of a special type of radioactive series of elements:

$$\frac{dx}{dt} = -\lambda_1 x$$

$$\frac{dy}{dt} = \lambda_1 x - \lambda_2 y$$

where λ_1 and λ_2 are constants. Solve the system subject to $x(0) = x_0$, $y(0) = y_0$.

62. *WRITING PROJECT* Write a short report on **Ricatti's differential equation**. Illustrate its method of solution.

63. *PROJECT PROBLEM* Try to find a substitution that will enable you to find two explicit solutions of the linear second-order equation

$$y'' - y' - e^{2x}y = 0.$$

64. *PROJECT PROBLEM* Solve

$$\frac{d^{n+1}y}{dx^{n+1}} - \frac{d^n y}{dx^n} = 1.$$

65. *PROJECT PROBLEM* The basic distinction between the so-called *elementary functions* and the *nonelementary functions* is simply a matter of familiarity. In other words, in practice we use functions such as x^n, e^x, $\sin x$, and $\cos x$ more often than we do functions such as erf(x), erfc(x), Si(x), Ci(x), S(x), and C(x).

(a) Write a short report defining the **error function**, **complementary error function**, **sine** and **cosine integral functions**, and **Fresnel integrals**.

(b) Show that the solution of the initial-value problem $y' - 2xy = -1$, $y(0) = \sqrt{\pi}/2$, is $y = (\sqrt{\pi}/2)e^{x^2}\operatorname{erfc}(x)$. Use tables or a computer to calculate $y(2)$. Use a CAS to graph the solution for $x \ge 0$.

(c) Show that the solution of the initial-value problem $x^3y' + 2x^2y = 10 \sin x$, $y(1) = 0$ is $y = 10x^{-2}[\mathrm{Si}(x) - \mathrm{Si}(1)]$. Use tables or a computer to calculate $y(2)$. Use a CAS to graph the solution for $x > 0$.

2.5 Applications

Introduction

In science and engineering, business and economics, and even psychology, we often wish to describe, or **model**, the behavior of some system or phenomenon in mathematical terms. This description starts with

(i) identifying the variables that are responsible for changing the system, and

(ii) a set of reasonable assumptions about the system.

These assumptions also include any empirical laws that are applicable to the system. The mathematical construct of all these assumptions, or the **mathematical model** of the system, is in many instances a differential equation or a system of differential equations. We expect a reasonable mathematical model of a system to have a solution that is consistent with the known behavior of the system.

A mathematical model of a physical system often involves the variable time. The solution of the model then gives the **state of the system**; in other words, for appropriate values of time t, the values of the dependent variable (or variables) describe the system in the past, present, and future.

2.5.1 Linear Equations

Growth and Decay

The initial-value problem

$$\frac{dx}{dt} = kx, \quad x(t_0) = x_0, \tag{1}$$

where k is a constant of proportionality, occurs in many physical theories involving either **growth** or **decay**. For example, in biology it is often observed that the rate at which certain bacteria grow is proportional to the number of bacteria present at any time. Over short intervals of time, the population of small animals, such as rodents, can be predicted fairly accurately by the solution of (1). In physics an initial-value problem such as (1) provides a model for approximating the remaining amount of a substance that is disintegrating, or decaying, through radioactivity. The differential equation in (1) could also determine the temperature of a cooling body. In chemistry the amount of a substance remaining during certain reactions is also described by (1).

The constant of proportionality k in (1) can be determined from the solution of the initial-value problem using a subsequent measurement of x at a time $t_1 > t_0$.

EXAMPLE 1 Bacterial Growth

A culture initially has N_0 number of bacteria. At $t = 1$ hour the number of bacteria is measured to be $(3/2)N_0$. If the rate of growth is proportional to the number of bacteria present, determine the time necessary for the number of bacteria to triple.

Solution We first solve the differential equation

$$\frac{dN}{dt} = kN \tag{2}$$

subject to $N(0) = N_0$. Then we use the empirical condition $N(1) = (3/2)N_0$ to determine the constant of proportionality k.

Now (2) is both separable and linear. When it is put in the form

$$\frac{dN}{dt} - kN = 0,$$

we can see by inspection that the integrating factor is e^{-kt}. Multiplying both sides of the equation by this term immediately gives

$$\frac{d}{dt}[e^{-kt}N] = 0.$$

Integrating both sides of the last equation yields

$$e^{-kt}N = c \quad \text{or} \quad N(t) = ce^{kt}.$$

At $t = 0$ it follows that $N_0 = ce^0 = c$ and so $N(t) = N_0e^{kt}$. At $t = 1$ we have $(3/2)N_0 = N_0e^k$ or $e^k = 3/2$. From the last equation we get $k = \ln(3/2) = 0.4055$. Thus

$$N(t) = N_0e^{0.4055t}.$$

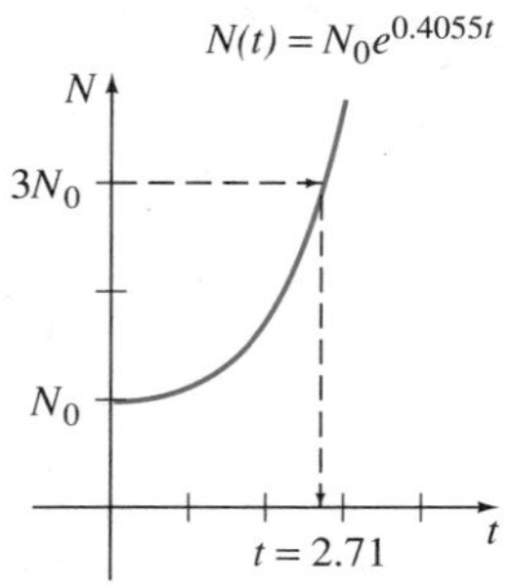

FIGURE 2.16

To find the time at which the number of bacteria have tripled, we solve $3N_0 = N_0e^{0.4055t}$ for t. It follows that $0.4055\ t = \ln 3$, and so

$$t = \frac{\ln 3}{0.4055} \approx 2.71 \text{ hours}.$$

See Figure 2.16. ●

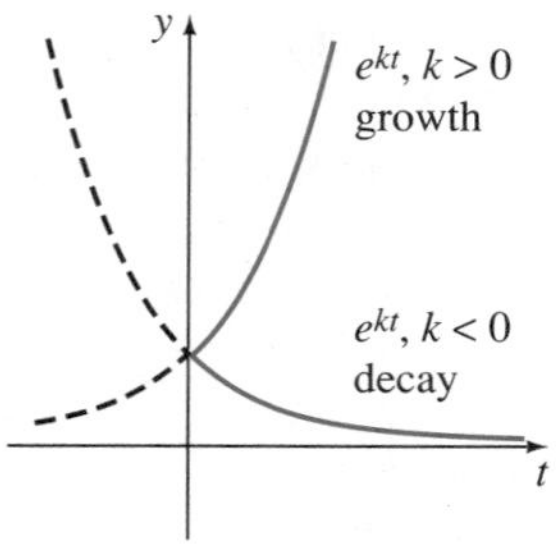

FIGURE 2.17

Notice in Example 1 that the actual number N_0 of bacteria present at time $t = 0$ played no part in determining the time required to triple the number in the culture. The time necessary for an initial population of, say, 100 or 1,000,000 bacteria to triple is still approximately 2.71 hours.

As shown in Figure 2.17, the exponential function e^{kt} increases as t increases for $k > 0$, and decreases as t increases for $k < 0$. Thus problems describing growth, such as population, bacteria, or even capital, are characterized by a positive value of k, whereas problems involving decay, as in radioactive disintegration, yield a negative k value. Accordingly, we say that k is either a **growth constant** ($k > 0$) or a **decay constant** ($k < 0$).

You should also verify that the growth and decay behavior just described is consistent with the fact that the differential equation $dx/dt = kx$ is autonomous and that a solution $x(t)$ moves away from 0 (that is, grows) because 0 is an unstable critical point of the equation for $k > 0$ and a solution $x(t)$ moves closer and closer to 0 as $t \to \infty$ (that is, decays) because 0 is an asymptotically stable critical point of the equation for $k < 0$.

Half-Life

In physics the **half-life** is a measure of the stability of a radioactive substance. The half-life is simply the time it takes for one-half of the atoms in an initial amount A_0 to disintegrate, or transmute, into the atoms of another element. The longer the half-life of a substance, the more stable it is. For example, the half-life of highly radioactive radium, Ra-226, is about 1700 years. In 1700 years one-half of a given quantity of Ra-226 is transmuted into radon, Rn-222. The most commonly occurring uranium isotope, U-238, has a half-life of approximately 4,500,000,000 years. In about 4.5 billion years, one-half of a quantity of U-238 is transmuted into lead, Pb-206.

EXAMPLE 2 Half-Life of Plutonium

A breeder reactor converts relatively stable uranium 238 into the isotope plutonium 239. After 15 years it is determined that 0.043% of the initial amount A_0 of the plutonium has disintegrated. Find the half-life of this isotope if the rate of disintegration is proportional to the amount remaining.

Solution Let $A(t)$ denote the amount of plutonium remaining at any time. As in Example 1, the solution of the initial-value problem

$$\frac{dA}{dt} = kA, \quad A(0) = A_0,$$

is $A(t) = A_0e^{kt}$. If 0.043% of the atoms of A_0 have disintegrated, then 99.957% of the substance remains. To find the decay constant k we use $0.99957A_0 = A(15)$, that is, $0.99957A_0 = A_0e^{15k}$. Solving for k then gives $k = (1/15)\ln(0.99957) = -0.00002867$. Hence

$$A(t) = A_0e^{-0.00002867t}.$$

Now the half-life is the corresponding value of time at which $A(t) = A_0/2$. Solving for t gives $A_0/2 = A_0e^{-0.00002867t}$ or $1/2 = e^{-0.00002867t}$. The last equation yields

$$t = \frac{\ln 2}{0.00002867} \approx 24{,}180 \text{ years.}$$

●

Carbon Dating

About 1950 the chemist Willard Libby devised a method of using radioactive carbon as a means of determining the approximate ages of fossils. The theory of **carbon dating** is based on the fact that the isotope carbon 14 is produced in the atmosphere by the action of cosmic radiation on nitrogen. The ratio of the amount of C-14 to ordinary carbon in the atmosphere appears to be a constant, and as a consequence the proportionate amount of the isotope present in all living organisms is the same as that in the atmosphere. When an organism dies, the absorption of C-14, by either breathing or eating, ceases. Thus, by comparing the proportionate amount of C-14 present, say, in a fossil with the constant ratio found in the atmosphere, it is possible to obtain a reasonable estimation of its age. The method is based on the knowledge that the half-life of the radioactive C-14 is approximately 5600 years. For his work Libby won the Nobel Prize for chemistry in 1960. Libby's method has been used to date wooden furniture in Egyptian tombs and the woven flax wrappings of the Dead Sea scrolls.

EXAMPLE 3 Age of a Fossil

A fossilized bone is found to contain 1/1000 the original amount of C-14. Determine the age of the fossil.

Solution The starting point is again $A(t) = A_0e^{kt}$. To determine the value of the decay constant k we use the fact that $A_0/2 = A(5600)$, or $A_0/2 = A_0e^{5600k}$. From $5600k = \ln(1/2) = -\ln 2$ we get $k = -(\ln 2)/5600 = -0.00012378$. Therefore

$$A(t) = A_0e^{-0.00012378t}.$$

With $A(t) = A_0/1000$ we have $A_0/1000 = A_0e^{-0.00012378t}$ so that $-0.00012378t = \ln(1/1000) = -\ln 1000$. Thus

$$t = \frac{\ln 1000}{0.00012378} \approx 55{,}800 \text{ years.}$$

The date found in Example 3 is really at the border of accuracy for this method. The usual carbon-14 technique is limited to about 9 half-lives of the isotope, or about 50,000 years. One reason is that the chemical analysis needed to obtain an accurate measurement of the remaining C-14 becomes somewhat formidable around the point of $A_0/1000$. Also, this analysis demands the destruction of a rather large sample of the specimen. If this measurement is accomplished indirectly, based on the actual radioactivity of the specimen, then it is very difficult to distinguish between the radiation from the fossil and the normal background radiation. But in recent developments, the use of a particle accelerator has enabled scientists to separate the C-14 from the stable C-12 directly. By computing the precise value of the ratio of C-14 to C-12, the accuracy of this method can be extended to 70,000–100,000 years. Other isotopic techniques such as using potassium 40 and argon 40 can give dates of several million years. Nonisotopic methods based on the use of amino acids are also sometimes possible.

Cooling

Newton's law of cooling states that the rate at which the temperature $T(t)$ changes in a cooling body is proportional to the difference between the temperature in the body and the constant temperature T_m of the surrounding medium. The foregoing sentence is a verbal description of a linear differential equation, that is,

$$\frac{dT}{dt} = k(T - T_m), \tag{3}$$

where k is a constant of proportionality.

Series Circuits

In a series circuit containing only a resistor and an inductor, Kirchhoff's second law states that the sum of the voltage drop across the inductor $(L(di/dt))$ and the

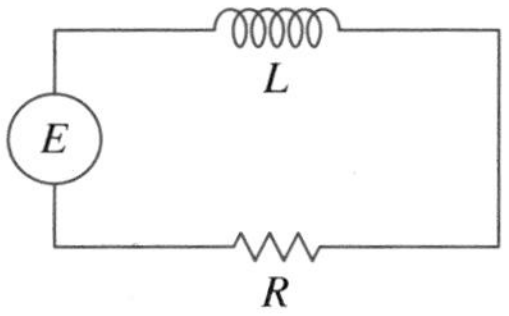

L-R Series Circuit

FIGURE 2.18

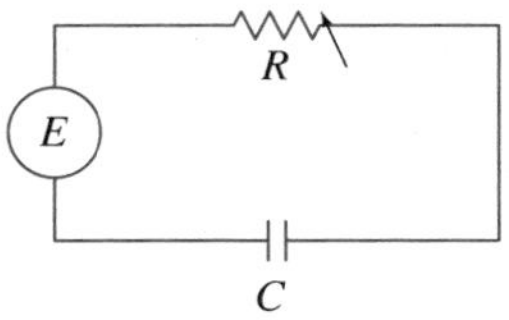

R-C Series Circuit

FIGURE 2.19

voltage drop across the resistor (iR) is the same as the impressed voltage ($E(t)$) on the circuit. See Figure 2.18.

Thus we obtain the linear differential equation for the current $i(t)$,

$$L\frac{di}{dt} + Ri = E(t), \tag{4}$$

where L and R are constants known as the inductance and the resistance, respectively. The current $i(t)$ is sometimes called the **response** of the system.

The voltage drop across a capacitor with capacitance C is given by $q(t)/C$, where q is the charge on the capacitor. Hence, for the series circuit shown in Figure 2.19, Kirchhoff's second law gives

$$Ri + \frac{1}{C}q = E(t). \tag{5}$$

But current i and charge q are related by $i = dq/dt$, so (5) becomes the linear differential equation

$$R\frac{dq}{dt} + \frac{1}{C}q = E(t). \tag{6}$$

EXAMPLE 4 Series Circuit

A 12-volt battery is connected to a series circuit in which the inductance is $\frac{1}{2}$ henry and the resistance is 10 ohms. Determine the current i if the initial current is zero.

Solution From (4) we see that we must solve

$$\frac{1}{2}\frac{di}{dt} + 10i = 12$$

subject to $i(0) = 0$. First, we multiply the differential equation by 2 and read off the integrating factor e^{20t}. We then obtain

$$\frac{d}{dt}[e^{20t}i] = 24e^{20t}.$$

Integrating each side of the last equation and solving for i gives

$$i = \tfrac{6}{5} + ce^{-20t}.$$

Now $i(0) = 0$ implies $0 = 6/5 + c$, or $c = -6/5$. Therefore the response is

$$i(t) = \frac{6}{5} - \frac{6}{5}e^{-20t}.$$

●

From (4) of Section 2.4 we can write a general solution of (4):

$$i(t) = \frac{e^{-(R/L)t}}{L}\int e^{(R/L)t}E(t)\,dt + ce^{-(R/L)t}. \tag{7}$$

In particular, when $E(t) = E_0$ is a constant, (7) becomes

$$i(t) = \frac{E_0}{R} + ce^{-(R/L)t}. \tag{8}$$

Note that as $t \to \infty$, the second term in equation (8) approaches zero. Such a term is usually called a **transient term**; any remaining terms are called the **steady-state** part of the solution. In this case E_0/R is also called the **steady-state current**; for large values of time it then appears that the current in the circuit is simply governed by Ohm's law ($E = iR$).

Mixture Problem

The mixing of two fluids sometimes gives rise to a linear first-order differential equation. In the next example we consider the mixture of two salt solutions with different concentrations.

EXAMPLE 5 Mixture of Two Salt Solutions

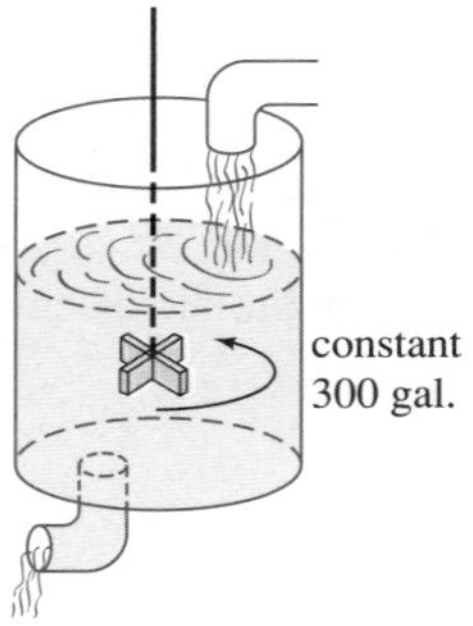

FIGURE 2.20

Initially 50 pounds of salt is dissolved in a large tank holding 300 gallons of water. A brine solution is pumped into the tank at a rate of 3 gallons per minute, and the well-stirred solution is then pumped out at the same rate. See Figure 2.20. If the concentration of the solution entering is 2 pounds per gallon, determine the amount of salt in the tank at any time. How much salt is present after 50 minutes? After a long time?

Solution Let $A(t)$ be the amount of salt (in pounds) in the tank at any time. For problems of this sort, the net rate at which $A(t)$ changes is given by

$$\frac{dA}{dt} = \left(\begin{matrix} \textit{rate of} \\ \textit{substance entering} \end{matrix}\right) - \left(\begin{matrix} \textit{rate of} \\ \textit{substance leaving} \end{matrix}\right) = R_1 - R_2. \tag{9}$$

Now the rate at which the salt enters the tank is, in pounds per minute,

$$R_1 = (3 \text{ gal/min}) \cdot (2 \text{ lb/gal}) = 6 \text{ lb/min},$$

whereas the rate at which salt is leaving is

$$R_2 = (3 \text{ gal/min}) \cdot \left(\frac{A}{300} \text{ lb/gal}\right) = \frac{A}{100} \text{ lb/min}.$$

Thus equation (9) becomes

$$\frac{dA}{dt} = 6 - \frac{A}{100}, \tag{10}$$

which we solve subject to the initial condition $A(0) = 50$.

Since the integrating factor is $e^{t/100}$, we can write (10) as

$$\frac{d}{dt}[e^{t/100}A] = 6e^{t/100}$$

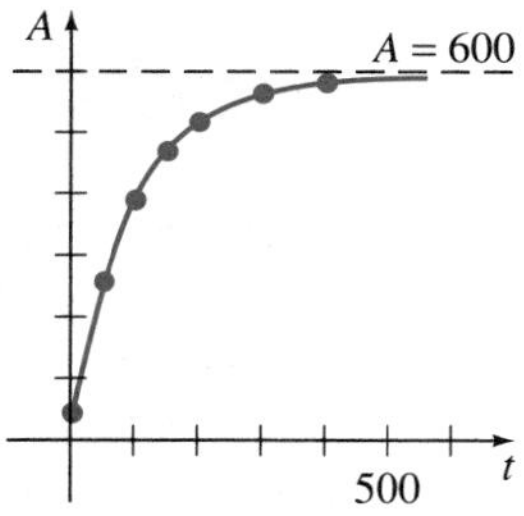

(a)

t (minutes)	A (lbs)
50	266.41
100	397.67
150	477.27
200	525.57
300	572.62
400	589.93

(b)

FIGURE 2.21

and therefore $$e^{t/100}A = 600e^{t/100} + c$$

$$A = 600 + ce^{-t/100}.$$

When $t = 0$, $A = 50$, so we find that $c = -550$. Finally, we obtain

$$A(t) = 600 - 550e^{-t/100}. \tag{11}$$

At $t = 50$ we find $A(50) = 266.41$ pounds. Also, as $t \to \infty$ it is seen from (11) and Figure 2.21 that $A \to 600$. Of course this is what we would expect; over a long period of time the number of pounds of salt in the solution must be

$$(300 \text{ gal})(2 \text{ lb/gal}) = 600 \text{ lb}.$$

This long-term behavior can also be deduced from the fact that 600 is an asymptotically stable critical point of equation (10). ●

In Example 5 we assumed that the rate at which the solution was pumped in was the same as the rate at which the solution was pumped out. This need not be the case; the mixed brine solution could be pumped out at a rate faster or slower than the rate at which the other solution is pumped in. For example, if the well-stirred solution in Example 5 is pumped out at a slower rate of 2 gallons per minute, the solution is accumulating at a rate of $(3 - 2)$ gal/min $=$ 1 gal/min. After t minutes there are $300 + t$ gallons of brine in the tank. The rate at which the salt is leaving is then

$$R_2 = (2 \text{ gal/min}) \left(\frac{A}{300 + t} \text{ lb/gal} \right).$$

Hence equation (9) becomes

$$\frac{dA}{dt} + \frac{2A}{300 + t} = 6.$$

You should verify that the solution of this equation subject to $A(0) = 50$ is

$$A(t) = 600 + 2t - (4.95 \times 10^7)(300 + t)^{-2}.$$

2.5.2 Nonlinear Equations

Logistic Equation

At the beginning of this section we saw that if a population P is described by $dP/dt = kP$, $k > 0$, then $P(t)$ exhibits unbounded exponential growth. In many instances this linear equation is an unrealistic model for the growth of a population; that is, what is actually observed differs substantially from what is predicted by the solution of the equation.

Around 1840 the Belgian mathematician-biologist P. F. Verhulst was concerned with mathematical models for predicting the human population of various countries. One of the equations he studied was

$$\frac{dP}{dt} = P(a - bP), \tag{12}$$

where $a > 0$, $b > 0$. Equation (12) came to be known as the **logistic equation**, and its solution is called the **logistic function**. The graph of a logistic function is called a **logistic curve**.

The differential equation $dP/dt = kP$ does not provide a very accurate model for population when the population itself is very large. Overcrowded conditions with the resulting detrimental effects on the environment, such as pollution and excessive and competitive demands for food and fuel, can have an inhibitive effect on population growth. If a, $a > 0$, is a constant average birth rate, let us assume that the average death rate is proportional to the population $P(t)$ at any time. Thus, if $(1/P)(dP/dt)$ is the rate of growth per individual in a population, then

$$\frac{1}{P}\frac{dP}{dt} = \begin{pmatrix} \text{average} \\ \text{birth rate} \end{pmatrix} - \begin{pmatrix} \text{average} \\ \text{death rate} \end{pmatrix} = a - bP,$$

where b is a positive constant of proportionality. Cross multiplying by P immediately gives (12).

As we shall now see, certain solutions of (12) are bounded as $t \to \infty$. If we rewrite (12) as $dP/dt = aP - bP^2$, the term $-bP^2$, $b > 0$, can be interpreted as an "inhibition" or "competition" term. Also, in most applications, the positive constant a is much larger than the constant b.

Logistic curves have proved to be quite accurate in predicting the growth patterns, in a limited space, of certain types of bacteria, protozoa, water fleas (*Daphnia*), and fruit flies (*Drosophila*). This differential equation also provides a reasonable model for describing the spread of an epidemic brought about by initially introducing an infected individual into a static population. The solution $x(t)$ represents the number of individuals infected with the disease at any time. Sociologists and even business analysts have borrowed this latter model to study the spread of information and the impact of advertising in centers of population.

Solution of Equation

One method of solving (12) is by separation of variables. Decomposing the left side of $dP/P(a - bP) = dt$ into partial fractions and integrating gives

$$\left[\frac{1/a}{P} + \frac{b/a}{a - bP}\right] dP = dt$$

$$\frac{1}{a}\ln|P| - \frac{1}{a}\ln|a - bP| = t + c$$

$$\ln\left|\frac{P}{a - bP}\right| = at + ac$$

$$\frac{P}{a - bP} = c_1 e^{at}.$$

It follows from the last equation that

$$P(t) = \frac{ac_1e^{at}}{1 + bc_1e^{at}} = \frac{ac_1}{bc_1 + e^{-at}}.$$

If $P(0) = P_0$, $P_0 \neq a/b$, we find $c_1 = P_0/(a - bP_0)$, and so, after substituting and simplifying, the solution becomes

$$P(t) = \frac{aP_0}{bP_0 + (a - bP_0)e^{-at}}. \tag{13}$$

Graphs of *P*(*t*)

We have already examined the graphs of solutions of the autonomous equation (12) for different initial conditions in Figure 2.7. Recall from Example 6 of Section 2.1, 0 is an unstable critical point, and a/b is an asymptotically stable critical point, and that $x(t) = a/b$ is an equilibrium solution of the logistic equation. Thus, for $0 < P_0 < a/b$, the basic graph of $P(t)$ is the same as that given in color in Figure 2.7, and thus we know that the population $P(t) \to a/b$ as $t \to \infty$. The dashed line $P = a/2b$ shown in Figure 2.22 corresponds to the ordinate of a point of inflection of the logistic curve. To see this, we differentiate (12) by the product rule

$$\begin{aligned}\frac{d^2P}{dt^2} = P\left(-b\frac{dP}{dt}\right) + (a - bP)\frac{dP}{dt} &= \frac{dP}{dt}(a - 2bP)\\ &= P(a - bP)(a - 2bP)\\ &= 2b^2P\left(P - \frac{a}{b}\right)\left(P - \frac{a}{2b}\right).\end{aligned}$$

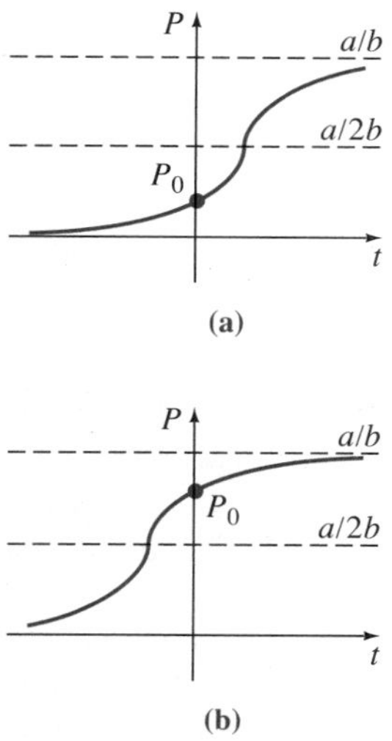

FIGURE 2.22

From calculus recall that the points where $d^2P/dt^2 = 0$ are possible points of inflection, but $P = 0$ and $P = a/b$ can obviously be ruled out. Hence $P = a/2b$ is the only possible ordinate value at which the concavity of the graph can change. For $0 < P < a/2b$ it follows that $P'' > 0$, and $a/2b < P < a/b$ implies $P'' < 0$. Thus, as we read from left to right, the graph changes from concave up to concave down at the point corresponding to $P = a/2b$. When the initial value satisfies $0 < P_0 < a/2b$, the graph of $P(t)$ assumes the shape of an S, as we see in Figure 2.22(a). For $a/2b < P_0 < a/b$ the graph is still S-shaped, but the point of inflection occurs at a negative value of t, as shown in Figure 2.22(b).

EXAMPLE 6 Logistic Growth

Suppose a student carrying a flu virus returns to an isolated college campus of 1000 students. If it is assumed that the rate at which the virus spreads is proportional not only to the number x of infected students but also to the number of students not infected, determine the number of infected students after 6 days if it is further observed that after 4 days $x(4) = 50$.

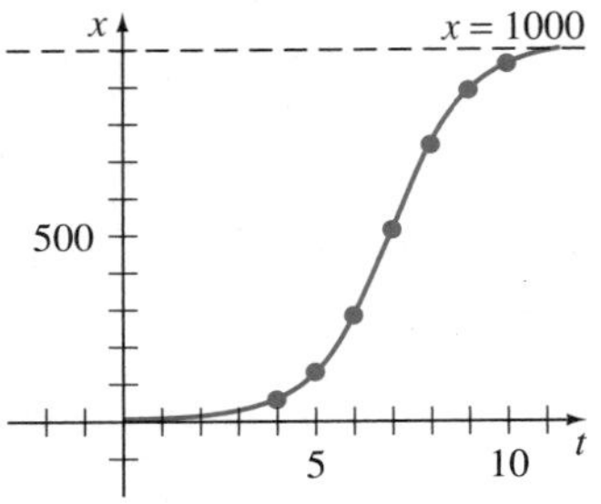

(a)

t (days)	x (number infected)
4	50 (observed)
5	124
6	276
7	507
8	735
9	882
10	953

(b)

FIGURE 2.23

Solution Assuming that no one leaves the campus throughout the duration of the disease, we must solve the initial-value problem

$$\frac{dx}{dt} = kx(1000 - x), \quad x(0) = 1.$$

By making the identifications $a = 1000k$ and $b = k$, we have immediately from (13) that

$$x(t) = \frac{1000k}{k + 999ke^{-1000kt}} = \frac{1000}{1 + 999e^{-1000kt}}.$$

Now, using the information $x(4) = 50$, we determine k from

$$50 = \frac{1000}{1 + 999e^{-4000k}}.$$

We find $-1000k = (1/4)\ln(19/999) = -0.9906$. Thus

$$x(t) = \frac{1000}{1 + 999e^{-0.9906t}}.$$

Finally $$x(6) = \frac{1000}{1 + 999e^{-5.9436}} = 276 \text{ students}.$$

Additional calculated values of $x(t)$ are given in the table in Figure 2.23(b). ●

Chemical Reactions

The disintegration of a radioactive substance, governed by equation (1) of this section, is said to be a **first-order reaction**. In chemistry a few reactions follow the same empirical law: If the molecules of a substance A decompose into smaller molecules, it is a natural assumption that the rate at which this decomposition takes place is proportional to the amount of the first substance that has not undergone conversion; that is, if $X(t)$ is the amount of substance A remaining at any time, then $dX/dt = kX$, where k is negative, since X is decreasing. An example of a first-order chemical reaction is the conversion of t-butyl chloride into t-butyl alcohol:

$$(CH_3)_3CCl + NaOH \rightarrow (CH_3)_3COH + NaCl.$$

Only the concentration of the t-butyl chloride controls the rate of reaction.

Now, in the reaction

$$CH_3Cl + NaOH \rightarrow CH_3OH + NaCl,$$

for every molecule of methyl chloride one molecule of sodium hydroxide is consumed, thus forming one molecule of methyl alcohol and one molecule of sodium chloride. In this case the rate at which the reaction proceeds is proportional to the product of the remaining concentrations of CH_3Cl and of NaOH. If X denotes the amount of CH_3OH formed and α and β are the given amounts of the first two chemicals A and B, then the instantaneous amounts not converted to chemical C are $\alpha - X$ and $\beta - X$, respectively. Hence the rate of formation of C is given by

$$\frac{dX}{dt} = k(\alpha - X)(\beta - X), \tag{14}$$

where k is a constant of proportionality. A reaction described by equation (14) is said to be of **second-order**.

EXAMPLE 7 Second-Order Chemical Reaction

A compound C is formed when two chemicals A and B are combined. The resulting reaction between the two chemicals is such that for each gram of A, 4 grams of B are used. It is observed that 30 grams of the compound C are formed in 10 minutes. Determine the amount of C at any time if the rate of the reaction is proportional to the amounts of A and B remaining and if initially there are 50 grams of A and 32 grams of B. How much of the compound C is present at 15 minutes? Interpret the solution as $t \to \infty$.

Solution Let $X(t)$ denote the number of grams of the compound C present at any time t. Clearly $X(0) = 0$ g and $X(10) = 30$ g.

If, for example, there are 2 grams of compound C, we must have used, say, a grams of A and b grams of B so that $a + b = 2$ and $b = 4a$. Thus we must use $a = 2/5 = 2(1/5)$ grams of chemical A and $b = 8/5 = 2(4/5)$ grams of B. In general, for X grams of C we must use

$$\frac{X}{5} \text{ grams of } A \quad \text{and} \quad \frac{4}{5}X \text{ grams of } B.$$

The amounts of A and B remaining at any time are then

$$50 - \frac{X}{5} \quad \text{and} \quad 32 - \frac{4}{5}X,$$

respectively.

Now we know that the rate at which chemical C is formed satisfies

$$\frac{dX}{dt} \propto \left(50 - \frac{X}{5}\right)\left(32 - \frac{4}{5}X\right).$$

To simplify the subsequent algebra, we factor 1/5 from the first term and 4/5 from the second, and then introduce the constant of proportionality:

$$\frac{dX}{dt} = k(250 - X)(40 - X).$$

By separation of variables and partial fractions, we can write

$$-\frac{1/210}{250 - X}\,dX + \frac{1/210}{40 - X}\,dX = k\,dt.$$

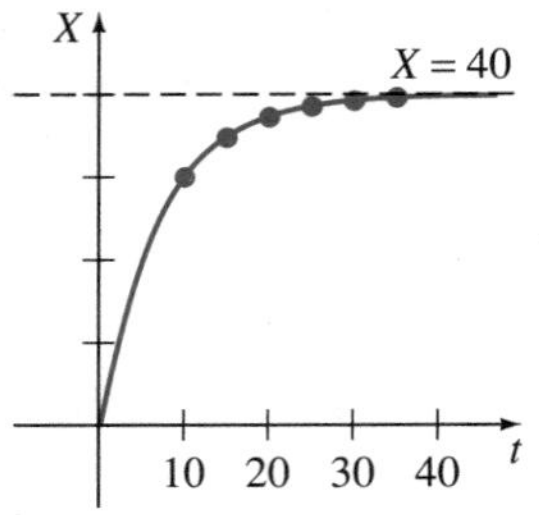

(a)

t (minutes)	x (grams)
10	30 (measured)
15	34.78
20	37.25
25	38.54
30	39.22
35	39.59

(b)

FIGURE 2.24

Integrating gives

$$\ln\left|\frac{250 - X}{40 - X}\right| = 210kt + c_1 \quad \text{or} \quad \frac{250 - X}{40 - X} = c_2 e^{210kt}. \tag{15}$$

When $t = 0$, $X = 0$, so it follows at this point that $c_2 = 25/4$. Using $X = 30$ g at $t = 10$, we find $210k = (1/10)\ln(88/25) = 0.1258$. With this information we solve the last equation in (15) for X:

$$X(t) = 1000\frac{1 - e^{-0.1258t}}{25 - 4e^{-0.1258t}}. \tag{16}$$

The behavior of X as a function of time is displayed in Figure 2.24. It is clear from the accompanying table and equation (16) that $X \to 40$ as $t \to \infty$. This means there are 40 grams of compound C formed, leaving

$$50 - \tfrac{1}{5}(40) = 42 \text{ g of } A \quad \text{and} \quad 32 - \tfrac{4}{5}(40) = 0 \text{ g of } B.$$ ●

Escape Velocity

The differential equation of a free-falling body of mass m near the surface of the Earth is given by

$$m\frac{d^2s}{dt^2} = -mg \quad \text{or simply} \quad \frac{d^2s}{dt^2} = -g,$$

where s represents the distance from the surface of the Earth to the object and the positive direction is considered to be upward. In other words, the underlying assumption here is that the distance s to the object is small when compared with the radius R of the Earth; put yet another way, the distance y from the center of the Earth to the object is approximately the same as R. If, on the other hand, the distance y to an object, such as a rocket or a space probe, is large compared to R, then we combine Newton's second law of motion and his universal law of gravitation to derive a differential equation in the variable y. The solution of this differential equation can be used to determine the minimum velocity, the so-called **escape velocity**, needed by a rocket to break free of the Earth's gravitational attraction.

EXAMPLE 8 Rocket Motion

A rocket is shot vertically upward from the ground as shown in Figure 2.25. If the positive direction is upward and air resistance is ignored, then the differential equation of motion after fuel burnout is

$$m\frac{d^2y}{dt^2} = -k\frac{mM}{y^2} \quad \text{or} \quad \frac{d^2y}{dt^2} = -k\frac{M}{y^2}, \tag{17}$$

where k is a constant of proportionality, y is the distance from the center of the Earth to the rocket, M is the mass of the Earth, and m is the mass of the rocket.

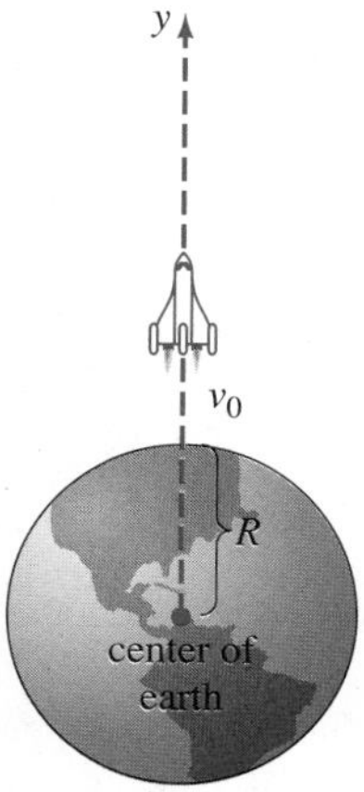

FIGURE 2.25

To determine the constant k, we use the fact that when $y = R$,

$$k\frac{mM}{R^2} = mg \quad \text{or} \quad k = \frac{gR^2}{M}.$$

Thus the last equation in (17) becomes

$$\frac{d^2y}{dt^2} = -g\frac{R^2}{y^2}. \tag{18}$$

Although this is not a first-order equation, if we write the acceleration as

$$\frac{d^2y}{dt^2} = \frac{dv}{dt} = \frac{dv}{dy}\frac{dy}{dt} = v\frac{dv}{dy},$$

then (18) becomes first-order in v; that is,

$$v\frac{dv}{dy} = -g\frac{R^2}{y^2}.$$

This last equation can be solved by separation of variables. From

$$\int v\,dv = -gR^2\int y^{-2}\,dy \quad \text{we get} \quad \frac{v^2}{2} = g\frac{R^2}{y} + c. \tag{19}$$

If we assume that the velocity is $v = v_0$ at burnout and that $y \approx R$ at that instant, we can obtain the (approximate) value of c. From (19) we find $c = -gR + v_0^2/2$. Substituting this value in (19) and multiplying the resulting equation by 2 yield

$$v^2 = 2g\frac{R^2}{y} - 2gR + v_0^2. \tag{20}$$

●

You might object, correctly, that in Example 8 we have not really solved the original equation for y. Actually the solution (20) gives quite a bit of information. Now that we have done the hard part, we leave the actual determination of the escape velocity of the rocket as an exercise. See Problem 39 in Exercises 2.5.

Falling Bodies and Air Resistance

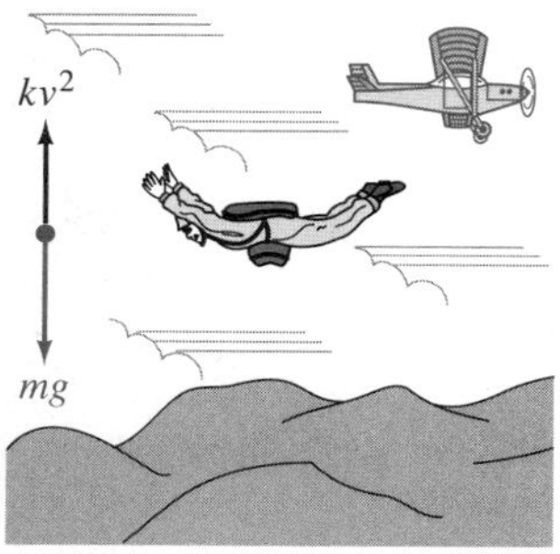

FIGURE 2.26

Under some circumstances a falling body of mass m encounters air resistance proportional to its instantaneous velocity v. If the positive direction is downward, then by equating the net force $mg - kv$ acting on the mass with Newton's second law $F = ma = m\,dv/dt$, we obtain a differential equation for the velocity of the body

$$m\frac{dv}{dt} = mg - kv. \tag{21}$$

Here k is a positive constant of proportionality. For high-speed motion through the air, such as the sky diver shown in Figure 2.26 falling before the parachute has opened, the air resistance is closer to the square of the instantaneous velocity. In this case the differential equation for the velocity is given by

$$m\frac{dv}{dt} = mg - kv^2. \tag{22}$$

Equations (21) and (22) are, in turn, a linear and a nonlinear differential equation. See Problem 40 in Exercises 2.5.

EXERCISES 2.5

Answers to odd-numbered problems begin on page AN-6.

2.5.1

1. The population of a certain community is known to increase at a rate proportional to the number of people present at any time. If the population has doubled in 5 years, how long will it take to triple? to quadruple?

2. Suppose it is known that the population of the community in Problem 1 is 10,000 after 3 years. What was the initial population? What will be the population in 10 years?

3. The population of a town grows at a rate proportional to the population at any time. Its initial population of 500 increases by 15% in 10 years. What will be the population in 30 years?

4. The population of bacteria in a culture grows at a rate proportional to the number of bacteria present at any time. After 3 hours it is observed that there are 400 bacteria present. After 10 hours there are 2000 bacteria present. What was the initial number of bacteria?

5. The radioactive isotope of lead, Pb-209, decays at a rate proportional to the amount present at any time and has a half-life of 3.3 hours. If 1 gram of lead is present initially, how long will it take for 90% of the lead to decay?

6. Initially there were 100 milligrams of a radioactive substance present. After 6 hours the mass decreased by 3%. If the rate of decay is proportional to the amount of the substance present at any time, find the amount remaining after 24 hours.

7. Determine the half-life of the radioactive substance described in Problem 6.

8. Show that the half-life of a radioactive substance is, in general,

$$t = \frac{(t_2 - t_1)\ln 2}{\ln(A_1/A_2)},$$

where $A_1 = A(t_1)$ and $A_2 = A(t_2)$, $t_1 < t_2$.

9. When a vertical beam of light passes through a transparent substance, the rate at which its intensity I decreases is proportional to $I(t)$, where t represents the thickness of the medium (in feet). In clear seawater the intensity 3 feet below the surface is 25% of the initial intensity I_0 of the incident beam. What is the intensity of the beam 15 feet below the surface?

10. When interest is compounded continuously, the amount of money S increases at a rate proportional to the amount present at any time: $dS/dt = rS$, where r is the annual rate of interest (see (26) of Section 1.2).

   (a) Find the amount of money accrued at the end of 5 years when \$5000 is deposited in a savings account drawing $5\frac{3}{4}\%$ annual interest compounded continuously.

   (b) In how many years will the initial sum deposited be doubled?

   (c) Use a hand calculator to compare the number obtained in part (a) with the value

$$S = 5000\left(1 + \frac{0.0575}{4}\right)^{5(4)}.$$

   This value represents the amount accrued when interest is compounded quarterly.

11. In a piece of burned wood, or charcoal, it was found that 85.5% of the C-14 had decayed. Use the information in Example 3 to determine the approximate age of the wood. (It is precisely these data that archaeologists used to date prehistoric paintings in a cave in Lascaux, France.)

12. A thermometer is taken from an inside room to the outside where the air temperature is 5°F. After 1 minute the thermometer reads 55°F, and after 5 minutes the reading is 30°F. What is the initial temperature of the room?

13. A thermometer is removed from a room where the air temperature is 70°F to the outside where the temperature is 10°F. After $\frac{1}{2}$ minute the thermometer reads 50°F. What is

the reading at $t = 1$ minute? How long will it take for the thermometer to reach 15°F?

14. Formula (3) also holds when an object absorbs heat from the surrounding medium. If a small metal bar whose initial temperature is 20°C is dropped into a container of boiling water, how long will it take for the bar to reach 90°C if it is known that its temperature increased 2° in 1 second? How long will it take the bar to reach 98°C?

15. A 30-volt electromotive force is applied to an L-R series circuit in which the inductance is 0.1 henry and the resistance is 50 ohms. Find the current $i(t)$ if $i(0) = 0$. Determine the current as $t \to \infty$.

16. Solve equation (6) under the assumption that $E(t) = E_0 \sin \omega t$ and $i(0) = i_0$.

17. A 100-volt electromotive force is applied to an R-C series circuit in which the resistance is 200 ohms and the capacitance is 10^{-4} farad. Find the charge $q(t)$ on the capacitor if $q(0) = 0$. Find the current $i(t)$.

18. A 200-volt electromotive force is applied to an R-C series circuit in which the resistance is 1000 ohms and the capacitance is 5×10^{-6} farad. Find the charge $q(t)$ on the capacitor if $i(0) = 0.4$. Determine the charge and current at $t = 0.005$ second. Determine the charge as $t \to \infty$.

19. An electromotive force

$$E(t) = \begin{cases} 120, & 0 \leq t \leq 20 \\ 0, & t > 20 \end{cases}$$

is applied to an L-R series circuit in which the inductance is 20 henry and the resistance is 2 ohms. Find the current $i(t)$ if $i(0) = 0$.

20. Suppose an R-C series circuit has a variable resistor. If the resistance at any time t is given by $R = k_1 + k_2 t$, where $k_1 > 0$ and $k_2 > 0$ are known constants, then (6) becomes

$$(k_1 + k_2 t)\frac{dq}{dt} + \frac{1}{C}q = E(t).$$

Show that if $E(t) = E_0$ and $q(0) = q_0$, then

$$q(t) = E_0 C + (q_0 - E_0 C)\left(\frac{k_1}{k_1 + k_2 t}\right)^{1/Ck_2}.$$

21. A tank contains 200 liters of fluid in which 30 g of salt is dissolved. Brine containing 1 g of salt per liter is then pumped into the tank at a rate of 4 liters per minute; the well-mixed solution is pumped out at the same rate. Find the number of grams of salt $A(t)$ in the tank at any time.

22. Solve Problem 21 assuming pure water is pumped into the tank.

23. A large tank is filled with 500 gallons of pure water. Brine containing 2 lb of salt per gallon is pumped into the tank at a rate of 5 gallons per minute. The well-mixed solution is pumped out at the same rate. Find the number of pounds of salt $A(t)$ in the tank at any time.

24. Solve Problem 23 under the assumption that the solution is pumped out at a faster rate of 10 gallons per minute. When is the tank empty?

25. A large tank is partially filled with 100 gallons of fluid in which 10 lb of salt is dissolved. Brine containing $\frac{1}{2}$ lb of salt per gallon is pumped into the tank at a rate of 6 gallons per minute. The well-mixed solution is then pumped out at a slower rate of 4 gallons per minute. Find the number of pounds of salt in the tank after 30 minutes.

26. *PROJECT PROBLEM*

(a) Look up the U.S. census population for the years 1790, 1800, 1810, . . . , 1950.

(b) Use the data obtained in part (a) to construct a population model of the form

$$\frac{dP}{dt} = kP, \quad P(0) = P_0.$$

(c) Construct a table comparing the population predicted by the model in part (b) with the actual population. Compute the error and percentage error for each entry pair.

27. *PROJECT PROBLEM* Suppose a medical examiner, after arriving at the scene of a homicide, found that the temperature of the body was 82°F. Make up additional, but plausible, data necessary to determine an approximate time of death of the victim using Newton's law of cooling (3).

28. *PROJECT PROBLEM* Mr. Jones puts two cups of coffee on the breakfast table at the same time. Mr. Jones immediately pours cream into his coffee from a pitcher that was sitting on the table for a long time. He then reads the morning paper for 5 minutes before taking his first sip. Mrs. Jones arrives at the table 5 minutes after the cups were set down, adds cream to her coffee, and takes a sip. Determine who drinks the hotter cup of coffee. Assume that both Mr. and Mrs. Jones add exactly the same amount of cream to their cups.

29. *WRITING PROJECT* Who of us has not, at one or more times, resolved to stay on a diet? Write a report on the article "A Linear Diet Model,"

Arthur C. Segal, *The College Mathematics Journal*, January, 1987, pp. 44–45, and elaborate on the author's meaning of "why so many dieters give up in frustration."

30. **WRITING PROJECT** Go to the library and find any edition of the text *Differential Equations and Their Applications*, M. Braun, Springer-Verlag. Prepare a report, written or oral, on *The Van Meegeren Art Forgeries*.

31. **PROJECT PROBLEM** In Example 5 the size of the tank containing the salt mixture was not given. Suppose, as discussed on page 59, that the rate at which brine is pumped into the tank is the same, but that the mixed solution is pumped out a rate of 2 gallons per minute. It stands to reason that since brine is accumulating in the tank at the rate of 1 gal/min any finite tank must eventually overflow. Now suppose that the tank has an open top and has a total capacity of 400 gallons.
   (a) When will the tank overflow?
   (b) What are the number of pounds of salt in the tank at the instant it overflows?
   (c) Assume that the tank is overflowing, that brine solution continues to be pumped in at a rate of 3 gallons per minute, and that the well-stirred solution continues to be pumped out at a rate of 2 gallons per minute. Devise a method for determining the number of pounds of salt in the tank at $t = 150$ minutes.
   (d) Determine the number of pounds of salt in the tank as $t \to \infty$. Does your answer agree with your intuition?
   (e) Use a computer graphing program to obtain the graph $A(t)$ on the interval $[0, \infty)$.

32. **PROJECT PROBLEM** As is usually the case when mathematical models are constructed, equation (10) in Example 5 is the product of several assumptions. In this problem we focus on one of these assumptions.

Although the liquid entering is "well-stirred" with the solution in the tank, it cannot instantly affect the concentration of the liquid leaving. In other words, it is unrealistic to evaluate the rate $A'(t)$ and the concentration $A(t)/300$ at the same time t. It would probably make more sense to assume that the concentration of the solution leaving the tank is an average concentration, say, $A(t - t_1)/300$, where $A(t - t_1)$ represents the amount of salt in the tank at some earlier instant t_1 in time. The resulting equation

$$\frac{dA}{dt} = 6 - 3\frac{A(t - t_1)}{300}$$

is called a **delay differential equation**. In such an equation the unknown function and its derivatives have different arguments.
   (a) Write a report elaborating on the concept of delay differential equations.
   (b) Set up a delay differential equation for Problem 22. Show how the **method of steps** can be used to solve this equation.

2.5.2

33. The number of supermarkets $C(t)$ throughout the country that are using a computerized checkout system is described by the initial-value problem

$$\frac{dC}{dt} = C(1 - 0.0005C), \quad C(0) = 1,$$

where $t > 0$. How many supermarkets are using the computerized method when $t = 10$? How many companies are estimated to adopt the new procedure over a long period of time?

34. The number of people $N(t)$ in a community who are exposed to a particular advertisement is governed by the logistic equation. Initially $N(0) = 500$, and it is observed that $N(1) = 1000$. If it is predicted that the limiting number of people in the community who will see the advertisement is 50,000, determine $N(t)$ at any time.

35. The population $P(t)$ at any time in a suburb of a large city is governed by the initial-value problem

$$\frac{dP}{dt} = P(10^{-1} - 10^{-7}P), \quad P(0) = 5000,$$

where t is measured in months. What is the limiting value of the population? At what time will the population be equal to one-half of this limiting value?

36. Two chemicals A and B are combined to form a chemical C. The rate or velocity of the reaction is proportional to the product of the instantaneous amounts of A and B not converted to chemical C. Initially there are 40 grams of A and 50 grams of B, and for each gram of B, 2 grams of A are used. It is observed that 10 grams of C are formed in 5 minutes. How much is formed in 20 minutes? What is the limiting amount of C after a long time? How much of chemicals A and B remains after a long time?

37. Obtain a solution of the equation

$$\frac{dX}{dt} = k(\alpha - X)(\beta - X)$$

governing second-order reactions in the two cases $\alpha \neq \beta$ and $\alpha = \beta$.

38. In a third-order chemical reaction the number of grams X of a compound obtained by combining three chemicals is

governed by

$$\frac{dX}{dt} = k(\alpha - X)(\beta - X)(\gamma - X).$$

Solve the equation under the assumption $\alpha \neq \beta \neq \gamma$.

39. **(a)** Use equation (20) to show that the escape velocity of the rocket is given by $v_0 = \sqrt{2gR}$. [*Hint:* Take $y \to \infty$ in (20) and assume $v > 0$ for all time t.]

(b) The result in part (a) holds for any body in the solar system. Use the values $g = 32$ ft/s and $R = 4000$ miles to show that the escape velocity from the Earth is (approximately) $v_0 = 25{,}000$ mi/hr.

(c) Find the escape velocity from the moon if the acceleration of gravity is 0.165 g and $R = 1080$ miles.

40. **(a)** Solve the differential equation (21) subject to $v(0) = v_0$.

(b) Solve the differential equation (22) subject to $v(0) = v_0$.

(c) Determine the limiting, or terminal, velocity of a falling mass m first using the model in (21), and then using the model in (22).

41. **(a)** Determine the differential equation for the velocity $v(t)$ of an object of mass m sinking in water that imparts a resistance proportional to the square of the instantaneous velocity and that also exerts an upward buoyant force whose magnitude is given by Archimedes' principle. Assume the positive direction is downward.

(b) Solve the differential equation in part (a).

(c) Determine the limiting, or terminal, velocity of the sinking mass.

42. ***PROJECT PROBLEM***

(a) Use the census data from 1790, 1850, and 1910 obtained in part (a) of Problem 26 to construct a population model of the form

$$\frac{dP}{dt} = P(a - bP), \quad P(0) = P_0.$$

(b) Construct a table comparing the population predicted by the model in part (a) with the actual population. Compute the error and percentage error for each entry pair.

43. ***PROJECT PROBLEM*** If a constant number of animals h have been removed or harvested per unit time, then the population $P(t)$ of animals is modeled by the initial-value problem

$$\frac{dP}{dt} = P(a - bP) - h, \quad P(0) = P_0,$$

where a, b, h, and P_0 are positive constants.

(a) Without solving, analyze the affect of h on the population by finding the critical points of the equation and phase portraits in the cases $0 < h < a^2/4b$, $h = a^2/4b$, and $h > a^2/4b$. Where do these cases come from? Use the phase portraits to sketch the behavior of solution curves for various values of P_0. Use the phase portraits and solution curves to determine the long term behavior of the population. Does the population ever become extinct?

(b) Find an explicit solution of

$$\frac{dP}{dt} = P(5 - P) - 4, \quad P(0) = P_0$$

and relate the solution back to one of the cases in part (a). Determine the long-term behavior of the population by considering $P_0 > 4$, $1 < P_0 < 4$, and $0 < P_0 < 1$. If the population becomes extinct in finite time, find that time.

44. ***PROJECT PROBLEM*** A sky diver weighing 160 pounds steps out the door of an airplane that is flying at an altitude of 12,000 feet. After falling freely for 15 seconds the parachute is opened. Assume that air resistance is proportional to v^2 while the parachute is unopened and is proportional to the velocity v after the parachute is opened. See Figure 2.27. For a person of this weight, typical values for k in the models

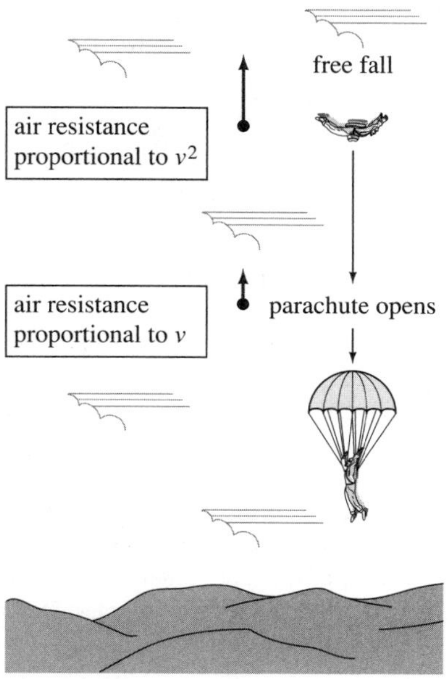

FIGURE 2.27

(21) and (22) are, respectively, $k = 7.857$ and $k = 0.0053$. Determine the time that it takes the sky diver to reach the ground. What is the sky diver's impact velocity?

45. ***WRITING PROJECT*** When all curves in a family $G(x, y, c_1) = 0$ intersect orthogonally all curves in another family $H(x, y, c_2) = 0$, the families are said to be **orthogonal trajectories** of each other. See Figure 2.28. (Do not confuse this with trajectories in a phase plane.)

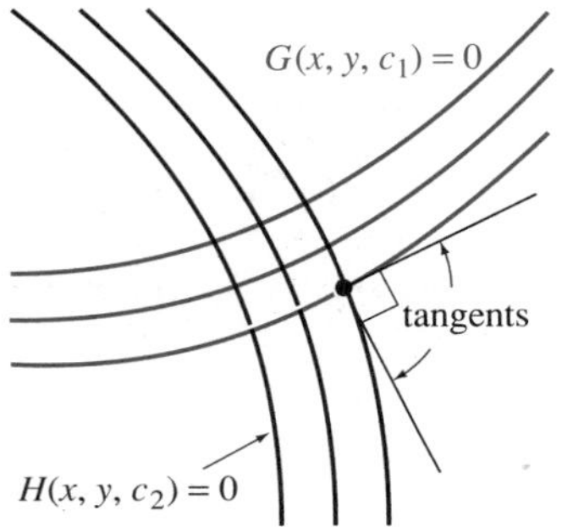

FIGURE 2.28

(a) Write a short report on orthogonal trajectories. Include several examples where orthogonal trajectories occur in the sciences.
(b) Illustrate your report by finding the orthogonal trajectories for the family $y = 1/(x + c_1)$. Use a computer graphing program to graph both families on the same axes.
(c) Find the orthogonal trajectories for the family $x^2 - xy + y^2 = c_1$. Use a computer program (such as a CAS) with a feature for graphing contour curves to graph both families on the same axes for $x > 0$, $y > 0$.
(d) If instructed to do so, include in your report a discussion of **isogonal trajectories** (also called oblique trajectories).

In Problems 46 and 47 the models that you are asked to derive are nonlinear second-order differential equations that can be solved by an appropriate substitution. See Example 6, Section 2.2.

46. ***PROJECT PROBLEM*** Consider a suspended wire hanging under its own weight. As Figure 2.29 shows, a physical model for this could be a long telephone wire strung between two poles. Determine the nonlinear second-order differential equation that describes the shape that the hanging wire assumes. Find an explicit solution of the differential equation subject to the initial conditions $y(0) = 1$, $y'(0) = 0$.

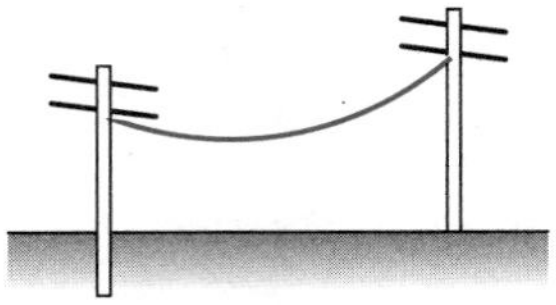

FIGURE 2.29

47. ***PROJECT PROBLEM*** In a naval exercise, a ship S_1 is pursued by a submarine S_2, as shown in Figure 2.30. Ship S_1 departs point (0, 0) at $t = 0$ and proceeds along a straight line course (the y-axis) at a constant speed v_1. The submarine S_2 keeps ship S_1 in visual contact, indicated by the straight dashed line L in the figure, while traveling at a constant speed v_2 along a curve C. Assume that S_2 starts at the point $(a, 0)$, $a > 0$, at $t = 0$ and that L is tangent to C. Determine the nonlinear second-order differential equation that describes the curve C. Find an explicit solution of the differential equation. For convenience, define $r = v_1/v_2$. Determine whether the paths of S_1 and S_2 will ever intersect by considering the cases $r > 1$, $r < 1$, and $r = 1$. $\left[\textit{Hint: } \dfrac{dt}{dx} = \dfrac{dt}{ds}\dfrac{ds}{dx}\right.$ where s is arc length measured along C.$\left.\right]$

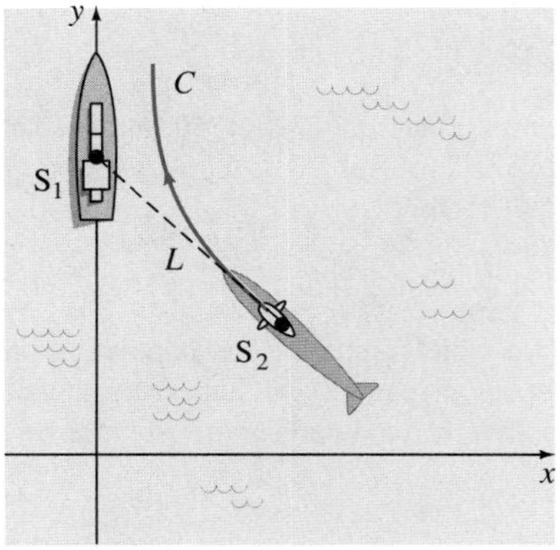

FIGURE 2.30

48. ***PROJECT PROBLEM*** In another naval exercise, a destroyer S_1 pursues a sub-

merged submarine S_2. Suppose, that S_1 at (9, 0) on the x-axis detects S_2 at (0, 0) and that S_2 simultaneously detects S_1. The captain of the destroyer S_1 assumes that the submarine will take immediate evasive action and conjectures that its likely new course is the straight line L indicated in Figure 2.31. When S_1 is at (3, 0) it changes from its straight-line course toward the origin to a pursuit curve C. Assume that the speed of the destroyer is, at all times, a constant 30 mi/hr and that the submarine's speed is a constant 15 mi/hr.

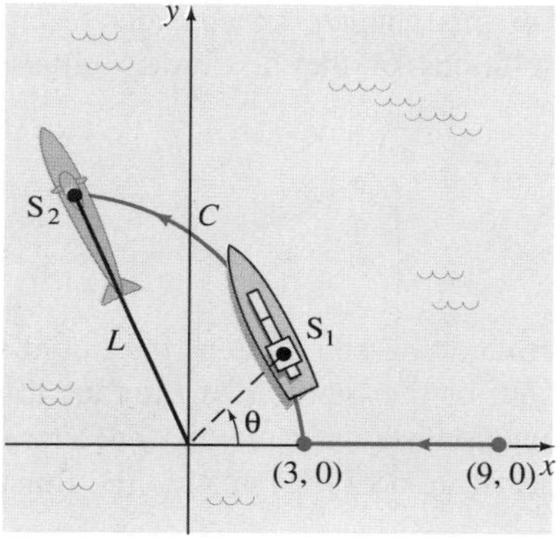

FIGURE 2.31

(a) Explain why the captain waits until S_1 reaches (3, 0) before ordering a course change to C.

(b) Using polar coordinates, find an equation $r = f(\theta)$ of the curve C.

(c) Explain why the time, measured from the initial detection, at which the destroyer intercepts the submarine must be less than $(1 + e^{2\pi/\sqrt{3}})/5$.

49. *PROJECT PROBLEM* As shown in Figure 2.32, a plane flying horizontally at a constant speed v_0 drops a relief food supply pack to persons on the ground. Assume that the origin is the point where the supply pack is released and that the positive y-axis points downward.

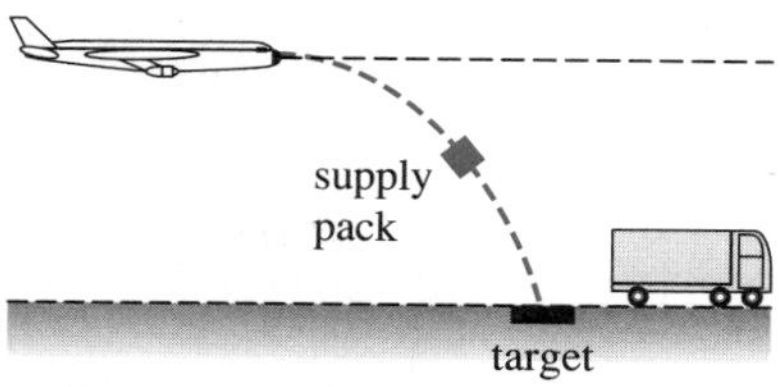

FIGURE 2.32

(a) Determine the system of differential equations and initial conditions that define the trajectory of the supply pack under the assumption that the horizontal and vertical components of the air resistance are proportional to $(dx/dt)^2$ and $(dy/dt)^2$, respectively. [*Hint:* Use vector methods.]

(b) Solve the differential equations in part (a).

(c) Suppose the plane flies at an altitude of 1000 ft and that its constant speed is 300 mi/hr., Assume that the constant of proportionality for air resistance is $k = 0.0053$ and that the supply pack weighs 256 lbs. Determine the horizontal distance the pack travels, measured from its point of release, when it hits the ground.

50. *PROJECT PROBLEM* A large snowball is shaped into the form of a sphere. Starting at some time, which we can designate as $t = 0$, the snowball begins to melt. Let V_0 and r_0 be the volume and radius, respectively, of the snowball at $t = 0$.

(a) Suppose "melting" is interpreted as the rate of change of volume V with respect to time t. To devise a mathematical model for the volume V of the snowball for any time $t > 0$, make the assumption that this rate is proportional to another quantity. Discuss what that quantity could be and solve the resulting differential equation.

(b) Determine the radius of the snowball as a function of time t. When does the snowball disappear?

2.6 NUMERICAL SOLUTIONS: EULER METHODS

Introduction

So far in this chapter we have primarily considered first-order differential equations whose solutions can be found by an appropriate technique, such as separation of variables, and that can be expressed in terms of the elementary functions studied in calculus. Since these techniques involve integration, we may obtain an integral that we are unable to evaluate. For example, an integrating factor for the linear differential equation

$$y' + 2xy = \ln x$$

is e^{x^2}, and the general solution on the interval $(0, \infty)$ is

$$y = e^{-x^2} \int e^{x^2} \ln x \, dx + ce^{-x^2}.$$

In this case the integral cannot be expressed in terms of elementary functions. In other instances, particularly involving nonlinear differential equations, we will be required to resort to approximation techniques to obtain a representation of the solution. In the remaining sections of this chapter we consider techniques for obtaining approximate numerical solutions of the first-order initial-value problem

$$y' = f(x, y), \quad y(x_0) = y_0. \tag{1}$$

Euler's Method

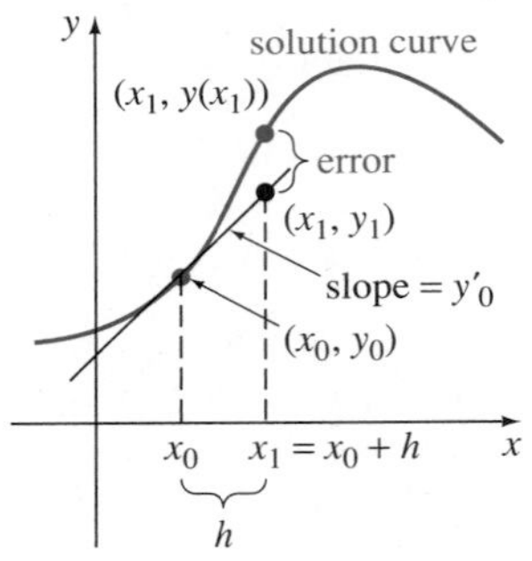

FIGURE 2.33

One of the simplest techniques for approximating solutions of the initial-value problem (1) is known as **Euler's Method**, or the **method of tangent lines**. It uses the fact that the derivative of a function $y(x)$ at a point x_0 determines a **linearization** of $y(x)$ at $x = x_0$. This linearization of $y(x)$ at x_0 is the function

$$L(x) = y'(x_0)(x - x_0) + y_0,$$

whose graph is a straight line tangent to the graph of $y = y(x)$ at the point (x_0, y_0). We now let h be a positive increment on the x-axis, as shown in Figure 2.33. Then for $x_1 = x_0 + h$ we have

$$L(x_1) = y'(x_0)(x_0 + h - x_0) + y_0 = y_0 + hy_0',$$

where $y_0' = y'(x_0) = f(x_0, y_0)$. Letting $y_1 = L(x_1)$ we get

$$y_1 = y_0 + hf(x_0, y_0).$$

The point (x_1, y_1) on the tangent line is an approximation to the point $(x_1, y(x_1))$ on the solution curve; that is, $L(x_1) \approx y(x_1)$ or $y_1 \approx y(x_1)$ is a *local linear approximation* of $y(x)$ at x_1. Of course, the accuracy of the approximation depends heavily on the size of the increment h. Usually we must choose this **step size** to be "reasonably small." If we now repeat this process, identifying the new starting point (x_1, y_1) with (x_0, y_0), we obtain the approximation

$$y(x_2) = y(x_0 + 2h) = y(x_1 + h) \approx y_2 = y_1 + hf(x_1, y_1).$$

In general it follows that

$$y_{n+1} = y_n + hf(x_n, y_n), \tag{2}$$

where $x_n = x_0 + nh$.

To illustrate Euler's Method we use the iteration scheme (2) on a differential equation for which we know the explicit solution; in this way we can compare the estimated values y_n with the true values $y(x_n)$.

EXAMPLE 1 Euler's Method

Consider the initial-value problem

$$y' = 0.2xy, \qquad y(1) = 1.$$

Use the Euler method to obtain an approximation to $y(1.5)$ using first $h = 0.1$ and then $h = 0.05$.

Solution We first identify $f(x, y) = 0.2xy$ so that (2) becomes

$$y_{n+1} = y_n + h(0.2x_n y_n).$$

Then for $h = 0.1$ we find

$$y_1 = y_0 + (0.1)(0.2x_0 y_0) = 1 + (0.1)[0.2(1)(1)] = 1.02,$$

which is an estimate to the value of $y(1.1)$. However, if we use $h = 0.05$, it takes *two* iterations to reach $x = 1.1$. We have

$$y_1 = 1 + (0.05)[0.2(1)(1)] = 1.01$$

$$y_2 = 1.01 + (0.05)[0.2(1.05)(1.01)] = 1.020605.$$

Here we note that $y_1 \approx y(1.05)$ and $y_2 \approx y(1.1)$. The remainder of the calculations are summarized in Tables 2.1 and 2.2. Each entry is rounded to four decimal places.

TABLE 2.1 Euler's Method with $h = 0.1$

x_n	y_n	True Value	Abs. Error	% Rel. Error
1.00	1.0000	1.0000	0.0000	0.00
1.10	1.0200	1.0212	0.0012	0.12
1.20	1.0424	1.0450	0.0025	0.24
1.30	1.0675	1.0714	0.0040	0.37
1.40	1.0952	1.1008	0.0055	0.50
1.50	1.1259	1.1331	0.0073	0.64

TABLE 2.2 Euler's Method with $h = 0.05$

x_n	y_n	True Value	Abs. Error	% Rel. Error
1.00	1.0000	1.0000	0.0000	0.00
1.05	1.0100	1.0103	0.0003	0.03
1.10	1.0206	1.0212	0.0006	0.06
1.15	1.0318	1.0328	0.0009	0.09
1.20	1.0437	1.0450	0.0013	0.12
1.25	1.0562	1.0579	0.0016	0.16
1.30	1.0694	1.0714	0.0020	0.19
1.35	1.0833	1.0857	0.0024	0.22
1.40	1.0980	1.1008	0.0028	0.25
1.45	1.1133	1.1166	0.0032	0.29
1.50	1.1295	1.1331	0.0037	0.32

In Example 1 the true values were calculated from the known solution $y = e^{0.1(x^2-1)}$. Also, the **absolute error** is defined to be

$$|\textit{true value} - \textit{approximation}|.$$

The **relative error** and the **percentage relative error** are, in turn.

$$\frac{|\textit{true value} - \textit{approximation}|}{|\textit{true value}|}$$

and $$\frac{|\textit{true value} - \textit{approximation}|}{|\textit{true value}|} \times 100 = \frac{\textit{absolute error}}{|\textit{true value}|} \times 100.$$

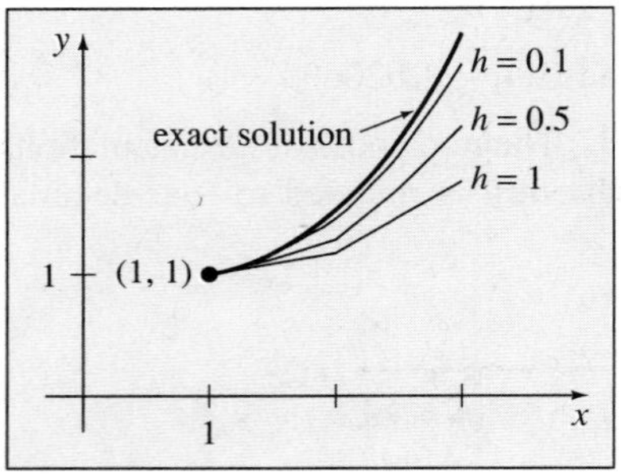

FIGURE 2.34

Computer software enables us to examine approximations to the graph of the solution $y(x)$ of an initial-value problem by plotting straight lines through the points (x_n, y_n) generated by Euler's method. (An alternative method for obtaining the graph of an approximate solution from a table of discrete values is examined in *Computer Lab Experiment* 6 for Sections 2.6–2.8.) In Figure 2.34 we have compared, on the interval [1, 3], the graph of the exact solution of the initial-value problem in Example 1 with the graphs obtained from Euler's method using the step sizes $h = 1$, $h = 0.5$, and $h = 0.1$. It is apparent from the figure that the approximation improves as the step size decreases.

Although we see that the percentage relative error in Tables 2.1 and 2.2 is growing, it does not appear to be that bad. But you should not be deceived by Example 1 and Figure 2.34. Watch what happens in the next example, when we simply change the coefficient 0.2 of the differential equation in Example 1 to the number 2.

EXAMPLE 2 Comparison of Exact/Approximate Values

Use the Euler method to obtain the approximate value of $y(1.5)$ for the solution of

$$y' = 2xy, \qquad y(1) = 1.$$

Solution You should verify that the exact or analytic solution is now $y = e^{x^2-1}$. Proceeding as in Example 1, we obtain the results shown in Tables 2.3 and 2.4.

In this case, with a step size $h = 0.1$, a 16% relative error in the calculation of the approximation to $y(1.5)$ is totally unacceptable. At the expense of doubling the number of calculations, a slight improvement in accuracy is obtained by halving the step size to $h = 0.05$.

TABLE 2.3 Euler's Method with $h = 0.1$

x_n	y_n	True Value	Abs. Error	% Rel. Error
1.00	1.0000	1.0000	0.0000	0.00
1.10	1.2000	1.2337	0.0337	2.73
1.20	1.4640	1.5527	0.0887	5.71
1.30	1.8154	1.9937	0.1784	8.95
1.40	2.2874	2.6117	0.3244	12.42
1.50	2.9278	3.4904	0.5625	16.12

TABLE 2.4 Euler's Method with $h = 0.05$

x_n	y_n	True Value	Abs. Error	% Rel. Error
1.00	1.0000	1.0000	0.0000	0.00
1.05	1.1000	1.1079	0.0079	0.72
1.10	1.2155	1.2337	0.0182	1.47
1.15	1.3492	1.3806	0.0314	2.27
1.20	1.5044	1.5527	0.0483	3.11
1.25	1.6849	1.7551	0.0702	4.00
1.30	1.8955	1.9937	0.0982	4.93
1.35	2.1419	2.2762	0.1343	5.90
1.40	2.4311	2.6117	0.1806	6.92
1.45	2.7714	3.0117	0.2403	7.98
1.50	3.1733	3.4904	0.3171	9.08

Errors in Numerical Methods

In choosing and using a numerical method for the solution of an initial-value problem, we must be aware of the various sources of errors. For some kinds of computation, the accumulation of errors might reduce the accuracy of an approximation to the point of being useless. On the other hand, depending on the use to which a numerical solution may be put, extreme accuracy may not be worth the added expense and complication.

One source of error always present in calculations is **round-off error**. This error results from the fact that any calculator or computer can represent numbers using only a finite number of digits. Suppose for the sake of illustration that we have a calculator that uses base 10 arithmetic and carries four digits, so that $\frac{1}{3}$ is represented in the calculator as 0.3333 and $\frac{1}{9}$ is represented as 0.1111. If we use this calculator to compute $(x^2 - \frac{1}{9})/(x - \frac{1}{3})$ for $x = 0.3334$, we obtain

$$\frac{(0.3334)^2 - 0.1111}{0.3334 - 0.3333} = \frac{0.1112 - 0.1111}{0.3334 - 0.3333} = 1.$$

With the help of a little algebra, however, we see that

$$\frac{x^2 - 1/9}{x - 1/3} = \frac{(x - 1/3)(x + 1/3)}{x - 1/3} = x + \frac{1}{3},$$

so that when $x = 0.3334$, $(x^2 - 1/9)/(x - 1/3) \approx 0.3334 + 0.3333 = 0.6667$. This example shows that the effects of round-off error can be quite serious unless some care is taken. One way to reduce the effect of round-off error is to minimize the number of calculations. Another technique on a computer is to use double-precision arithmetic to check the results. In general, round-off error is unpredictable and difficult to analyze, and we will neglect it in the error analysis that follows. We will concentrate on investigating the error made by using a formula or algorithm to approximate the values of the solution.

Truncation Errors for Euler's Method

When iterating Euler's formula

$$y_{n+1} = y_n + hf(x_n, y_n),$$

we obtain a sequence of values $y_1, y_2\, y_3, \ldots$. Usually the value y_1 will not agree with $y(x_1)$, the actual solution evaluated at x_1, because the algorithm gives only a straight-line approximation to the solution. See Figure 2.33. The error is called the **local truncation error**, **formula error**, or **discretization error**. It occurs at each step; that is, if we assume that y_n is accurate, then y_{n+1} will contain local truncation error.

To derive a formula for the local truncation error for Euler's method, we use Taylor's formula with remainder. If a function $y(x)$ possesses $k + 1$ derivatives that are continuous on an open interval containing a and x, then

$$y(x) = y(a) + y'(a)\frac{(x-a)}{1!} + \cdots + y^{(k)}(a)\frac{(x-a)^k}{k!} + y^{(k+1)}(c)\frac{(x-a)^{(k+1)}}{(k+1)!}, \tag{3}$$

where c is some point between a and x. Setting $k = 1$, $a = x_n$, and $x = x_{n+1} = x_n + h$, we get

$$y(x_{n+1}) = y(x_n) + y'(x_n)\frac{h}{1!} + y''(c)\frac{h^2}{2!}$$

or

$$y_{n+1} = y_n + hf(x_n, y_n) + y''(c)\frac{h^2}{2!}.$$

Euler's method is this formula without the last term; hence, the local truncation error in y_{n+1} is

$$y''(c)\frac{h^2}{2!}, \quad \text{where} \quad x_n < c < x_{n+1}.$$

Unfortunately, the value of c is usually unknown (it exists theoretically) and so the exact error cannot be calculated, but an upper bound on the absolute value of the error is

$$M\frac{h^2}{2}, \quad \text{where} \quad M = \max_{x_n < x < x_{n+1}} |y''(x)|.$$

When discussing errors arising from the use of numerical methods it is helpful to use the notation $O(h^n)$. To define this concept we let $e(h)$ denote the error in a numerical calculation depending on h. Then $e(h)$ is said to be of order h^n, denoted by $O(h^n)$, if there exists a constant C and a positive integer n such that $|e(h)| \leq Ch^n$ for h sufficiently small. Thus, the local truncation error for the Euler method is $O(h^2)$. We note that in general if a numerical method has order h^n and h is halved, the new error is approximately $C(h/2)^n = Ch^n/2^n$; that is, the error is reduced by a factor of $1/2^n$.

EXAMPLE 3 Bound for Local Truncation Errors

Find a bound for the local truncation errors for Euler's method applied to

$$y' = 2xy, \qquad y(1) = 1.$$

Solution This differential equation was studied in Example 2 and its analytic solution is $y(x) = e^{x^2 - 1}$.

The local truncation error is

$$y''(c)\frac{h^2}{2} = (2 + 4c^2)e^{(c^2 - 1)}\frac{h^2}{2},$$

where c is between x_n and $x_n + h$. In particular, for $h = 0.1$, we can get an upper bound on the local truncation error for y_1 by replacing c by 1.1:

$$[2 + (4)(1.1)^2]e^{((1.1)^2 - 1)}\frac{(0.1)^2}{2} = 0.0422.$$

From Table 2.3 we see that the error after the first step is 0.0337, less than the value given by the bound.

Similarly we can get a bound for the local truncation error for any of the five steps given in Table 2.3 by replacing c by 1.5 (this value of c gives the largest value of $y''(c)$ for any of the steps and may be too generous for the first few steps). Doing this gives

$$[2 + (4)(1.5)^2]e^{((1.5)^2 - 1)}\frac{(0.1)^2}{2} = 0.1920 \qquad \textbf{(4)}$$

as an upper bound for the local truncation error in each step. ●

Note in Example 3 that if h is halved to 0.05, then the error bound (4) would be 0.0480, about one fourth as much. This is expected because the local truncation error for Euler's method is $O(h^2)$.

In the above analysis we assumed that the value of y_n is correct in the calculation of y_{n+1}, but it is not because it contains local truncation errors from previous steps. The total error in y_{n+1} is an accumulation of the errors in each of the previous steps. This total error is called the **global truncation error**. A complete analysis of the global truncation error is beyond the scope of this text, but it can be shown that the global truncation error for the Euler method is $O(h)$.

We expect that for Euler's method if the step size is halved, then the error would be approximately halved as well. This is borne out in Example 2, where the absolute error at $x = 1.50$ with $h = 0.1$ is 0.5625 and the absolute error with $h = 0.05$ is 0.3171, approximately half as large. See Tables 2.3 and 2.4.

In general it can be shown that if a method for the numerical solution of a differential equation has local truncation error $O(h^{\alpha+1})$, then the global truncation error is $O(h^{\alpha})$.

Improved Euler's Method

While the Euler formula is attractive for its simplicity, it is seldom used in serious calculations. In the remainder of this section and in subsequent sections we study methods that give significantly greater accuracy than Euler's method.

The formula

$$y_{n+1} = y_n + h\frac{f(x_n, y_n) + f(x_{n+1}, y^*_{n+1})}{2}$$

where

$$y^*_{n+1} = y_n + hf(x_n, y_n) \tag{5}$$

is known as the **improved Euler formula**, or **Heun's formula**. Euler's formula is used to obtain the initial estimate y^*_{n+1}. The values $f(x_n, y_n)$ and $f(x_{n+1}, y^*_{n+1})$ are approximations to the slopes of the solution curve at $(x_n, y(x_n))$ and $(x_{n+1}, y(x_{n+1}))$, and, consequently, the quotient

$$\frac{f(x_n, y_n) + f(x_{n+1}, y^*_{n+1})}{2}$$

can be interpreted as an average slope on the interval from x_n to x_{n+1}. The value of y_{n+1} is then computed in a manner similar to that used in Euler's method, except that an average slope over the interval is used instead of the slope at $(x_n, y(x_n))$. We say that the value of y^*_{n+1} *predicts* a value of $y(x_n)$, whereas

$$y_{n+1} = y_n + h\frac{f(x_n, y_n) + f(x_{n+1}, y^*_{n+1})}{2}$$

corrects this estimate.

EXAMPLE 4 Improved Euler's Method

Use the improved Euler formula to obtain the approximate value of $y(1.5)$ for the solution of the initial-value problem in Example 2. Compare the results for $h = 0.1$ and $h = 0.05$.

Solution For $n = 0$ and $h = 0.1$, we first compute

$$y^*_1 = y_0 + (0.1)(2x_0y_0) = 1.2.$$

Then from (5)

$$y_1 = y_0 + (0.1)\frac{2x_0y_0 + 2x_1y_1^*}{2} = 1 + (0.1)\frac{2(1)(1) + 2(1.1)(1.2)}{2} = 1.232.$$

The comparative values of the calculations for $h = 0.1$ and $h = 0.5$ are given in Tables 2.5 and 2.6, respectively.

TABLE 2.5 Improved Euler's Method with $h = 0.1$

x_n	y_n	True Value	Abs. Error	% Rel. Error
1.00	1.0000	1.0000	0.0000	0.00
1.10	1.2320	1.2337	0.0017	0.14
1.20	1.5479	1.5527	0.0048	0.31
1.30	1.9832	1.9937	0.0106	0.53
1.40	2.5908	2.6117	0.0209	0.80
1.50	3.4509	3.4904	0.0394	1.13

TABLE 2.6 Improved Euler's Method with $h = 0.05$

x_n	y_n	True Value	Abs. Error	% Rel. Error
1.00	1.0000	1.0000	0.0000	0.00
1.05	1.1077	1.1079	0.0002	0.02
1.10	1.2332	1.2337	0.0004	0.04
1.15	1.3798	1.3806	0.0008	0.06
1.20	1.5514	1.5527	0.0013	0.08
1.25	1.7531	1.7551	0.0020	0.11
1.30	1.9909	1.9937	0.0029	0.14
1.35	2.2721	2.2762	0.0041	0.18
1.40	2.6060	2.6117	0.0057	0.22
1.45	3.0038	3.0117	0.0079	0.26
1.50	3.4795	3.4904	0.0108	0.31

A brief word of caution is in order here. We cannot compute all the values of y_n^* first and then substitute these values in the first formula of (5). In other words, we cannot use the data in Table 2.3 to help construct the values in Table 2.5. Why not?

Truncation Errors for the Improved Euler Method

The local truncation error for the improved Euler method is $O(h^3)$. The derivation of this result is similar to the derivation of the local truncation error for Euler's method and is left for you. See Problem 16. Since the local truncation for the improved Euler method is $O(h^3)$, the global truncation error is $O(h^2)$. This can be seen in Example 4; when the step size is halved from $h = 0.1$ to $h = 0.05$, the absolute error at $x = 1.50$ is reduced from 0.0394 to 0.0108, a reduction of approximately $(1/2)^2 = 1/4$.

ODE Solvers

As already noted, Euler's method uses local linear approximations to generate the sequence of points (x_n, y_n). If we connect these points with straight line segments, we obtain a polygonal curve that approximates the actual solution curve. This is illustrated in Figure 2.34 using step sizes $h = 1$, $h = 0.5$, and $h = 0.1$. It is apparent from the figure that the approximation improves as the step size decreases. For h sufficiently small, the polygonal curve will appear to be smooth and will hopefully be close to the actual solution curve. This polygonal approximation is well-suited to computers, and a program which does this is sometimes referred to as an **ODE solver**. Frequently, an ODE solver is part of a larger, more versatile software package. The ODE solver which is included as a supplement to this text is briefly described in Section 1.1.

EXERCISES 2.6

Answers to odd-numbered problems begin on page AN-7.

1. Consider the initial-value problem

$$y' = (x + y - 1)^2, \quad y(0) = 2.$$

(a) Solve the initial-value problem in terms of elementary functions. [*Hint*: Let $u = x + y - 1$.]

(b) Use the Euler formula with $h = 0.1$ and $h = 0.05$ to obtain approximate values of the solution of the initial-value problem at $x = 0.5$. Compare the approximate values with the exact values computed using the solution from part (a).

2. Repeat the calculations of Problem 1(b) using the improved Euler formula.

Given the initial-value problems in Problems 3–12, use the Euler formula to obtain a four-decimal approximation to the indicated value. First use (a) $h = 0.1$ and then (b) $h = 0.05$.

3. $y' = 2x - 3y + 1$, $y(1) = 5$; $y(1.5)$

4. $y' = 4x - 2y$, $y(0) = 2$; $y(0.5)$

5. $y' = 1 + y^2$, $y(0) = 0$; $y(0.5)$

6. $y' = x^2 + y^2$, $y(0) = 1$; $y(0.5)$

7. $y' = e^{-y}$, $y(0) = 0$; $y(0.5)$

8. $y' = x + y^2$, $y(0) = 0$; $y(0.5)$

9. $y' = (x - y)^2$, $y(0) = 0.5$; $y(0.5)$

10. $y' = xy + \sqrt{y}$, $y(0) = 1$; $y(0.5)$

11. $y' = xy^2 - \dfrac{y}{x}$, $y(1) = 1$; $y(1.5)$

12. $y' = y - y^2$, $y(0) = 0.5$; $y(0.5)$

13. As parts (a)–(e) of this problem, repeat the calculations of Problems 3, 5, 7, 9, and 11 using the improved Euler formula.

14. As parts (a)–(e) of this problem, repeat the calculations of Problems 4, 6, 8, 10, and 12 using the improved Euler formula.

15. Although it may not be obvious from the differential equation, its solution could "behave badly" near a point x at which we wish to approximate $y(x)$. Numerical procedures may give widely differing results near this point. Let $y(x)$ be the solution of the initial-value problem

$$y' = x^2 + y^3, \quad y(1) = 1.$$

(a) Use an ODE solver to obtain a graph of the solution on the interval [1, 1.4].

(b) Using the step size $h = 0.1$, compare the results obtained from the Euler formula with the results from the improved Euler formula in the approximation of $y(1.4)$.

16. In this problem we show that the local truncation error for the improved Euler method is $O(h^3)$.

(a) Use Taylor's formula with remainder to show that

$$y''(x_n) = \frac{y'(x_{n+1}) - y'(x_n)}{h} - \frac{1}{2}hy'''(c).$$

[*Hint*: Set $k = 2$ and differentiate the Taylor polynomial (3).]

(b) Use the Taylor polynomial with $k = 2$ to show that

$$y(x_{n+1}) = y(x_n) + h\frac{y'(x_n) + y'(x_{n+1})}{2} + O(h^3).$$

17. Consider the initial-value problem $y' = 2y$, $y(0) = 1$. The analytic solution is $y(x) = e^{2x}$.

(a) Approximate $y(0.1)$ using one step and the Euler method.

(b) Find a bound for the local truncation error in y_1.

(c) Compare the actual error in y_1 with your error bound.

(d) Approximate $y(0.1)$ using two steps and the Euler method.

(e) Verify that the global truncation error for Euler's method is $O(h)$ by comparing the errors in parts (a) and (d).

18. Repeat Problem 17 using the improved Euler method. Its global truncation error is $O(h^2)$.

19. Repeat Problem 17 using the initial-value problem $y' = -2y + x$, $y(0) = 1$. The analytic solution is $y(x) = \frac{1}{2}x - \frac{1}{4} + \frac{5}{4}e^{-2x}$.

20. Repeat Problem 19 using the improved Euler method. Its global truncation error is $O(h^2)$.

21. Consider the initial-value problem $y' = 2x - 3y + 1$, $y(1) = 5$. The analytic solution is $y(x) = \frac{1}{9} + \frac{2}{3}x + \frac{38}{9}e^{-3(x-1)}$.

(a) Find a formula involving c and h for the local truncation error in the nth step if the Euler method is used.

(b) Find a bound for the local truncation error in each step if $h = 0.1$ is used to approximate $y(1.5)$.

(c) Approximate $y(1.5)$ using $h = 0.1$ and $h = 0.05$ with the Euler method. See Problem 3.

(d) Calculate the errors in part (c) and verify that the global truncation error of the Euler method is $O(h)$.

22. Repeat Problem 21 using the improved Euler method, which has global truncation error $O(h^2)$. See Problem 13(a). You may need to keep more than four decimal places to see the effect of reducing the order of error.

23. Repeat Problem 21 for the initial-value problem $y' = e^{-y}$, $y(0) = 0$. The analytic solution is $y(x) = \ln(x + 1)$. Approximate $y(0.5)$. See Problem 7.

24. Repeat Problem 23 using the improved Euler method, which has global truncation error $O(h^2)$. See Problem 13(a). You may need to keep more than four decimal places to see the effect of reducing the order of error.

25. ***PROJECT PROBLEM*** In this problem we examine a relationship between numerical integration techniques and numerical methods for solving $y' = f(x, y)$, $y(a) = y_0$.

(a) Define $I(x) = \int_a^x f(t)dt$. Use Euler's method to approximate $I(a + h)$. For $f(t) \geq 0$ on $[a, a + h]$ interpret geometrically the numerical integration technique developed.

(b) Repeat part (a) using the improved Euler method. Show that this method corresponds to the trapezoidal rule.

(c) Develop a numerical method for solving $y' = f(x, y)$, $y(a) = y_0$ that is suggested by Simpson's rule. Compare this method for accuracy with the improved Euler method.

26. ***WRITING PROJECT*** Write a short report on the **midpoint method** for solving the initial-value problem $y' = f(x, y)$, $y(x_0) = y_0$:

$$y_{n+1} = y_n + hf(x_n + \tfrac{1}{2}h, y_n + \tfrac{1}{2}hf(x_n, y_n)).$$

Compare the Euler, improved Euler, and midpoint methods for accuracy.

2.7 NUMERICAL SOLUTIONS: RUNGE-KUTTA METHODS

Introduction

Probably one of the most popular as well as most accurate numerical procedures used in obtaining approximate solutions to the initial-value problem $y' = f(x, y)$, $y(x_0) = y_0$ is the **fourth-order Runge-Kutta method**. As the name suggests, there are Runge-Kutta methods of different orders. These methods are derived using the Taylor series expansion with remainder for $y(x_n + h)$:

$$y(x_{n+1}) = y(x_n + h) = y(x_n) + hy'(x_n) + \frac{h^2}{2!}y''(x_n) + \frac{h^3}{3!}y'''(x_n) + \cdots + \frac{h^{k+1}}{(k+1)!}y^{(k+1)}(c),$$

where c is a number between x_n and $x_n + h$. In the case when $k = 1$, and the remainder $\frac{h^2}{2}y''(c)$ is small, we obtain the familiar iteration formula

$$y_{n+1} = y_n + hy_n' = y_n + hf(x_n, y_n).$$

In other words, the basic Euler method is a **first-order** Runge-Kutta procedure.

We consider now the **second-order Runge-Kutta** procedure. This consists of finding constants a, b, α, and β so that the formula

$$y_{n+1} = y_n + ak_1 + bk_2, \tag{1}$$

where

$$k_1 = hf(x_n, y_n)$$

$$k_2 = hf(x_n + \alpha h, y_n + \beta k_1)$$

agrees with a Taylor polynomial of degree 2. It can be shown that this can be done whenever the constants satisfy

$$a + b = 1, \quad b\alpha = \tfrac{1}{2}, \quad \text{and} \quad b\beta = \tfrac{1}{2}. \tag{2}$$

This is a system of three equations in four unknowns and has infinitely many solutions. Observe that when $a = b = 1/2$ and $\alpha = \beta = 1$, (1) reduces to the improved Euler formula. Since the formula agrees with a Taylor polynomial of degree 2, the local truncation error for this method is $O(h^3)$ and the global truncation error is $O(h^2)$.

Notice that the sum $ak_1 + bk_2$ in (1) is a weighted average of k_1 and k_2 since $a + b = 1$. The numbers k_1 and k_2 are multiples of approximations to the slope of the solution curve $y(x)$ at two different points in the interval from x_n to x_{n+1}.

Fourth-Order Runge-Kutta Formula

The **fourth-order Runge-Kutta procedure** consists of finding appropriate constants so that the formula

$$y_{n+1} = y_n + ak_1 + bk_2 + ck_3 + dk_4,$$

where

$$k_1 = hf(x_n, y_n)$$

$$k_2 = hf(x_n + \alpha_1 h, y_n + \beta_1 k_1)$$

$$k_3 = hf(x_n + \alpha_2 h, y_n + \beta_2 k_1 + \beta_3 k_2)$$

$$k_4 = hf(x_n + \alpha_3 h, y_n + \beta_4 k_1 + \beta_5 k_2 + \beta_6 k_3)$$

agrees with a Taylor polynomial of degree 4. This results in 11 equations in 13 unknowns. The most commonly used set of values for the constants yields the following result:

$$
\begin{aligned}
y_{n+1} &= y_n + \tfrac{1}{6}(k_1 + 2k_2 + 2k_3 + k_4), \\
k_1 &= hf(x_n, y_n) \\
k_2 &= hf(x_n + \tfrac{1}{2}h, y_n + \tfrac{1}{2}k_1) \\
k_3 &= hf(x_n + \tfrac{1}{2}h, y_n + \tfrac{1}{2}k_2) \\
k_4 &= hf(x_n + h, y_n + k_3).
\end{aligned} \tag{3}
$$

You are advised to look carefully at the formulas in (3);, note that k_2 depends on k_1, k_3 depends on k_2, and k_4 depends on k_3. Also k_2 and k_3 involve approximations to the slope at the midpoint of the interval between x_n and x_{n+1}.

EXAMPLE 1 Runge-Kutta Method

Use the Runge-Kutta method with $h = 0.1$ to obtain an approximation to $y(1.5)$ for the solution of

$$y' = 2xy, \quad y(1) = 1.$$

Solution For the sake of illustration let us compute the case when $n = 0$. From (3) we find

$$
\begin{aligned}
k_1 &= (0.1)f(x_0, y_0) = (0.1)(2x_0 y_0) = 0.2 \\
k_2 &= (0.1)f\left(x_0 + \frac{1}{2}(0.1), y_0 + \frac{1}{2}(0.2)\right) \\
&= (0.1)2\left(x_0 + \frac{1}{2}(0.1)\right)\left(y_0 + \frac{1}{2}(0.2)\right) = 0.231 \\
k_3 &= (0.1)f\left(x_0 + \frac{1}{2}(0.1), y_0 + \frac{1}{2}(0.231)\right) \\
&= (0.1)2\left(x_0 + \frac{1}{2}(0.1)\right)\left(y_0 + \frac{1}{2}(0.231)\right) = 0.234255 \\
k_4 &= (0.1)f(x_0 + 0.1, y_0 + 0.234255) \\
&= (0.1)2(x_0 + 0.1)(y_0 + 0.234255) = 0.2715361
\end{aligned}
$$

and therefore

$$
\begin{aligned}
y_1 &= y_0 + \frac{1}{6}(k_1 + 2k_2 + 2k_3 + k_4) \\
&= 1 + \frac{1}{6}(0.2 + 2(0.231) + 2(0.234255) + 0.2715361) = 1.23367435.
\end{aligned}
$$

The remaining calculations are summarized in Table 2.7 whose entries are rounded to four decimal places.

TABLE 2.7 Runge-Kutta Method with $h = 0.1$

x_n	y_n	True Value	Abs. Error	% Rel. Error
1.00	1.0000	1.0000	0.0000	0.00
1.10	1.2337	1.2337	0.0000	0.00
1.20	1.5527	1.5527	0.0000	0.00
1.30	1.9937	1.9937	0.0000	0.00
1.40	2.6116	2.6117	0.0001	0.00
1.50	3.4902	3.4904	0.0001	0.00

Inspection of Table 2.7 shows why the fourth-order Runge-Kutta method is so popular. If four-decimal-place accuracy is all that we desire, there is no need to use a smaller step size. Table 2.8 compares the results of applying the Euler, improved Euler, and fourth-order Runge-Kutta methods to the initial-value problem $y' = 2xy$, $y(1) = 1$. (See Examples 2 and 3 in Section 2.6.)

TABLE 2.8 $y' = 2xy$, $y(1) = 1$

Comparison of Numerical Methods with $h = 0.1$

x_n	Euler	Improved Euler	Three-Term Taylor	Runge-Kutta	True Value
1.00	1.0000	1.0000	1.0000	1.0000	1.0000
1.10	1.2000	1.2320	1.2300	1.2337	1.2337
1.20	1.4640	1.5479	1.5427	1.5527	1.5527
1.30	1.8154	1.9832	1.9728	1.9937	1.9937
1.40	2.2874	2.5908	2.5721	2.6116	2.6117
1.50	2.9278	3.4509	3.4188	3.4902	3.4904

Comparison of Numerical Methods with $h = 0.05$

x_n	Euler	Improved Euler	Three-Term Taylor	Runge-Kutta	True Value
1.00	1.0000	1.0000	1.0000	1.0000	1.0000
1.05	1.1000	1.1077	1.1075	1.1079	1.1079
1.10	1.2155	1.2332	1.2327	1.2337	1.2337
1.15	1.3492	1.3798	1.3788	1.3806	1.3806
1.20	1.5044	1.5514	1.5499	1.5527	1.5527
1.25	1.6849	1.7531	1.7509	1.7551	1.7551
1.30	1.8955	1.9909	1.9879	1.9937	1.9937
1.35	2.1419	2.2721	2.2681	2.2762	2.2762
1.40	2.4311	2.6060	2.6006	2.6117	2.6117
1.45	2.7714	3.0038	2.9967	3.0117	3.0117
1.50	3.1733	3.4795	3.4702	3.4903	3.4904

Truncation Errors for the Runge-Kutta Method

Since the first equation in (3) agrees with a Taylor polynomial of degree 4, the local truncation error for this method is

$$y^{(5)}(c)\frac{h^5}{5!} \quad \text{or} \quad O(h^5),$$

and the global truncation error is thus $O(h^4)$. It is now obvious why this is called the *fourth-order* Runge-Kutta method.

EXAMPLE 2 Bound for Local/Global Truncation Errors

Analyze the local and global truncation errors for the fourth-order Runge-Kutta method applied to $y' = 2xy$, $y(1) = 1$.

Solution By differentiating the known solution $y(x) = e^{x^2-1}$, we get

$$y^{(5)}(c)\frac{h^5}{5!} = (120c + 160c^3 + 32c^5)e^{c^2-1}\frac{h^5}{5!}. \tag{4}$$

Thus with $c = 1.5$, (4) yields a bound of 0.00028 on the local truncation error for each of the five steps when $h = 0.1$. Note that in Table 2.7 the actual error in y_1 is much less than this bound.

Table 2.9 gives the approximations to the solution of the initial-value problem at $x = 1.5$ which are obtained from the fourth-order Runge-Kutta method. By computing the value of the exact solution at $x = 1.5$ we can find the error in these approximations. Because the method is so accurate, many decimal places must be used in the numerical solution to see the effect of halving the step size. Note that when h is halved, from $h = 0.1$ to $h = 0.05$, the error is divided by a factor of about $2^4 = 16$, as expected.

TABLE 2.9 Runge-Kutta Method

h	Approximation	Error
0.1	3.49021064	$1.323210889 \times 10^{-4}$
0.05	3.49033382	$9.137760898 \times 10^{-6}$

Adaptive Methods

We have seen that the accuracy of a numerical method can be improved by decreasing the step size h. Of course, this enhanced accuracy is usually obtained at a cost, namely, increased computation time and greater possibility of round-off error. In general, over the interval of approximation, there may be subintervals where a relatively large step size suffices and other subintervals where a smaller step size is necessary in order to keep the truncation error within a desired limit. Numerical methods that use a variable step size are called **adaptive methods**. One of the more popular of these methods for approximating solutions of differential equations is called the **Runge-Kutta-Fehlberg algorithm**. See Problem 22 and *Computer Lab Experiment* 8 for Sections 2.6–2.8.

EXERCISES 2.7

Answers to odd-numbered problems begin on page AN-9.

1. Use the fourth-order Runge-Kutta method with $h = 0.1$ to obtain a four-decimal-place approximation to the solution of the initial-value problem
$$y' = (x + y - 1)^2, \quad y(0) = 2$$
at $x = 0.5$. Compare the approximate values with the exact values obtained in Problem 1 in Exercises 2.6.

2. Solve the equations in (2) using the assumption $a = \frac{1}{4}$. Use the resulting second-order Runge-Kutta method to obtain a four-decimal-place approximation to the solution of the initial-value problem
$$y' = (x + y - 1)^2, \quad y(0) = 2$$
at $x = 0.5$. Compare the approximate values with the values obtained in Problem 2 in Exercises 2.6.

Given the initial-value problems in Problems 3–12, use the Runge-Kutta method with $h = 0.1$ to obtain a four-decimal-place approximation to the indicated value.

3. $y' = 2x - 3y + 1$, $y(1) = 5$; $y(1.5)$
4. $y' = 4x - 2y$, $y(0) = 2$; $y(0.5)$
5. $y' = 1 + y^2$, $y(0) = 0$; $y(0.5)$
6. $y' = x^2 + y^2$, $y(0) = 1$; $y(0.5)$
7. $y' = e^{-y}$, $y(0) = 0$; $y(0.5)$
8. $y' = x + y^2$, $y(0) = 0$; $y(0.5)$
9. $y' = (x - y)^2$, $y(0) = 0.5$; $y(0.5)$
10. $y' = xy + \sqrt{y}$, $y(0) = 1$; $y(0.5)$
11. $y' = xy^2 - \dfrac{y}{x}$, $y(1) = 1$; $y(1.5)$
12. $y' = y - y^2$, $y(0) = 0.5$; $y(0.5)$

13. If air resistance is proportional to the square of the instantaneous velocity, then the velocity v of a mass m dropped from a height h is determined from
$$m\frac{dv}{dt} = mg - kv^2, \quad k > 0.$$
Let $v(0) = 0$, $k = 0.125$, $m = 5$ slugs, and $g = 32$ ft/s^2.
   (a) Use the Runge-Kutta method with $h = 1$ to find an approximation to the velocity of the falling mass at $t = 5$ seconds.
   (b) Use an ODE solver to graph the solution of the initial-value problem.
   (c) Use separation of variables to solve the initial-value problem and find the true value $v(5)$.

14. A mathematical model for the area A (in cm^2) that a colony of bacteria (*B. dendroides*) occupies is given by
$$\frac{dA}{dt} = A(2.128 - 0.0432A).*$$
Suppose that the initial area is 0.24 cm^2.
   (a) Use the Runge-Kutta method with $h = 0.5$ to complete the following table.

t (days)	1	2	3	4	5
A (observed)	2.78	13.53	36.30	47.50	49.40
A (approximated)					

   (b) Use an ODE solver to graph the solution of the initial-value problem. Estimate the values $A(1)$, $A(2)$, $A(3)$, $A(4)$, and $A(5)$ from the graph.
   (c) Use separation of variables to solve the initial-value problem and compute the values $A(1)$, $A(2)$, $A(3)$, $A(4)$, and $A(5)$.

15. Consider the initial-value problem
$$y' = x^2 + y^3, \quad y(1) = 1.$$
(See Problem 15 in Exercises 2.6.)
   (a) Compare the results obtained from using the Runge-Kutta formula over the interval [1, 1.4] with step sizes $h = 0.1$ and $h = 0.05$.
   (b) Use an ODE solver to obtain a graph of the solution on the interval [1, 1.4].

16. Consider the initial-value problem $y' = 2y$, $y(0) = 1$. The analytic solution is $y(x) = e^{2x}$.
   (a) Approximate $y(0.1)$ using one step and the fourth-order Runge-Kutta method.
   (b) Find a bound for the local truncation error in y_1.
   (c) Compare the actual error in y_1 with your error bound.
   (d) Approximate $y(0.1)$ using two steps and the fourth-order Runge-Kutta method.
   (e) Verify that the global truncation error for the fourth-order Runge-Kutta method is $O(h^4)$ by comparing the errors in parts (a) and (d).

---

* See. V. A. Kostitzin, *Mathematical Biology* (London: Harrap, 1939).

17. Repeat Problem 16 using the initial-value problem $y' = -2y + x$, $y(0) = 1$. The analytic solution is $y(x) = \frac{1}{2}x - \frac{1}{4} + \frac{5}{4}e^{-2x}$.

18. Consider the initial-value problem $y' = 2x - 3y + 1$, $y(1) = 5$. The analytic solution is $y(x) = \frac{1}{9} + \frac{2}{3}x + \frac{38}{9}e^{-3(x-1)}$.
   (a) Find a formula involving c and h for the local truncation error in the nth step if the fourth-order Runge-Kutta method is used.
   (b) Find a bound for the local truncation error in each step if $h = 0.1$ is used to approximate $y(1.5)$.
   (c) Approximate $y(1.5)$ using the fourth-order Runge-Kutta method with $h = 0.1$ and $h = 0.05$. See Problem 3. You will need to carry more than six decimal places to see the effect of reducing the order of error.

19. Repeat Problem 18 for the initial-value problem $y' = e^{-y}$, $y(0) = 0$. The analytic solution is $y(x) = \ln(x + 1)$. Approximate $y(0.5)$. See Problem 7.

20. *PROJECT PROBLEM* A count of the number of evaluations of the function f used in solving the initial-value problem $y' = f(x, y)$, $y(x_0) = y_0$ is used as a measure of the computational complexity of a numerical method. Determine the number of evaluations of f required for each step of the Euler, improved Euler, and Runge-Kutta methods. By considering some specific examples, compare the accuracy of these methods when used with comparable computational complexities.

21. *PROJECT PROBLEM* In the examples in the last two sections we checked the accuracy of the numerical methods by comparing the results with the exact or analytic solution. You might wonder how we can confirm the accuracy of a numerical technique when an exact solution cannot be found. To examine this question we begin by considering a problem whose exact solution can be determined.
   (a) Solve the initial-value problem $y' = 15y - 20x$, $y(0) = 1$.
   (b) Use the fourth-order Runge-Kutta method to approximate the solution of the problem in part (a). Compare the results.
   (c) If you did not know the exact solution of the problem in part (a), determine how you might discover that the numerical method is not giving an accurate result.
   (d) Approximate the solutions of the following initial-value problems and discuss the accuracy of your solution:

$$y' = x^2 + y^3,\ y(0) = 1 \quad \text{at} \quad x = 0.5;$$

$$y' = x + \sin y,\ y(0) = 1 \quad \text{at} \quad x = 1.$$

22. *WRITING PROJECT* Write a report on the **Runge-Kutta-Fehlberg method** for approximating the solution of a first-order initial-value problem. Discuss the formulas used and the method by which the step size is determined at each step. Contrast the Runge-Kutta-Fehlberg method with the Runge-Kutta methods of orders four and five in terms of accuracy and number of function evaluations required.

2.8 NUMERICAL SOLUTIONS: MULTISTEP METHODS

Introduction

The Euler and Runge-Kutta formulas discussed in the preceding sections are examples of **single-step** methods. In these methods each successive value y_{n+1} is computed based only on information about the immediately preceding value y_n. A **multistep** or **continuing** method, on the other hand, uses the values from several previously computed steps to obtain the value of y_{n+1}. There are a large number of multistep formulas that can be applied to obtain approximations to solutions of differential equations. Although it is not our intention to survey this vast field of numerical methods, we will consider two methods of special interest here. Both of these, like the improved Euler formula, are **predictor-corrector methods**. That is, one formula is used to predict a value y^*_{n+1}, which in turn is used to obtain a corrected value y_{n+1}.

Adams-Bashforth/Adams-Moulton Method

One of the most popular multistep methods is the **Adams-Bashforth/Adams-Moulton Method**. The predictor in this method is the Adams-Bashforth formula

$$y^*_{n+1} = y_n + \frac{h}{24}(55y'_{n-1} - 59y'_{n-1} + 37y'_{n-2} - 9y'_{n-3}), \tag{1}$$

$$y'_n = f(x_n, y_n)$$

$$y'_{n-1} = f(x_{n-1}, y_{n-1})$$

$$y'_{n-2} = f(x_{n-2}, y_{n-2})$$

$$y'_{n-3} = f(x_{n-3}, y_{n-3}),$$

for $n \geq 3$. The value of y^*_{n+1} is then substituted into the Adams-Moulton corrector

$$\begin{aligned} y_{n+1} &= y_n + \frac{h}{24}(9y'_{n+1} + 19y'_n - 5y'_{n-1} + y'_{n-2}), \\ y'_{n+1} &= f(x_{n+1}, y^*_{n+1}). \end{aligned} \tag{2}$$

Notice that formula (1) requires that we know the values of y_0, y_1, y_2, and y_3 in order to obtain y_4. The value of y_0 is, of course, the given initial condition. Since the local truncation error of the Adams-Bashforth/Adams-Moulton method is $O(h^5)$, the values of y_1, y_2, and y_3 are generally computed by a method with the same error property, such as the fourth-order Runge-Kutta formula.

EXAMPLE 1 Adams-Bashforth/Adams-Moulton Method

Use the Adams-Bashforth/Adams-Moulton method with $h = 0.2$ to obtain an approximation to $y(0.8)$ for the solution of

$$y' = x + y - 1, \quad y(0) = 1.$$

Solution With a step size of $h = 0.2$, $y(0.8)$ will be approximated by y_4. To get started we use the Runge-Kutta method with $x_0 = 0$, $y_0 = 1$, and $h = 0.2$ to obtain

$$y_1 = 1.02140000, \quad y_2 = 1.09181796, \quad y_3 = 1.22210646.$$

Now with the identifications $x_0 = 0$, $x_1 = 0.2$, $x_2 = 0.4$, $x_3 = 0.6$, and $f(x, y) = x + y - 1$, we find

$$y'_0 = f(x_0, y_0) = (0) + (1) - 1 = 0$$

$$y'_1 = f(x_1, y_1) = (0.2) + (1.02140000) - 1 = 0.22140000$$

$$y'_2 = f(x_2, y_2) = (0.4) + (1.09181796) - 1 = 0.49181796$$

$$y'_3 = f(x_3, y_3) = (0.6) + (1.22210646) - 1 = 0.82210646.$$

With the foregoing values the predictor (1) then gives

$$y_4^* = y_3 + \frac{0.2}{24}(55y_3' - 59y_2' + 37y_1' - 9y_0') = 1.42535975.$$

To use the corrector (2) we first need

$$y_4' = f(x_4, y_4^*) = (0.8) + (1.42535975) - 1 = 1.22535975.$$

Finally, (2) yields

$$y_4 = y_3 + \frac{0.2}{24}(9y_4' + 19y_3' - 5y_2' + y_1') = 1.42552788.$$

●

You should verify that the exact value of $y(0.8)$ in Example 1 is $y(0.8) = 1.42554093$.

Stability of Numerical Methods

An important consideration when using numerical methods to approximate the solution of an initial-value problem is the stability of the method. Simply stated, a numerical method is **stable** if small changes in the initial condition result in only small changes in the computed solution. A numerical method is said to be **unstable** if it is not stable. The reason that stability considerations are important is that in each step after the first step of a numerical technique we are essentially starting over again with a new initial-value problem, where the initial condition is the approximate solution value computed in the preceding step. Due to the presence of round-off error this value will almost certainly vary at least slightly from the true value of the solution. Besides round-off error, another common source of error occurs in the initial condition itself; in physical applications the data are often obtained by imprecise measurements.

One possible method for detecting instability in the numerical solution of a specific initial-value problem is to compare the approximate solutions obtained when decreasing step sizes are used. If the numerical method is unstable, the error may actually increase with smaller step sizes. Another way of checking stability is to observe what happens to solutions when the initial condition is slightly perturbed (for example, change $y(0) = 1$ to $y(0) = 0.999$).

For a more detailed and precise discussion of stability consult a numerical analysis text. In general, all of the methods we have discussed in this chapter have good stability characteristics. For an example of a method that is sometimes unstable see Problem 9 in Exercises 2.8.

Advantages and Disadvantages of Multistep Methods

Many considerations enter into the choice of a method to solve a differential equation numerically. Single-step methods, particularly the Runge-Kutta method, are often chosen because of their accuracy and the fact that they are easy to program. However, a major drawback to them is that the right-hand side of the differential equation must be evaluated many times at each step. For instance, the fourth-order Runge-Kutta method requires four function evaluations for each step. See Problem 20 in Exercises 2.7. On the other hand, if the function evaluations in the previous step have been calculated and stored, a multistep method requires only one new function evaluation for each step. This can lead to great savings in time and expense.

As an example, to solve $y' = f(x, y)$, $y(x_0) = y_0$ numerically using n steps by the fourth-order Runge-Kutta method requires $4n$ function evaluations. The Adams-Bashforth multistep method requires 16 function evaluations for the Runge-Kutta fourth-order starter and $n - 4$ for the n Adams-Bashforth steps, giving a total of $n + 12$ function evaluations for this method. In general the Adams-Bashforth multistep method requires slightly more than a quarter of the number of function evaluations required for the fourth-order Runge-Kutta method. If the evaluation of $f(x, y)$ is complicated, the multistep method will be more efficient.

Another issue involved with multistep methods is how many times the corrector formula should be repeated in each step. Each time the corrector is used, another function evaluation is done, and so the accuracy is increased at the expense of losing an advantage of the multistep method. In practice, the corrector is calculated once, and if the value of y_{n+1} is changed by a large amount, the entire problem is restarted using a smaller step size. This is often the basis of the variable step size methods, whose discussion is beyond the scope of this text.

EXERCISES 2.8

Answers to odd-numbered problems begin on page AN-11.

1. Find the exact solution of the initial-value problem in Example 1. Compare the exact values of $y(0.2)$, $y(0.4)$, $y(0.6)$, and $y(0.8)$ with the approximations y_1, y_2, y_3, and y_4.

2. Write a computer program for the Adams-Bashforth/Adams-Moulton method.

In Problems 3 and 4 use the Adams-Bashforth/Adams-Moulton method to approximate $y(0.8)$, where $y(x)$ is the solution of the given initial-value problem. Use $h = 0.2$ and the Runge-Kutta method to compute y_1, y_2, and y_3.

3. $y' = 2x - 3y + 1$, $y(0) = 1$

4. $y' = 4x - 2y$, $y(0) = 2$

In Problems 5–8 use the Adams-Bashforth/Adams-Moulton method to approximate $y(1.0)$, where $y(x)$ is the solution of the given initial-value problem. Use $h = 0.2$ and $h = 0.1$ and the Runge-Kutta method to compute y_1, y_2, and y_3.

5. $y' = 1 + y^2$, $y(0) = 0$

6. $y' = y + \cos x$, $y(0) = 1$

7. $y' = (x - y)^2$, $y(0) = 0$

8. $y' = xy + \sqrt{y}$, $y(0) = 1$

9. ***WRITING PROJECT*** A multistep method that is frequently discussed in differential equations and numerical analysis texts is **Milne's method**, also referred to as the **Milne-Simpson method**.

   (a) Write a report on Milne's method, including derivations of the predictor and corrector formulas and a discussion of stability.

   (b) Write a computer program to implement Milne's method. Use your program to obtain an approximation to $y(2)$, where $y' = -10y + 5$, $y(1) = 1$, and $h = 0.1$.

   (c) Consider the differential equation $y' + ay = 0$ for various values of the parameter a. Use your program to approximate a range of values of a for which Milne's method is stable.

CHAPTER 3

HIGHER-ORDER DIFFERENTIAL EQUATIONS

3.1 PRELIMINARY THEORY: LINEAR EQUATIONS

Introduction

We turn now to the solution of differential equations of order two or higher. In this and the nine sections that follow we examine some of the underlying theory and methods for solving *linear* equations. As we did in Section 2.4, we set conditions on the differential equation under which we can obtain a general solution, that is, a solution that contains all solutions of the equation on some interval. In the last section of this chapter, we shall briefly examine *nonlinear* higher-order equations.

3.1.1 Initial-Value and Boundary-Value Problems

Initial-Value Problem

For a linear nth-order differential equation, the problem

$$\textit{Solve:} \quad a_n(x)\frac{d^ny}{dx^n} + a_{n-1}(x)\frac{d^{n-1}y}{dx^{n-1}} + \cdots + a_1(x)\frac{dy}{dx} + a_0(x)y = g(x)$$

$$\textit{Subject to:} \quad y(x_0) = y_0, \quad y'(x_0) = y_1, \ldots, y^{(n-1)}(x_0) = y_{n-1} \tag{1}$$

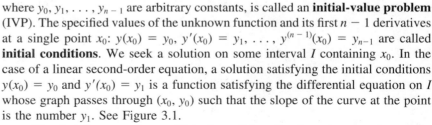

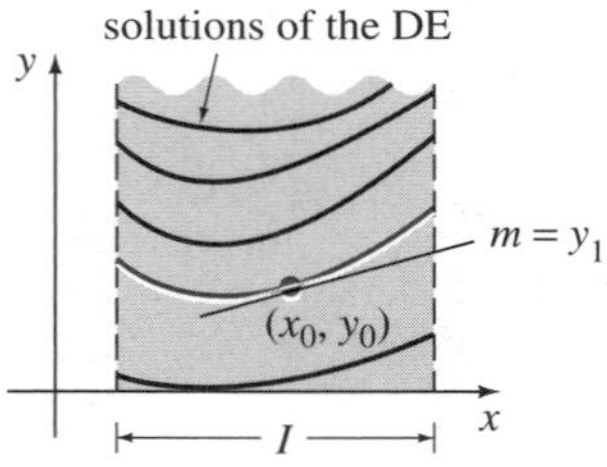

FIGURE 3.1

where $y_0, y_1, \ldots, y_{n-1}$ are arbitrary constants, is called an **initial-value problem** (IVP). The specified values of the unknown function and its first $n - 1$ derivatives at a single point x_0: $y(x_0) = y_0$, $y'(x_0) = y_1, \ldots, y^{(n-1)}(x_0) = y_{n-1}$ are called **initial conditions**. We seek a solution on some interval I containing x_0. In the case of a linear second-order equation, a solution satisfying the initial conditions $y(x_0) = y_0$ and $y'(x_0) = y_1$ is a function satisfying the differential equation on I whose graph passes through (x_0, y_0) such that the slope of the curve at the point is the number y_1. See Figure 3.1.

The first theorem gives sufficient conditions for the existence of a unique solution of (1).

THEOREM 3.1 Existence of a Unique Solution

Let $a_n(x), a_{n-1}(x), \ldots, a_1(x), a_0(x)$ and $g(x)$ be continuous on an interval I and let $a_n(x) \neq 0$ for every x in this interval. If $x = x_0$ is any point in this interval, then a solution $y(x)$ of the initial-value problem (1) exists on the interval and is unique.

EXAMPLE 1 Unique Solution of an IVP

The initial-value problem

$$3y''' + 5y'' - y' + 7y = 0, \qquad y(1) = 0, \quad y'(1) = 0, \quad y''(1) = 0$$

possesses the trivial solution $y = 0$. Since the third-order equation is linear with constant coefficients, it follows that all the conditions of Theorem 3.1 are fulfilled. Hence $y = 0$ is the *only* solution on any interval containing $x = 1$. ●

EXAMPLE 2 Unique Solution of an IVP

You should verify that the function $y = 3e^{2x} + e^{-2x} - 3x$ is a solution of the initial-value problem

$$y'' - 4y = 12x, \qquad y(0) = 4, \quad y'(0) = 1.$$

Now the differential equation is linear, the coefficients as well as $g(x) = 12x$ are continuous, and $a_2(x) = 1 \neq 0$ on any interval containing $x = 0$. We conclude from Theorem 3.1 that the given function is the unique solution. ●

The requirements in Theorem 3.1 that $a_i(x)$, $i = 0, 1, 2, \ldots, n$ be continuous and $a_n(x) \neq 0$ for every x in I are both important. Specifically, if $a_n(x) = 0$ for some x in the interval, then the solution of a linear initial-value problem may not be unique or even exist. For example, you should verify that the function $y = cx^2 + x + 3$ is a solution of the initial-value problem

$$x^2y'' - 2xy' + 2y = 6, \qquad y(0) = 3, \quad y'(0) = 1$$

on the interval $(-\infty, \infty)$ for any choice of the parameter c. In other words, there is no unique solution of the problem. Although most of the conditions of Theorem 3.1 are satisfied, the obvious difficulties are that $a_2(x) = x^2$ is zero at $x = 0$ and that the initial conditions are also imposed at $x = 0$.

Boundary-Value Problem

Another type of problem consists of solving a differential equation of order two or greater in which the dependent variable y or its derivatives are specified at *different points*. A problem such as

$$\textit{Solve:} \quad a_2(x)\frac{d^2y}{dx^2} + a_1(x)\frac{dy}{dx} + a_0(x)y = g(x)$$

$$\textit{Subject to:} \quad y(a) = y_0, \quad y(b) = y_1$$

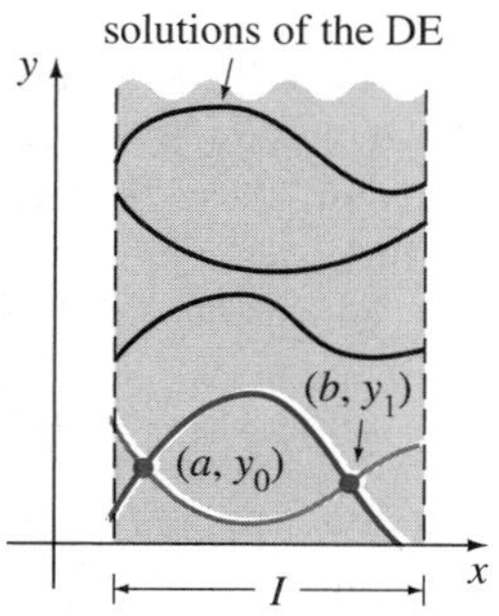

FIGURE 3.2

is called a **boundary-value problem** (BVP). The prescribed values $y(a) = y_0$ and $y(b) = y_1$ are called **boundary conditions**. A solution of the foregoing problem is a function satisfying the differential equation on some interval I, containing a and b, whose graph passes through the two points (a, y_0) and (b, y_1). See Figure 3.2.

For a second-order differential equation, other pairs of boundary conditions could be

$$y'(a) = y_0, \qquad y(b) = y_1;$$

$$y(a) = y_0, \qquad y'(b) = y_1;$$

or

$$y'(a) = y_0, \qquad y'(b) = y_1,$$

where y_0 and y_1 denote arbitrary constants. These three pairs of conditions are just special cases of the general boundary conditions

$$\alpha_1 y(a) + \beta_1 y'(a) = \gamma_1$$

$$\alpha_2 y(b) + \beta_2 y'(b) = \gamma_2.$$

The next examples show that even when the conditions of Theorem 3.1 are fulfilled, a boundary-value problem may have (i) several solutions (as suggested in Figure 3.2), (ii) a unique solution, or (iii) no solution at all.

EXAMPLE 3 A BVP can have Many, One, or No Solutions

In Example 5 of Section 1.1 we saw that the two-parameter family of solutions of the differential equation $x'' + 16x = 0$ is

$$x = c_1 \cos 4t + c_2 \sin 4t. \tag{2}$$

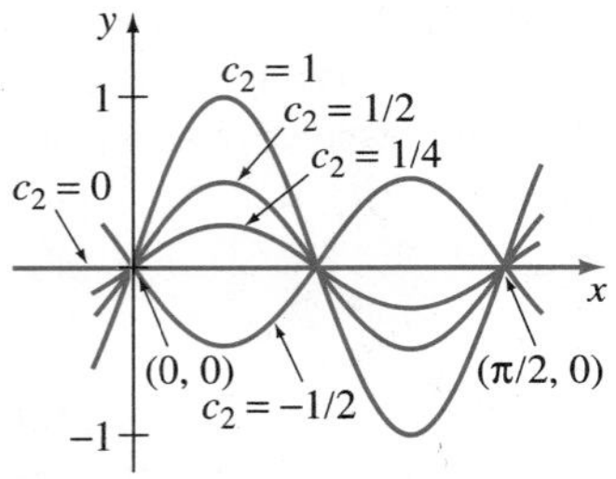

FIGURE 3.3

(a) Suppose we now wish to determine that solution of the equation that further satisfies the boundary conditions $x(0) = 0$, $x(\pi/2) = 0$. Observe that the first condition $0 = c_1 \cos 0 + c_2 \sin 0$ implies $c_1 = 0$, so that $x = c_2 \sin 4t$. But when $t = \pi/2$, $0 = c_2 \sin 2\pi$ is satisfied for any choice of c_2 since $\sin 2\pi = 0$. Hence the boundary-value problem

$$x'' + 16x = 0, \qquad x(0) = 0, \quad x(\pi/2) = 0 \tag{3}$$

has infinitely many solutions. Figure 3.3 shows the graphs of some of the members of the one-parameter family $x = c_2 \sin 4t$ that pass through the two points (0, 0) and ($\pi/2$, 0).

(b) If the boundary-value problem in (3) is changed to

$$x'' + 16x = 0, \qquad x(0) = 0, \quad x(\pi/8) = 0 \tag{4}$$

then $x(0) = 0$ still requires $c_1 = 0$ in the solution (2). But applying $x(\pi/8) = 0$ to $x = c_2 \sin 4t$ demands that $0 = c_2 \sin(\pi/2) = c_2 \cdot 1$. Hence $x = 0$ is a solution of this new boundary-value problem. Indeed, it can be proved that $x = 0$ is the *only* solution of (4).

(c) Finally, if we change the problem to

$$x'' + 16x = 0, \qquad x(0) = 0, \quad x(\pi/2) = 1, \tag{5}$$

we find again that $c_1 = 0$ from $x(0) = 0$, but that applying $x(\pi/2) = 1$ to $x = c_2 \sin 4t$ leads to the contradiction $1 = c_2 \sin 2\pi = c_2 \cdot 0 = 0$. Hence the boundary-value problem (5) has no solution. ●

3.1.2 Homogeneous Equations

A linear nth-order differential equation of the form

$$a_n(x)\frac{d^n y}{dx^n} + a_{n-1}(x)\frac{d^{n-1} y}{dx^{n-1}} + \cdots + a_1(x)\frac{dy}{dx} + a_0(x)\,y = 0 \tag{6}$$

is said to be **homogeneous**, whereas an equation of the form

$$a_n(x)\frac{d^n y}{dx^n} + a_{n-1}(x)\frac{d^{n-1}y}{dx^{n-1}} + \cdots + a_1(x)\frac{dy}{dx} + a_0(x)\,y = g(x) \tag{7}$$

with $g(x)$ not identically zero, is said to be **nonhomogeneous**. For example, $2y'' + 3y' - 5y = 0$ is a homogeneous linear second-order differential equation, whereas $x^3y''' + 6y' + 10y = e^x$ is a nonhomogeneous linear third-order differential equation. The word *homogeneous* in this context does not refer to coefficients that are homogeneous functions as in Section 2.2.

We shall see that in order to solve a nonhomogeneous equation (7), we must first be able to solve the associated homogeneous equation (6).

Note To avoid needless repetition throughout the remainder of this text we shall, as a matter of course, make the following important assumptions when stating definitions and theorems about the linear equations (6) and (7). On some common interval I,

- the coefficients $a_i(x)$, $i = 0, 1, 2, \ldots, n$ are continuous,
- the right-hand member $g(x)$ is continuous, and
- $a_n(x) \neq 0$ for every x in the interval.

Superposition Principle

In the next theorem we see that the sum, or **superposition**, of two or more solutions of a homogeneous linear differential equation is also a solution.

THEOREM 3.2 Superposition Principle—Homogeneous Equations

Let $y_1, y_2, \ldots, y_k$ be solutions of the homogeneous linear nth-order differential equation (6) on an interval I. Then the linear combination

$$y = c_1y_1(x) + c_2y_2(x) + \cdots + c_ky_k(x),$$

where the c_i, $i = 1, 2, \ldots, k$ are arbitrary constants, is also a solution on the interval.

Proof We prove the case when $n = k = 2$. Let $y_1(x)$ and $y_2(x)$ be solutions of $a_2(x)y'' + a_1(x)y' + a_0(x)y = 0$. If we define $y = c_1y_1(x) + c_2y_2(x)$, then

$$a_2(x)[c_1y_1'' + c_2y_2''] + a_1(x)[c_1y_1' + c_2y_2'] + a_0(x)[c_1y_1 + c_2y_2]$$

$$= c_1\underbrace{[a_2(x)y_1'' + a_1(x)y_1' + a_0(x)y_1]}_{\text{zero}} + c_2\underbrace{[a_2(x)y_2'' + a_1(x)y_2' + a_0(x)y_2]}_{\text{zero}}$$

$$= c_1 \cdot 0 + c_2 \cdot 0 = 0. \qquad \square$$

COROLLARIES

(A) A constant multiple $y = c_1y_1(x)$ of a solution $y_1(x)$ of a homogeneous linear differential equation is also a solution.
(B) A homogeneous linear differential equation always possesses the trivial solution $y = 0$.

EXAMPLE 4 Superposition/Homogeneous DE

The functions $y_1 = x^2$ and $y_2 = x^2 \ln x$ are both solutions of the homogeneous linear equation $x^2y''' - 2xy' + 4y = 0$ on the interval $(0, \infty)$. By the superposition principle, the linear combination

$$y = c_1x^2 + c_2x^2 \ln x$$

is also a solution of the equation on the interval. ●

The function $y = e^{7x}$ is a solution of $y'' - 9y' + 14y = 0$. Since the differential equation is linear and homogeneous, the constant multiple $y = ce^{7x}$ is also a solution. For various values of c we see that $y = 9e^{7x}$, $y = 0$, $y = -\sqrt{5}\, e^{7x}$ . . . are all solutions of the equation.

Linear Dependence and Linear Independence

The next two concepts are basic to the study of linear differential equations.

DEFINITION 3.1 Linear Dependence/Independence

A set of functions $f_1(x), f_2(x), \ldots, f_n(x)$ is said to be **linearly dependent** on an interval I if there exist constants $c_1, c_2, \ldots, c_n$, not all zero, such that

$$c_1f_1(x) + c_2f_2(x) + \cdots + c_nf_n(x) = 0$$

for every x in the interval. If the set of functions is not linearly dependent on the interval, it is said to be **linearly independent**.

In other words, a set of functions is linearly independent on an interval if the only constants for which

$$c_1f_1(x) + c_2f_2(x) + \cdots + c_nf_n(x) = 0,$$

for every x in the interval, are $c_1 = c_2 = \cdots = c_n = 0$.

It is easy to understand these definitions in the case of two functions $f_1(x)$ and $f_2(x)$. If the functions are linearly dependent on an interval, then there exist constants c_1 and c_2 that are not both zero such that for every x in the interval,

$c_1 f_1(x) + c_2 f_2(x) = 0$. Therefore, if we assume that $c_1 \neq 0$, it follows that

$$f_1(x) = -\frac{c_2}{c_1} f_2(x);$$

that is, *if two functions are linearly dependent, then one is simply a constant multiple of the other.* Conversely, if $f_1(x) = c_2 f_2(x)$ for some constant c_2, then

$$(-1) \cdot f_1(x) + c_2 f_2(x) = 0$$

for every x on some interval. Hence the functions are linearly dependent since at least one of the constants (namely, $c_1 = -1$) is not zero. We conclude that *two functions are linearly independent when neither is a constant multiple of the other* on an interval. For example, the functions $f_1(x) = \sin 2x$ and $f_2(x) = \sin x \cos x$ are linearly dependent on $(-\infty, \infty)$ because $f_1(x)$ is a constant multiple of $f_2(x)$. Recall from the double angle formula for the sine that $\sin 2x = 2 \sin x \cos x$. On the other hand, the functions $f_1(x) = x$ and $f_2(x) = |x|$ are linearly independent on $(-\infty, \infty)$. Inspection of Figure 3.4 should convince you that neither function is a constant multiple of the other on the interval.

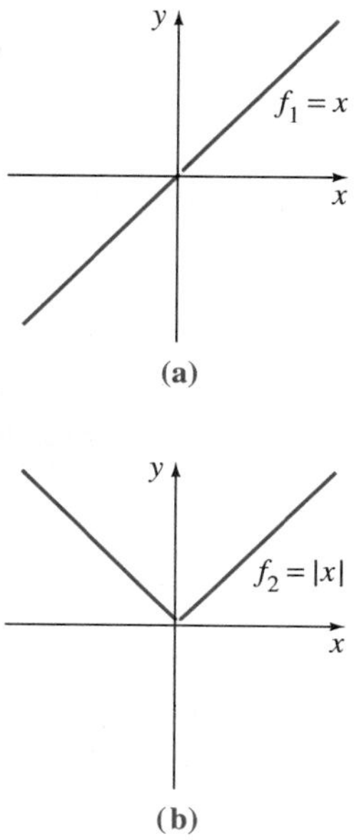

FIGURE 3.4

It follows from the preceding discussion that the ratio $f_2(x)/f_1(x)$ is not a constant on an interval on which $f_1(x)$ and $f_2(x)$ are linearly independent. This little fact will be used in the next section.

EXAMPLE 5 Linearly Dependent Functions

The functions $f_1(x) = \cos^2 x, f_2(x) = \sin^2 x, f_3(x) = \sec^2 x, f_4(x) = \tan^2 x$ are linearly dependent on the interval $(-\pi/2, \pi/2)$ since

$$c_1 \cos^2 x + c_2 \sin^2 x + c_3 \sec^2 x + c_4 \tan^2 x = 0,$$

when $c_1 = c_2 = 1, c_3 = -1, c_4 = 1$. We used here $\cos^2 x + \sin^2 x = 1$ and $1 + \tan^2 x = \sec^2 x$. ●

A set of functions $f_1(x), f_2(x), \ldots, f_n(x)$ is linearly dependent on an interval if at least one function can be expressed as a linear combination of the remaining functions.

EXAMPLE 6 Linearly Dependent Functions

The functions $f_1(x) = \sqrt{x} + 5, f_2(x) = \sqrt{x} + 5x, f_3(x) = x - 1, f_4(x) = x^2$ are linearly dependent on the interval $(0, \infty)$ since f_2 can be written as a linear combination of f_1, f_3, and f_4. Observe that

$$f_2(x) = 1 \cdot f_1(x) + 5 \cdot f_3(x) + 0 \cdot f_4(x)$$

for every x in the interval $(0, \infty)$. ●

Solutions of Differential Equations

We are primarily interested in linearly independent functions or, more to the point, linearly independent solutions of a linear differential equation. Although we could always appeal directly to Definition 3.1, it turns out that the question of whether n solutions $y_1, y_2, \ldots, y_n$ of a homogeneous linear nth-order differential equation (6) are linearly independent can be settled somewhat mechanically using a determinant.

DEFINITION 3.2 Wronskian

Suppose each of the functions $f_1(x), f_2(x), \ldots, f_n(x)$ possess at least $n - 1$ derivatives. The determinant

$$W(f_1, f_2, \ldots, f_n) = \begin{vmatrix} f_1 & f_2 & \cdots & f_n \\ f_1' & f_2' & \cdots & f_n' \\ \vdots & \vdots & & \vdots \\ f_1^{(n-1)} & f_2^{(n-1)} & \cdots & f_n^{(n-1)} \end{vmatrix},$$

where the primes denote derivatives, is called the **Wronskian** of the functions.

THEOREM 3.3 Criterion for Linearly Independent Solutions

Let $y_1, y_2, \ldots, y_n$ be n solutions of the homogeneous linear nth-order differential equation (6) on an interval I. Then the set of solutions is **linearly independent** on I if and only if

$$W(y_1, y_2, \ldots, y_n) \neq 0$$

for every x in the interval.

It follows from Theorem 3.3 that when $y_1, y_2, \ldots, y_n$ are n solutions of (6) on an interval I, the Wronskian $W(y_1, y_2, \ldots, y_n)$ is either identically zero or never zero on the interval.

A set of n linearly independent solutions of a homogeneous linear nth-order differential equation is given a special name.

DEFINITION 3.3 Fundamental Set of Solutions

Any set $y_1, y_2, \ldots, y_n$ of n linearly independent solutions of the homogeneous linear nth-order differential equation (6) on an interval I is said to be a **fundamental set of solutions** on the interval.

The basic question of whether a fundamental set of solutions exists for a linear equation is answered in the next theorem.

THEOREM 3.4 Existence of a Fundamental Set

There exists a fundamental set of solutions for the homogeneous linear nth-order differential equation (6) on an interval I.

Analogous to the fact that any vector in three dimensions can be expressed as a linear combination of the *linearly independent* vectors **i**, **j**, **k**, any solution of an nth-order homogeneous linear differential equation on an interval I can be expressed as a linear combination of n linearly independent solutions on I. In other words, n linearly independent solutions $y_1, y_2, \ldots, y_n$ are the basic building blocks for the general solution of the equation.

THEOREM 3.5 General Solution—Homogeneous Equations

Let $y_1, y_2, \ldots, y_n$ be a fundamental set of solutions of the homogeneous linear nth-order differential equation (6) on an interval I. Then the **general solution** of the equation on the interval is

$$y = c_1y_1(x) + c_2y_2(x) + \cdots + c_ny_n(x),$$

where c_i, $i = 1, 2, \ldots, n$ are arbitrary constants.

Theorem 3.5 states that if $Y(x)$ is any solution of (6) on the interval then constants $C_1, C_2, \ldots, C_n$ can always be found so that

$$Y(x) = C_1y_1(x) + C_2y_2(x) + \cdots + C_ny_n(x).$$

We will prove the case when $n = 2$.

Proof Let Y be a solution and y_1 and y_2 be linearly independent solutions of $a_2y'' + a_1y' + a_0y = 0$ on an interval I. Suppose $x = t$ is a point in I for which $W(y_1(t), y_2(t)) \neq 0$. Suppose also that $Y(t) = k_1$ and $Y'(t) = k_2$. If we now examine the equations

$$C_1y_1(t) + C_2y_2(t) = k_1$$

$$C_1y_1'(t) + C_2y_2'(t) = k_2,$$

it follows that we can determine C_1 and C_2 uniquely, provided that the determinant of the coefficients satisfies

$$\begin{vmatrix} y_1(t) & y_2(t) \\ y_1'(t) & y_2'(t) \end{vmatrix} \neq 0.$$

But this determinant is simply the Wronskian evaluated at $x = t$ and, by assumption, $W \neq 0$. If we define $G(x) = C_1 y_1(x) + C_2 y_2(x)$, we observe **(i)** $G(x)$ satisfies the differential equation since it is a superposition of two known solutions; **(ii)** $G(x)$ satisfies the initial conditions

$$G(t) = C_1 y_1(t) + C_2 y_2(t) = k_1$$

$$G'(t) = C_1 y_1'(t) + C_2 y_2'(t) = k_2;$$

(iii) $Y(x)$ satisfies the *same* linear equation and the *same* initial conditions. Since the solution of this linear initial-value problem is unique (Theorem 3.1), we have $Y(x) = G(x)$ or $Y(x) = C_1 y_1(x) + C_2 y_2(x)$. ○

EXAMPLE 7 General Solution of a Homogeneous DE

The functions $y_1 = e^{3x}$ and $y_2 = e^{-3x}$ are both solutions of the homogeneous linear equation $y'' - 9y = 0$ on the interval $(-\infty, \infty)$. By inspection, the solutions are linearly independent on the x-axis. This fact can be corroborated by observing that the Wronskian

$$W(e^{3x}, e^{-3x}) = \begin{vmatrix} e^{3x} & e^{-3x} \\ 3e^{3x} & -3e^{-3x} \end{vmatrix} = -6 \neq 0$$

for every x. We conclude that y_1 and y_2 form a fundamental set of solutions, and consequently

$$y = c_1 e^{3x} + c_2 e^{-3x}$$

is the general solution of the equation on the interval. ●

EXAMPLE 8 A Solution Obtained from a General Solution

The function $y = 4 \sinh 3x - 5e^{3x}$ is a solution of the differential equation in Example 7. (Verify this.) In view of Theorem 3.5, we must be able to obtain this solution from the general solution $y = c_1 e^{3x} + c_2 e^{-3x}$. Observe that if we choose $c_1 = 2$ and $c_2 = -7$, then $y = 2e^{3x} - 7e^{-3x}$ can be rewritten as

$$y = 2e^{3x} - 2e^{-3x} - 5e^{-3x} = 4\left(\frac{e^{3x} - e^{-3x}}{2}\right) - 5e^{-3x}.$$

The last expression is recognized as $y = 4 \sinh 3x - 5e^{-3x}$. ●

EXAMPLE 9 General Solution of a Homogeneous DE

The functions $y_1 = e^{x}$, $y_2 = e^{2x}$, and $y_3 = e^{3x}$ satisfy the third-order equation

$$\frac{d^3y}{dx^3} - 6\frac{d^2y}{dx^2} + 11\frac{dy}{dx} - 6y = 0.$$

Since $$W(e^x, e^{2x}, e^{3x}) = \begin{vmatrix} e^x & e^{2x} & e^{3x} \\ e^x & 2e^{2x} & 3e^{3x} \\ e^x & 4e^{2x} & 9e^{3x} \end{vmatrix} = 2e^{6x} \neq 0$$

for every real value of x, y_1, y_2, and y_3 form a fundamental set of solutions on $(-\infty, \infty)$. We conclude that

$$y = c_1 e^x + c_2 e^{2x} + c_3 e^{3x}$$

is the general solution of the differential equation on the interval. ●

3.1.3 Nonhomogeneous Equations

Any function y_p, free of arbitrary parameters, that satisfies (7) is said to be a **particular solution** or **particular integral** of the equation. For example, it is straightforward task to show that the constant function $y_p = 3$ is a particular solution of the nonhomogeneous equation $y'' + 9y = 27$.

Now if $y_1, y_2, \ldots, y_k$ are solutions of (6) on an interval I and y_p is any particular solution of (7) on I, then the linear combination

$$y = c_1 y_1(x) + c_2 y_2(x) + \cdots + c_k y_k(x) + y_p \tag{8}$$

is also a solution of the nonhomogeneous equation (7). If you think about it, this makes sense because the linear combination $c_1 y_1(x) + c_2 y_2(x) + \cdots + c_k y_k(x)$ is mapped into 0 by $a_n(x)y^{(n)} + a_{n-1}(x)y^{(n-1)} + \cdots + a_1(x)y' + a_0(x)y$, whereas y_p is mapped into $g(x)$. If we use $k = n$ linearly independent solutions of the nth-order equation (6), then the expression in (8) becomes the general solution of (7).

THEOREM 3.6 General Solution—Nonhomogeneous Equations

Let y_p be any particular solution of the nonhomogeneous linear nth-order differential equation (7) on an interval I and let $y_1, y_2, \ldots, y_n$ be a fundamental set of solutions of the associated homogeneous differential equation (6) on I. Then the **general solution** of the equation on the interval is

$$y = c_1 y_1(x) + c_2 y_2(x) + \cdots + c_n y_n(x) + y_p,$$

where the c_i, $i = 1, 2, \ldots, n$ are arbitrary constants.

The proof this result, which follows from Theorem 3.5, is left as an exercise.

Complementary Function

We see in Theorem 3.6 that the general solution of a nonhomogeneous linear equation consists of a sum of two functions:

$$y = c_1 y_1(x) + c_2 y_2(x) + \cdots + c_n y_n(x) + y_p = y_c(x) + y_p(x).$$

The linear combination $y_c(x) = c_1 y_1(x) + c_2 y_2(x) + \cdots + c_n y_n(x)$, which is the general solution of (6), is called the **complementary function** for equation (7).

In other words, to solve a nonhomogeneous linear differential equation we first solve the associated homogeneous equation and then we must find any particular solution of the nonhomogeneous equation. The general solution of the nonhomogeneous equation is then

$$y = \textit{complementary function} + \textit{any particular solution.}$$

EXAMPLE 10 General Solution of a Nonhomogeneous DE

By substitution, the function $y_p = -\frac{11}{12} - \frac{1}{2}x$ is readily shown to be a particular solution of the nonhomogeneous equation

$$\frac{d^3y}{dx^3} - 6\frac{d^2y}{dx^2} + 11\frac{dy}{dx} - 6y = 3x. \tag{9}$$

In order to write the general solution of (9), we must also be able to solve the associated homogeneous equation

$$\frac{d^3y}{dx^3} - 6\frac{d^2y}{dx^2} + 11\frac{dy}{dx} - 6y = 0.$$

But in Example 9 we saw that the general solution of this latter equation on the interval $(-\infty, \infty)$ was $y_c = c_1e^x + c_2e^{2x} + c_3e^{3x}$. Hence the general solution of (9) on the interval is

$$y = y_c + y_p = c_1e^x + c_2e^{2x} + c_3e^{3x} - \frac{11}{12} - \frac{1}{2}x.$$

●

Another Superposition Principle

The last theorem of this discussion will be useful in Section 3.4 when we consider a method for finding particular solutions of nonhomogeneous equations.

THEOREM 3.7 Superposition Principle—Nonhomogeneous Equations

Let $y_{p_1}, y_{p_2}, \ldots, y_{p_k}$ be k particular solutions of the nonhomogeneous linear nth-order differential equation (7) on an interval I corresponding, in turn, to k distinct functions $g_1, g_2, \ldots, g_k$. That is, suppose y_{p_i} denotes a particular solution of the corresponding differential equation

$$a_n(x)y^{(n)} + a_{n-1}(x)y^{(n-1)} + \cdots + a_1(x)y' + a_0(x)y = g_i(x), \tag{10}$$

where $i = 1, 2, \ldots, k$. Then

$$y_p = y_{p_1}(x) + y_{p_2}(x) + \cdots + y_{p_k}(x) \tag{11}$$

is a particular solution of

$$a_n(x)y^{(n)} + a_{n-1}(x)y^{(n-1)} + \cdots + a_1(x)y' + a_0(x)y$$
$$= g_1(x) + g_2(x) + \cdots + g_k(x). \tag{12}$$

We leave the proof of this result when $k = 2$ as an exercise. See Problem 40 in Exercises 3.1.

EXAMPLE 11 Superposition/Nonhomogeneous DE

You should verify that

$y_{p_1} = -4x^2$ is a particular solution of $y'' - 3y' + 4y = -16x^2 + 24x - 8$,

$y_{p_2} = e^{2x}$ is a particular solution of $y'' - 3y' + 4y = 2e^{2x}$, and

$y_{p_3} = xe^x$ is a particular solution of $y'' - 3y' + 4y = 2xe^x - e^x$.

It follows from Theorem 3.7 that the superposition of y_{p_1}, y_{p_2}, and y_{p_3},

$$y = y_{p_1} + y_{p_2} + y_{p_3} = -4x^2 + e^{2x} + xe^x,$$

is a solution of

$$y'' - 3y' + 4y = \underbrace{-16x^2 + 24x - 8}_{g_1(x)} + \underbrace{2e^{2x}}_{g_2(x)} + \underbrace{2xe^x - e^x}_{g_3(x)}.$$

Note If the y_{p_i} are particular solutions of (10) for $i = 1, 2, \ldots, k$, then it also follows that the linear combination

$$y_p = c_1 y_{p_1} + c_2 y_{p_2} + \cdots + c_k y_{p_k},$$

where the c_i are constants, is a particular solution of (12) when the right-hand member of the equation is the linear combination

$$c_1 g_1(x) + c_2 g_2(x) + \cdots + c_k g_k(x)$$

Before we actually start solving homogeneous and nonhomogeneous linear differential equations, we need one additional bit of theory presented in the next section.

EXERCISES 3.1

Answers to odd-numbered problems begin on page AN-11.

3.1.1

1. Given that $y = c_1e^x + c_2e^{-x}$ is a two-parameter family of solutions of $y'' - y = 0$ on the interval $(-\infty, \infty)$, find a member of the family satisfying the initial conditions $y(0 = 0)$, $y'(0) = 1$.

2. Find a solution of the differential equation in Problem 1 satisfying the boundary conditions $y(0) = 0$, $y(1) = 1$.

3. Given that $y = c_1e^{4x} + c_2e^{-x}$ is a two-parameter family of solutions of $y'' - 3y' - 4y = 0$ on the interval $(-\infty, \infty)$, find a member of the family satisfying the initial conditions $y(0) = 1$, $y'(0) = 2$.

4. Given that $y = c_1 + c_2 \cos x + c_3 \sin x$ is a three-parameter family of solutions of $y''' + y' = 0$ on the interval $(-\infty, \infty)$, find a member of the family satisfying the initial conditions $y(\pi) = 0$, $y'(\pi) = 2$, $y''(\pi) = -1$.

5. Given that $y = c_1x + c_2x \ln x$ is a two-parameter family of solutions of $x^2y'' - xy' + y = 0$ on the interval $(-\infty, \infty)$, find a member of the family satisfying the initial conditions $y(1) = 3$, $y'(1) = -1$.

6. Given that $y = c_1 + c_2x^2$ is a two-parameter family of solutions of $xy'' - y' = 0$ on the interval $(-\infty, \infty)$, show that constants c_1 and c_2 cannot be found so that a member of the family satisfies the initial conditions $y(0) = 0$, $y'(0) = 1$. Explain why this does not violate Theorem 3.1.

7. Find two members of the family of solutions of $xy'' - y' = 0$ given in Problem 6 satisfying the initial conditions $y(0) = 0$, $y'(0) = 0$.

8. Find a member of the family of solutions of $xy'' - y' = 0$ given in Problem 6 satisfying the boundary conditions $y(0) = 1$, $y'(1) = 6$. Does Theorem 3.1 guarantee that this solution is unique?

9. Given that $y = c_1e^x \cos x + c_2e^x \sin x$ is a two-parameter family of solutions of $y'' - 2y' + 2y = 0$ on the interval $(-\infty, \infty)$, determine whether a member of the family can be found that satisfies the boundary conditions

(a) $y(0) = 1, \quad y'(0) = 0$ **(b)** $y(0) = 1, \quad y(\pi) = -1$
(c) $y(0) = 1, \quad y(\pi/2) = 1$ **(d)** $y(0) = 0, \quad y(\pi) = 0$.

10. Given that $y = c_1x^2 + c_2x^4 + 3$ is a two-parameter family of solutions of $x^2y'' - 5xy' + 8y = 24$ on the interval $(-\infty, \infty)$, determine whether a member of the family can be found that satisfies the boundary conditions

(a) $y(-1) = 0, \quad y(1) = 4$ **(b)** $y(0) = 1, \quad y(1) = 2$
(c) $y(0) = 3, \quad y(1) = 0$ **(d)** $y(1) = 3, \quad y(2) = 15$.

In Problems 11 and 12 find an interval around $x = 0$ for which the given initial-value problem has a unique solution.

11. $(x - 2)y'' + 3y = x, \quad y(0) = 0, \quad y'(0) = 1$

12. $y'' + (\tan x)y = e^x, \quad y(0) = 1, \quad y'(0) = 0$

13. **(a)** Given that $x = c_1 \cos \omega t + c_2 \sin \omega t$ is a two-parameter family of solutions of $x'' + \omega^2 x = 0$ on the interval $(-\infty, \infty)$, show that a solution satisfying the initial-conditions $x(0) = x_0$, $x'(0) = x_1$ is given by

$$x(t) = x_0 \cos \omega t + \frac{x_1}{\omega} \sin \omega t.$$

(b) Use the two-parameter family $x = c_1 \cos \omega t + c_2 \sin \omega t$ to show that a solution of the differential equation satisfying $x(t_0) = x_0$, $x'(t_0) = x_1$ is the solution of the initial-value problem in part (a) shifted by an amount t_0:

$$x(t) = x_0 \cos \omega(t - t_0) + \frac{x_1}{\omega} \sin \omega(t - t_0).$$

14. *PROJECT PROBLEM* In this problem you are asked to prove that the solution given in part (b) of Problem 13 is unique without invoking Theorem 3.1.

(a) Start by defining $X(t) = x_1(t) - x_2(t)$, where $x_1(t)$ and $x_2(t)$ represent two different solutions of the initial value problem. Show that

$$X'' + \omega^2 X = 0, \qquad X(t_0) = 0, \quad X'(t_0) = 0.$$

(b) We now wish to show that the preceding initial-value problem possesses only one solution. Show that by multiplying the differential equation $X'' + \omega^2 X = 0$ by X' and integrating on the interval $[t_0, t]$ we obtain

$$[X'(t)]^2 + \omega^2[X(t)]^2 = 0$$

for *every* t. What conclusion can you draw from this?

(c) Put the results of parts (a) and (b) together to conclude that there is only one solution of the initial-value problem in part (b) of Problem 13, namely, the one that is given.

3.1.2

In Problems 15–22 determine whether the given functions are linearly independent or dependent on $(-\infty, \infty)$.

15. $f_1(x) = x, \quad f_2(x) = x^2, \quad f_3(x) = 4x - 3x^2$

16. $f_1(x) = 0, \quad f_2(x) = x, \quad f_3(x) = e^x$

17. $f_1(x) = 5, \quad f_2(x) = \cos^2 x, \quad f_3(x) = \sin^2 x$

18. $f_1(x) = \cos 2x, \quad f_2(x) = 1, \quad f_3(x) = \cos^2 x$

19. $f_1(x) = x, \quad f_2(x) = x - 1, \quad f_3(x) = x + 3$

20. $f_1(x) = 2 + x, \quad f_2(x) = 2 + |x|$

21. $f_1(x) = 1 + x, \quad f_2(x) = x, \quad f_3(x) = x^2$

22. $f_1(x) = e^x, \quad f_2(x) = e^{-x}, \quad f_3(x) = \sinh x$

In Problems 23–30 verify that the given functions form a fundamental set of solutions of the differential equation on the indicated interval. Form the general solution.

23. $y'' - y' - 12y = 0; \quad e^{-3x}, e^{4x}, (-\infty, \infty)$

24. $y'' - 4y = 0; \quad \cosh 2x, \sinh 2x, (-\infty, \infty)$

25. $y'' - 2y' + 5y = 0$; $e^x \cos 2x$, $e^x \sin 2x$, $(-\infty, \infty)$

26. $4y'' - 4y' + y = 0$; $e^{x/2}$, $xe^{x/2}$, $(-\infty, \infty)$

27. $x^2y'' - 6xy' + 12y = 0$; x^3, x^4, $(0, \infty)$

28. $x^2y'' + xy' + y = 0$; $\cos(\ln x)$, $\sin(\ln x)$, $(0, \infty)$

29. $x^3y''' + 6x^2y'' + 4xy' - 4y = 0$; x, x^{-2}, $x^{-2} \ln x$, $(0, \infty)$

30. $y^{(4)} + y'' = 0$; 1, x, $\cos x$, $\sin x$, $(-\infty, \infty)$

31. *PROJECT PROBLEM*

(a) Verify that $y_1 = x^3$ and $y_2 = |x|^3$ are linearly independent solutions of the differential equation $x^2y'' - 4xy' + 6y = 0$ on the interval $(-\infty, \infty)$.

(b) Show that $W(y_1, y_2) = 0$ for every real number x. Does this result violate Theorem 3.3? Explain.

(c) Verify that $Y_1 = x^3$ and $Y_2 = x^2$ are also linearly independent solutions of the differential equation in part (a) on the interval $(-\infty, \infty)$.

(d) Find a solution of the differential equation satisfying $y(0) = 0$, $y'(0) = 0$.

(e) By the superposition principle, Theorem 3.2, both linear combinations

$$y = c_1y_1 + c_2y_2 \quad \text{and} \quad Y = c_1Y_1 + c_2Y_2$$

are solutions of the differential equation. Discuss whether one, both, or neither of the linear combinations is a general solution of the differential equation on the interval $(-\infty, \infty)$.

32. *PROJECT PROBLEM*

(a) Let y_1 and y_2 denote two solutions of $y'' + P(x)y' + Q(x)y = 0$, where P and Q are continuous on an interval I. Show that the Wronskian of y_1 and y_2 satisfies the first-order equation

$$\frac{dW}{dx} + P(x)W = 0.$$

(b) Derive **Abel's formula,** $W = ce^{-\int P(x)\,dx}$, where c is a constant.

(c) Why does the result in part (b) prove the statement following Theorem 3.3?

(d) Let y_1 and y_2 denote two solutions of $x^2y'' + 3xy' - 24y = 0$. Find the Wronskian $W(y_1, y_2)$.

(e) By inspection find a solution $y_1(x)$ of $(1 - x^2)y'' - 2xy' + 2y = 0$. (Do not think profound thoughts here.) Use $y_1(x)$ and the result in part (b) to find a second linearly independent solution $y_2(x)$. Give the general solution of the differential equation and an appropriate interval.

3.1.3

In Problems 33–36 verify that the given two-parameter family of functions is the general solution of the nonhomogeneous differential equation on the indicated interval.

33. $y'' - 7y' + 10y = 24e^x$
$y = c_1e^{2x} + c_2e^{5x} + 6e^x$, $(-\infty, \infty)$

34. $y'' + y = \sec x$
$y = c_1 \cos x + c_2 \sin x + x \sin x + (\cos x) \ln(\cos x)$, $(-\pi/2, \pi/2)$

35. $y'' - 4y' + 4y = 2e^{2x} + 4x - 12$
$y = c_1e^{2x} + c_2xe^{2x} + x^2e^{2x} + x - 2$, $(-\infty, \infty)$

36. $2x^2y'' + 5xy' + y = x^2 - x$
$y = c_1x^{-1/2} + c_2x^{-1} + \frac{1}{15}x^2 - \frac{1}{6}x$, $(0, \infty)$

37. Given that $y_{p_1} = 3e^{2x}$ and $y_{p_2} = x^2 + 3x$ are particular solutions of

$$y'' - 6y' + 5y = -9e^{2x}$$

and
$$y'' - 6y' + 5y = 5x^2 + 3x - 16,$$

respectively. Find particular solutions of

$$y'' - 6y' + 5y = 5x^2 + 3x - 16 - 9e^{2x}$$

and
$$y'' - 6y' + 5y = -10x^2 - 6x + 32 + e^{2x}.$$

38. (a) By inspection determine a particular solution of $y'' + 2y = 10$.

(b) By inspection determine a particular solution of $y'' + 2y = -4x$.

(c) Find a particular solution of $y'' + 2y = -4x + 10$.

(d) Find a particular solution of $y'' + 2y = 8x + 5$.

39. *PROJECT PROBLEM* In this problem you are asked to prove Theorem 3.6 in the case when $n = 2$. To get started suppose that Y and y_p are any two solutions of $a_2(x)y'' + a_1(x)y' + a_0(x)y = g(x)$. See what happens when $u(x) = Y(x) - y_p(x)$ is substituted into this differential equation. Then reread Theorem 3.5.

40. Prove Theorem 3.7 in the case when $k = 2$.

3.2 REDUCTION OF ORDER

Introduction

It is one of the more interesting facts of mathematical life in the study of linear *second-order* differential equations that we can construct a second solution y_2 of

$$a_2(x)y'' + a_1(x)y' + a_0(x)y = 0 \tag{1}$$

on an interval I from a known nontrivial solution y_1. We seek a second solution $y_2(x)$ of (1) so that y_1 and y_2 are linearly independent on I. Recall, if y_1 and y_2 are linearly independent, then their ratio y_2/y_1 is nonconstant on I, that is, $y_2/y_1 = u(x)$ or $y_2(x) = u(x)y_1(x)$. The idea is to find the function $u(x)$ by substituting $y_2(x) = u(x)y_1(x)$ into the given differential equation. The method is called **reduction of order** since we must solve a linear *first-order* equation to find u.

EXAMPLE 1 A Second Solution by Reduction of Order

Given that $y_1 = e^x$ is a solution of $y'' - y = 0$ on the interval $(-\infty, \infty)$, use reduction of order to find a second solution y_2.

Solution If $y = u(x)y_1(x) = u(x)e^x$, then the product rule gives

$$y' = ue^x + e^x u', \quad y'' = ue^x + 2e^x u' + e^x u''$$

and so

$$y'' - y = e^x(u'' + 2u') = 0.$$

Since $e^x \neq 0$, the last equation requires $u'' + 2u' = 0$. If we make the substitution $w = u'$, this linear second-order equation in u becomes $w' + 2w = 0$, which is a linear first-order equation in w. Using the integrating factor e^{2x}, we can write

$$\frac{d}{dx}[e^{2x} w] = 0.$$

After integrating we get $w = c_1 e^{-2x}$ or $u' = c_1 e^{-2x}$. Integrating again

then yields

$$u = -\frac{c_1}{2} e^{-2x} + c_2.$$

Thus

$$y = u(x)e^x = -\frac{c_1}{2} e^{-x} + c_2 e^x. \tag{2}$$

By picking $c_2 = 0$ and $c_1 = -2$, we obtain the desired second solution, $y_2 = e^{-x}$. Because $W(e^x, e^{-x}) \neq 0$ for every x, the solutions are linearly independent on $(-\infty, \infty)$. ●

Since we have shown that $y_1 = e^x$ and $y_2 = e^{-x}$ are linearly independent solutions of a linear second-order equation the expression in (2) is actually the general solution of $y'' - y = 0$ on $(-\infty, \infty)$.

General Case

Suppose we divide by $a_2(x)$ in order to put equation (1) in the **standard form**

$$y'' + P(x)y' + Q(x)y = 0, \tag{3}$$

where $P(x)$ and $Q(x)$ are continuous on some interval I. Let us suppose further that $y_1(x)$ is a known solution of (3) on I and that $y_1(x) \neq 0$ for every x in the interval. If we define $y = u(x)y_1(x)$, it follows that

$$y' = uy_1' + y_1u', \quad y'' = uy_1'' + 2y_1'u' + y_1u''$$

$$y'' + Py' + Qy = u\underbrace{[y_1'' + Py_1' + Qy_1]}_{\text{zero}} + y_1u'' + (2y_1' + Py_1)u' = 0.$$

This implies that we must have

$$y_1u'' + (2y_1' + Py_1)u' = 0 \quad \text{or} \quad y_1w' + (2y_1' + Py_1)w = 0, \tag{4}$$

where we have let w $= u'$. Observe that the last equation in (4) is both linear and separable. Separating variables and integrating, we obtain

$$\frac{dw}{w} + 2\frac{y_1'}{y_1}dx + P\,dx = 0$$

$$\ln|wy_1^2| = -\int P\,dx + c \quad \text{or} \quad wy_1^2 = c_1e^{-\int P\,dx}.$$

We solve the last equation for w, use $w = u'$, and integrate again:

$$u = c_1\int \frac{e^{-\int P\,dx}}{y_1^2}dx + c_2.$$

By choosing $c_1 = 1$ and $c_2 = 0$ we find from $y = u(x)y_1(x)$ that a second solution of equation (3) is

$$y_2 = y_1(x)\int \frac{e^{-\int P(x)\,dx}}{y_1^2(x)}dx. \tag{5}$$

It makes a good review of differentiation to verify that the function $y_2(x)$ defined in (5) satisfies equation (3) and that y_1 and y_2 are linearly independent on any interval on which $y_1(x)$ is not zero. See Problem 29 in Exercises 3.2.

EXAMPLE 2 A Second Solution by Formula (5)

The function $y_1 = x^2$ is a solution of $x^2y'' - 3xy' + 4y = 0$. Find the general solution on the interval $(0, \infty)$.

Solution Since the equation has the alternative form

$$y'' - \frac{3}{x}y' + \frac{4}{x^2}y = 0,$$

we find from (5) $$y_2 = x^2 \int \frac{e^{3\int dx/x}}{x^4}\,dx \qquad \leftarrow e^{3\int dx/x} = e^{\ln x^3} = x^3$$

$$= x^2 \int \frac{dx}{x} = x^2 \ln x.$$

The general solution on $(0, \infty)$ is given by $y = c_1y_1 + c_2y_2$; that is,

$$y = c_1x^2 + c_2x^2 \ln x.$$

Remark We have derived and illustrated how to use (5) because this formula appears again in the next section and in Section 3.6. We use (5) simply to save time in obtaining a desired result. Your instructor will tell you whether you should memorize (5) or whether you should know the first principles of reduction of order.

EXERCISES 3.2

Answers to odd-numbered problems begin on page AN-11.

In Problems 1–24 find a second solution of each differential equation. Use reduction of order or formula (5) as instructed. Assume an appropriate interval of validity.

1. $y'' + 5y' = 0; \quad y_1 = 1$
2. $y'' - y' = 0; \quad y_1 = 1$
3. $y'' - 4y' + 4y = 0; \quad y_1 = e^{2x}$
4. $y'' + 2y' + y = 0; \quad y_1 = xe^{-x}$
5. $y'' + 16y = 0; \quad y_1 = \cos 4x$
6. $y'' + 9y = 0; \quad y_1 = \sin 3x$
7. $y'' - y = 0; \quad y_1 = \cosh x$
8. $y'' - 25y = 0; \quad y_1 = e^{5x}$
9. $9y'' - 12y' + 4y = 0; \quad y_1 = e^{2x/3}$
10. $6y'' + y' - y = 0; \quad y_1 = e^{x/3}$
11. $x^2y'' - 7xy' + 16y = 0; \quad y_1 = x^4$
12. $x^2y'' + 2xy' - 6y = 0; \quad y_1 = x^2$
13. $xy'' + y' = 0; \quad y_1 = \ln x$
14. $4x^2y'' + y = 0; \quad y_1 = x^{1/2} \ln x$
15. $(1 - 2x - x^2)y'' + 2(1 + x)y' - 2y = 0; \quad y_1 = x + 1$
16. $(1 - x^2)y'' - 2xy' = 0; \quad y_1 = 1$
17. $x^2y'' - xy' + 2y = 0; \quad y_1 = x \sin(\ln x)$
18. $x^2y'' - 3xy' + 5y = 0; \quad y_1 = x^2 \cos(\ln x)$
19. $(1 + 2x)y'' + 4xy' - 4y = 0; \quad y_1 = e^{-2x}$
20. $(1 + x)y'' + xy' - y = 0; \quad y_1 = x$
21. $x^2y'' - xy' + y = 0; \quad y_1 = x$
22. $x^2y'' - 20y = 0; \quad y_1 = x^{-4}$
23. $x^2y'' - 5xy' + 9y = 0; \quad y_1 = x^3 \ln x$
24. $x^2y'' + xy' + y = 0; \quad y_1 = \cos(\ln x)$

In Problems 25–28 use the method of reduction of order to find a solution of the given nonhomogeneous equation. The indicated function $y_1(x)$ is a solution of the associated homogeneous equation. Determine a second solution of the homogeneous equation and a particular solution of the nonhomogeneous equation.

25. $y'' - 4y = 2; \quad y_1 = e^{-2x}$
26. $y'' + y' = 1; \quad y_1 = 1$
27. $y'' - 3y' + 2y = 5e^{3x}; \quad y_1 = e^x$
28. $y'' - 4y' + 3y = x; \quad y_1 = e^x$
29. **(a)** Verify by direct substitution that (5) satisfies (3).
    (b) Show that $W(y_1(x), y_2(x)) = u'(y_1)^2 = e^{-\int P(x)\,dx}$.
30. ***PROJECT PROBLEM*** Use the method of reduction of order to find two additional solutions y_2 and y_3 of $y''' - 4y'' + 5y' - 2y = 0$ given that $y_1 = e^x$ is a solution.

3.3 HOMOGENEOUS LINEAR EQUATIONS WITH CONSTANT COEFFICIENTS

Introduction

We have seen that the linear first-order equation $dy/dx + ay = 0$, where a is a constant, has the exponential solution $y = c_1e^{-ax}$ on the interval $(-\infty, \infty)$. Therefore, it is natural to seek to determine whether exponential solutions exist on $(-\infty, \infty)$ for higher-order linear equations

$$a_ny^{(n)} + a_{n-1}y^{(n-1)} + \cdots + a_2y'' + a_1y' + a_0y = 0, \tag{1}$$

where the coefficients a_i, $i = 0, 1, \ldots, n$ are constants. The surprising fact is that all solutions of (1) are exponential functions or constructed out of exponential functions.

3.3.1 Method of Solution

We begin by considering the special case of the second-order equation

$$ay'' + by' + cy = 0. \tag{2}$$

If we try a solution of the form $y = e^{mx}$, then $y' = me^{mx}$ and $y'' = m^2e^{mx}$ so that equation (2) becomes

$$am^2e^{mx} + bme^{mx} + ce^{mx} = 0 \quad \text{or} \quad e^{mx}(am^2 + bm + c) = 0.$$

Because e^{mx} is never zero for real values of x, it is apparent that the only way that this exponential function can satisfy the differential equation is to choose m so that it is a root of the quadratic equation

$$am^2 + bm + c = 0. \tag{3}$$

This latter equation is called the **auxiliary equation**, or **characteristic equation**, of the differential equation (2). We consider three cases, namely, the solutions of the auxiliary equation corresponding to distinct real roots, real but equal roots, and a conjugate pair of complex roots.

CASE I **Distinct Real Roots** Under the assumption that the auxiliary equation (3) has two unequal real roots m_1 and m_2, we find two solutions, $y_1 = e^{m_1x}$ and $y_2 = e^{m_2x}$. We see that these functions are linearly independent on $(-\infty, \infty)$ and hence form a fundamental set. It follows that the general solution of (2) on this interval is

$$y = c_1e^{m_1x} + c_2e^{m_2x}. \tag{4}$$

CASE II **Repeated Real Roots** When $m_1 = m_2$ we necessarily obtain only one exponential solution, $y_1 = e^{m_1x}$. From the quadratic formula we find that $m_1 = -b/2a$ since the only way to have $m_1 = m_2$ is to have $b^2 - 4ac = 0$. It follows from the discussion in Section 3.2 that a second solution of the equation is

$$y_2 = e^{m_1x}\int \frac{e^{2m_1x}}{e^{2m_1x}}\,dx = e^{m_1x}\int dx = xe^{m_1x}. \tag{5}$$

In (5) we have used the fact that $-b/a = 2m_1$. The general solution is then

$$y = c_1e^{m_1x} + c_2xe^{m_1x}. \tag{6}$$

CASE III **Conjugate Complex Roots** If m_1 and m_2 are complex, then we can write $m_1 = \alpha + i\beta$ and $m_2 = \alpha - i\beta$, where α and $\beta > 0$ are real and $i^2 = -1$. Formally, there is no difference between this case and Case I, and hence

$$y = C_1e^{(\alpha + i\beta)x} + C_2e^{(\alpha - i\beta)x}.$$

However, in practice we prefer to work with real functions instead of complex exponentials. To this end we use Euler's formula:

$$e^{i\theta} = \cos\theta + i\sin\theta,$$

where θ is any real number.* It follows from this formula that

$$e^{i\beta x} = \cos\beta x + i\sin\beta x \quad \text{and} \quad e^{-i\beta x} = \cos\beta x - i\sin\beta x, \tag{7}$$

where we have used $\cos(-\beta x) = \cos\beta x$ and $\sin(-\beta x) = -\sin\beta x$. Note that by first adding and then subtracting the two equations in (7), we obtain, respectively.

$$e^{i\beta x} + e^{-i\beta x} = 2\cos\beta x \quad \text{and} \quad e^{i\beta x} - e^{-i\beta x} = 2i\sin\beta x.$$

Since $y = C_1e^{(\alpha + i\beta)x} + C_2e^{(\alpha - i\beta)x}$ is a solution of (2) for any choice of the constants C_1 and C_2, the choices $C_1 = C_2 = 1$ and $C_1 = 1$, $C_2 = -1$ give, in turn, two solutions:

$$y_1 = e^{(\alpha + i\beta)x} + e^{(\alpha - i\beta)x} \quad \text{and} \quad y_2 = e^{(\alpha + i\beta)x} - e^{(\alpha - i\beta)x}.$$

But
$$y_1 = e^{\alpha x}(e^{i\beta x} + e^{-i\beta x}) = 2e^{\alpha x}\cos\beta x$$

and
$$y_2 = e^{\alpha x}(e^{i\beta x} - e^{-i\beta x}) = 2ie^{\alpha x}\sin\beta x.$$

Hence from Corollary (A) of Theorem 3.2, the last two results show that the *real* functions $e^{\alpha x}\cos\beta x$ and $e^{\alpha x}\sin\beta x$ are solutions of (2). Moreover, these solutions form a fundamental set on $(-\infty, \infty)$. Consequently, the general solution is

$$\begin{aligned} y &= c_1e^{\alpha x}\cos\beta x + c_2e^{\alpha x}\sin\beta x \\ &= e^{\alpha x}(c_1\cos\beta x + c_2\sin\beta x). \end{aligned} \tag{8}$$

---

* A formal derivation of Euler's formula can be obtained from the Maclaurin series $e^x = \sum_{n=0}^{\infty} \frac{x^n}{n!}$ by substituting $x = i\theta$, using $i^2 = -1$, $i^3 = -i, \ldots$ and then separating the series into real and imaginary parts. The plausibility thus established, we then adopt $\cos\theta + i\sin\theta$ as the *definition* of $e^{i\theta}$.

EXAMPLE 1 Second-Order DEs

Solve the following differential equations:

(a) $2y'' - 5y' - 3y = 0$ (b) $y'' - 10y' + 25y = 0$

(c) $y'' + y' + y = 0$

Solution We give the auxiliary equations, roots, and the corresponding general solutions.

(a) $2m^2 - 5m - 3 = (2m + 1)(m - 3) = 0, \quad m_1 = -\frac{1}{2}, \quad m_2 = 3,$
$y = c_1 e^{-x/2} + c_2 e^{3x}$

(b) $m^2 - 10m + 25 = (m - 5)^2 = 0, \quad m_1 = m_2 = 5,$
$y = c_1 e^{5x} + c_2 x e^{5x}$

(c) $m^2 + m + 1 = 0, \quad m_1 = -\frac{1}{2} + \frac{\sqrt{3}}{2} i, \quad m_2 = -\frac{1}{2} - \frac{\sqrt{3}}{2} i,$
$y = e^{-x/2}\left(c_1 \cos \frac{\sqrt{3}}{2} x + c_2 \sin \frac{\sqrt{3}}{2} x\right)$ ●

EXAMPLE 2 An Initial-Value Problem

Solve the initial-value problem $y'' - 4y' + 13y = 0$, $y(0) = -1$, $y'(0) = 2$.

Solution The roots of the auxiliary equation $m^2 - 4m + 13 = 0$ are $m_1 = 2 + 3i$ and $m_2 = 2 - 3i$ so that

$$y = e^{2x}(c_1 \cos 3x + c_2 \sin 3x).$$

Applying the condition $y(0) = -1$ we see from $-1 = e^0(c_1 \cos 0 + c_2 \sin 0)$ that $c_1 = -1$. Differentiating $y = e^{2x}(c_1 \cos 3x + c_2 \sin 3x)$ and then using $y'(0) = 2$ give $2 = 3c_2 - 2$ or $c_2 = \frac{4}{3}$. Hence the solution is

$$y = e^{2x}(-\cos 3x + \tfrac{4}{3} \sin 3x).$$ ●

The two differential equations $y'' + k^2 y = 0$ and $y'' - k^2 y = 0$, k real, are important in applied mathematics. For the former equation, the auxiliary equation $m^2 + k^2 = 0$ has imaginary roots $m_1 = ki$ and $m_2 = -ki$. It follows from (8) with $\alpha = 0$ and $\beta = k$ that its general solution is

$$y = c_1 \cos kx + c_2 \sin kx. \tag{9}$$

The auxiliary equation of the second equation $m^2 - k^2 = 0$ has distinct real roots $m_1 = k$ and $m_2 = -k$. Hence its general solution is

$$y = c_1 e^{kx} + c_2 e^{-kx}. \tag{10}$$

Notice that if we choose $c_1 = c_2 = \frac{1}{2}$ and then $c_1 = \frac{1}{2}$, $c_2 = -\frac{1}{2}$ in (10) we get the particular solutions $y = (e^{kx} + e^{-kx})/2 = \cosh kx$ and $y = (e^{kx} - e^{-kx})/2 = \sinh kx$. Because $\cosh kx$ and $\sinh kx$ are linearly independent on any interval of the x-axis, an alternative form for the general solution of $y'' - k^2 y = 0$ is

$$y = c_1 \cosh kx + c_2 \sinh kx.$$

Higher-Order Equations

In general, to solve an nth-order differential equation

$$a_ny^{(n)} + a_{n-1}y^{(n-1)} + \cdots + a_2y'' + a_1y' + a_0y = 0, \tag{11}$$

where the a_i, $i = 0, 1, \ldots, n$, are real constants, we must solve an nth-degree polynominal equation

$$a_nm^n + a_{n-1}m^{n-1} + \cdots + a_2m^2 + a_1m + a_0 = 0. \tag{12}$$

If all the roots of (12) are real and distinct, then the general solution of (11) is

$$y = c_1e^{m_1x} + c_2e^{m_2x} + \cdots + c_ne^{m_nx}.$$

It is somewhat harder to summarize the analogues of Cases II and III because the roots of an auxiliary equation of degree greater than two can occur in many combinations. For example, a fifth-degree equation could have five distinct real roots, or three distinct real and two complex roots, or one real and four complex roots, or five real but equal roots, or five real roots but two of them equal, and so on. When m_1 is a root of multiplicity k of an nth-degree auxiliary equation (that is, k roots are equal to m_1), it can be shown that the linearly independent solutions are

$$e^{m_1x}, xe^{m_1x}, x^2e^{m_1x}, \ldots, x^{k-1}e^{m_1x},$$

and the general solution must contain the linear combination

$$c_1e^{m_1x} + c_2xe^{m_1x} + c_3x^2e^{m_1x} + \cdots + c_kx^{k-1}e^{m_1x}.$$

Lastly, it should be remembered that when the coefficients are real, complex roots of an auxiliary equation always appear in conjugate pairs. Thus, for example, a cubic polynomial equation can have at most two complex roots.

EXAMPLE 3 Third-Order DE

Solve $\quad y''' + 3y'' - 4y = 0.$

Solution It should be apparent from inspection of $m^3 + 3m^2 - 4 = 0$ that one root is $m_1 = 1$. Now if we divide $m^3 + 3m^2 - 4$ by $m - 1$, we find

$$m^3 + 3m^2 - 4 = (m - 1)(m^2 + 4m + 4) = (m - 1)(m + 2)^2,$$

and so the other roots are $m_2 = m_3 = -2$. Thus the general solution is

$$y = c_1e^x + c_2e^{-2x} + c_3xe^{-2x}.$$

EXAMPLE 4 Fourth-Order DE

Solve $\quad \dfrac{d^4y}{dx^4} + 2\dfrac{d^2y}{dx^2} + y = 0.$

Solution The auxiliary equation $m^4 + 2m^2 + 1 = (m^2 + 1)^2 = 0$ has roots $m_1 = m_3 = i$ and $m_2 = m_4 = -i$. Thus from Case II the solution is

$$y = C_1 e^{ix} + C_2 e^{-ix} + C_3 x e^{ix} + C_4 x e^{-ix}.$$

By Euler's formula the grouping $C_1 e^{ix} + C_2 e^{-ix}$ can be rewritten as

$$c_1 \cos x + c_2 \sin x$$

after a relabeling of constants. Similarly, $x(C_3 e^{ix} + C_4 e^{-ix})$ can be expressed as $x(c_3 \cos x + c_4 \sin x)$. Hence the general solution is

$$y = c_1 \cos x + c_2 \sin x + c_3 x \cos x + c_4 x \sin x.$$

Example 4 illustrates a special case when the auxiliary equation has repeated complex roots. In general, if $m_1 = \alpha + i\beta$ is a complex root of multiplicity k of an auxiliary equation with real coefficients, then its conjugate $m_2 = \alpha - i\beta$ is also a root of multiplicity k. From the $2k$ complex-valued solutions

$$e^{(\alpha + i\beta)x}, xe^{(\alpha + i\beta)x}, x^2 e^{(\alpha + i\beta)x}, \ldots, x^{k-1} e^{(\alpha + i\beta)x}$$

$$e^{(\alpha - i\beta)x}, xe^{(\alpha - i\beta)x}, x^2 e^{(\alpha - i\beta)x}, \ldots, x^{k-1} e^{(\alpha - i\beta)x}$$

we conclude, with the aid of Euler's formula, that the general solution of the corresponding differential equation must then contain a linear combination of the $2k$ real linearly independent solutions

$$e^{\alpha x} \cos \beta x, xe^{\alpha x} \cos \beta x, x^2 e^{\alpha x} \cos \beta x, \ldots, x^{k-1} e^{\alpha x} \cos \beta x$$

$$e^{\alpha x} \sin \beta x, xe^{\alpha x} \sin \beta x, x^2 e^{\alpha x} \sin \beta x, \ldots, x^{k-1} e^{\alpha x} \sin \beta x.$$

In Example 4 we identify $k = 2$, $\alpha = 0$, and $\beta = 1$.

Of course the most difficult aspect of solving constant-coefficient differential equations is finding roots of auxiliary equations of degree greater than two. For example, to solve $3y''' + 5y'' + 10y' - 4y = 0$ we must solve $3m^3 + 5m^2 + 10m - 4 = 0$. Something we can try is to test the auxiliary equation for rational roots. Recall, if $m_1 = p/q$ is a rational root (expressed in lowest terms) of an auxiliary equation $a_n m^n + \cdots + a_1 m + a_0 = 0$ with integer coefficients, then p is a factor of a_0 and q is a factor of a_n. For our specific cubic auxiliary equation, all the factors of $a_0 = -4$ and $a_n = 3$ are p: $\pm 1, \pm 2, \pm 4$ and q: $\pm 1, \pm 3$, so the possible rational roots are p/q: $\pm 1, \pm 2, \pm 4, \pm 1/3, \pm 2/3, \pm 4/3$. Each of these numbers can then be tested, say, by synthetic division. In this way we discover both the root $m_1 = \frac{1}{3}$ and the factorization

$$3m^3 + 5m^2 + 10m - 4 = (m - \tfrac{1}{3})(3m^2 + 6m + 12).$$

The quadratic formula then yields the remaining roots $m_2 = -1 + \sqrt{3}\, i$ and $m_3 = -1 - \sqrt{3}\, i$. Therefore the general solution of $3y''' + 5y'' + 10y' - 4y = 0$ is

$$y = c_1 e^{x/3} + e^{-x}(c_2 \cos \sqrt{3}x + c_3 \sin \sqrt{3}x).$$

Use of Computers

Finding roots, or approximations of roots, of polynomial equations is a routine problem on a computer. Moreover, some computer algebra systems are capable of giving explicit solutions of linear homogeneous constant-coefficient differential equations. For example, using *Mathematica*,

DSolve [y″[x] + 2y′[x] + 2y[x] = = 0, y[x], x]

yields $$y[x] \;\text{->}\; \frac{\text{C[2] Cos [x]} - \text{C[1] Sin [x]}}{\text{E}^{\text{x}}} \qquad \textbf{(13)}$$

Translated into standard symbols, this means that $y = c_2e^{-x}\cos x + c_1e^{-x}\sin x$ is a solution of $y'' + 2y' + 2y = 0$. Note that the minus sign in front of C[1] in (13) is superfluous. In the classic text *Differential Equations*, by Ralph Palmer Agnew,* (used by the author as a student) the following statement is made:

> *It is not reasonable to expect students in this course to have computing skill and equipment necessary for efficient solving of equations such as*
>
> $$4.317\frac{d^4y}{dx^4} + 2.179\frac{d^3y}{dx^3} + 1.416\frac{d^2y}{dx^2} + 1.295\frac{dy}{dx} + 3.169y = 0. \qquad \textbf{(14)}$$

While it is debatable whether computing skills have improved in the intervening years, it is certain that technology has. If you have access to a computer algebra system, equation (14) could be considered reasonable. After simplification and some relabeling of the output, *Mathematica* yields the general solution

$$y = c_1e^{-0.728852x}\cos(0.618605x) + c_2e^{-0.728852x}\sin(0.618605x)$$

$$+\, c_3e^{0.476478x}\cos(0.759081x) + c_4e^{0.476478x}\sin(0.759081x).$$

We note in passing that the DSolve command in *Mathematica*, like most aspects of any CAS, has its limitations.

3.3.2 Trajectories and the Phase Plane

Every differential equation of the form given in (2), that is, every homogeneous linear second-order equation with constant coefficients, is autonomous. In order to examine and discuss the various phase portraits of autonomous second-order differential equations, it is convenient to relabel the dependent variable as x and the independent variable as t. Thus (2), written as

$$ax'' + bx' + cx = 0 \qquad \textbf{(15)}$$

has the form $F(x, x', x'') = 0$. As in Section 1.2, we wish to reserve the symbol y to denote dx/dt. Also, we assume throughout that both a and c are not zero.

---

* McGraw-Hill Book Company, Inc., New York, 1960.

Recall, there are three possible ways of obtaining a phase portrait of an autonomous equation of form (15):

(i) If $x = f(t)$ is an explicit real general solution of (15), we can then compute $y = dx/dt = g(t)$. The equations $x = f(t)$, $y = g(t)$, $-\infty < t < \infty$, describe the trajectories parametrically.

(ii) If we let $dx/dt = y$ and solve for $x'' = dy/dt$, then an equivalent form of (15) is given by the system of two first-order equations

$$\begin{aligned} \frac{dx}{dt} &= y \\ \frac{dy}{dt} &= -\frac{c}{a}x - \frac{b}{a}y. \end{aligned} \tag{16}$$

We can use this system and software to plot curves (y versus x) through selected initial points (x_0, y_0).

(iii) We can divide the second equation in (16) by the first and solve, if possible, the resulting first-order differential equation in x and y. The one-parameter family of solution curves of that equation is a one-parameter Cartesian equation of the family of trajectories for (15). (See Example 7, Section 2.2.)

Since equation (15) is autonomous, the equivalent system (16) is also said to be autonomous. The system (16) is just a special case of a more general autonomous system of two first-order equations

$$\begin{aligned} \frac{dx}{dt} &= f(x, y) \\ \frac{dy}{dt} &= g(x, y) \end{aligned} \tag{17}$$

that will be studied in greater detail in Chapter 4. Points where both dx/dt and dy/dt are zero in an autonomous system of two equations are said to be **critical points of the system**. In other words, (x_1, y_1) is a critical point of the system (17) if $f(x_1, y_1) = 0$ and $g(x_1, y_1) = 0$. As in the discussion of autonomous first-order equations, critical points correspond to *constant solutions*, in this case constant solutions of the system of equations. A constant solution of the system (17) is called an **equilibrium solution**. Note that when we set the right sides in (16) equal to zero and solve the system of algebraic equations $y = 0$, $(c/a)\,x + (b/a)y = 0$ $(a \neq 0,\ c \neq 0)$, we find the solution to be $x = 0$, $y = 0$. Thus the only critical point of the system (16) is the origin (0, 0). In general,

A critical point of an autonomous second-order differential equation $F(x, x', x'') = 0$ is a critical point of its equivalent autonomous system.

Thus for $a \neq 0$, $c \neq 0$ we are guaranteed that (0, 0) is the *only* critical point of the homogeneous linear equation (15). The restriction $c \neq 0$ guarantees that the critical point (0, 0) of (15) is **isolated**. This simply means that there is some circle around the critical point that is free of all other critical points. See Problem 79 in Exercises 3.3.

A Gallery of Phase Portraits

The phase portrait of a differential equation of the form given in (15) depends on the nature of the roots of its auxiliary equation. Within the three cases for the general solution of (15) we can distinguish eight subcases. For example, in Case I, where the roots m_1 and m_2 are real and unequal, we have three possibilities: The roots can be both negative, both positive, or have opposite algebraic signs. (Remember, $c \neq 0$ in (15) guarantees that $m = 0$ is not a root of its auxiliary equation.) All of the subcases are summarized in the table that follows.

CASE I **Distinct Real Roots** $m_1 \neq m_2$	*CASE II* **Repeated Real Roots** $m_1 = m_2$	*CASE III* **Conjugate Complex Roots** $m_1 = \alpha + i\beta$, $m_2 = \alpha - i\beta$
(i) Both negative: $m_1 < m_2 < 0$	(i) Negative: $m_1 < 0$	(i) Complex: $\alpha < 0$, $\beta \neq 0$
(ii) Both positive: $m_1 > m_2 > 0$	(ii) Positive: $m_1 > 0$	(ii) Complex: $\alpha > 0$, $\beta \neq 0$
(iii) Opposite signs: $m_1 < 0 < m_2$	—	(iii) Pure imaginary: $\alpha = 0$, $\beta \neq 0$

The next three figures show the phase portrait of the differential equation (y versus x) along with a typical solution curve (x versus t) corresponding to the different subcases listed. In order to properly understand the nature of a phase portrait it is recommended that you verify that the graph of $x(t)$ is consistent with the pattern of the trajectories. For example, in Figure 3.5(b), observe that $x(t) \to 0$ and $x'(t) \to 0$ as $t \to \infty$; in Figure 3.5(a), if $(x(t), y(t))$ denotes a point on a trajectory, then clearly $x(t) \to 0$ and $y(t) \to 0$ as $t \to \infty$. The arrows on all the trajectories in Figure 3.5(a) indicate that points $(x(t), y(t))$ approach (0, 0) as $t \to \infty$. In Figure 3.5(c) the arrows indicate that points on all trajectories move away, or recede, from (0, 0) as $t \to \infty$. Correspondingly, notice that the solution $x(t)$ in Figure 3.5(d) becomes unbounded as $t \to \infty$.

As indicated in the captions of the phase portraits, it is standard practice to give the critical point (0, 0) a descriptive name depending on the behavior of the trajectories near this point. The *saddle point* derives its name from the resemblance of the phase portrait with the level curves of a hyperbolic paraboloid. (Recall that the graph of that surface has a saddle point.) We will say more about these names in Section 4.4.

CASE I General solution $x(t) = c_1e^{m_1t} + c_2e^{m_2t}$

(i) $m_1 < m_2 < 0$

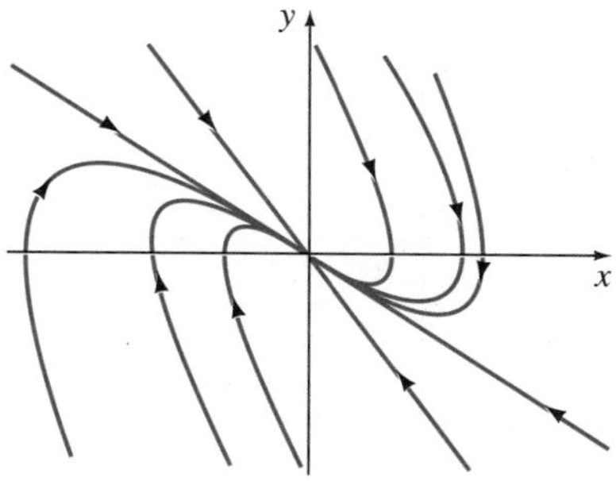

(a) The point (0, 0) is called a **node.**

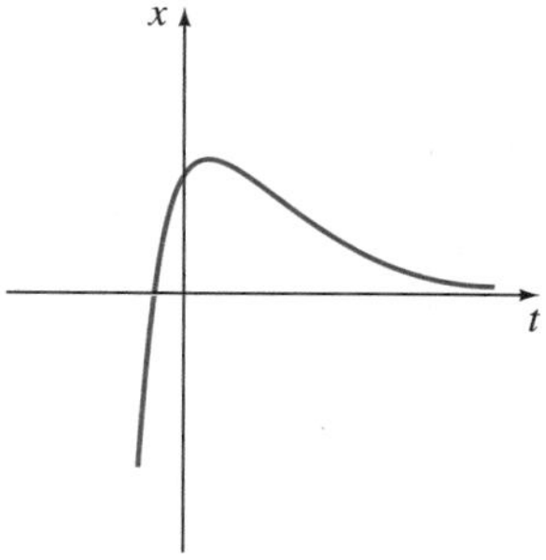

(b)

(ii) $m_1 > m_2 > 0$

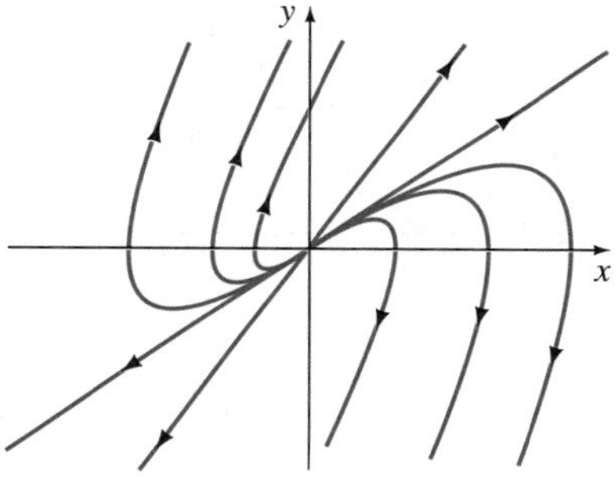

(c) The point (0, 0) is called a **node.**

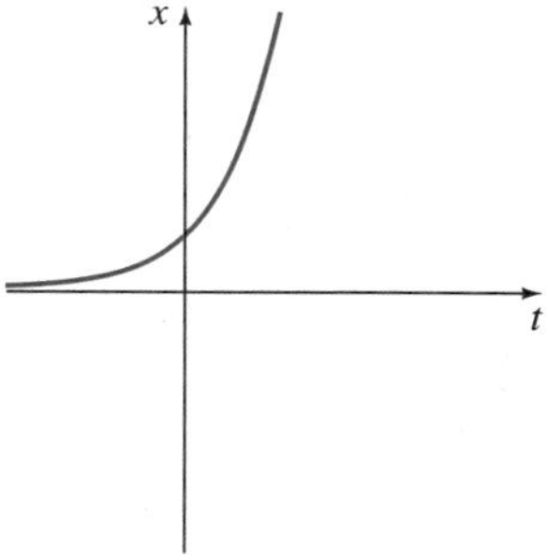

(d)

(iii) $m_1 < 0 < m_2$

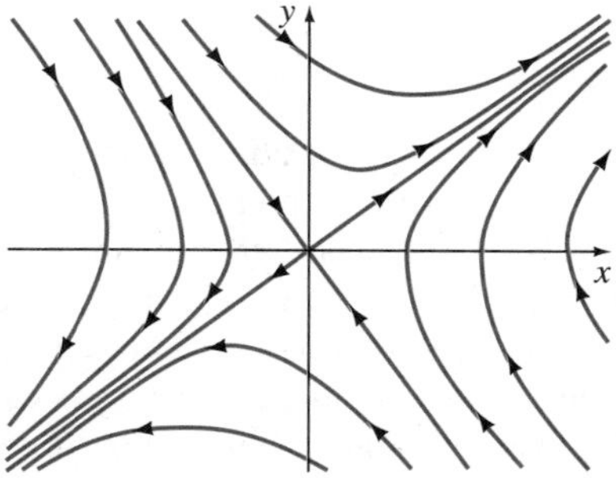

(e) The point (0, 0) is called a **saddle point.**

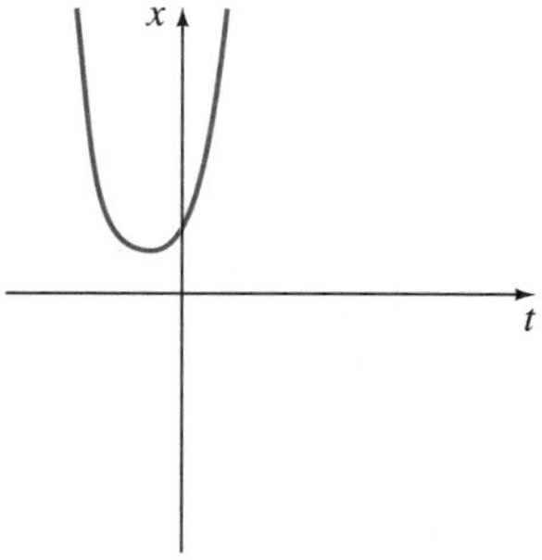

(f)

FIGURE 3.5

CASE II General solution $x(t) = c_1e^{m_1t} + c_2te^{m_1t}$

(i) $m_1 < 0$

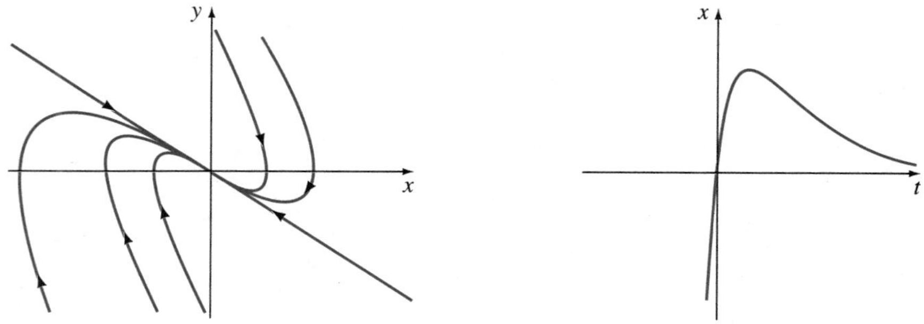

(a) The point (0, 0) is called a **node.** (b)

(ii) $m_1 > 0$

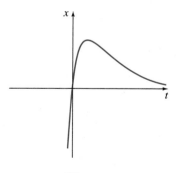

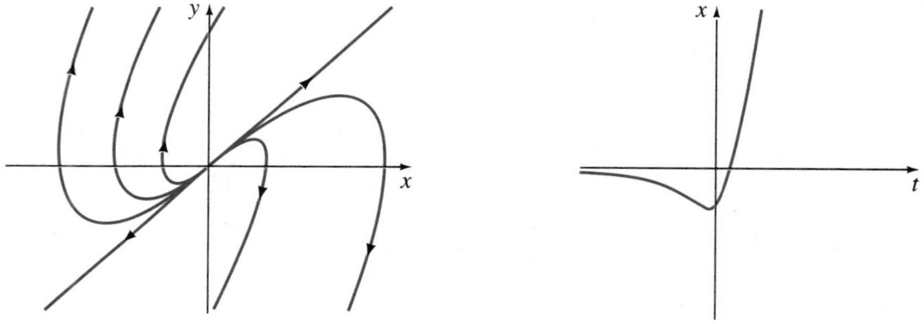

(c) The point (0, 0) is called a **node.** (d)

FIGURE 3.6

Stability

As in the discussion of critical points of first-order differential equations in Section 2.1, critical points for autonomous second-order equations $F(x, x', x'') = 0$ are classified as either stable or unstable. A critical point (x_1, y_1) of $F(x, x', x'') = 0$ is said to be **stable** provided a solution $(x(t), y(t))$ of the equivalent autonomous system stays arbitrarily close to (x_1, y_1) for all $t > t_0$ whenever an initial point (x_0, y_0) can be chosen sufficiently close to (x_1, y_1). (We shall elaborate on the notions of ''arbitrarily close'' and ''sufficiently close'' in Section 4.4.) If, in addition, $(x(t), y(t))$ approaches a stable critical point (x_1, y_1), that is, $(x(t), y(t)) \to (x_1, y_1)$ as $t \to \infty$, then (x_1, y_1) is called an **asymptotically stable** critical point. A critical point is **unstable** if it is not stable. In Figure 3.5 the node shown in (a) is asymptotically stable, the node in (c) is unstable, and the saddle point in (e) is unstable even though it is apparent that *some* solutions approach (0, 0) as $t \to \infty$. The nodes in (a) and (c) of Figure 3.6 are asymptotically stable and unstable, respectively. In Figure 3.7(a)

CASE III General solution $x(t) = c_1 e^{\alpha t} \cos \beta t + c_2 e^{\alpha t} \sin \beta t$

(i) $\alpha < 0, \beta \neq 0$

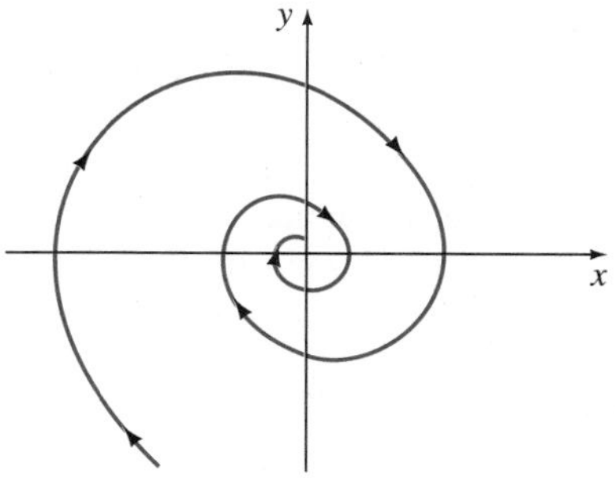

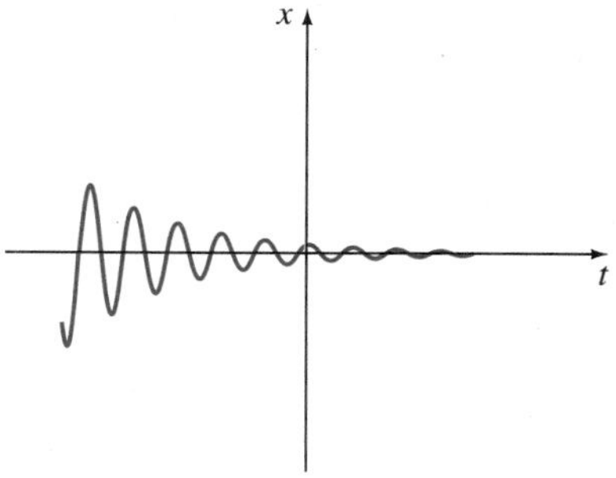

(**a**) The point (0, 0) is called a **spiral point**

(**b**)

(ii) $\alpha > 0, \beta \neq 0$

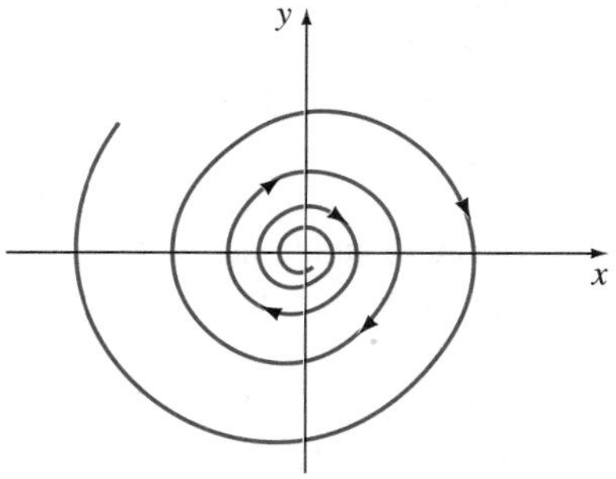

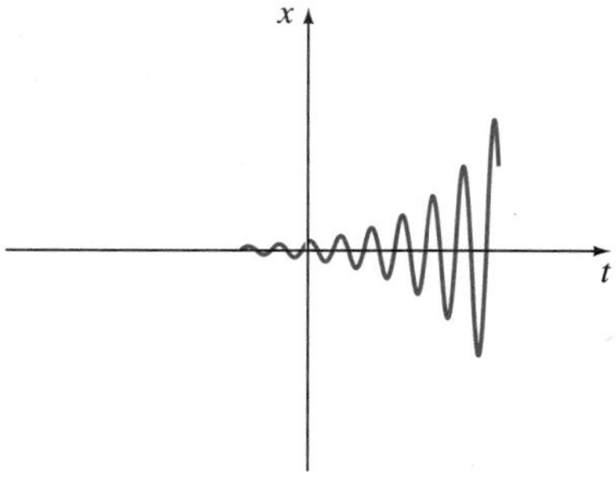

(**c**) The point (0, 0) is called a **spiral point.**

(**d**)

(iii) $\alpha = 0, \beta \neq 0$

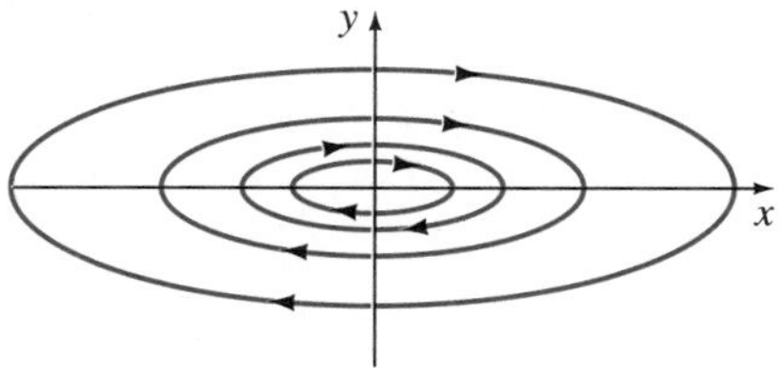

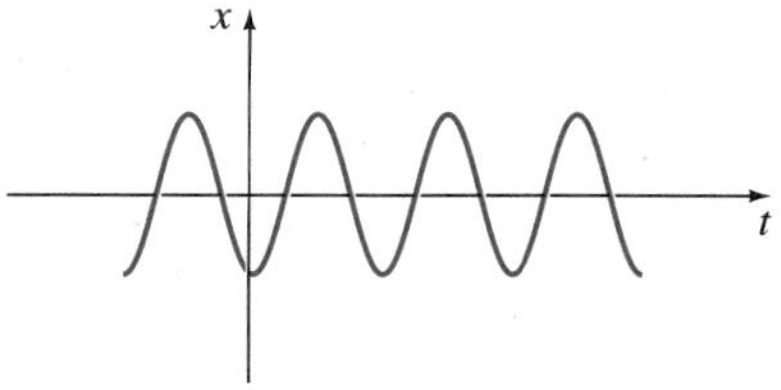

(**e**) The point (0, 0) is called a **center.**

(**f**)

FIGURE 3.7

the spiral point is asymptotically stable, in Figure 3.7(c) the spiral point is unstable, and in Figure 3.7(e) the center is stable. Recall, the closed trajectories reflect the fact that solutions of (16) (and the corresponding solutions of (15)) are periodic; that is, there exists a positive number T such that $x(t + T) = x(t)$, $y(t + T) = y(t)$. The center in Figure 3.7(e) is, however, not asymptotically stable because, regardless of how close an ellipse in that family of trajectories may be to the origin, a point $(x(t), y(t))$ on the ellipse must move around that curve in a periodic manner; in other words, $(x(t), y(t))$ remains nearby but cannot approach the critical point (0, 0) as $t \to \infty$.

Words, Words, Words

If you explore the topic of stability you will find that there is a vast amount of terminology, and often texts will differ on what things are called. For example, a critical point is variously called an **equilibrium point**, **singular point**, **stationary point**, or **rest point**. A spiral point is called a **focus**, **focal point**, and **vortex point**. Trajectories are called **paths** and **orbits**. Some texts use the word ''orbit'' only to refer to the trajectories around a center. Occasionally a center is also called a **vortex point**. The direction of the arrowheads in Figures 3.5(a), 3.6(a), and 3.7(a), indicating that points $(x(t), y(t))$ on all trajectories are drawn into the origin as $t \to \infty$ in a manner reminiscent of iron filings pulled to a magnet, is why asymptotically stable nodes and spiral points are also known as **attractors**. Similarly, the unstable nodes in Figures 3.5(c) and 3.6(c) as well as the unstable spiral point in Figure 3.7(c) certainly look like points that could be called **repellers**. Asymptotically stable nodes and spiral points are also called **sinks** and their unstable counterparts are **sources**. The phrase ''a critical point is stable but not asymptotically stable'' is probably considered awkward by some, and so the terms **neutrally stable** and **weakly stable** are sometimes used instead.

EXERCISES 3.3

Answers to odd-numbered problems begin on page AN-12.

3.3.1

In Problems 1–36 find the general solution of the given differential equation.

1. $4y'' + y' = 0$ **2.** $2y'' - 5y' = 0$

3. $y'' - 36y = 0$ **4.** $y'' - 8y = 0$

5. $y'' + 9y = 0$ **6.** $3y'' + y = 0$

7. $y'' - y' - 6y = 0$ **8.** $y'' - 3y' + 2y = 0$

9. $\dfrac{d^2y}{dx^2} + 8\dfrac{dy}{dx} + 16y = 0$ **10.** $\dfrac{d^2y}{dx^2} - 10\dfrac{dy}{dx} + 25y = 0$

11. $y'' + 3y' - 5y = 0$ **12.** $y'' + 4y' - y = 0$

13. $12y'' - 5y' - 2y = 0$ **14.** $8y'' + 2y' - y = 0$

15. $y'' - 4y' + 5y = 0$ **16.** $2y'' - 3y' + 4y = 0$

17. $3y'' + 2y' + y = 0$ **18.** $2y'' + 2y' + y = 0$

19. $y''' - 4y'' - 5y' = 0$ **20.** $4y''' + 4y'' + y' = 0$

21. $y''' - y = 0$ **22.** $y''' + 5y'' = 0$

23. $y''' - 5y'' + 3y' + 9y = 0$

24. $y''' + 3y'' - 4y' - 12y = 0$

25. $y''' + y'' - 2y = 0$ **26.** $y''' - y'' - 4y = 0$

27. $y''' + 3y'' + 3y' + y = 0$

28. $y''' - 6y'' + 12y' - 8y = 0$

29. $\frac{d^4y}{dx^4} + \frac{d^3y}{dx^3} + \frac{d^2y}{dx^2} = 0$ 30. $\frac{d^4y}{dx^4} - 2\frac{d^2y}{dx^2} + y = 0$

31. $16\frac{d^4y}{dx^4} + 24\frac{d^2y}{dx^2} + 9y = 0$

32. $\frac{d^4y}{dx^4} - 7\frac{d^2y}{dx^2} - 18y = 0$

33. $\frac{d^5y}{dx^5} - 16\frac{dy}{dx} = 0$

34. $\frac{d^5y}{dx^5} - 2\frac{d^4y}{dx^4} + 17\frac{d^3y}{dx^3} = 0$

35. $\frac{d^5y}{dx^5} + 5\frac{d^4y}{dx^4} - 2\frac{d^3y}{dx^3} - 10\frac{d^2y}{dx^2} + \frac{dy}{dx} + 5y = 0$

36. $2\frac{d^5y}{dx^5} - 7\frac{d^4y}{dx^4} + 12\frac{d^3y}{dx^3} + 8\frac{d^2y}{dx^2} = 0$

In Problems 37–52 solve the given differential equation subject to the indicated initial conditions.

37. $y'' + 16y = 0, \quad y(0) = 2, \ y'(0) = -2$

38. $y'' - y = 0, \quad y(0) = y'(0) = 1$

39. $y'' + 6y' + 5y = 0, \quad y(0) = 0, \ y'(0) = 3$

40. $y'' - 8y' + 17y = 0, \quad y(0) = 4, \ y'(0) = -1$

41. $2y'' - 2y' + y = 0, \quad y(0) = -1, \ y'(0) = 0$

42. $y'' - 2y' + y = 0, \quad y(0) = 5, \ y'(0) = 10$

43. $y'' + y' + 2y = 0, \quad y(0) = y'(0) = 0$

44. $4y'' - 4y' - 3y = 0, \quad y(0) = 1, \ y'(0) = 5$

45. $y'' - 3y' + 2y = 0, \quad y(1) = 0, \ y'(1) = 1$

46. $y'' + y = 0, \quad y(\pi/3) = 0, \ y'(\pi/3) = 2$

47. $y''' + 12y'' + 36y' = 0, \quad y(0) = 0, \ y'(0) = 1,$ $y''(0) = -7$

48. $y''' + 2y'' - 5y' - 6y = 0, \quad y(0) = y'(0) = 0,$ $y''(0) = 1$

49. $y''' - 8y = 0, \quad y(0) = 0, \ y'(0) = -1, \ y''(0) = 0$

50. $\frac{d^4y}{dx^4} = 0, \quad y(0) = 2, \ y'(0) = 3, \ y''(0) = 4,$ $y'''(0) = 5$

51. $\frac{d^4y}{dx^4} - 3\frac{d^3y}{dx^3} + 3\frac{d^2y}{dx^2} - \frac{dy}{dx} = 0, \quad y(0) = y'(0) = 0,$ $y''(0) = y'''(0) = 1$

52. $\frac{d^4y}{dx^4} - y = 0, \quad y(0) = y'(0) = y''(0) = 0, \ y'''(0) = 1$

In Problems 53–56 solve the given differential equation subject to the indicated boundary conditions.

53. $y'' - 10y' + 25y = 0, \quad y(0) = 1, \ y(1) = 0$

54. $y'' + 4y = 0, \quad y(0) = 0, \ y(\pi) = 0$

55. $y'' + y = 0, \quad y'(0) = 0, \ y'(\pi/2) = 2$

56. $y'' - y = 0, \quad y(0) = 1, \ y'(1) = 0$

57. The roots of an auxiliary equation are $m_1 = 4$, $m_2 = m_3 = -5$. What is the corresponding differential equation?

58. The roots of an auxiliary equation are $m_1 = -1/2$, $m_2 = 3 + i$, $m_3 = 3 - i$. What is the corresponding differential equation?

In Problems 59 and 60 find the general solution of the given equation if it is known that y_1 is a solution.

59. $y''' - 9y'' + 25y' - 17y = 0; \quad y_1 = e^x$

60. $y''' + 6y'' + y' - 34y = 0; \quad y_1 = e^{-4x}\cos x$

In Problems 61–64 use a computer either as an aid in solving the auxiliary equation or as a means of directly obtaining the general solution of the given differential equation. If you use a CAS to obtain the general solution, simplify the output and write the solution in terms of real functions.

61. $y''' - 6y'' + 2y' + y = 0$

62. $6.11y''' + 8.59y'' + 7.93y' + 0.778y = 0$

63. $3.15y^{(4)} - 5.34y'' + 6.33y' - 2.03y = 0$

64. $y^{(4)} + 2y'' - y' + 2y = 0$

65. **PROJECT PROBLEM** Consider the second-order differential equation with constant coefficients $y'' + by' + cy = 0$.

(a) If $y(x)$ is a solution of the equation, show that $y(x \pm k)$, k a constant, is also a solution.

(b) Suppose that $y(x)$ is a solution of the differential equation that satisfies the initial conditions $y(0) = y_0$, $y'(0) = y_1$. Find a solution of the differential equation that satisfies the initial conditions $y(x_0) = y_0$, $y'(x_0) = y_1$.

(c) Find conditions on the coefficients b and c so that $\lim_{x\to\infty} y(x) = 0$.

(d) Find conditions on the coefficients b and c so that the equation possesses a nontrivial solution satisfying the boundary conditions $y(0) = 0$, $y(1) = 0$.

66. *WRITING PROJECT*

(a) Write a short report on how to find the roots of a complex number $\alpha + i\beta$, where α and β are real and $i^2 = -1$. Illustrate this by expressing $i^{1/2}$ and $(-i)^{1/2}$ in the form $\alpha + i\beta$.

(b) Use part (a) to find the general solution of the differential equation $y^{(4)} + y = 0$.

3.3.2

In Problems 67–74 the origin (0, 0) is a critical point of the given autonomous differential equation. Classify (0, 0) as a node, saddle point, spiral point, or center. Classify (0, 0) as unstable, stable (but not asymptotically stable), or asymptotically stable.

67. $x'' - 9x = 0$

68. $x'' + x = 0$

69. $x'' - 7x' + 10x = 0$

70. $x'' + 8x' + 7x = 0$

71. $x'' - 2x' + 37x = 0$

72. $x'' - 6x' + 9x = 0$

73. $x'' + 10x' + 25x = 0$

74. $3x'' + 2x' + 3x = 0$

In Problems 75 and 76 do not solve the given differential equation, but use computer software to plot the trajectory of the equation corresponding to the given initial point. Plot the corresponding solution of the initial-value problem either by hand or by using an ODE solver as instructed. Indicate the initial point on each graph.

75. $x'' - x = 0$

(a) $x(0) = 0, x'(0) = 1$ **(b)** $x(0) = 1, x'(0) = 0$, **(c)** $x(0) = 1, x'(0) = -1$

76. $x'' + 2x' + x = 0$

(a) $x(0) = 2, x'(0) = 1$ **(b)** $x(0) = -1, x'(0) = 1$ **(c)** $x(0) = -4, x'(0) = -2$

77. Without graphing, describe the phase portrait of $x'' = 0$.

78. Four of the trajectories in the phase portrait of $x'' - 9x' + 18x = 0$ are straight line segments.

(a) Find Cartesian equations of these lines.

(b) Find four initial-value problems whose solutions correspond to these straight-line trajectories.

79. Consider the autonomous equation $x'' + x' = 0$. Explain why (0, 0) is not an isolated critical point.

80. Consider equation (15) in the form $x'' + bx' + cx = 0$, where b and c are real constants and $c \neq 0$. Give conditions on b and c such that (0, 0) is

(a) a node, which is either unstable or asymptotically stable,

(b) a saddle point,

(c) a spiral point, which is either unstable or asymptotically stable, or

(d) a center.

81. (a) Let $a \neq 0$, b, $c \neq 0$, and $d \neq 0$ be constants. Show that the critical point of $ax'' + bx' + cx = d$ is $(d/c, 0)$.

(b) Show that the change of variables $u(t) = x(t) - d/c$ gives $au'' + bu' + cu = 0$. Briefly discuss the significance of this result.

3.4 UNDETERMINED COEFFICIENTS

To solve a nonhomogeneous linear differential equation

$$a_n y^{(n)} + a_{n-1} y^{(n-1)} + \cdots + a_1 y' + a_0 y = g(x), \qquad (1)$$

we must do two things:

(i) Find the complementary function y_c.

(ii) Find *any* particular solution y_p of the nonhomogeneous equation.

Then, as discussed in Section 3.1, the general solution of (1) on an interval is $y = y_c + y_p$.

The complementary function y_c is the solution of the associated homogeneous equation $a_n y^{(n)} + a_{n-1} y^{(n-1)} + \cdots + a_1 y' + a_0 y = 0$. We saw how to solve these kinds of equations in the last section when the coefficients were constants. The first of two ways we shall consider for obtaining a particular solution y_p is called the **method of undetermined coefficients**. The underlying idea in this method is a conjecture, or educated guess, about the form of y_p motivated by the kinds of functions that comprise $g(x)$. Basically straightforward, the method regrettably is limited to nonhomogeneous linear equations such as (1) where

- the coefficients a_i, $i = 0, 1, \ldots, n$ are constants, and
- $g(x)$ is a constant k, a polynomial function, an exponential function $e^{\alpha x}$, sine or cosine functions $\sin \beta x$, $\cos \beta x$, or finite sums and products of these functions.

Note Strictly speaking $g(x) = k$ (a constant) is a polynomial function. Since a constant function is probably not the first thing that comes to mind when you think of polynomial functions, for emphasis we shall continue to use the redundancy "constant functions, polynomials, ..."

The following are some examples of the types of functions $g(x)$ that are appropriate for this discussion:

$$g(x) = 10, \quad g(x) = x^2 - 5x, \quad g(x) = 15x - 6 + 8e^{-x}$$

$$g(x) = \sin 3x - 5x \cos 2x, \quad g(x) = e^x \cos x + (3x^2 - 1)e^{-x}$$

and so on. That is, $g(x)$ is a linear combination of functions of the type

$$k \text{ (constant)}, \quad x^n, \quad x^n e^{\alpha x}, \quad x^n e^{\alpha x} \cos \beta x, \quad \text{and} \quad x^n e^{\alpha x} \sin \beta x,$$

where n is a nonnegative integer and α and β are real numbers. The method of undetermined coefficients is not applicable to equations of form (1) when

$$g(x) = \ln x, \quad g(x) = \frac{1}{x}, \quad g(x) = \tan x, \quad g(x) = \sin^{-1} x,$$

and so on. Differential equations in which $g(x)$ is a function of this last kind will be considered in Section 3.5.

The set of functions that consists of constants, polynomials, exponentials $e^{\alpha x}$, sines, and cosines has the remarkable property that derivatives of their sums and products are again sums and products of constants, polynomials, exponentials $e^{\alpha x}$, sines, and cosines. Since the linear combination of derivatives $a_n y_p^{(n)} + a_{n-1} y_p^{(n-1)} + \cdots + a_1 y_p' + a_0 y_p$ must be identical to $g(x)$, it seems reasonable to assume that *y_p has the same form as $g(x)$*.

The next two examples illustrate the basic method.

EXAMPLE 1 General Solution Using Undetermined Coefficients

Solve $$y'' + 4y' - 2y = 2x^2 - 3x + 6. \qquad \textbf{(2)}$$

Solution **Step 1.** We first solve the associated homogeneous equation $y'' + 4y' - 2y = 0$. From the quadratic formula we find that the roots of the auxiliary equation $m^2 + 4m - 2 = 0$ are $m_1 = -2 - \sqrt{6}$ and $m_2 = -2 + \sqrt{6}$. Hence the complementary function is

$$y_c = c_1 e^{-(2+\sqrt{6})x} + c_2 e^{(-2+\sqrt{6})x}.$$

Step 2. Now since the function $g(x)$ is a quadratic polynomial, let us assume a particular that is also in the form of a quadratic polynomial:

$$y_p = Ax^2 + Bx + C.$$

We seek to determine *specific* coefficients A, B, and C for which y_p is a solution of (2). Substituting y_p and the derivatives

$$y_p' = 2Ax + B \quad \text{and} \quad y_p'' = 2A$$

into the given differential equation (2), we get

$$y_p'' + 4y_p' - 2y_p = 2A + 8Ax + 4B - 2Ax^2 - 2Bx - 2C = 2x^2 - 3x + 6.$$

Since the last equation is supposed to be an identity, the coefficients of like powers of x must be equal:

equal

$$\boxed{-2A}\,x^2 + \boxed{8A - 2B}\,x + \boxed{2A + 4B - 2C} = 2x^2 - 3x + 6.$$

That is, $\quad -2A = 2, \quad 8A - 2B = -3, \quad 2A + 4B - 2C = 6.$

Solving this system of equations leads to the values $A = -1$, $B = -5/2$, and $C = -9$. Thus a particular solution is

$$y_p = -x^2 - \frac{5}{2}x - 9.$$

Step 3. The general solution of the given equation is

$$y = y_c + y_p = c_1 e^{-(2+\sqrt{6})x} + c_2 e^{(-2+\sqrt{6})x} - x^2 - \frac{5}{2}x - 9.$$

●

EXAMPLE 2 Particular Solution Using Undetermined Coefficients

Find a particular solution of $\quad y'' - y' + y = 2 \sin 3x.$

Solution A natural first guess for a particular solution would be $A \sin 3x$. But since successive differentiations of $\sin 3x$ produce $\sin 3x$ *and* $\cos 3x$, we are prompted instead to assume a particular solution that includes both of these terms:

$$y_p = A \cos 3x + B \sin 3x.$$

Differentiating y_p and substituting the results into the differential equation give, after regrouping,

$$y_p'' - y_p' + y_p = (-8A - 3B) \cos 3x + (3A - 8B) \sin 3x = 2 \sin 3x$$

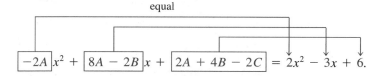

or

$$\underbrace{\boxed{-8A - 3B}}\cos 3x + \boxed{3A - 8B}\sin 3x = 0\cos 3x + 2\sin 3x.$$

equal

From the resulting system of equations,

$$-8A - 3B = 0, \quad 3A - 8B = 2,$$

we get $A = 6/73$ and $B = -16/73$. A particular solution of the equation is

$$y_p = \frac{6}{73}\cos 3x - \frac{16}{73}\sin 3x. \quad \bullet$$

As we mentioned, the form that we assume for the particular solution y_p is an educated guess; it is not a blind guess. This educated guess must take into consideration not only the types of functions that make up $g(x)$, but also, as we shall see in Example 4, the functions that make up the complementary function y_c.

EXAMPLE 3 Forming y_p by Superposition

Solve $\quad y'' - 2y' - 3y = 4x - 5 + 6xe^{2x}. \quad$ **(3)**

Solution **Step 1.** First, the solution of the associated homogeneous equation $y'' - 2y' - 3y = 0$ is found to be $y_c = c_1e^{-x} + c_2e^{3x}$.

Step 2. Next, the presence of $4x - 5$ in $g(x)$ suggests that the particular solution includes a linear polynomial. Furthermore, since the derivative of the product xe^{2x} produces $2xe^{2x}$ and e^{2x}, we also assume that the particular solution includes both xe^{2x} and e^{2x}. In other words, g is the sum of two basic kinds of functions:

$$g(x) = g_1(x) + g_2(x) = \textit{polynomial} + \textit{exponentials}.$$

Correspondingly, the superposition principle for nonhomogeneous equations (Theorem 3.7) suggests that we seek a particular solution

$$y_p = y_{p_1} + y_{p_2},$$

where $y_{p_1} = Ax + B$ and $y_{p_2} = Cxe^{2x} + De^{2x}$. Substituting

$$y_p = Ax + B + Cxe^{2x} + De^{2x}$$

into the given equation (3) and grouping like terms give

$$y_p'' - 2y_p' - 3y_p = -3Ax - 2A - 3B - 3Cxe^{2x} + (2C - 3D)e^{2x} = 4x - 5 + 6xe^{2x}. \quad \textbf{(4)}$$

From this identity we obtain the four equations

$$-3A = 4, \quad -2A - 3B = -5, \quad -3C = 6, \quad 2C - 3D = 0.$$

The last equation in this system results from the interpretation that the coefficient of e^{2x} in the right member of (4) is zero. Solving, we find $A = -4/3$,

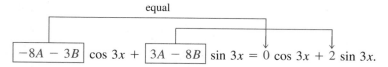

$B = 23/9$, $C = -2$, and $D = -4/3$. Consequently,

$$y_p = -\frac{4}{3}x + \frac{23}{9} - 2xe^{2x} - \frac{4}{3}e^{2x}.$$

Step 3. The general solution of the equation is

$$y = c_1 e^{-x} + c_2 e^{3x} - \frac{4}{3}x + \frac{23}{9} - \left(2x + \frac{4}{3}\right)e^{2x}.$$

In light of the superposition principle (Theorem 3.7) we can also approach Example 3 from the viewpoint of solving two simpler problems. You should verify that substituting

$$y_{p_1} = Ax + B \qquad \text{into} \quad y'' - 2y' - 3y = 4x - 5$$

and

$$y_{p_2} = Cxe^{2x} + De^{2x} \quad \text{into} \quad y'' - 2y' - 3y = 6xe^{2x}$$

yields in turn $y_{p_1} = -(4/3)x + 23/9$ and $y_{p_2} = -(2x + 4/3)e^{2x}$. A particular solution of (3) is then $y_p = y_{p_1} + y_{p_2}$.

The next example illustrates that sometimes the "obvious" assumption for the form of y_p is not a correct assumption.

EXAMPLE 4 A Glitch in the Method

Find a particular solution of $y'' - 5y' + 4y = 8e^x$.

Solution Differentiation of e^x produces no new functions. Thus, proceeding as we did in the earlier examples, we can reasonably assume a particular solution of the form $y_p = Ae^x$. But substitution of this expression into the differential equation yields the contradictory statement

$$0 = 8e^x,$$

and so we have clearly made the wrong guess for y_p.

The difficulty here is apparent upon examining the complementary function $y_c = c_1e^x + c_2e^{4x}$. Observe that our assumption Ae^x is already present in y_c. This means that e^x is a solution of the associated homogeneous differential equation, and a constant multiple Ae^x when substituted into the differential equation necessarily produces zero.

What then should be the form of y_p? Inspired by Case II of Section 3.3, let's see whether we can find a particular solution of the form

$$y_p = Axe^x.$$

Substituting $y_p' = Axe^x + Ae^x$ and $y_p'' = Axe^x + 2Ae^x$ into the differential equation and simplifying gives

$$y_p'' - 5y_p' + 4y_p = -3Ae^x = 8e^x.$$

From the last equality we see that the value of A is now determined as $A = -8/3$. Therefore a particular solution of the given equation is

$$y_p = -\frac{8}{3}xe^x.$$

The difference in the procedures used in Examples 1–3 and in Example 4 suggests that we consider two cases. The first case reflects the situation in Examples 1–3.

CASE I No function in the assumed particular solution is a solution of the associated homogeneous differential equation.

In Table 3.1 we illustrate some specific examples of $g(x)$ in (1) along with the corresponding form of the particular solution. We are, of course, taking for granted that no function in the assumed particular solution y_p is duplicated by a function in the complementary function y_c.

TABLE 3.1 Trial Particular Solutions

$g(x)$	Form of y_p
1. 1 (any constant)	A
2. $5x + 7$	$Ax + B$
3. $3x^3 - 2$	$Ax^2 + Bx + C$
4. $x^3 - x + 1$	$Ax^3 + Bx^2 + Cx + D$
5. $\sin 4x$	$A \cos 4x + B \sin 4x$
6. $\cos 4x$	$A \cos 4x + B \sin 4x$
7. e^{5x}	Ae^{5x}
8. $(9x - 2)e^{5x}$	$(Ax + B)e^{5x}$
9. x^2e^{5x}	$(Ax^2 + Bx + C)e^{5x}$
10. $e^{3x} \sin 4x$	$Ae^{3x} \cos 4x + Be^{3x} \sin 4x$
11. $5x^2 \sin 4x$	$(Ax^2 + Bx + C) \cos 4x + (Dx^2 + Ex + F) \sin 4x$
12. $xe^{3x} \cos 4x$	$(Ax + B)e^{3x} \cos 4x + (Cx + D)e^{3x} \sin 4x$

EXAMPLE 5 Forms of Particular Solutions—Case I

Determine the form of a particular solution of

$$\textbf{(a)}\ y'' - 8y' + 25y = 5x^3e^{-x} - 7e^{-x} \quad \text{and} \quad \textbf{(b)}\ y'' + 4y = x \cos x.$$

Solution (a) We can write

$$g(x) = (5x^3 - 7)e^{-x}.$$

Using entry 9 in Table 3.1 as a model, we assume a particular solution of the form

$$y_p = (Ax^3 + Bx^2 + Cx + D)e^{-x}.$$

Note that there is no duplication between the terms in y_p and the terms in the complementary function $y_c = e^{4x}(c_1 \cos 3x + c_2 \sin 3x)$.

(b) The function $g(x) = x \cos x$ is similar to entry 11 in Table 3.1, except of course we use a linear rather than a quadratic polynomial and $\cos x$ and $\sin x$ instead of $\cos 4x$ and $\sin 4x$ in the form of y_p:

$$y_p = (Ax + B) \cos x + (Cx + D) \sin x.$$

Again observe that there is no duplication of terms between y_p and $y_c = c_1 \cos 2x + c_2 \sin 2x$. ●

If $g(x)$ consists of a sum of, say, m terms of the kind listed in the table, then (as in Example 3) the assumption for a particular solution y_p consists of the sum of the trial forms $y_{p_1}, y_{p_2}, \ldots, y_{p_m}$ corresponding to these terms:

$$y_p = y_{p_1} + y_{p_2} + \cdots + y_{p_m}.$$

The foregoing sentence can be put another way.

> ***Form Rule for Case I*** *The form of y_p is a linear combination of all linearly independent functions that are generated by repeated differentiations of $g(x)$.*

EXAMPLE 6 Forming y_p by Superposition—Case I

Determine the form of a particular solution of

$$y'' - 9y' + 14y = 3x^2 - 5 \sin 2x + 7xe^{6x}.$$

Solution

Corresponding to $3x^2$ we assume: $\quad y_{p_1} = Ax^2 + Bx + C$

Corresponding to $-5 \sin 2x$ we assume: $\quad y_{p_2} = D \cos 2x + E \sin 2x$

Corresponding to $7xe^{6x}$ we assume: $\quad y_{p_3} = (Fx + G)e^{6x}$

The assumption for the particular solution is then

$$y_p = y_{p_1} + y_{p_2} + y_{p_3} = Ax^2 + Bx + C + D \cos 2x + E \sin 2x + (Fx + G)e^{6x}.$$

No term in this assumption duplicates a term in $y_c = c_1e^{2x} + c_2e^{7x}$. ●

CASE II A function in the assumed particular solution is also a solution of the associated homogeneous differential equation.

The next example is similar to Example 4.

EXAMPLE 7 Particular Solution—Case II

Find a particular solution of $y'' - 2y' + y = e^x$.

Solution The complementary function is $y_c = c_1e^x + c_2xe^x$. As in Example 4, the assumption $y_p = Ae^x$ will fail since it is apparent from y_c that e^x is a solution of the associated homogeneous equation $y'' - 2y' + y = 0$. Moreover, we will not be able to find a particular solution of the form $y_p = Axe^x$ since the term xe^x is also duplicated in y_c. We next try

$$y_p = Ax^2e^x.$$

Substituting into the given differential equation yields

$$2Ae^x = e^x \quad \text{and so} \quad A = \frac{1}{2}.$$

Thus a particular solution is $y_p = \frac{1}{2}x^2e^x$. ●

Suppose again that $g(x)$ consists of m terms of the kind given in Table 3.2 and suppose further that the usual assumption for a particular solution is

$$y_p = y_{p_1} + y_{p_2} + \cdots + y_{p_m}$$

where the y_{p_i}, $i = 1, 2, \ldots, m$ are the trial particular solution forms corresponding to these terms. Under the circumstances described in Case II, we can make up the following general rule.

Multiplication Rule for Case II *If any y_{p_i} contains terms that duplicate terms in y_c, then that y_{p_i} must be multiplied by x^n, where n is the smallest positive integer that eliminates that duplication.*

EXAMPLE 8 An Initial-Value Problem

Solve the initial-value problem

$$y'' + y = 4x + 10 \sin x, \qquad y(\pi) = 0, \quad y'(\pi) = 2.$$

Solution The solution of the associated homogeneous equation $y'' + y = 0$ is $y_c = c_1 \cos x + c_2 \sin x$. Now since $g(x) = 4x + 10 \sin x$ is the sum of a linear polynomial and a sine function, our normal assumption for y_p, from entries 2 and 5 of Table 3.1, would be the sum of $y_{p_1} = Ax + B$ and $y_{p_2} = C \cos x + D \sin x$:

$$y_p = Ax + B + C \cos x + D \sin x. \tag{5}$$

But there is an obvious duplication of the terms $\cos x$ and $\sin x$ in this assumed form and two terms in the complementary function. This duplication can be eliminated by simply multiplying y_{p_2} by x. Instead of (5) we now use

$$y_p = Ax + B + Cx \cos x + Dx \sin x. \tag{6}$$

Differentiating this expression and substituting the results into the differential equation give

$$y_p'' + y_p = Ax + B - 2C \sin x + 2D \cos x = 4x + 10 \sin x$$

and so

$$A = 4, \quad B = 0, \quad -2C = 10, \quad 2D = 0.$$

The solutions of the system are immediate: $A = 4$, $B = 0$, $C = -5$, and $D = 0$. Therefore from (6) we obtain

$$y_p = 4x - 5x \cos x.$$

The general solution of the given equation is

$$y = y_c + y_p = c_1 \cos x + c_2 \sin x + 4x - 5x \cos x.$$

We now apply the prescribed initial conditions to the general solution of the equation. First, $y(\pi) = c_1 \cos \pi + c_2 \sin \pi + 4\pi - 5\pi \cos \pi = 0$ yields $c_1 = 9\pi$ since $\cos \pi = -1$ and $\sin \pi = 0$. Next, from the derivative

$$y' = -9\pi \sin x + c_2 \cos x + 4 + 5x \sin x - 5 \cos x$$

and $$y'(\pi) = -9\pi \sin \pi + c_2 \cos \pi + 4 + 5\pi \sin \pi - 5 \cos \pi = 2$$

we find $c_2 = 7$. The solution of the initial value is then

$$y = 9\pi \cos x + 7 \sin x + 4x - 5x \cos x.$$

EXAMPLE 9 Using the Multiplication Rule

Solve $$y'' - 6y' + 9y = 6x^2 + 2 - 12e^{3x}.$$

Solution The complementary function is $y_c = c_1e^{3x} + c_2xe^{3x}$. And so, based on entries 3 and 7 of Table 3.1, the usual assumption for a particular solution would be

$$y_p = \underbrace{Ax^2 + Bx + C}_{y_{p_1}} + \underbrace{De^{3x}}_{y_{p_2}}.$$

Inspection of these functions shows that the one term in y_{p_2} is duplicated in y_c. If we multiply y_{p_2} by x, we note that the term xe^{3x} is still part of y_c. But multiplying y_{p_2} by x^2 eliminates all duplications. Thus the operative form of a particular solution is

$$y_p = Ax^2 + Bx + C + Dx^2e^{3x}.$$

Differentiating this last form, substituting into the differential equation, and collecting like terms give

$$y_p'' - 6y_p' + 9y_p = 9Ax^2 + (-12A + 9B)x + 2A - 6B + 9C + 2De^{3x} = 6x^2 + 2 - 12e^{3x}.$$

It follows from this identity that $A = 2/3$, $B = 8/9$, $C = 2/3$, and $D = -6$. Hence the general solution $y = y_c + y_p$ is

$$y = c_1e^{3x} + c_2xe^{3x} + \frac{2}{3}x^2 + \frac{8}{9}x + \frac{2}{3} - 6x^2e^{3x}.$$

EXAMPLE 10 Third-Order DE—Case I

Solve $y''' + y'' = e^x \cos x$.

Solution From the characteristic equation $m^3 + m^2 = 0$ we find $m_1 = m_2 = 0$ and $m_3 = -1$. Hence the complementary solution of the equation is $y_c = c_1 + c_2x + c_3e^{-x}$. With $g(x) = e^x \cos x$ we see from entry 10 of Table 3.1 that we should assume

$$y_p = Ae^x \cos x + Be^x \sin x.$$

Since there are no functions in y_p that duplicate functions in the complementary solution, we proceed in the usual manner. From

$$y_p''' + y_p'' = (-2A + 4B)e^x \cos x + (-4A - 2B)e^x \sin x = e^x \cos x$$

we get $-2A + 4B = 1, \quad -4A - 2B = 0.$

This system gives $A = -1/10$ and $B = 1/5$, so that a particular solution is

$$y_p = -\frac{1}{10}e^x \cos x + \frac{1}{5}e^x \sin x.$$

The general solution of the equation is

$$y = y_c + y_p = c_1 + c_2x + c_3e^{-x} - \frac{1}{10}e^x \cos x + \frac{1}{5}e^x \sin x.$$

●

EXAMPLE 11 Fourth-Order DE—Case II

Determine the form of a particular solution of $y^{(4)} + y''' = 1 - e^{-x}$.

Solution Comparing the complementary function

$$y_c = c_1 + c_2x + c_3x^2 + c_4e^{-x}$$

with our normal assumption for a particular solution

$$y_p = \underbrace{A}_{y_{p_1}} + \underbrace{Be^{-x}}_{y_{p_2}},$$

we see that the duplications between y_c and y_p are eliminated when y_{p_1} is multiplied by x^3 and y_{p_2} is multiplied by x. Thus the correct assumption for a particular solution is

$$y_p = Ax^3 + Bxe^{-x}.$$

●

EXERCISES 3.4

Answers to odd-numbered problems begin on page AN-13.

In Problems 1–26 solve the given differential equation by undetermined coefficients.

1. $y'' + 3y' + 2y = 6$

2. $4y'' + 9y = 15$

3. $y'' - 10y' + 25y = 30x + 3$

4. $y'' + y' - 6y = 2x$

5. $\frac{1}{4}y'' + y' + y = x^2 - 2x$

6. $y'' - 8y' + 20y = 100x^2 - 26xe^x$

7. $y'' + 3y = -48x^2e^{3x}$

8. $4y'' - 4y' - 3y = \cos 2x$

9. $y'' - y' = -3$

10. $y'' + 2y' = 2x + 5 - e^{-2x}$

11. $y'' - y' + \frac{1}{4}y = 3 + e^{x/2}$

12. $y'' - 16y = 2e^{4x}$

13. $y'' + 4y = 3 \sin 2x$

14. $y'' + 4y = (x^2 - 3) \sin 2x$

15. $y'' + y = 2x \sin x$

16. $y'' - 5y' = 2x^3 - 4x^2 - x + 6$

17. $y'' - 2y' + 5y = e^x \cos 2x$

18. $y'' - 2y' + 2y = e^{2x}(\cos x - 3 \sin x)$

19. $y'' + 2y' + y = \sin x + 3 \cos 2x$

20. $y'' + 2y' - 24y = 16 - (x + 2)e^{4x}$

21. $y''' - 6y'' = 3 - \cos x$

22. $y''' - 2y'' - 4y' + 8y = 6xe^{2x}$

23. $y''' - 3y'' + 3y' - y = x - 4e^x$

24. $y''' - y'' - 4y' + 4y = 5 - e^x + e^{2x}$

25. $y^{(4)} + 2y'' + y = (x - 1)^2$

26. $y^{(4)} - y'' = 4x + 2xe^{-x}$

In Problems 27–36 solve the given differential equation subject to the indicated initial conditions.

27. $y'' + 4y = -2, \quad y\left(\frac{\pi}{8}\right) = \frac{1}{2}, \quad y'\left(\frac{\pi}{8}\right) = 2$

28. $2y'' + 3y' - 2y = 14x^2 - 4x - 11,$
$y(0) = 0, \quad y'(0) = 0$

29. $5y'' + y' = -6x, \quad y(0) = 0, \quad y'(0) = -10$

30. $y'' + 4y' + 4y = (3 + x)e^{-2x}, \quad y(0) = 2, \quad y'(0) = 5$

31. $y'' + 4y' + 5y = 35e^{-4x}, \quad y(0) = -3, \quad y'(0) = 1$

32. $y'' - y = \cosh x, \quad y(0) = 2, \quad y'(0) = 12$

33. $\dfrac{d^2x}{dt^2} + \omega^2 x = F_0 \sin \omega t, \quad x(0) = 0, \quad x'(0) = 0$

34. $\dfrac{d^2x}{dt^2} + \omega^2 x = F_0 \cos \gamma t, \quad x(0) = 0, \quad x'(0) = 0$

35. $y''' - 2y'' + y' = 2 - 24e^x + 40e^{5x},$
$y(0) = \frac{1}{2}, \quad y'(0) = \frac{5}{2}, \quad y''(0) = -\frac{9}{2}$

36. $y''' + 8y = 2x - 5 + 8e^{-2x}, \quad y(0) = -5,$
$y'(0) = 3, \quad y''(0) = -4$

37. ***PROJECT PROBLEM*** The purpose of this problem is to further motivate the form of the particular solution under Case II of this section, in other words, why certain parts of a particular solution must be multiplied by x^n.

Consider $ay'' + by' + cy = de^{kx}$, where a, b, c, and d are known constants and k is a root of the auxiliary equation of the differential equation. Use the method discussed in Section 3.2 to find a particular solution of the equation having the form $y_p = u(x)e^{kx}$. Carefully consider two cases: **(i)** k is a distinct root of the auxiliary equation, and **(ii)** k is root of multiplicity two.

38. ***PROJECT PROBLEM*** The differentiation symbol or operator d/dx is often abbreviated by the symbol D. A linear differential equation with constant coefficients such as $y'' - 4y' + 4y = e^{2x}$ is then written as $(D^2 - 4D + 4)y = e^{2x}$. The polynomial $D^2 - 4D + 4$ or $(D - 2)^2$ in the symbol D is called a second-order linear differential operator with constant coefficients.

(a) Write a short report on the **exponential shift** in the form

$$(D - a)^n e^{ax} y = e^{ax} D^n y.$$

Give a justification of the last result and illustrate its utility in finding a particular solution of $y'' - 4y' + 4y = e^{2x}$.

(b) At the direction of your instructor, write a longer report discussing the utility of the general exponential shift $P(D - a)e^{ax}y = e^{ax}P(D)y$, where P is a polynomial in the symbol D.

3.5 VARIATION OF PARAMETERS

The procedure that we used in Section 2.4 to find a particular solution y_p of a linear first-order differential equation

$$\frac{dy}{dx} + P(x)y = f(x) \tag{1}$$

on an interval is applicable to linear higher-order equations as well. To adapt the method of **variation of parameters** to a linear second-order differential equation

$$a_2(x)y'' + a_1(x)y' + a_0(x)y = g(x) \tag{2}$$

we begin as we did in Section 2.4 by putting the differential equation in the **standard form**

$$y'' + P(x)y' + Q(x)y = f(x) \tag{3}$$

by dividing through by the lead coefficient $a_2(x)$. Here we assume that $P(x)$, $Q(x)$, and $f(x)$ are continuous on some interval I. Equation (3) is the analogue of (1). As we have already seen in Section 3.3, there is absolutely no difficulty in obtaining the complementary function y_c of (2) when the coefficients are constants.

Suppose y_1 and y_2 form a fundamental set of solutions on I of the associated homogeneous form of (3); that is,

$$y_1'' + P(x)y_1' + Q(x)y_1 = 0 \quad \text{and} \quad y_2'' + P(x)y_2' + Q(x)y_2 = 0.$$

Now we ask: Can two functions u_1 and u_2 be found so that

$$y_p = u_1(x)y_1(x) + u_2(x)y_2(x)$$

is a particular solution of (3)? Our assumption for y_p is the same as $y_c = c_1y_1 + c_2y_2$, but we have replaced the parameters c_1 and c_2 by the variables u_1 and u_2. Because we seek to determine two unknown functions, reason dictates that we need two equations. As we shall see one of these equations results from substituting $y_p = u_1y_1 + u_2y_2$ into the given differential equation (3). The other equation that we impose is

$$y_1u_1' + y_2u_2' = 0. \tag{4}$$

This equation is an assumption that is made to simplify the first derivative and, thereby, the second derivative of y_p. Using the product rule to differentiate y_p, we get

$$y_p' = u_1y_1' + y_1u_1' + u_2y_2' + y_2u_2' = u_1y_1' + u_2y_2' + \underbrace{y_1u_1' + y_2u_2'}_{\text{zero from (4)}} \tag{5}$$

and so

$$y_p' = u_1y_1' + u_2y_2'.$$

Continuing, we find $y_p'' = u_1 y_1'' + y_1' u_1' + u_2 y_2'' + y_2' u_2'$.

Substitution of these results in (3) yields

$$y_p'' + Py_p' + Qy_p = u_1 y_1'' + y_1' u_1' + u_2 y_2'' + y_2' u_2' + Pu_1 y_1' + Pu_2 y_2' + Qu_1 y_1 + Qu_2 y_2$$

$$= u_1 \underbrace{[y_1'' + Py_1' + Qy_1]}_{\text{zero}} + u_2 \underbrace{[y_2'' + Py_2' + Qy_2]}_{\text{zero}} + y_1' u_1' + y_2' u_2' = f(x).$$

In other words, u_1 and u_2 must be functions that also satisfy the condition

$$y_1' u_1' + y_2' u_2' = f(x). \qquad \textbf{(6)}$$

Equations (4) and (6) constitute a linear system of equations for determining the derivatives u_1' and u_2'. By Cramer's rule, the solution of

$$y_1 u_1' + y_2 u_2' = 0$$

$$y_1' u_1' + y_2' u_2' = f(x)$$

can be expressed in terms of determinants:

$$u_1' = \frac{W_1}{W} \quad \text{and} \quad u_2' = \frac{W_2}{W}, \qquad \textbf{(7)}$$

where

$$W = \begin{vmatrix} y_1 & y_2 \\ y_1' & y_2' \end{vmatrix}, \quad W_1 = \begin{vmatrix} 0 & y_2 \\ f(x) & y_2' \end{vmatrix}, \quad \text{and} \quad W_2 = \begin{vmatrix} y_1 & 0 \\ y_1' & f(x) \end{vmatrix}. \qquad \textbf{(8)}$$

The determinant W is recognized as the Wronskian of y_1 and y_2. By linear independence of y_1 and y_2 on I, we know that $W(y_1(x), y_2(x)) \neq 0$ for every x in the interval.

Summary of the Method

Usually it is not a good idea to memorize formulas in lieu of understanding a procedure. However, the foregoing procedure is too long and complicated to use each time we wish to solve a differential equation. In this case it is more efficient to simply use the formulas in (7). Thus, to solve $a_2 y'' + a_1 y' + a_0 y = g(x)$, first find the complementary function $y_c = c_1 y_1 + c_2 y_2$ and then compute the Wronskian $W(y_1(x), y_2(x))$. By dividing by a_2, we put the equation into the standard form $y'' + Py' + Qy = f(x)$ to determine $f(x)$. We find u_1 and u_2 by integrating $u_1' = W_1/W$ and $u_2' = W_2/W$, where W_1 and W_2 are defined in (8). A particular solution is $y_p = u_1 y_1 + u_2 y_2$. The general solution of the equation is then $y = y_c + y_p$.

EXAMPLE 1 General Solution Using Variation of Parameters

Solve $\quad y'' - 4y' + 4y = (x + 1)e^{2x}.$

Solution From the auxiliary equation $m^2 - 4m + 4 = (m - 2)^2 = 0$ we have $y_c = c_1e^{2x} + c_2xe^{2x}$. With the identifications $y_1 = e^{2x}$ and $y_2 = xe^{2x}$, we next compute the Wronskian:

$$W(e^{2x}, xe^{2x}) = \begin{vmatrix} e^{2x} & xe^{2x} \\ 2e^{2x} & 2xe^{2x} + e^{2x} \end{vmatrix} = e^{4x}.$$

Since the given differential equation is already in form (3) (that is, the coefficient of y'' is 1), we identify $f(x) = (x + 1)e^{2x}$. From (8) we obtain

$$W_1 = \begin{vmatrix} 0 & xe^{2x} \\ (x+1)e^{2x} & 2xe^{2x} + e^{2x} \end{vmatrix} = -(x+1)xe^{4x}, \quad W_2 = \begin{vmatrix} e^{2x} & 0 \\ 2e^{2x} & (x+1)e^{2x} \end{vmatrix} = (x+1)e^{4x},$$

and so from (7),

$$u_1' = -\frac{(x+1)xe^{4x}}{e^{4x}} = -x^2 - x, \quad u_2' = \frac{(x+1)e^{4x}}{e^{4x}} = x + 1.$$

It follows that $\quad u_1 = -\dfrac{x^3}{3} - \dfrac{x^2}{2}$ and $u_2 = \dfrac{x^2}{2} + x.$

Hence $\quad y_p = \left(-\dfrac{x^3}{3} - \dfrac{x^2}{2}\right)e^{2x} + \left(\dfrac{x^2}{2} + x\right)xe^{2x} = \left(\dfrac{x^3}{6} + \dfrac{x^2}{2}\right)e^{2x}$

and $\quad y = y_c + y_p = c_1e^{2x} + c_2xe^{2x} + \left(\dfrac{x^3}{6} + \dfrac{x^2}{2}\right)e^{2x}.$ ●

EXAMPLE 2 General Solution Using Variation of Parameters

Solve $\quad 4y'' + 36y = \csc 3x.$

Solution We first put the equation in the standard form (3) by dividing by 4:

$$y'' + 9y = \frac{1}{4}\csc 3x.$$

Since the roots of the auxiliary equation $m^2 + 9 = 0$ are $m_1 = 3i$ and $m_2 = -3i$, the complementary function is $y_c = c_1 \cos 3x + c_2 \sin 3x$. Using $y_1 = \cos 3x$, $y_2 = \sin 3x$, and $f(x) = (1/4) \csc x$, we obtain

$$W(\cos 3x, \sin 3x) = \begin{vmatrix} \cos 3x & \sin 3x \\ -3\sin 3x & 3\cos 3x \end{vmatrix} = 3$$

$$W_1 = \begin{vmatrix} 0 & \sin 3x \\ \frac{1}{4}\csc 3x & 3\cos 3x \end{vmatrix} = -\frac{1}{4}, \quad W_2 = \begin{vmatrix} \cos 3x & 0 \\ -3\sin 3x & \frac{1}{4}\csc 3x \end{vmatrix} = \frac{1}{4}\frac{\cos 3x}{\sin 3x}$$

Integrating $\quad u_1' = \dfrac{W_1}{W} = -\dfrac{1}{12} \quad$ and $\quad u_2' = \dfrac{W_2}{W} = \dfrac{1}{12}\dfrac{\cos 3x}{\sin 3x}$

gives $\quad u_1 = -\dfrac{1}{12}x \quad$ and $\quad u_2 = \dfrac{1}{36}\ln|\sin 3x|.$

Thus a particular solution is

$$y_p = -\frac{1}{12}x\cos 3x + \frac{1}{36}(\sin 3x)\ln|\sin 3x|.$$

The general solution of the equation is

$$y = y_c + y_p = c_1\cos 3x + c_2\sin 3x - \frac{1}{12}x\cos 3x + \frac{1}{36}(\sin 3x)\ln|\sin 3x|. \quad \textbf{(9)}$$

●

Equation (9) represents the general solution of the differential equation on, say, the interval $(0, \pi/6)$.

Constants of Integration

When computing the indefinite integrals of u_1' and u_2', we need not introduce any constants. This is because

$$\begin{aligned} y = y_c + y_p &= c_1y_1 + c_2y_2 + (u_1 + a_1)y_1 + (u_2 + b_1)y_2 \\ &= (c_1 + a_1)y_1 + (c_2 + b_1)y_2 + u_1y_1 + u_2y_2 \\ &= C_1y_1 + C_2y_2 + u_1y_1 + u_2y_2. \end{aligned}$$

EXAMPLE 3 General Solution Using VP

Solve $\quad y'' - y = \dfrac{1}{x}.$

Solution The auxiliary equation $m^2 - 1 = 0$ yields $m_1 = -1$ and $m_2 = 1$. Therefore $y_c = c_1e^x + c_2e^{-x}$. Now $W(e^x, e^{-x}) = -2$ and

$$u_1' = -\frac{e^{-x}(1/x)}{-2}, \qquad u_1 = \frac{1}{2}\int_{x_0}^{x}\frac{e^{-t}}{t}\,dt$$

$$u_2' = \frac{e^x(1/x)}{-2}, \qquad u_2 = -\frac{1}{2}\int_{x_0}^{x}\frac{e^{t}}{t}\,dt.$$

It is well known that the integrals defining u_1 and u_2 cannot be expressed in terms of elementary functions. Hence we write

$$y_p = \frac{1}{2}e^x\int_{x_0}^{x}\frac{e^{-t}}{t}\,dt - \frac{1}{2}e^{-x}\int_{x_0}^{x}\frac{e^{t}}{t}\,dt,$$

and so

$$y = y_c + y_p = c_1 e^x + c_2 e^{-x} + \frac{1}{2} e^x \int_{x_0}^{x} \frac{e^{-t}}{t}\, dt - \frac{1}{2} e^{-x} \int_{x_0}^{x} \frac{e^t}{t}\, dt.$$

In Example 3 we can integrate on any interval $x_0 \le t \le x$ not containing the origin.

Higher-Order Equations

The method we have just examined for nonhomogeneous second-order differential equations can be generalized to nth-order linear equations that have been put into the form

$$y^{(n)} + P_{n-1}(x)y^{(n-1)} + \cdots + P_1(x)y' + P_0(x)y = f(x). \quad \textbf{(10)}$$

If $y = c_1y_1 + c_2y_2 + \cdots + c_ny_n$ is the complementary function for (10), then a particular solution is

$$y_p = u_1(x)y_1(x) + u_2(x)y_2(x) + \cdots + u_n(x)y_n(x),$$

where the u_k', $k = 1, 2, \ldots, n$, are determined by the n equations

$$\begin{aligned}
y_1u_1' + \quad y_2u_2' + \cdots + \quad y_nu_n' &= 0\\
y_1'u_1' + \quad y_2'u_2' + \cdots + \quad y_n'u_n' &= 0\\
&\vdots\\
y_1^{(n-1)}u_1' + y_2^{(n-1)}u_2' + \cdots + y_n^{(n-1)}u_n' &= f(x).
\end{aligned}$$

The first $n - 1$ equations in this system, like (4), are assumptions made to simplify the first $n - 1$ derivatives of y_p. The last equation of the system results from substituting the nth derivative of y_p and the simplified lower derivatives into (10). In this case, Cramer's rule gives

$$u_k' = \frac{W_k}{W}, \qquad k = 1, 2, \ldots, n,$$

where W is the Wronskian of $y_1, y_2, \ldots, y_n$, and W_k is the determinant obtained by replacing the kth column of the Wronskian by the column

$$\begin{matrix} 0 \\ 0 \\ \vdots \\ 0 \\ f(x). \end{matrix}$$

When $n = 2$ we get (7).

Remarks **(i)** Variation of parameters has a distinct advantage over the method of undetermined coefficients in that it will *always* yield a particular solution y_p, provided the related homogeneous equation can be solved. The present method is not limited to a function $f(x)$, which is a combination of the four types of functions listed on page 123. Also, variation of parameters, unlike undetermined coefficients, is applicable to differential equations with variable coefficients.

(ii) In the problems that follow do not hesitate to simplify the form of y_p. Depending on how the antiderivatives of u_1' and u_2' are found, you may not obtain the same y_p as given in the answer section. For example, in Problem 3, both $y_p = \frac{1}{2}\sin x - \frac{1}{2}x\cos x$ and $y_p = \frac{1}{4}\sin x - \frac{1}{2}x\cos x$ are valid answers. In either case, the general solution $y = y_c + y_p$ simplifies to $y = c_1\cos x + c_2\sin x - \frac{1}{2}x\cos x$. Why?

(iii) In Problems 25–28 you are asked to solve initial-value problems. Be sure to apply the initial conditions to the general solution $y = y_c + y_p$. Students often make the mistake of applying the initial conditions to only the complementary function y_c since it is that part of the solution that contains the constants.

EXERCISES 3.5

Answers to odd-numbered problems begin on page AN-13.

In Problems 1–24 solve each differential equation by variation of parameters. State an interval on which the general solution is defined.

1. $y'' + y = \sec x$
2. $y'' + y = \tan x$
3. $y'' + y = \sin x$
4. $y'' + y = \sec x \tan x$
5. $y'' + y = \cos^2 x$
6. $y'' + y = \sec^2 x$
7. $y'' - y = \cosh x$
8. $y'' - y = \sinh 2x$
9. $y'' - 4y = e^{2x}/x$
10. $y'' - 9y = 9x/e^{3x}$
11. $y'' + 3y' + 2y = 1/(1 + e^x)$
12. $y'' - 3y' + 2y = e^{3x}/(1 + e^x)$
13. $y'' + 3y' + 2y = \sin e^x$
14. $y'' - 2y' + y = e^x \arctan x$
15. $y'' - 2y' + y = e^x/(1 + x^2)$
16. $y'' - 2y' + 2y = e^x \sec x$
17. $y'' + 2y' + y = e^{-x}\ln x$
18. $y'' + 10y' + 25y = e^{-10x}/x^2$
19. $3y'' - 6y' + 30y = e^x \tan 3x$
20. $4y'' - 4y' + y = e^{x/2}\sqrt{1 - x^2}$
21. $y''' + y' = \tan x$
22. $y''' + 4y' = \sec 2x$
23. $y''' - 2y'' - y' + 2y = e^{3x}$
24. $2y''' - 6y'' = x^2$

In Problems 25–28 solve each differential equation by variations of parameters subject to the initial conditions $y(0) = 1$, $y'(0) = 0$.

25. $4y'' - y = xe^{x/2}$
26. $2y'' + y' - y = x + 1$
27. $y'' + 2y' - 8y = 2e^{-2x} - e^{-x}$
28. $y'' - 4y' + 4y = (12x^2 - 6x)e^{2x}$
29. Given that $y_1 = x^{-1/2}\cos x$ and $y_2 = x^{-1/2}\sin x$, form a fundamental set of solutions of $x^2y'' + xy' + (x^2 - \frac{1}{4})y = 0$ on $(0, \infty)$. Find the general solution of
$$x^2y'' + xy' + \left(x^2 - \frac{1}{4}\right)y = x^{3/2}.$$
30. Given that $y_1 = \cos(\ln x)$ and $y_2 = \sin(\ln x)$ are known linearly independent solutions of $x^2y'' + xy' + y = 0$ on $(0, \infty)$:
   (a) Find a particular solution of
   $$x^2y'' + xy' + y = \sec(\ln x).$$
   (b) Give the general solution of the equation and state an interval of validity. [*Hint:* It is *not* $(0, \infty)$. Why?]

31. **PROJECT PROBLEM** Show how the methods of undetermined coefficients and variation of parameters can be combined to solve the differential equation

$$y'' + 2y' + y = 4x^2 - 3 + \frac{e^{-x}}{x}.$$

32. **PROJECT PROBLEM**

(a) Find a solution of the form $y_1 = e^{mx}$ for

$$xy'' - (5x + 1)y' + 5y = 0.$$

(b) Find the general solution of

$$xy'' - (5x + 1)y' + 5y = x^2,\ x > 0.$$

PROJECT PROBLEM

(a) Suppose y_1 and y_2 form a fundamental set of solutions of $y'' + P(x)y' + Q(x)y = 0$. Show that a particular solution of $y'' + P(x)y' + Q(x)y = f(x)$ on the interval $[x_0, x]$ is

$$y_p(x) = \int_{x_0}^{x} G(x, t) f(t)\, dt,$$

where

$$G(x, t) = \frac{\begin{vmatrix} y_1(t) & y_2(t) \\ y_1(x) & y_2(x) \end{vmatrix}}{W(t)}$$

is called a **Green's function** for the nonhomogeneous equation. In this case $W(t)$ denotes the Wronskian of $y_1(t)$ and $y_2(t)$.

(b) Verify that the particular solution obtained in Example 3 has the form given in part (a).

(c) Verify that $y_p(x)$ defined in part (a) satisfies $y_p(x_0) = 0$ and $y_p'(x_0) = 0$.

(d) Use a Green's function to solve the initial-value problem

$$y'' + \omega^2 y = \sin \omega x, \qquad y(0) = 0, \quad y'(0) = 0.$$

(e) Show how the particular solution obtained in part (d) can be used to solve the initial-value problem

$$y'' + \omega^2 y = \sin \omega x, \qquad y(0) = 1, \quad y'(0) = -1.$$

[*Hint*: First consider the problem $y'' + \omega^2 y = 0$, $y(0) = 1$, $y'(0) = -1$, then think about what "superposition" means.]

3.6 CAUCHY–EULER EQUATION

Introduction

Up to now we have solved linear differential equations of order two or higher only when the equation had constant coefficients. We shall see in Chapter 6 that when a linear differential equation has variable coefficients, the best that we can *usually* expect is to find an infinite series solution. However, the type of equation studied in this section is an exception to this rule; it is an equation with variable coefficients whose general solution can always be expressed in terms of powers of x, sines, cosines, logarithms, and exponentials. The method of solution for this equation is quite similar to that for constant-coefficient equations.

Cauchy–Euler or Equidimensional Equation

Any differential equation of the form

$$a_n x^n \frac{d^n y}{dx^n} + a_{n-1} x^{n-1} \frac{d^{n-1} y}{dx^{n-1}} + \cdots + a_1 x \frac{dy}{dx} + a_0 y = g(x)$$

where the coefficients $a_n, a_{n-1}, \ldots, a_0$ are constants, is known diversely as a **Cauchy–Euler equation**, an **Euler–Cauchy equation**, an **Euler equation**, or an **equidimensional equation**. The observable characteristic of this type of equation is that the degree $k = n, n - 1, \ldots, 1, 0$ of the monomial coefficients x^k matches the order k of differentiation d^ky/dx^k:

$$\underbrace{a_n x^n \frac{d^n y}{dx^n}}_{\text{same}} + \underbrace{a_{n-1} x^{n-1} \frac{d^{n-1} y}{dx^{n-1}}}_{\text{same}} + \cdots.$$

To start the discussion, we confine our attention to solving the homogeneous second-order equation

$$ax^2 \frac{d^2y}{dx^2} + bx\frac{dy}{dx} + cy = 0.$$

The solution of higher-order equations follows analogously. Also, we can solve the nonhomogeneous equation $ax^2y'' + bxy' + cy = g(x)$ by variation of parameters, once we have determined the complementary function $y_c(x)$.

Note The coefficient of d^2y/dx^2 is zero at $x = 0$. Hence in order to guarantee that the fundamental results of Theorem 3.1 are applicable to the Cauchy-Euler equation, we confine our attention to finding the general solution on the interval $(0, \infty)$. Solutions on the interval $(-\infty, 0)$ can be obtained by substituting $t = -x$ in the differential equation.

Method of Solution

We try a solution of the form $y = x^m$, where m is to be determined. The first and second derivatives are, respectively,

$$\frac{dy}{dx} = mx^{m-1} \quad \text{and} \quad \frac{d^2y}{dx^2} = m(m-1)x^{m-2}.$$

Consequently

$$\begin{aligned} ax^2\frac{d^2y}{dx^2} + bx\frac{dy}{dx} + cy &= ax^2 \cdot m(m-1)x^{m-2} + bx \cdot mx^{m-1} + cx^m \\ &= am(m-1)x^m + bmx^m + cx^m = x^m(am(m-1) + bm + c). \end{aligned}$$

Thus $y = x^m$ is a solution of the differential equation whenever m is a solution of the **auxiliary equation**

$$am(m-1) + bm + c = 0 \quad \text{or} \quad am^2 + (b-a)m + c = 0. \qquad \textbf{(1)}$$

There are three different cases to be considered, depending on whether the roots of this quadratic equation are real and distinct, real and equal, or complex. In the last case, the roots appear as a conjugate pair.

CASE I **Distinct Real Roots** Let m_1 and m_2 denote the real roots of (1) such that $m_1 \neq m_2$. Then $y_1 = x^{m_1}$ and $y_2 = x^{m_2}$ form a fundamental set of solutions. Hence the general solution is

$$y = c_1x^{m_1} + c_2x^{m_2}. \tag{2}$$

EXAMPLE 1 C-E Equation: Distinct Roots

Solve $\quad x^2\dfrac{d^2y}{dx^2} - 2x\dfrac{dy}{dx} - 4y = 0.$

Solution Rather than just memorizing equation (1), it is preferable to assume $y = x^m$ as the solution a few times in order to understand the origin and the difference between this new form of the auxiliary equation and that obtained in Section 3.3. Differentiate twice

$$\frac{dy}{dx} = mx^{m-1}, \quad \frac{d^2y}{dx^2} = m(m-1)x^{m-2}$$

and substitute back into the differential equation:

$$x^2\frac{d^2y}{dx^2} - 2x\frac{dy}{dx} - 4y = x^2 \cdot m(m-1)x^{m-2} - 2x \cdot mx^{m-1} - 4x^m$$

$$= x^m(m(m-1) - 2m - 4) = x^m(m^2 - 3m - 4) = 0$$

if $m^2 - 3m - 4 = 0$. Now $(m+1)(m-4) = 0$ implies $m_1 = -1$, $m_2 = 4$ so that

$$y = c_1x^{-1} + c_2x^4.$$

CASE II **Repeated Real Roots** If the roots of (1) are repeated, that is, $m_1 = m_2$, then we obtain only one solution, namely, $y = x^{m_1}$. When the roots of the quadratic equation $am^2 + (b-a)m + c = 0$ are equal, the discriminant of the coefficients is necessarily zero. It follows from the quadratic formula that the root must be $m_1 = -(b-a)/2a$.

Now we can construct a second solution y_2, using (5) of Section 3.2. We first write the Cauchy–Euler equation in the form

$$\frac{d^2y}{dx^2} + \frac{b}{ax}\frac{dy}{dx} + \frac{c}{ax^2}y = 0$$

and make the identifications $P(x) = b/ax$ and $\int (b/ax)\,dx = (b/a)\ln x$. Thus

$$y_2 = x^{m_1}\int \frac{e^{-(b/a)\ln x}}{x^{2m_1}}\,dx$$

$$= x^{m_1}\int x^{-b/a} \cdot x^{-2m_1}\,dx \qquad \leftarrow e^{-(b/a)\ln x} = e^{\ln x^{-b/a}} = x^{-b/a}$$

$$= x^{m_1}\int x^{-b/a} \cdot x^{(b-a)/a}\,dx \qquad \leftarrow -2m_1 = (b-a)/a$$

$$= x^{m_1}\int \frac{dx}{x} = x^{m_1}\ln x.$$

The general solution is then

$$y = c_1x^{m_1} + c_2x^{m_1}\ln x. \tag{3}$$

EXAMPLE 2 C-E Equation: Repeated Roots

Solve $\quad 4x^2\dfrac{d^2y}{dx^2} + 8x\dfrac{dy}{dx} + y = 0.$

Solution The substitution $y = x^m$ yields

$$4x^2\frac{d^2y}{dx^2} + 8x\frac{dy}{dx} + y = x^m(4m(m-1) + 8m + 1) = x^m(4m^2 + 4m + 1) = 0$$

when $4m^2 + 4m + 1 = 0$ or $(2m + 1)^2 = 0$. Since $m_1 = -1/2$, the general solution is

$$y = c_1x^{-1/2} + c_2x^{-1/2}\ln x.$$

●

For higher-order equations, if m_1 is a root of multiplicity k, then it can be shown that

$$x^{m_1}, \quad x^{m_1}\ln x, \quad x^{m_1}(\ln x)^2, \quad \ldots, \quad x^{m_1}(\ln x)^{k-1}$$

are k linearly independent solutions. Correspondingly, the general solution of the differential equation must then contain a linear combination of these k solutions.

CASE III **Conjugate Complex Roots** If the roots of (1) are the conjugate pair $m_1 = \alpha + i\beta$, $m_2 = \alpha - i\beta$, where α and $\beta > 0$ are real, then a solution is

$$y = C_1x^{\alpha + i\beta} + C_2x^{\alpha - i\beta}.$$

But, as in the case of equations with constant coefficients, when the roots of the auxiliary equation are complex, we wish to write the solution in terms of real functions only. We note the identity

$$x^{i\beta} = (e^{\ln x})^{i\beta} = e^{i\beta\ln x},$$

which, by Euler's formula, is the same as

$$x^{i\beta} = \cos(\beta\ln x) + i\sin(\beta\ln x).$$

Similarly, $\quad x^{-i\beta} = \cos(\beta\ln x) - i\sin(\beta\ln x).$

Adding and subtracting the last two results yield

$$x^{i\beta} + x^{-i\beta} = 2\cos(\beta\ln x) \quad \text{and} \quad x^{i\beta} - x^{-i\beta} = 2i\sin(\beta\ln x),$$

respectively. From the fact that $y = C_1x^{\alpha + i\beta} + C_2x^{\alpha - i\beta}$ is a solution for any values of the constants, we see, in turn, for $C_1 = C_2 = 1$ and $C_1 = 1$, $C_2 = -1$ that

$$y_1 = x^\alpha(x^{i\beta} + x^{-i\beta}) \quad \text{and} \quad y_2 = x^\alpha(x^{i\beta} - x^{-i\beta})$$

or $\qquad y_1 = 2x^{\alpha}\cos(\beta \ln x) \quad \text{and} \quad y_2 = 2ix^{\alpha}\sin(\beta \ln x)$

are also solutions. Since $W(x^{\alpha}\cos(\beta \ln x), x^{\alpha}\sin(\beta \ln x)) = \beta x^{2\alpha - 1} \neq 0, \beta > 0,$ on the interval $(0, \infty)$, we conclude that

$$y_1 = x^{\alpha}\cos(\beta \ln x) \quad \text{and} \quad y_2 = x^{\alpha}\sin(\beta \ln x)$$

constitute a fundamental set of real solutions of the differential equation. Hence, the general solution is

$$y = x^{\alpha}[c_1\cos(\beta \ln x) + c_2\sin(\beta \ln x)]. \tag{4}$$

EXAMPLE 3 An Initial-Value Problem

Solve the initial-value problem

$$x^2\frac{d^2y}{dx^2} + 3x\frac{dy}{dx} + 3y = 0, \qquad y(1) = 1, \quad y'(1) = -5.$$

Solution We have

$$x^2\frac{d^2y}{dx^2} + 3x\frac{dy}{dx} + 3y = x^m(m(m-1) + 3m + 3) = x^m(m^2 + 2m + 3) = 0$$

when $m^2 + 2m + 3 = 0$. From the quadratic formula we find $m_1 = -1 + \sqrt{2}i$ and $m_2 = -1 - \sqrt{2}i$. If we make the identifications $\alpha = -1$ and $\beta = \sqrt{2}$, we see from (4) that the general solution of the differential equation is

$$y = x^{-1}[c_1\cos(\sqrt{2}\ln x) + c_2\sin(\sqrt{2}\ln x)].$$

By applying the conditions $y(1) = 1$, $y'(1) = -5$ to the foregoing solution, we find, in turn, $c_1 = 1$ and $c_2 = -2\sqrt{2}$. Thus the solution to the initial-value problem is

$$y = x^{-1}[\cos(\sqrt{2}\ln x) - 2\sqrt{2}\sin(\sqrt{2}\ln x)].$$

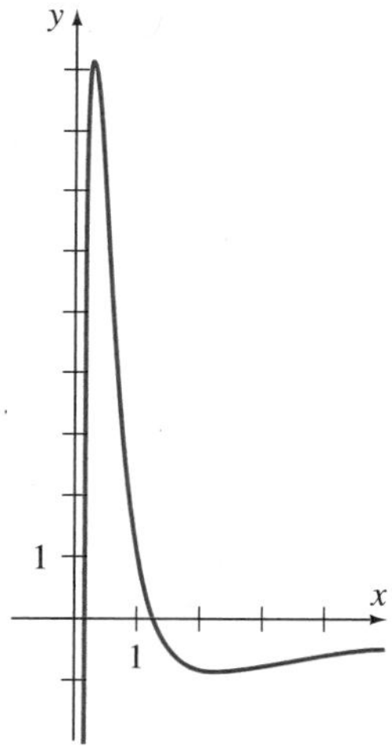

FIGURE 3.8

The graph of this solution, obtained with the aid of computer software, is given in Figure 3.8. ●

The next example illustrates the solution of a third-order Cauchy–Euler equation.

EXAMPLE 4 Third-Order Cauchy–Euler Equation

Solve $\qquad x^3\frac{d^3y}{dx^3} + 5x^2\frac{d^2y}{dx^2} + 7x\frac{dy}{dx} + 8y = 0.$

Solution The first three derivatives of $y = x^m$ are

$$\frac{dy}{dx} = mx^{m-1}, \quad \frac{d^2y}{dx^2} = m(m-1)x^{m-2}, \quad \frac{d^3y}{dx^3} = m(m-1)(m-2)x^{m-3},$$

so that the given differential equation becomes

$$x^3\frac{d^3y}{dx^3} + 5x^2\frac{d^2y}{dx^2} + 7x\frac{dy}{dx} + 8y = x^3m(m-1)(m-2)x^{m-3} + 5x^2m(m-1)x^{m-2} + 7xmx^{m-1} + 8x^m$$

$$= x^m(m(m-1)(m-2) + 5m(m-1) + 7m + 8)$$

$$= x^m(m^3 + 2m^2 + 4m + 8) = x^m(m+2)(m^2+4) = 0.$$

In this case we see that $y = x^m$ will be a solution of the differential equation for $m_1 = -2$, $m_2 = 2i$, and $m_3 = -2i$. Hence the general solution is

$$y = c_1x^{-2} + c_2\cos(2\ln x) + c_3\sin(2\ln x).$$ ●

Because the method of undetermined coefficients is applicable only to differential equations with constant coefficients it cannot be applied *directly* to a nonhomogeneous Cauchy–Euler equation (see Problems 35–40 in Exercises 3.6). In our last example, the method of variation of parameters is employed.

EXAMPLE 5 Using Variation of Parameters

Solve the nonhomogeneous equation

$$x^2y'' - 3xy' + 3y = 2x^4e^x.$$

Solution The substitution $y = x^m$ leads to the auxiliary equation

$$m(m-1) - 3m + 3 = 0 \quad \text{or} \quad (m-1)(m-3) = 0.$$

Thus $$y_c = c_1x + c_2x^3.$$

Before using variation of parameters to find a particular solution $y_p = u_1y_1 + u_2y_2$, recall that the formulas $u_1' = W_1/W$ and $u_2' = W_2/W$, where W_1, W_2, and W are the determinants defined on page 134, were derived under the assumption that the differential equation has been put into the special form $y'' + P(x)y' + Q(x)y = f(x)$. Therefore we divide the given equation by x^2, and from

$$y'' - \frac{3}{x}y' + \frac{3}{x^2}y = 2x^2e^x$$

we make the identification $f(x) = 2x^2e^x$. Now with $y_1 = x$, $y_2 = x^3$, and

$$W = \begin{vmatrix} x & x^3 \\ 1 & 3x^2 \end{vmatrix} = 2x^3, \quad W_1 = \begin{vmatrix} 0 & x^3 \\ 2x^2e^x & 3x^2 \end{vmatrix} = -2x^5e^x, \quad W_2 = \begin{vmatrix} x & 0 \\ 1 & 2x^2e^x \end{vmatrix} = 2x^3e^x$$

we find $$u_1' = -\frac{2x^5e^x}{2x^3} = -x^2e^x \quad \text{and} \quad u_2' = \frac{2x^3e^x}{2x^3} = e^x.$$

The integral of the latter function is immediate, but in the case of u_1' we integrate by parts twice. The results are $u_1 = -x^2e^x + 2xe^x - 2e^x$ and $u_2 = e^x$. Hence

$$y_p = u_1y_1 + u_2y_2 = (-x^2e^x + 2xe^x - 2e^x)x + e^xx^3 = 2x^2e^x - 2xe^x.$$

Finally we have $y = y_c + y_p = c_1x + c_2x^3 + 2x^2e^x - 2xe^x$. ●

EXERCISES 3.6

Answers to odd-numbered problems begin on page AN-14.

In Problems 1–22 solve the given differential equation.

1. $x^2y'' - 2y = 0$
2. $4x^2y'' + y = 0$
3. $xy'' + y' = 0$
4. $xy'' - y' = 0$
5. $x^2y'' + xy' + 4y = 0$
6. $x^2y'' + 5xy' + 3y = 0$
7. $x^2y'' - 3xy' - 2y = 0$
8. $x^2y'' + 3xy' - 4y = 0$
9. $25x^2y'' + 25xy' + y = 0$
10. $4x^2y'' + 4xy' - y = 0$
11. $x^2y'' + 5xy' + 4y = 0$
12. $x^2y'' + 8xy' + 6y = 0$
13. $x^2y'' - xy' + 2y = 0$
14. $x^2y'' - 7xy' + 41y = 0$
15. $3x^2y'' + 6xy' + y = 0$
16. $2x^2y'' + xy' + y = 0$
17. $x^3y''' - 6y = 0$
18. $x^3y''' + xy' - y = 0$
19. $x^3\dfrac{d^3y}{dx^3} - 2x^2\dfrac{d^2y}{dx^2} - 2x\dfrac{dy}{dx} + 8y = 0$
20. $x^3\dfrac{d^3y}{dx^3} - 2x^2\dfrac{d^2y}{dx^2} + 4x\dfrac{dy}{dx} - 4y = 0$
21. $x\dfrac{d^4y}{dx^4} + 6\dfrac{d^3y}{dx^3} = 0$
22. $x^4\dfrac{d^4y}{dx^4} + 6x^3\dfrac{d^3y}{dx^3} + 9x^2\dfrac{d^2y}{dx^2} + 3x\dfrac{dy}{dx} + y = 0$

In Problems 23–26 solve the given differential equation subject to the indicated initial conditions.

23. $x^2y'' + 3xy' = 0, \quad y(1) = 0, \; y'(1) = 4$
24. $x^2y'' - 5xy' + 8y = 0, \quad y(2) = 32, \; y'(2) = 0$
25. $x^2y'' + xy' + y = 0, \quad y(1) = 1, \; y'(1) = 2$
26. $x^2y'' - 3xy' + 4y = 0, \quad y(1) = 5, \; y'(1) = 3$

In Problems 27 and 28 solve the given differential equation subject to the indicated initial conditions. [*Hint:* Let $t = -x$.]

27. $4x^2y'' + y = 0, \quad y(-1) = 2, \; y'(-1) = 4$
28. $x^2y'' - 4xy' + 6y = 0, \quad y(-2) = 8, \; y'(-2) = 0$

Solve Problems 29–34 by variation of parameters.

29. $xy'' + y' = x$
30. $xy'' - 4y' = x^4$
31. $2x^2y'' + 5xy' + y = x^2 - x$
32. $x^2y'' - 2xy' + 2y = x^4e^x$
33. $x^2y'' - xy' + y = 2x$
34. $x^2y'' - 2xy' + 2y = x^3 \ln x$

Any Cauchy–Euler differential equation can be reduced to an equation with constant coefficients by means of the substitution $x = e^t$. In Problems 35–40 solve the given differential equation by means of this substitution and the procedures outlined in Sections 3.3 and 3.4.

35. $x^2\dfrac{d^2y}{dx^2} + 10x\dfrac{dy}{dx} + 8y = x^2$
36. $x^2y'' - 4xy' + 6y = \ln x^2$
37. $x^2y'' - 3xy' + 13y = 4 + 3x$
38. $2x^2y'' - 3xy' - 3y = 1 + 2x + x^2$
39. $x^2y'' + 9xy' - 20y = 5/x^3$
40. $x^3\dfrac{d^3y}{dx^3} - 3x^2\dfrac{d^2y}{dx^2} + 6x\dfrac{dy}{dx} - 6y = 3 + \ln x^3$

3.7 APPLICATIONS: INITIAL-VALUE PROBLEMS

Introduction

In this section we are going to consider several physical systems that change with time t and whose mathematical model is a linear second-order differential with constant coefficients

$$a_2\frac{d^2y}{dt^2} + a_1\frac{dy}{dt} + a_0y = g(t).$$

The values of the variables $y(t)$ and $y'(t)$ at a specific time describe the **state** of the system. The function g is variously called the **input function**, **excitation function**, **forcing function**, or **driving function**. A solution of the differential equation on an interval containing t_0 that satisfies the prescribed conditions $y(t_0) = y_0$ and $y'(t_0) = y_1$ is known as the **output** or **response** of the system.

3.7.1 Spring/Mass Systems: Free Undamped Motion

Hooke's Law

Suppose, as in Figure 3.9(b), that a mass m_1 is attached to a flexible spring suspended from a rigid support. When m_1 is replaced with a different mass m_2, the amount of stretch, or elongation, of the spring will of course be different.

By Hooke's law, the spring itself exerts a restoring force F opposite to the direction of elongation and proportional to the amount of elongation s. Simply stated, $F = ks$, where k is a constant of proportionality called the **spring constant**. Although masses with different weights stretch a spring by different amounts, the spring is essentially characterized by the number k. For example, if a mass weighing 10 lb stretches a spring $\frac{1}{2}$ ft, then $10 = k(\frac{1}{2})$ implies $k = 20$ lb/ft. Necessarily then, a mass weighing, say, 8 lb stretches the same spring $\frac{2}{5}$ ft.

Newton's Second Law

After a mass m is attached to a spring, it stretches the spring by an amount s and attains a position of equilibrium at which its weight W is balanced by the restoring force ks. Recall, weight is defined by $W = mg$, where mass is measured in slugs, kilograms, or grams, and $g = 32$ ft/s², 9.8 m/s², or 980 cm/s², respectively. An indicated in Figure 3.10(b), the condition of equilibrium is $mg = ks$ or $mg - ks = 0$. If the mass is displaced by an amount x from its equilibrium

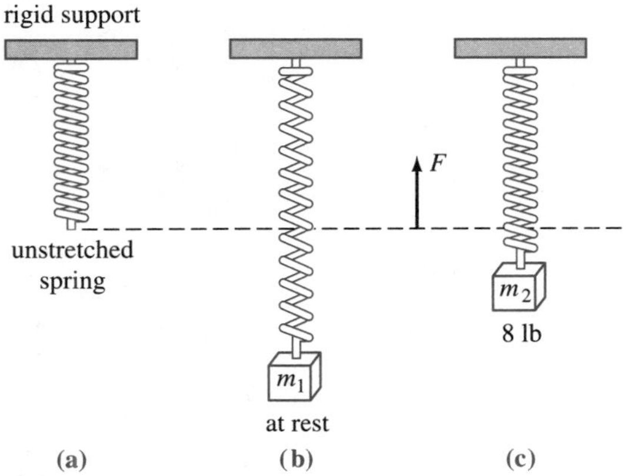

FIGURE 3.9

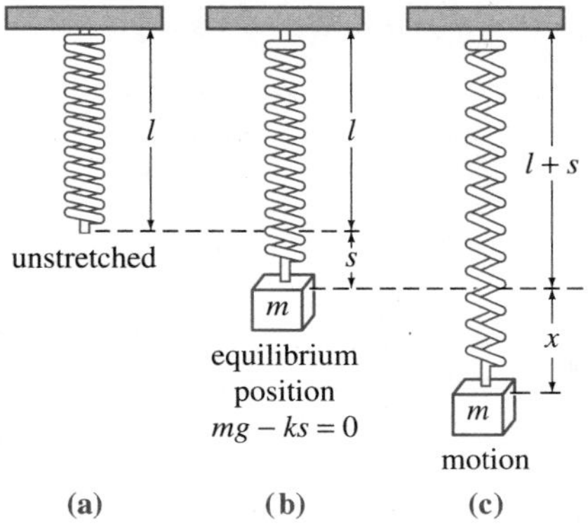

FIGURE 3.10

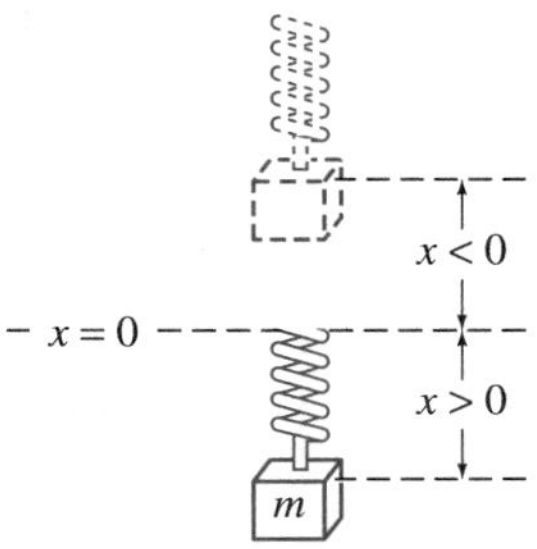

FIGURE 3.11

position, the restoring force of the spring is then $k(x + s)$. Assuming that there are no retarding forces acting on the system, and assuming that the mass vibrates free of other external forces—**free motion**—we can equate Newton's second law with the net or resultant force of the restoring force and the weight:

$$m\frac{d^2x}{dt^2} = -k(s + x) + mg = -kx + \underbrace{mg - ks}_{\text{zero}} = -kx. \tag{1}$$

The negative sign in (1) indicates that the restoring force of the spring acts opposite to the direction of motion. Furthermore, we can adopt the convention that displacements measured *below* the equilibrium position are positive. See Figure 3.11.

Differential Equation of Free Undamped Motion

By dividing (1) by the mass m, we obtain the second-order differential equation $d^2x/dt^2 + (k/m)x = 0$, or

$$\frac{d^2x}{dt^2} + \omega^2 x = 0, \tag{2}$$

where $\omega^2 = k/m$. Equation (2) is said to describe **simple harmonic motion**, or **free undamped motion**. Two obvious initial conditions associated with (2) are $x(0) = \alpha$, the amount of initial displacement, and $x'(0) = \beta$, the initial velocity of the mass. For example, if $\alpha > 0$, $\beta < 0$, the mass starts from a point *below* the equilibrium position with an imparted *upward* velocity. If $\alpha < 0$, $\beta = 0$, the mass is released from *rest* from a point $|\alpha|$ units *above* the equilibrium position, and so on.

Solution and Equation of Motion

To solve equation (2) we note that the solutions of the auxiliary equation $m^2 + \omega^2 = 0$ are the complex numbers $m_1 = \omega i$, $m_2 = -\omega i$. Thus from (8) of Section 3.3, we find the general solution of (2) to be

$$x(t) = c_1 \cos \omega t + c_2 \sin \omega t. \tag{3}$$

The **period** of free vibrations described by (3) is $T = 2\pi/\omega$ and the **frequency** is $f = 1/T = \omega/2\pi$. For example, for $x(t) = 2 \cos 3t - 4 \sin 3t$, the period is $2\pi/3$ and the frequency is $3/2\pi$. The former number means that the graph of $x(t)$ repeats every $2\pi/3$ units; the latter number means that there are 3 cycles of the graph every 2π units or, equivalently, the mass undergoes $3/2\pi$ complete vibrations per unit time. In addition, it can be shown that the period $2\pi/\omega$ is the time interval between two successive maxima of $x(t)$. Keep in mind that a maximum of $x(t)$ is a positive displacement corresponding to the mass attaining a maximum distance *below* the equilibrium position, whereas a minimum of $x(t)$ is a negative displacement corresponding to the mass attaining a maximum height *above* the equilibrium position. We refer to either case as an **extreme displacement** of the mass. Finally, when the initial conditions are used to determine the constants c_1 and c_2 in (3), we say that the resulting particular solution is the **equation of motion**.

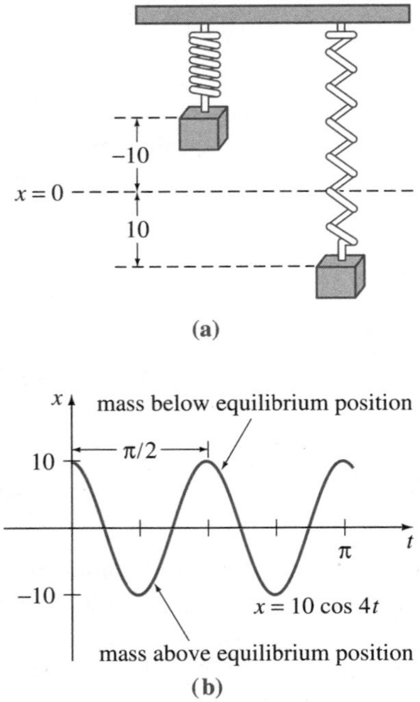

FIGURE 3.12

EXAMPLE 1 Interpretation of an IVP

Solve and interpret the initial-value problem

$$\frac{d^2x}{dt^2} + 16x = 0, \qquad x(0) = 10, \quad x'(0) = 0.$$

Solution The problem is equivalent to pulling a mass on a spring down 10 units below the equilibrium position, holding it until $t = 0$, and then releasing it from rest. Applying the initial conditions to the solution

$$x(t) = c_1 \cos 4t + c_2 \sin 4t$$

gives $x(0) = 10 = c_1 \cdot 1 + c_2 \cdot 0$ so that $c_1 = 10$. Hence

$$x(t) = 10 \cos 4t + c_2 \sin 4t.$$

From $x'(t) = -40 \sin 4t + 4c_2 \cos 4t$ we see that $x'(0) = 0 = 4c_2 \cdot 1$ and so $c_2 = 0$. Therefore, the equation of motion is $x(t) = 10 \cos 4t$.

The solution clearly shows that once the system is set in motion, it stays in motion with the mass bouncing back and forth 10 units on either side of the equilibrium position $x = 0$. As shown in Figure 3.12(b), the period of oscillation is $2\pi/4 = \pi/2$ seconds. ●

EXAMPLE 2 Free Undamped Motion

A mass weighing 2 lb stretches a spring 6 inches. At $t = 0$ the mass is released from a point 8 inches below the equilibrium position with an upward velocity of 4/3 ft/s. Determine the equation of free motion.

Solution Because we are using the engineering system of units, the measurements given in terms of inches must be converted into feet: 6 inches $= \frac{1}{2}$ foot; 8 inches $= \frac{2}{3}$ foot. In addition, we must convert the units of weight given in pounds into units of mass. From $m = W/g$ we have $m = \frac{2}{32} = \frac{1}{16}$ slug. Also, from Hooke's law, $2 = k(\frac{1}{2})$ implies that the spring constant is $k = 4$ lb/ft. Hence (1) gives

$$\frac{1}{16}\frac{d^2x}{dt^2} = -4x \quad \text{or} \quad \frac{d^2x}{dt^2} + 64x = 0.$$

The initial displacement and initial velocity are $x(0) = \frac{2}{3}$, $x'(0) = -\frac{4}{3}$, where the negative sign in the last condition is a consequence of the fact that the mass is given an initial velocity in the negative or upward direction.

Now $\omega^2 = 64$ or $\omega = 8$, so that the general solution of the differential equation is

$$x(t) = c_1 \cos 8t + c_2 \sin 8t. \tag{4}$$

Applying the initial conditions to $x(t)$ and $x'(t)$ gives $c_1 = \frac{2}{3}$ and $c_2 = -\frac{1}{6}$. Thus the equation of motion is

$$x(t) = \frac{2}{3}\cos 8t - \frac{1}{6}\sin 8t. \tag{5}$$

Alternative Form of $x(t)$

When $c_1 \neq 0$ and $c_2 \neq 0$, the actual **amplitude** A of free vibrations is not obvious from inspection of equation (3). For example, although the mass in Example 2 is initially displaced $\frac{2}{3}$ ft beyond the equilibrium position, the amplitude of vibrations is a number larger than $\frac{2}{3}$. Hence it is often convenient to convert a solution of form (3) to the simpler form

$$x(t) = A \sin(\omega t + \phi), \tag{6}$$

where $A = \sqrt{c_1^2 + c_2^2}$ and ϕ is a **phase angle** defined by

$$\left.\begin{aligned} \sin\phi &= \frac{c_1}{A} \\ \cos\phi &= \frac{c_2}{A} \end{aligned}\right\} \tan\phi = \frac{c_1}{c_2}. \tag{7}$$

To verify this we expand (6) by the addition formula for the sine function:

$$A \sin \omega t \cos \phi + A \cos \omega t \sin \phi = (A \sin \phi) \cos \omega t + (A \cos \phi) \sin \omega t. \tag{8}$$

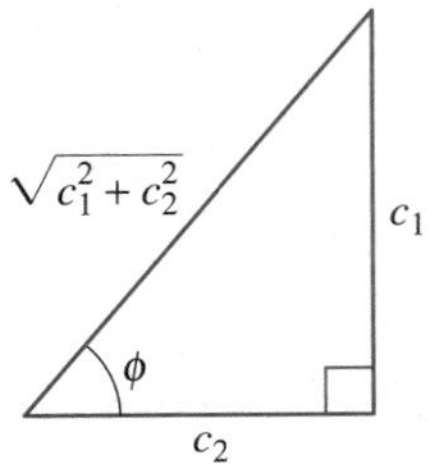

FIGURE 3.13

It follows from Figure 3.13 that if ϕ is defined by

$$\sin \phi = \frac{c_1}{\sqrt{c_1^2 + c_2^2}} = \frac{c_1}{A}, \quad \cos \phi = \frac{c_2}{\sqrt{c_1^2 + c_2^2}} = \frac{c_2}{A},$$

then (8) becomes

$$A\frac{c_1}{A}\cos \omega t + A\frac{c_2}{A}\sin \omega t = c_1 \cos \omega t + c_2 \sin \omega t = x(t).$$

EXAMPLE 3 **Alternative Form of Solution (5)**

In view of the foregoing discussion, we can write the solution (5) in Example 2 as

$$x(t) = \frac{2}{3}\cos 8t - \frac{1}{6}\sin 8t \quad \text{alternatively as} \quad x(t) = A\sin(8t + \phi).$$

The amplitude is given by

$$A = \sqrt{\left(\frac{2}{3}\right)^2 + \left(-\frac{1}{6}\right)^2} = \frac{\sqrt{17}}{6} \approx 0.69 \text{ ft}.$$

You should exercise some care when computing the phase angle ϕ defined by (7). With $c_1 = \frac{2}{3}$ and $c_2 = -\frac{1}{6}$ we find $\tan \phi = -4$, and a calculator then gives $\tan^{-1}(-4) = -1.326$ radians.* But this angle is located in the fourth quadrant and therefore contradicts the fact that $\sin \phi > 0$ and $\cos \phi < 0$ (recall, $c_1 > 0$ and $c_2 < 0$). Hence we must take ϕ to be the second-quadrant angle $\phi = \pi + (-1.326) = 1.816$ radians.

Thus we have

$$x(t) = \frac{\sqrt{17}}{6}\sin(8t + 1.816). \tag{9}$$

Form (6) is very useful since it is easy to find values of time for which the graph of $x(t)$ crosses the positive t-axis (the line $x = 0$). We observe that $\sin(\omega t + \phi) = 0$ when $\omega t + \phi = n\pi$, where n is a nonegative integer.

The Aging Spring

In the preceeding model we assumed an ideal world, a world in which the physical characteristics of the spring/mass system do not change over time. In the nonideal world, however, it seems reasonable to expect that when such a spring/mass system is in motion for a long period the spring would weaken; in other words, the spring "constant" would decay with time. In one model for the **aging spring**, the spring constant k is replaced by the function $K(t) = ke^{-\alpha t}$, $k > 0$, $\alpha > 0$. We leave it to you to investigate the behavior of aging springs in *Computer Lab Experiments* 2 and 3 for this section. The differential equation

* The range of the inverse tangent is $-\pi/2 < \tan^{-1}x < \pi/2$.

$mx'' + ke^{-\alpha t}x = 0$ cannot be solved by the methods considered earlier in this chapter. Nevertheless, we can obtain two linearly independent solutions using the methods in Chapter 6. See Problem 24, Exercises 6.4.

Remarks **(i)** The model in (1) has the form $md^2x/dt^2 + F(x) = 0$, where $F(x) = kx$. Since x denotes the displacement of the mass from its equilibrium position, $F(x) = kx$ is simply Hooke's law; that is, the force exerted by the spring tends to restore the mass to the equilibrium position. A spring acting under a linear restoring force $F(x) = kx$ is naturally referred to as a **linear spring**. But springs are seldom perfectly linear. Depending on how it is constructed and what material is used, a spring can range from ''mushy'' or soft to ''stiff'' or hard, so that its restorative force may vary from something below or above that given by the linear law. In the case of free motion, if we assume that a non-aging spring possesses some nonlinear characteristics, then another reasonable assumption for the restorative force acting on the system is given by the nonlinear function $F(x) = kx + k_1x^3$. A spring whose mathematical model is $md^2x/dt^2 + kx + k_1x^3 = 0$ is called a **nonlinear spring**. Nonlinear springs will be discussed in Section 3.11.

(ii) In general, any spring/mass system whose mathematical model is $md^2x/dt^2 + F(x) = 0$ is a said to be a **conservative system**. For example, the trajectories of the phase portrait of (1) are defined by the one-parameter family of solutions of

$$\frac{dy}{dx} = -\frac{kx}{my},$$

where $y = dx/dt$ is the velocity of the mass. See Example 7, Section 2.2. Separating variables in this first-order equation and integrating then gives

$$\frac{1}{2}my^2 + \frac{1}{2}kx^2 = E,$$

where E is a constant. This last equation expresses the fact that the law of conservation of energy holds in the undamped spring/mass system since the sum of the kinetic energy of the mass ($\frac{1}{2}my^2$) and the potential energy of the spring ($\frac{1}{2}kx^2$) is a constant (E). Thus the trajectories defined by this equation represent various energy states of the system. Recall, that since all the trajectories are closed curves (ellipses), (0, 0) is a stable center of (1) and all solutions are periodic.

3.7.2 Spring/Mass Systems: Free Damped Motion

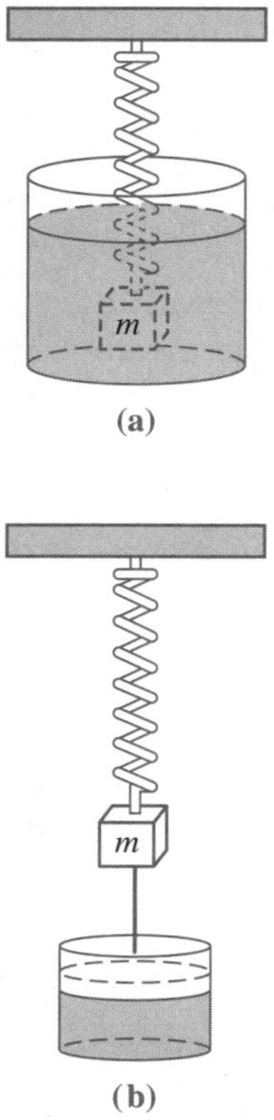

FIGURE 3.14

The discussion of free harmonic motion is somewhat unrealistic since the motion described by equation (1) assumes that no retarding forces are acting on the moving mass. Unless the mass is suspended in a perfect vacuum, there will be at least a resisting force due to the surrounding medium. For example, as Figure 3.14 shows, the mass m could be suspended in a viscous medium or connected to a dashpot damping device.

Differential Equation of Free Damped Motion

In the study of mechanics, damping forces acting on a body are considered to be proportional to a power of the instantaneous velocity. In particular, we shall assume throughout the subsequent discussion that this force is given by a constant multiple of dx/dt. When no other external forces are impressed on the system, it follows from Newton's second law that

$$m\frac{d^2x}{dt^2} = -kx - \beta\frac{dx}{dt}, \tag{10}$$

where β is a positive *damping constant* and the negative sign is a consequence of the fact that the damping force acts in a direction opposite to the motion.

Dividing (10) by the mass m, we find the differential equation of **free damped motion** is $d^2x/dt^2 + (\beta/m)\, dx/dt + (k/m)x = 0$ or

$$\frac{d^2x}{dt^2} + 2\lambda\frac{dx}{dt} + \omega^2 x = 0 \tag{11}$$

where

$$2\lambda = \beta/m, \quad \omega^2 = k/m. \tag{12}$$

The symbol 2λ is used only for algebraic convenience since the auxiliary equation is $m^2 + 2\lambda m + \omega^2 = 0$ and the corresponding roots are then

$$m_1 = -\lambda + \sqrt{\lambda^2 - \omega^2}, \qquad m_2 = -\lambda - \sqrt{\lambda^2 - \omega^2}.$$

We can now distinguish three possible cases depending on the algebraic sign of $\lambda^2 - \omega^2$. Since each solution contains the *damping factor* $e^{-\lambda t}$, $\lambda > 0$, the displacements of the mass become negligible for large time.

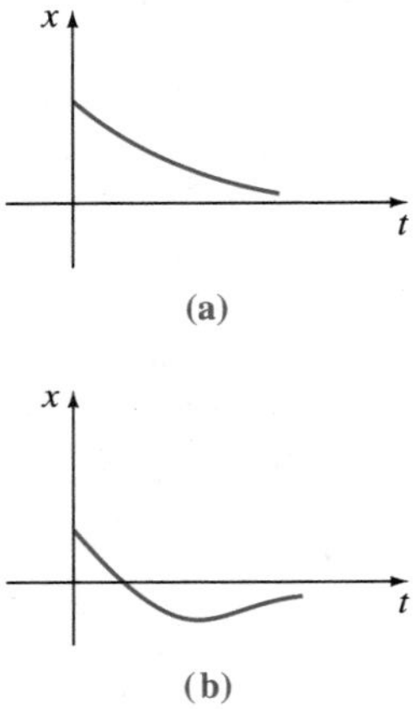

FIGURE 3.15

CASE I $\lambda^2 - \omega^2 > 0$. In this situation the system is said to be **overdamped**, since the damping coefficient β is large when compared to the spring constant k. The corresponding solution of (11) is

$$x(t) = c_1 e^{m_1 t} + c_2 e^{m_2 t}$$

or

$$x(t) = e^{-\lambda t}\left(c_1 e^{\sqrt{\lambda^2 - \omega^2}\,t} + c_2 e^{-\sqrt{\lambda^2 - \omega^2}\,t}\right). \tag{13}$$

This equation represents a smooth and nonoscillatory motion. Figure 3.15 shows two possible graphs of $x(t)$.

CASE II $\lambda^2 - \omega^2 = 0$. The system is said to be **critically damped**, since any slight decrease in the damping force would result in oscillatory motion. The general solution of (11) is

$$x(t) = c_1 e^{m_1 t} + c_2 t e^{m_1 t}$$

or

$$x(t) = e^{-\lambda t}(c_1 + c_2 t). \tag{14}$$

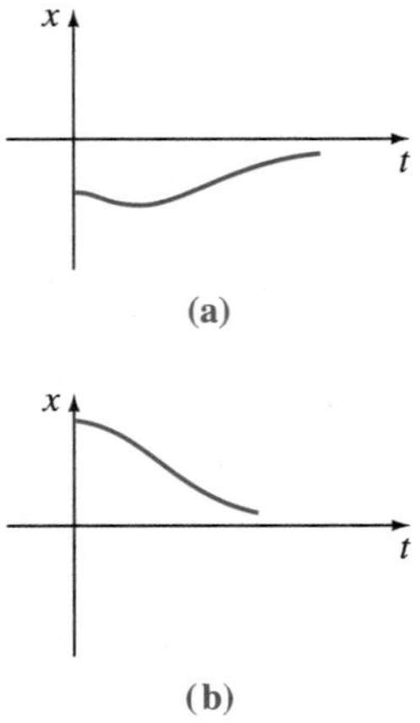

FIGURE 3.16

Some graphs of typical motion are given in Figure 3.16. Notice that the motion is quite similar to that of an overdamped system. It is also apparent from (14) that the mass can pass through the equilibrium position at most one time.

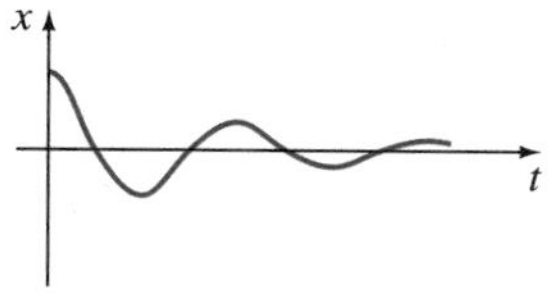

FIGURE 3.17

CASE III $\lambda^2 - \omega^2 < 0.$ In this case the system is said to be **underdamped**, since the damping coefficient is small compared to the spring constant. The roots m_1 and m_2 are now complex:

$$m_1 = -\lambda + \sqrt{\omega^2 - \lambda^2}\, i, \quad m_2 = -\lambda - \sqrt{\omega^2 - \lambda^2}\, i,$$

and so the general solution of equation (11) is

$$x(t) = e^{-\lambda t}(c_1 \cos \sqrt{\omega^2 - \lambda^2}t + c_2 \sin \sqrt{\omega^2 - \lambda^2}t). \tag{15}$$

As indicated in Figure 3.17 the motion described by (15) is oscillatory; but because of the coefficient $e^{-\lambda t}$, the amplitudes of vibration $\to 0$ as $t \to \infty$.

EXAMPLE 4 Overdamped Motion

It is readily verified that the solution of the initial-value problem

$$\frac{d^2x}{dt^2} + 5\frac{dx}{dt} + 4x = 0, \quad x(0) = 1, \quad x'(0) = 1,$$

is

$$x(t) = \frac{5}{3}e^{-t} - \frac{2}{3}e^{-4t}. \tag{16}$$

The problem can be interpreted as representing the overdamped motion of a mass on a spring. The mass starts from a position 1 unit *below* the equilibrium position with a *downward* velocity of 1 ft/s.

To graph $x(t)$ we find the value of t for which the function has an extremum, that is, the value of time for which the first derivative (velocity) is zero. Differentiating (16) gives

$$x'(t) = -\frac{5}{3}e^{-t} + \frac{8}{3}e^{-4t}$$

so that $x'(t) = 0$ implies $e^{3t} = 8/5$ or $t = (1/3)\ln(8/5) = 0.157$. It follows from the first derivative test, as well as our physical intuition, that $x(0.157) = 1.069$ ft is actually a maximum. In other words, the mass attains an extreme displacement of 1.069 ft below the equilibrium position.

We should also check to see whether the graph crosses the t-axis, that is, whether the mass passes through the equilibrium position. This cannot happen in this instance since the equation $x(t) = 0$, or $e^{3t} = 2/5$, has the physically irrelevant solution $t = (1/3)\ln(2/5) = -0.305$.

The graph of $x(t)$, along with some other pertinent data, is given in Figure 3.18. ●

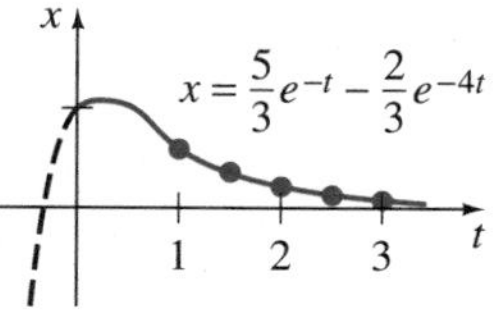

(a)

t	$x(t)$
1	0.601
1.5	0.370
2	0.225
2.5	0.137
3	0.083

(b)

FIGURE 3.18

EXAMPLE 5 Critically Damped Motion

An 8-lb weight stretches a spring 2 ft. Assuming that a damping force numerically equal to two times the instantaneous velocity acts on the system, determine the equation of motion if the weight is released from the equilibrium position with an upward velocity of 3 ft/s.

Solution From Hooke's law we see that $8 = k(2)$ gives $k = 4$ lb/ft and that $W = mg$ gives $m = 8/32 = 1/4$ slug. The differential equation of motion is then

$$\frac{1}{4}\frac{d^2x}{dt^2} = -4x - 2\frac{dx}{dt} \quad \text{or} \quad \frac{d^2x}{dt^2} + 8\frac{dx}{dt} + 16x = 0. \tag{17}$$

The auxiliary equation for (17) is $m^2 + 8m + 16 = (m + 4)^2 = 0$ so that $m_1 = m_2 = -4$. Hence the system is critically damped and

$$x(t) = c_1e^{-4t} + c_2te^{-4t}. \tag{18}$$

Applying the initial conditions $x(0) = 0$ and $x'(0) = -3$, we find, in turn, that $c_1 = 0$ and $c_2 = -3$. Thus the equation of motion is

$$x(t) = -3te^{-4t}. \tag{19}$$

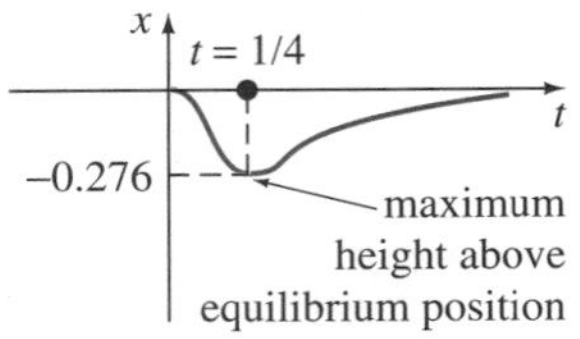

FIGURE 3.19

To graph $x(t)$ we proceed as in Example 4:

$$x'(t) = -3(-4te^{-4t} + e^{-4t}) = -3e^{-4t}(1 - 4t).$$

Clearly $x'(t) = 0$ when $t = 1/4$. The corresponding extreme displacement is $x(\frac{1}{4}) = -3(\frac{1}{4})e^{-1} = -0.276$ ft. As shown in Figure 3.19, we interpret this value to mean that the weight reaches a maximum height of 0.276 ft above the equilibrium position. ●

EXAMPLE 6 Underdamped Motion

A 16-lb weight is attached to a 5-ft-long spring. At equilibrium the spring measures 8.2 ft. If the weight is pushed up and released from rest at a point 2 ft above the equilibrium position, find the displacements $x(t)$ if it is further known that the surrounding medium offers a resistance numerically equal to the instantaneous velocity.

Solution The elongation of the spring after the weight is attached is $8.2 - 5 = 3.2$ ft, so it follows from Hooke's law that $16 = k(3.2)$ or $k = 5$ lb/ft. In addition, $m = 16/32 = 1/2$ slug, so that the differential equation is given by

$$\frac{1}{2}\frac{d^2x}{dt^2} = -5x - \frac{dx}{dt} \quad \text{or} \quad \frac{d^2x}{dt^2} + 2\frac{dx}{dt} + 10x = 0. \tag{20}$$

Proceeding, we find that the roots of $m^2 + 2m + 10 = 0$ are $m_1 = -1 + 3i$ and $m_2 = -1 - 3i$, which then implies the system is underdamped and

$$x(t) = e^{-t}(c_1 \cos 3t + c_2 \sin 3t). \tag{21}$$

Finally, the initial conditions $x(0) = -2$ and $x'(0) = 0$ yield $c_1 = -2$ and $c_2 = -2/3$, so the equation of motion is

$$x(t) = e^{-t}\left(-2\cos 3t - \frac{2}{3}\sin 3t\right). \tag{22}$$ ●

Alternative Form of the Solution

In a manner identical to the procedure used on page 149, we can write any solution

$$x(t) = e^{-\lambda t}(c_1 \cos \sqrt{\omega^2 - \lambda^2}t + c_2 \sin \sqrt{\omega^2 - \lambda^2}t)$$

in the alternative form

$$x(t) = Ae^{-\lambda t} \sin(\sqrt{\omega^2 - \lambda^2}t + \phi), \tag{23}$$

where $A = \sqrt{c_1^2 + c_2^2}$ and the phase angle ϕ is determined from the equations

$$\sin \phi = \frac{c_1}{A}, \quad \cos \phi = \frac{c_2}{A}, \quad \tan \phi = \frac{c_1}{c_2}.$$

The coefficient $Ae^{-\lambda t}$ is sometimes called the **damped amplitude** of vibrations. Because (23) is not a periodic function, the number $2\pi/\sqrt{\omega^2 - \lambda^2}$ is called the **quasi period** and $\sqrt{\omega^2 - \lambda^2}/2\pi$ is the **quasi frequency**. The quasi period is the time interval between two successive maxima of $x(t)$. You should verify, for the equation of motion in Example 6, that $A = 2\sqrt{10}/3$ and $\phi = 4.931$. Therefore an equivalent form of (22) is

$$x(t) = \frac{2\sqrt{10}}{3} e^{-t} \sin(3t + 4.931).$$

Remarks **(i)** We noted on page 151 that an undamped spring/mass system whose model is the autonomous equation $md^2x/dt^2 + F(x) = 0$ is an example of a conservative system. When damping forces are taken into account, such a system is nonconservative; in other words, energy is not conserved.

(ii) When $\beta > 0$, (0, 0) is always a stable critical point of the linear differential equation (11). In the cases of overdamped and critically damped spring/mass systems, (0, 0) is an asymptotically stable node. For an underdamped system, (0, 0) is an asymptotically stable spiral point. See Figures 3.5(a), 3.6(a), and 3.7(a).

3.7.3 Spring/Mass Systems: Driven Motion

Differential Equation of Driven Motion with Damping

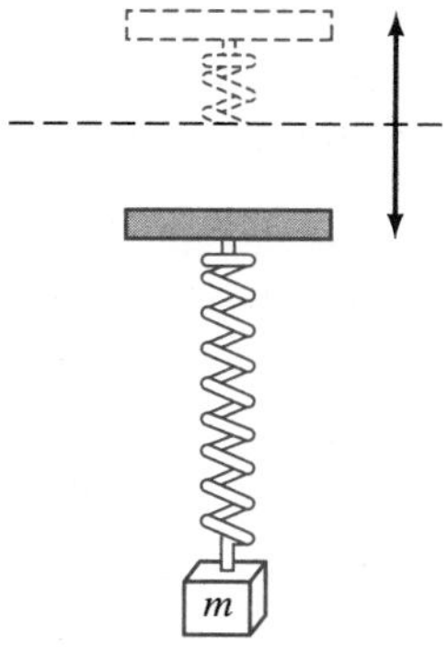

FIGURE 3.20

Suppose we now take into consideration an external force $f(t)$ acting on a vibrating mass on a spring. For example, $f(t)$ could represent a driving force causing an oscillatory vertical motion of the support of the spring. See Figure 3.20. The inclusion of $f(t)$ in the formulation of Newton's second law gives the differential equation of **driven** or **forced motion**:

$$m\frac{d^2x}{dt^2} = -kx - \beta\frac{dx}{dt} + f(t). \tag{24}$$

Dividing (24) by m gives

$$\frac{d^2x}{dt^2} + 2\lambda\frac{dx}{dt} + \omega^2 x = F(t), \tag{25}$$

where $F(t) = f(t)/m$ and, as in the preceding section, $2\lambda = \beta/m$, $\omega^2 = k/m$. To solve the latter nonhomogeneous equation we can use either the method of undetermined coefficients or variation of parameters.

EXAMPLE 7 Interpretation of an Initial-Value Problem

Interpret and solve the initial-value problem

$$\frac{1}{5}\frac{d^2x}{dt^2} + 1.2\frac{dx}{dt} + 2x = 5\cos 4t, \qquad x(0) = \frac{1}{2}, \quad x'(0) = 0. \tag{26}$$

Solution We can interpret the problem to represent a vibrational system consisting of a mass ($m = 1/5$ slug or kilogram) attached to a spring ($k = 2$ lb/ft or N/m). The mass is released from rest $\frac{1}{2}$ unit (foot or meter) below the equilibrium position. The motion is damped ($\beta = 1.2$) and is being driven by an external periodic ($T = \pi/2$ seconds) force beginning at $t = 0$. Intuitively we would expect that even with damping the system would remain in motion until such time as the forcing function is "turned off," in which case the amplitudes diminish. However, as the problem is given, $f(t) = 5\cos 4t$ will remain "on" forever.

We first multiply the differential equation in (26) by 5 and solve

$$\frac{dx^2}{dt^2} + 6\frac{dx}{dt} + 10x = 0$$

by the usual methods. Since $m_1 = -3 + i$, $m_2 = -3 - i$, it follows that

$$x_c(t) = e^{-3t}(c_1 \cos t + c_2 \sin t).$$

Using the method of undetermined coefficients, we assume a particular solution of the form $x_p(t) = A\cos 4t + B\sin 4t$. Now

$$x_p' = -4A\sin 4t + 4B\cos 4t, \quad x_p'' = -16A\cos 4t - 16B\sin 4t$$

so that

$$x_p'' + 6x_p' + 10x_p = (-6A + 24B)\cos 4t + (-24A - 6B)\sin 4t = 25\cos 4t.$$

The resulting system of equations

$$-6A + 24B = 25, \quad -24A - 6B = 0$$

yields $A = -25/102$ and $B = 50/51$. It follows that

$$x(t) = e^{-3t}(c_1\cos t + c_2\sin t) - \frac{25}{102}\cos 4t + \frac{50}{51}\sin 4t. \tag{27}$$

When we set $t = 0$ in the above equation, we obtain $c_1 = 38/51$. By differentiating the expression and then setting $t = 0$, we also find that $c_2 = -86/51$. Therefore the equation of motion is

$$x(t) = e^{-3t}\left(\frac{38}{51}\cos t - \frac{86}{51}\sin t\right) - \frac{25}{102}\cos 4t + \frac{50}{51}\sin 4t. \tag{28}$$

Transient and Steady-State Terms

Notice that the complementary function

$$x_c(t) = e^{-3t}\left(\frac{38}{51}\cos t - \frac{86}{51}\sin t\right)$$

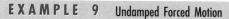

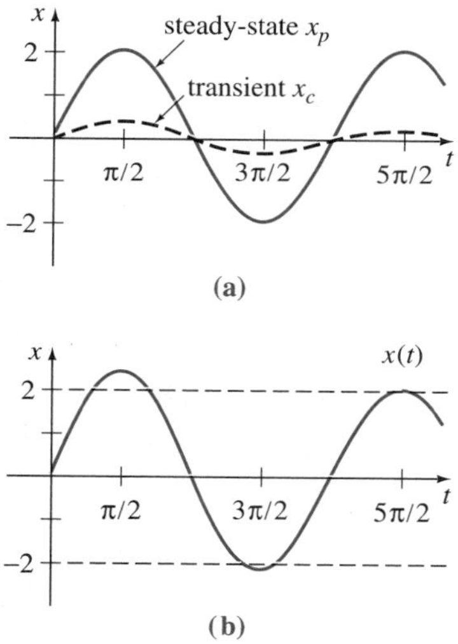

FIGURE 3.21

in (28) possesses the property that $\lim_{t\to\infty} x_c(t) = 0$. Since $x_c(t)$ becomes negligible (namely, $\to 0$) as $t \to \infty$, it is said to be a **transient term**, or **transient solution**. Thus for large time the displacements of the weight in the preceding problem are closely approximated by the particular solution $x_p(t)$. This latter function is also called the **steady-state solution**. When F is a periodic function, such as $F(t) = F_0 \sin \gamma t$ or $F(t) = F_0 \cos \gamma t$, the general solution of (25) consists of

$$x(t) = \textit{transient} + \textit{steady-state}.$$

EXAMPLE 8 Transient/Steady-State Solutions

It is readily shown that the solution of the initial-value problem

$$\frac{d^2x}{dt^2} + 2\frac{dx}{dt} + 2x = 4\cos t + 2\sin t, \qquad x(0) = 0, \quad x'(0) = 3,$$

is

$$x = x_c + x_p = \underbrace{e^{-t}\sin t}_{\text{transient}} + \underbrace{2\sin t}_{\text{steady-state}}.$$

Inspection of Figure 3.21 shows that the effect of the transient term on the solution is, in this case, negligible for about $t > 2\pi$.

Differential Equation of Driven Motion Without Damping

With a periodic impressed force and no damping force, there is no transient term in the solution of a problem. Also, we shall see that a periodic impressed force with a frequency near, or the same as, the frequency of free undamped vibrations can cause a severe problem in any oscillatory mechanical system.

EXAMPLE 9 Undamped Forced Motion

Solve the initial-value problem

$$\frac{d^2x}{dt^2} + \omega^2 x = F_0 \sin \gamma t, \qquad x(0) = 0, \quad x'(0) = 0, \tag{29}$$

where F_0 is a constant and $\gamma \neq \omega$.

Solution The complementary function is $x_c(t) = c_1 \cos \omega t + c_2 \sin \omega t$. To obtain a particular solution we assume $x_p(t) = A \cos \gamma t + B \sin \gamma t$, so that

$$x_p'' + \omega^2 x_p = A(\omega^2 - \gamma^2) \cos \gamma t + B(\omega^2 - \gamma^2) \sin \gamma t = F_0 \sin \gamma t$$

Equating coefficients immediately gives $A = 0$ and $B = F_0/(\omega^2 - \gamma^2)$. Therefore

$$x_p(t) = \frac{F_0}{\omega^2 - \gamma^2} \sin \gamma t.$$

Applying the given initial conditions to the general solution

$$x(t) = c_1 \cos \omega t + c_2 \sin \omega t + \frac{F_0}{\omega^2 - \gamma^2} \sin \gamma t$$

yields $c_1 = 0$ and $c_2 = -\gamma F_0/\omega(\omega^2 - \gamma^2)$. Thus the solution is

$$x(t) = \frac{F_0}{\omega(\omega^2 - \gamma^2)}(-\gamma \sin \omega t + \omega \sin \gamma t), \quad \gamma \neq \omega. \tag{30}$$

●

Pure Resonance

Although equation (30) is not defined for $\gamma = \omega$, it is interesting to observe that its limiting value as $\gamma \to \omega$ can be obtained by applying L'Hôpital's rule. This limiting process is analogous to "tuning in" the frequency of the driving force $(\gamma/2\pi)$ to the frequency of free vibrations $(\omega/2\pi)$. Intuitively we expect that over a length of time we should be able to substantially increase the amplitudes of vibration. For $\gamma = \omega$ we define the solution to be

$$x(t) = \lim_{\gamma \to \omega} F_0 \frac{-\gamma \sin \omega t + \omega \sin \gamma t}{\omega(\omega^2 - \gamma^2)} = F_0 \lim_{\gamma \to \omega} \frac{\frac{d}{d\gamma}(-\gamma \sin \omega t + \omega \sin \gamma t)}{\frac{d}{d\gamma}(\omega^3 - \omega\gamma^2)}$$

$$= F_0 \lim_{\gamma \to \omega} \frac{-\sin \omega t + \omega t \cos \gamma t}{-2\omega\gamma}$$

$$= F_0 \frac{-\sin \omega t + \omega t \cos \omega t}{-2\omega^2}$$

$$= \frac{F_0}{2\omega^2} \sin \omega t - \frac{F_0}{2\omega} t \cos \omega t. \tag{31}$$

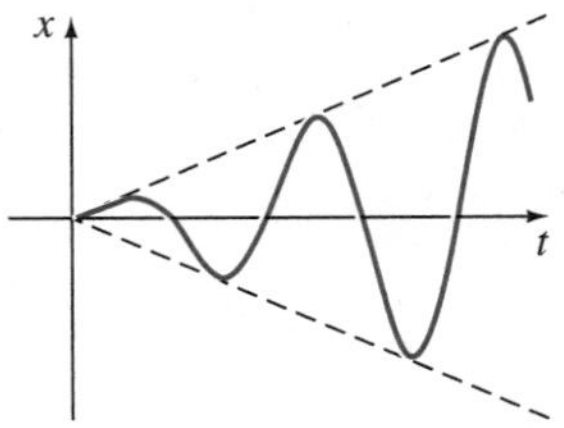

FIGURE 3.22

As suspected, when $t \to \infty$, the displacements become large; in fact, $|x(t_n)| \to \infty$ when $t_n = n\pi/\omega$, $n = 1, 2, \ldots$. The phenomenon we have just described is known as **pure resonance**. The graph given in Figure 3.22 shows typical motion in this case.

In conclusion it should be noted that there is no actual need to use a limiting process on (30) to obtain the solution for $\gamma = \omega$. Alternatively, equation (31) follows by solving the initial-value problem

$$\frac{d^2x}{dt^2} + \omega^2 x = F_0 \sin \omega t, \qquad x(0) = 0, \quad x'(0) = 0,$$

directly by conventional methods.

If the displacements of a spring/mass system were actually described by a function such as (31) it would necessarily fail. Large oscillations of the mass would eventually force the spring beyond its elastic limit. One might argue too that the resonating model presented in Figure 3.22 is completely unrealistic, because it ignores the retarding effects of ever-present damping forces. Although it is true that pure resonance cannot occur when the smallest amount of damping is taken in consideration, large and equally destructive amplitudes of vibration (although bounded as $t \to \infty$) can occur. See Problem 35 in Exercises 3.7.

3.7.4 Analogous Systems

L-R-C Series Circuits

As mentioned in the introduction to this chapter, many different physical systems can be described by a linear second-order differential equation similar to the differential equation of forced motion with damping:

$$m\frac{d^2x}{dt^2} + \beta\frac{dx}{dt} + kx = f(t). \tag{32}$$

If $i(t)$ denotes current in an ***L-R-C* series electrical circuit** shown in Figure 3.23(a), then the voltage drops across the inductor, resistor, and capacitor are as shown in Figure 3.23(b). By Kirchhoff's second law, the sum of these voltages equals the voltage $E(t)$ impressed on the circuit; that is,

$$L\frac{di}{dt} + Ri + \frac{1}{C}q = E(t). \tag{33}$$

But the charge $q(t)$ on the capacitor is related to the current $i(t)$ by $i = dq/dt$ and so (33) becomes the linear second-order differential equation

$$L\frac{d^2q}{dt^2} + R\frac{dq}{dt} + \frac{1}{C}q = E(t). \tag{34}$$

The nomenclature used in the analysis of circuits is similar to that used to describe spring-mass systems.

If $E(t) = 0$, the **electrical vibrations** of the circuit are said to be **free**. Since the auxiliary equation for (34) is $Lm^2 + Rm + 1/C = 0$, there will be three forms of the solution when $R \neq 0$, depending on the value of the discriminant $R^2 - 4L/C$. We say that the circuit is

overdamped if $\quad R^2 - 4L/C > 0,$

critically damped if $\quad R^2 - 4L/C = 0,$

and **underdamped** if $\quad R^2 - 4L/C < 0.$

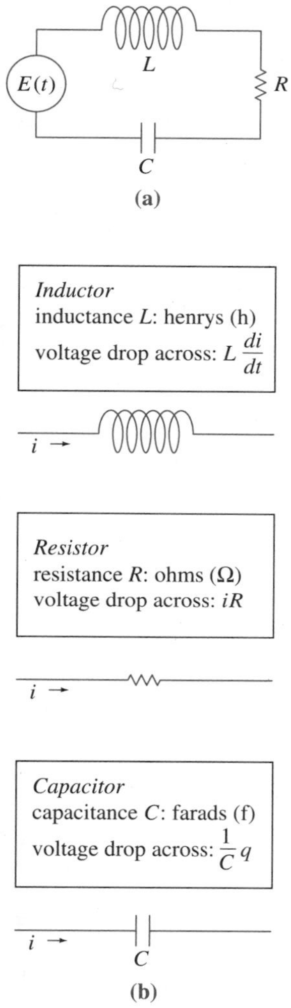

FIGURE 3.23

In each of these three cases, the general solution of (34) contains the factor $e^{-Rt/2L}$ and so $q(t) \to 0$ as $t \to \infty$. In the underdamped case when $q(0) = q_0$, the charge on the capacitor oscillates as it decays; in other words, the capacitor is charging and discharging as $t \to \infty$. When $E(t) = 0$ and $R = 0$, the circuit is said to be undamped and the electrical vibrations do not approach zero as t increases without bound; the response of the circuit is **simple harmonic**.

EXAMPLE 10 Underdamped Series Circuit

Find the charge $q(t)$ on the capacitor in an L-R-C series circuit when $L = 0.25$ henry, $R = 10$ ohms, $C = 0.001$ farad, $E(t) = 0$, $q(0) = q_0$ coulombs, and $i(0) = 0$.

Solution Since $1/C = 1000$, equation (34) becomes

$$\frac{1}{4}q'' + 10q' + 1000q = 0 \quad \text{or} \quad q'' + 40q' + 4000q = 0.$$

Solving this homogeneous equation in the usual manner, we find that the circuit is underdamped and

$$q(t) = e^{-20t}(c_1 \cos 60t + c_2 \sin 60t).$$

Applying the initial conditions, we find $c_1 = q_0$ and $c_2 = q_0/3$. Thus

$$q(t) = q_0 e^{-20t}\left(\cos 60t + \frac{1}{3}\sin 60t\right).$$

Using (23), we can write the foregoing solution as

$$q(t) = \frac{q_0\sqrt{10}}{3}e^{-20t}\sin(60t + 1.249).$$

When there is an impressed voltage $E(t)$ on the circuit, the electrical vibrations are said to be **forced**. In the case when $R \neq 0$, the complementary function $q_c(t)$ of (34) is called a **transient solution**. If $E(t)$ is periodic or a constant, then the particular solution $q_p(t)$ of (34) is a **steady-state solution**.

EXAMPLE 11 Steady-State Current

Find the steady-state solution $q_p(t)$ and the **steady-state current** in an L-R-C series circuit when the impressed voltage is $E(t) = E_0 \sin \gamma t$.

Solution The steady-state solution $q_p(t)$ is a particular solution of the differential equation

$$L\frac{d^2q}{dt^2} + R\frac{dq}{dt} + \frac{1}{C}q = E_0 \sin \gamma t.$$

Using the method of undetermined coefficients, we assume a particular solution of the form $q_p(t) = A \sin \gamma t + B \cos \gamma t$. Substituting this expression into the differential equation, simplifying, and equating coefficients give

$$A = \frac{E_0(L\gamma - 1/C\gamma)}{-\gamma\left[L^2\gamma^2 - \frac{2L}{C} + \frac{1}{C^2\gamma^2} + R^2\right]}, \quad B = \frac{E_0 R}{-\gamma\left[L^2\gamma^2 - \frac{2L}{C} + \frac{1}{C^2\gamma^2} + R^2\right]}.$$

It is convenient to express A and B in terms of some new symbols. If

$$X = L\gamma - \frac{1}{C\gamma}, \quad \text{then} \quad X^2 = L^2\gamma^2 - \frac{2L}{C} + \frac{1}{C^2\gamma^2}.$$

If

$$Z = \sqrt{X^2 + R^2}, \quad \text{then} \quad Z^2 = L^2\gamma^2 - \frac{2L}{C} + \frac{1}{C^2\gamma^2} + R^2.$$

Therefore $A = E_0X/(-\gamma Z^2)$ and $B = E_0R/(-\gamma Z^2)$, so the steady-state charge is

$$q_p(t) = -\frac{E_0X}{\gamma Z^2}\sin\gamma t - \frac{E_0R}{\gamma Z^2}\cos\gamma t.$$

Now the steady-state current is given by $i_p(t) = q_p'(t)$:

$$i_p(t) = \frac{E_0}{Z}\left(\frac{R}{Z}\sin\gamma t - \frac{X}{Z}\cos\gamma t\right). \tag{35}$$

●

The quantities $X = L\gamma - 1/C\gamma$ and $Z = \sqrt{X^2 + R^2}$ defined in Example 11 are called, respectively, the **reactance** and **impedance** of the circuit. Both the reactance and the impedance are measured in ohms.

Twisted Shaft

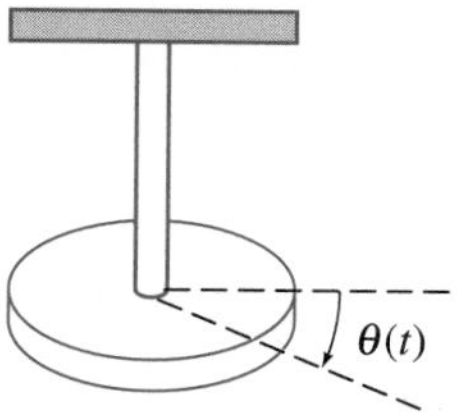

FIGURE 3.24

The differential equation governing the torsional motion of a weight suspended from the end of an elastic shaft is

$$I\frac{d^2\theta}{dt^2} + c\frac{d\theta}{dt} + k\theta = T(t). \tag{36}$$

As shown in Figure 3.24 the function $\theta(t)$ represents the amount of twist of the weight at any time.

By comparing equations (25) and (34) with (36), we see that, with the exception of terminology, there is absolutely no difference between the mathematics of vibrating springs, simple series circuits, and torsional vibrations.

EXERCISES 3.7

Answers to odd-numbered problems begin on page AN-14.

3.7.1

1. A 4-lb weight is attached to a spring whose spring constant is 16 lb/ft. What is the period of simple harmonic motion?

2. A 20-kg mass is attached to a spring. If the frequency of simple harmonic motion is $2/\pi$ vibrations/second, what is the spring constant k? What is the frequency of simple harmonic motion if the original mass is replaced with an 80-kg mass?

3. A 24-lb weight, attached to the end of a spring, stretches it 4 in. Find the equation of motion if the weight is released from rest from a point 3 in. above the equilibrium position.

4. Determine the equation of motion if the weight in Problem 3 is released from the equilibrium position with an initial downward velocity of 2 ft/s.

5. A 20-lb weight stretches a spring 6 in. The weight is released from rest 6 in. below the equilibrium position.
   (a) Find the position of the weight at $t = \pi/12, \pi/8, \pi/6, \pi/4, 9\pi/32$ seconds.
   (b) What is the velocity of the weight when $t = 3\pi/16$ second? In which direction is the weight heading at this instant?
   (c) At what times does the weight pass through the equilibrium position?

6. A force of 400 newtons stretches a spring 2 m. A mass of 50 kg is attached to the end of the spring and released from the equilibrium position with an upward velocity of 10 m/s. Find the equation of motion.

7. Another spring whose constant is 20 N/m is suspended from the same rigid support but parallel to the spring/mass system in Problem 6. A mass of 20 kg is attached to the second spring and both masses are released from the equilibrium position with an upward velocity of 10 m/s.
   (a) Which mass exhibits the greater amplitude of motion?
   (b) Which mass is moving faster at $t = \pi/4$ second? at $\pi/2$ seconds?
   (c) At what times are the two masses in the same position? Where are the masses at these times? In which directions are they moving?

8. A 32-lb weight stretches a spring 2 ft. Determine the amplitude and period of motion if the weight is released 1 ft above the equilibrium position with an initial upward velocity of 2 ft/s. How many complete vibrations will the weight have completed at the end of 4π seconds?

9. An 8-lb weight attached to a spring exhibits simple harmonic motion. Determine the equation of motion if the spring constant is 1 lb/ft and if the weight is released 6 in. below the equilibrium position with a downward velocity of 3/2 ft/s. Express the solution in form (6).

10. A mass weighing 10 lb stretches a spring 1/4 ft. This mass is removed and replaced with a mass of 1.6 slugs, which is released 1/3 ft above the equilibrium position with a downward velocity of 5/4 ft/s. Express the solution in form (6). At what times does the mass attain a displacement below the equilibrium position numerically equal to one-half the amplitude?

11. A 64-lb weight attached to the end of a spring stretches it 0.32 ft. From a position 8 in. above the equilibrium position the weight is given a downward velocity of 5 ft/s.
   (a) Find the equation of motion.
   (b) What are the amplitude and period of motion?
   (c) How many complete vibrations will the weight have completed at the end of 3π seconds?
   (d) At what time does the weight pass through the equilibrium position heading downward for the second time?
   (e) At what time does the weight attain its extreme displacement on either side of the equilibrium position?
   (f) What is the position of the weight at $t = 3$ seconds?
   (g) What is the instantaneous velocity at $t = 3$ seconds?
   (h) What is the acceleration at $t = 3$ seconds?
   (i) What is the instantaneous velocity at the times when the weight passes through the equilibrium position?
   (j) At what times is the weight 5 in. below the equilibrium position?
   (k) At what times is the weight 5 in. below the equilibrium position heading in the upward direction?

12. A mass of 1 slug is suspended from a spring whose characteristic spring constant is 9 lb/ft. Initially the mass starts from a point 1 ft above the equilibrium position with an upward velocity of $\sqrt{3}$ ft/s. Find the times for which the mass is heading downward at a velocity of 3 ft/s.

13. Under some circumstances when two parallel springs, with constants k_1 and k_2, support a single weight W, the **effective spring constant** of the system is given by $k = 4k_1k_2/(k_1 + k_2)$. A 20-lb weight stretches one spring 6 in. and another spring 2 in. The springs are attached to a common rigid support and then to a metal plate. As shown in Figure 3.25 the 20-lb weight is attached to the center of the plate in the double spring arrangement. Determine the effective spring constant of this system. Find the equation of motion if the weight is released from the equilibrium position with a downward velocity of 2 ft/s.

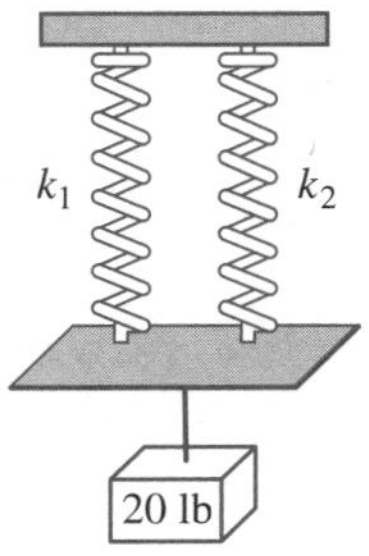

FIGURE 3.25

14. A certain weight stretches one spring 1/3 ft and another spring 1/2 ft. The two springs are attached to a common rigid support in a manner indicated in Problem 13 and Figure 3.25. The first weight is set aside, and an 8-lb weight is attached to the double spring arrangement and the system is set in motion. If the period of motion is $\pi/15$ second, determine the numerical value of the first weight.

3.7.2

In Problems 15–18 the given figure represents the graph of an equation of motion for a mass on a spring. The spring/mass system is damped. Use the graph to determine

(a) whether the initial displacement of the mass is above or below the equilibrium position; and
(b) whether the mass is initially released from rest, heading downward, or heading upward.

15.
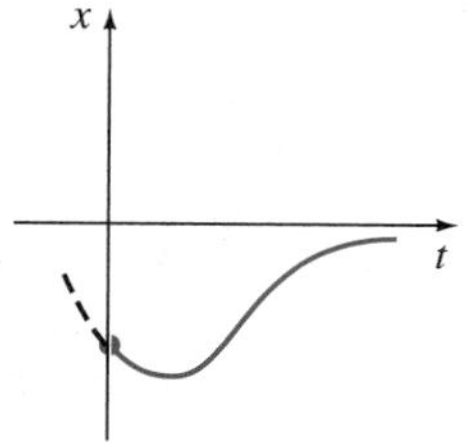

FIGURE 3.26

16.
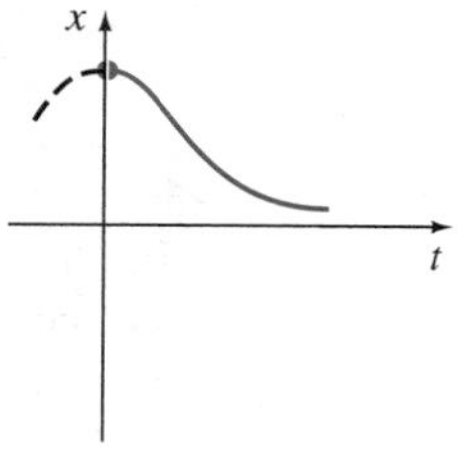

FIGURE 3.27

17.
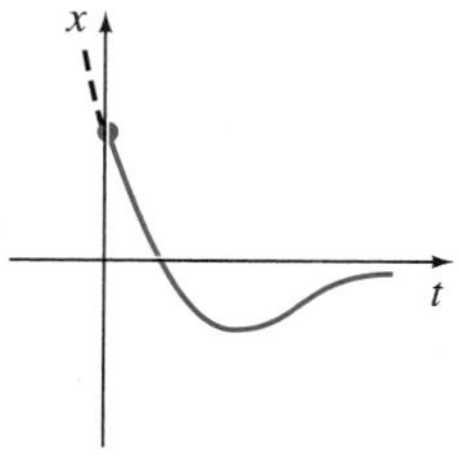

FIGURE 3.28

18.
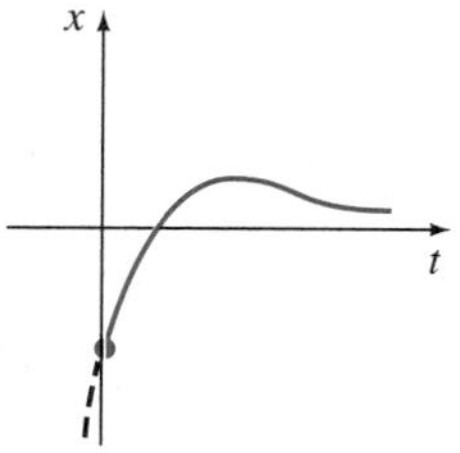

FIGURE 3.29

19. A 4-lb weight is attached to a spring whose constant is 2 lb/ft. The medium offers a resistance to the motion of the weight numerically equal to the instantaneous velocity. If the weight is released from a point 1 ft above the equilibrium position with a downward velocity of 8 ft/s, determine the time that the weight passes through the equilibrium position. Find the time for which the weight attains its extreme displacement from the equilibrium position. What is the position of the weight at this instant?

20. A 4-ft spring measures 8 ft long after an 8-lb weight is attached to it. The medium through which the weight moves offers a resistance numerically equal to $\sqrt{2}$ times the instantaneous velocity. Find the equation of motion if the weight is released from the equilibrium position with a downward velocity of 5 ft/s. Find the time for which the weight attains its extreme displacement from the equilibrium position. What is the position of the weight at this instant?

21. A 1-kg mass is attached to a spring whose constant is 16 N/m and the entire system is then submerged in a liquid that imparts a damping force numerically equal to 10 times the instantaneous velocity. Determine the equations of motion if

(a) the weight is released from rest 1 m below the equilibrium position; and
(b) the weight is released 1 m below the equilibrium position with an upward velocity of 12 m/s.

22. In parts (a) and (b) of Problem 21 determine whether the weight passes through the equilibrium position. In each case find the time at which the weight attains its extreme displacement from the equilibrium position. What is the position of the weight at this instant?

23. A force of 2 lb stretches a spring 1 ft. A 3.2-lb weight is attached to the spring and the system is then immersed in a medium that imparts a damping force numerically equal to 0.4 times the instantaneous velocity.

(a) Find the equation of motion if the weight is released from rest 1 ft above the equilibrium position.
(b) Express the equation of motion in the form given in (23).
(c) Find the first time for which the weight passes through the equilibrium position heading upward.

24. After a 10-lb weight is attached to a 5-ft spring, the spring measures 7 ft long. The 10-lb weight is removed and replaced with an 8-lb weight and the entire system is placed in a medium offering a resistance numerically equal to the instantaneous velocity.

(a) Find the equation of motion if the weight is released 1/2 ft below the equilibrium position with a downward velocity of 1 ft/s.
(b) Express the equation of motion in the form given in (23).
(c) Find the times for which the weight passes through the equilibrium position heading downward.
(d) Graph the equation of motion.

25. A 10-lb weight attached to a spring stretches it 2 ft. The weight is attached to a dashpot damping device that offers

a resistance numerically equal to $\beta (\beta > 0)$ times the instantaneous velocity. Determine the values of the damping constant β so that the subsequent motion is **(a)** overdamped, **(b)** critically damped, and **(c)** underdamped.

26. A 24-lb weight stretches a spring 4 ft. The subsequent motion takes place in a medium offering a resistance numerically equal to $\beta (\beta > 0)$ times the instantaneous velocity. If the weight starts from the equilibrium position with an upward velocity of 2 ft/s, show that if $\beta > 3\sqrt{2}$, the equation of motion is

$$x(t) = \frac{-3}{\sqrt{\beta^2 - 18}} e^{-2\beta t/3} \sinh \frac{2}{3} \sqrt{\beta^2 - 18}\,t.$$

3.7.3

27. A 16-lb weight stretches a spring 8/3 ft. Initially the weight starts from rest 2 ft below the equilibrium position and the subsequent motion takes place in a medium that offers a damping force numerically equal to 1/2 the instantaneous velocity. Find the equation of motion if the weight is driven by an external force equal to $f(t) = 10 \cos 3t$.

28. A mass of 1 slug is attached to a spring whose constant is 5 lb/ft. Initially the mass is released 1 ft below the equilibrium position with a downward velocity of 5 ft/s, and the subsequent motion takes place in a medium that offers a damping force numerically equal to 2 times the instantaneous velocity.

(a) Find the equation of motion if the mass is driven by an external force equal to $f(t) = 12 \cos 2t + 3 \sin 2t$.

(b) Graph the transient and steady-state solutions on the same coordinate axes.

(c) Graph the equation of motion.

29. A mass of 1 slug, when attached to a spring, stretches it 2 ft and then comes to rest in the equilibrium position. Starting at $t = 0$, an external force equal to $f(t) = 8 \sin 4t$ is applied to the system. Find the equation of motion if the surrounding medium offers a damping force numerically equal to 8 times the instantaneous velocity.

30. In Problem 29 determine the equation of motion if the external force is $f(t) = e^{-t} \sin 4t$. Analyze the displacements for $t \to \infty$.

31. When a mass of 2 kilograms is attached to a spring whose constant is 32 N/m, it comes to rest in the equilibrium position. Starting at $t = 0$, a force equal to $f(t) = 68e^{-2t} \cos 4t$ is applied to the system. Find the equation of motion in the absence of damping.

32. In Problem 31 write the equation of motion in the form $x(t) = A \sin(\omega t + \phi) + Be^{-2t} \sin(4t + \theta)$. What is the amplitude of vibrations after a very long time?

33. A mass m is attached to the end of a spring whose constant is k. After the mass reaches equilibrium, its support begins to oscillate vertically about a horizontal line L according to a formula $h(t)$. The value of h represents the distance in feet measured from L. See Figure 3.30.

(a) Determine the differential equation of motion if the entire system moves through a medium offering a damping force numerically equal to $\beta(dx/dt)$.

(b) Solve the differential equation in part (a) if the spring is stretched 4 ft by a weight of 16 lb, and $\beta = 2$, $h(t) = 5 \cos t$, $x(0) = x'(0) = 0$.

34. A mass of 100 g is attached to a spring whose constant is 1600 dynes/cm. After the mass reaches equilibrium, its support oscillates according to the formula $h(t) = \sin 8t$, where h represents displacement from its original position. See Problem 33 and Figure 3.30.

(a) In the absence of damping, determine the equation of motion if the mass starts from rest from the equilibrium position.

(b) At what times does the mass pass through the equilibrium position?

(c) At what times does the mass attain its extreme displacements?

(d) What are the maximum and minimum displacements?

(e) Graph the equation of motion.

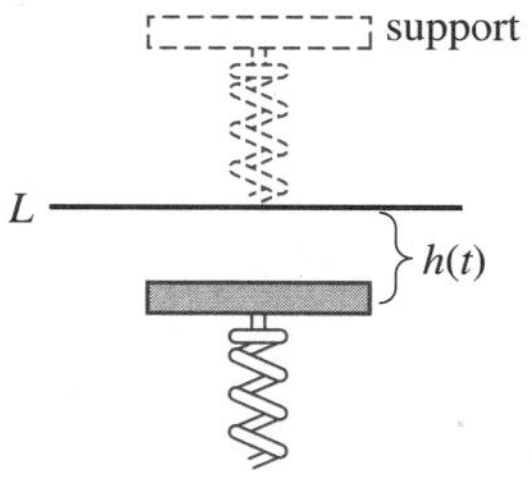

FIGURE 3.30

35. *PROJECT PROBLEM*

(a) Show that the general solution of

$$\frac{d^2x}{dt^2} + 2\lambda \frac{dx}{dt} + \omega^2 x = F_0 \sin \gamma t$$

is $x(t) = Ae^{-\lambda t}\sin(\sqrt{\omega^2 - \lambda^2}t + \phi)$

$$+ \frac{F_0}{\sqrt{(\omega^2 - \gamma^2)^2 + 4\lambda^2\gamma^2}}\sin(\gamma t + \theta),$$

where $A = \sqrt{c_1^2 + c_2^2}$ and the phase angles ϕ and θ are, respectively, defined by $\sin\phi = c_1/A$, $\cos\phi = c_2/A$ and

$$\sin\theta = \frac{-2\lambda\gamma}{\sqrt{(\omega^2 - \gamma^2)^2 + 4\lambda^2\gamma^2}},$$

$$\cos\theta = \frac{\omega^2 - \gamma^2}{\sqrt{(\omega^2 - \gamma^2)^2 + 4\lambda^2\gamma^2}}.$$

(b) The solution in part (a) has the form $x(t) = x_c(t) + x_p(t)$. Inspection shows that $x_c(t)$ is transient and hence, for large values of time, the solution is approximated by $x_p(t) = g(\gamma)\sin(\gamma t + \theta)$, where

$$g(\gamma) = \frac{F_0}{\sqrt{(\omega^2 - \gamma^2)^2 + 4\lambda^2\gamma^2}}.$$

Although the amplitude $g(\gamma)$ of $x_p(t)$ is bounded as $t \to \infty$, show that the maximum oscillations will occur at the value $\gamma_1 = \sqrt{\omega^2 - 2\lambda^2}$. What is the maximum value of g? The number $\sqrt{\omega^2 - 2\lambda^2}/2\pi$ is said to be the **resonance frequency** of the system.

(c) When $F_0 = 2$, $m = 1$, and $k = 4$, g becomes

$$g(\gamma) = \frac{2}{\sqrt{(4 - \gamma^2)^2 + \beta^2\gamma^2}}.$$

Construct a table of the values of γ_1 and $g(\gamma_1)$ corresponding to the damping coefficients $\beta = 2$, $\beta = 1$, $\beta = \frac{3}{4}$, $\beta = \frac{1}{2}$, and $\beta = \frac{1}{4}$. Use a computer graphing program to obtain the graphs of g corresponding to these damping coefficients. Use the same coordinate axes. This family of graphs is called the **resonance curve** or **frequency response curve** of the system. What is γ_1 approaching as $\beta \to 0$? What is happening to the resonance curve as $\beta \to 0$?

36. *PROJECT PROBLEM* Consider a driven undamped spring/mass system described by the initial-value problem

$$\frac{d^2x}{dt^2} + \omega^2 x = F_0 \sin^n \gamma t, \qquad x(0) = 0, \quad x'(0) = 0.$$

(a) Demonstrate for $n = 2$ that there is a single frequency $\gamma_1/2\pi$ at which the system is in pure resonance.

(b) Demonstrate for $n = 3$ that there are two frequencies $\gamma_1/2\pi$ and $\gamma_2/2\pi$ at which the system is in pure resonance.

(c) Suppose $\omega = 1$ and $F_0 = 1$. Use an ODE solver to obtain the graph of the solution of the initial-value problem for $n = 2$ and $\gamma = \gamma_1$ in part (a). Obtain the graph of the solution of the initial-value problem for $n = 3$ corresponding, in turn, to $\gamma = \gamma_1$ and $\gamma = \gamma_2$ in part (b).

37. *WRITING PROJECT* Write a report on the 1940 collapse of the Tacoma Narrows bridge. Elaborate on the early theories that the collapse of the bridge was due to a resonance phenomenon. Conclude your report with a discussion of some modern theories (c. 1990), which provide an alternative explanation for the bridge's structural failure.

38. *WRITING PROJECT* In late 1959 and early 1960 two commercial plane crashes occurred that involved the relatively new propjet called the *Electra*. Write a report on how it was discovered that a resonance phenomenon was the probable cause for the crashes. Start your research with the book *Loud and Clear*, by Robert J. Serling (New York; Dell, 1970).

3.7.4

39. Find the charge on the capacitor in an *L-R-C* series circuit at $t = 0.01$ second when $L = 0.05$ henry, $R = 2$ ohms, $C = 0.01$ farad, $E(t) = 0$ volts, $q(0) = 5$ coulombs, and $i(0) = 0$ amperes. Determine the first time at which the charge on the capacitor is equal to zero.

40. Find the charge on the capacitor in an *L-R-C* series circuit when $L = 1/4$ henry, $R = 20$ ohms, $C = 1/300$ farad, $E(t) = 0$ volts, $q(0) = 4$ coulombs, and $i(0) = 0$ amperes. Is the charge on the capacitor ever equal to zero?

In Problems 41 and 42 find the charge on the capacitor and the current in the given *L-R-C* series circuit. Find the maximum charge on the capacitor.

41. $L = 5/3$ henrys, $R = 10$ ohms, $C = 1/30$ farad, $E(t) = 300$ volts, $q(0) = 0$ coulombs, $i(0) = 0$ amperes.

42. $L = 1$ henry, $R = 100$ ohms, $C = 0.0004$ farad, $E(t) = 30$ volts, $q(0) = 0$ coulombs, $i(0) = 2$ amperes.

43. Find the steady-state charge and the steady-state current in an *L-R-C* series circuit when $L = 1$ henry, $R = 2$ ohms, $C = 0.25$ farad, and $E(t) = 50\cos t$ volts.

44. Show that the amplitude of the steady-state current in the L-R-C series circuit in Example 11 is given by E_0/Z, where Z is the impedance of the circuit.

45. Show that the steady-state current in an L-R-C series circuit when $L = 1/2$ henry, $R = 20$ ohms, $C = 0.001$ farad, and $E(t) = 100 \sin 60t$ volts is given by $i_p(t) = (4.160) \sin(60t - 0.588)$. [*Hint:* Use Problem 44.]

46. Find the steady-state current in an L-R-C series circuit when $L = 1/2$ henry, $R = 20$ ohms, $C = 0.001$ farad, and $E(t) = 100 \sin 60t + 200 \cos 40t$ volts.

47. Find the charge on the capacitor in an L-R-C series circuit when $L = 1/2$ henry, $R = 10$ ohms, $C = 0.01$ farad, $E(t) = 150$ volts, $q(0) = 1$ coulomb, and $i(0) = 0$ amperes. What is the charge on the capacitor after a long time?

48. Show that if L, R, C, and E_0 are constant, then the amplitude of the steady-state current in Example 11 is a maximum when $\gamma = 1/\sqrt{LC}$. What is the maximum amplitude?

49. Show that if L, R, E_0, and γ are constant, then the amplitude of the steady-state current in Example 11 is a maximum when the capacitance is $C = 1/L\gamma^2$.

50. Find the charge on the capacitor and the current in an L-C circuit when $L = 0.1$ henry, $C = 0.1$ farad, $E(t) = 100 \sin \gamma t$ volts, $q(0) = 0$ coulombs, and $i(0) = 0$ amperes.

51. Find the charge on the capacitor and the current in an L-C circuit when $E(t) = E_0 \cos \gamma t$ volts, $q(0) = q_0$ coulombs, and $i(0) = i_0$ amperes.

52. In Problem 51 find the current when the circuit is in resonance.

3.8 Applications: Boundary-Value Problems

In the preceding section we examined some second-order mathematical models involving prescribed initial conditions, that is, side conditions that are specified on the unknown function and its first derivative at a single point. But often the mathematical description of a physical system demands that we solve a differential equation subject to boundary conditions, that is, conditions specified on the unknown function, or on one of its derivatives, or even on a linear combination of the unknown function and one of its derivatives, at two (or more) different points.

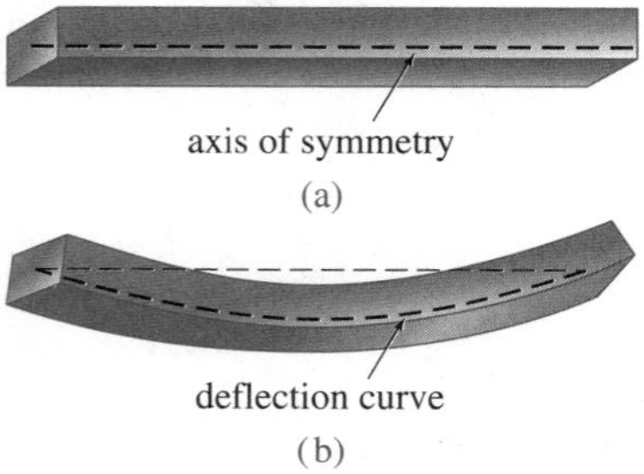

FIGURE 3.31

Deflection of a Beam

Many structures are constructed using girders or beams, and these beams deflect or distort under their own weight or under the influence of some external force. As we shall now see this deflection $y(x)$ can be modeled by a relatively simple linear fourth-order differential equation.

To begin, let us assume that a beam of length L is homogeneous and has uniform cross sections along its length. In the absence of any load on the beam (including its weight), a curve joining the centroids of all its cross sections is a straight line called the **axis of symmetry**. See Figure 3.31(a). If a load is applied to the beam in a vertical plane containing the axis of symmetry, then, as shown in Figure 3.31(b), the beam undergoes a distortion and the curve connecting the centroids of all cross sections is called the **deflection curve** or **elastic curve**. The deflection curve approximates the shape of the beam. Now suppose that the x-axis coincides with the axis of symmetry, and that the deflection $y(x)$, measured from this axis, is positive if downward. In the theory of elasticity, it is argued that a good model for the bending moment $M(x)$ at a point x along the beam is related to the load per unit length $w(x)$ by the equation

$$\frac{d^2M}{dx^2} = w(x). \tag{1}$$

In addition, the bending moment $M(x)$ is proportional to the curvature κ of the elastic curve

$$M(x) = EI\kappa, \tag{2}$$

where E and I are constants; E is Young's modulus of elasticity of the material of the beam and I is the moment of inertia of a cross-section of the beam (about an axis known as the neutral axis). The product EI is called the **flexural rigidity** of the beam.

Now, from calculus, curvature is given by $\kappa = y''/[1 + (y')^2]^{3/2}$. When the deflection $y(x)$ is small, the slope $y' \approx 0$ and so $[1 + (y')^2]^{3/2} \approx 1$. If we let $\kappa = y''$, equation (2) becomes $M = EIy''$. The second derivative of this last expression is

$$\frac{d^2M}{dx^2} = EI\frac{d^2}{dx^2}y'' = EI\frac{d^4y}{dx^4}. \tag{3}$$

Using the given result in (1) to replace d^2M/dx^2 in (3), we see that the deflection $y(x)$ satisfies the fourth-order differential equation

$$EI\frac{d^4y}{dx^4} = w(x). \tag{4}$$

Boundary conditions associated with equation (4) depend on how the ends of the beam are supported. A cantilever beam is **embedded** or **clamped** at one end and **free** at the other. A diving board, an outstretched arm, an airplane wing, and a balcony are common examples of such beams, but even trees, flagpoles, skyscrapers, and the George Washington monument can act as cantilever beams, because they are embedded at one end and are subject to the bending force of the wind. For a cantilever beam, the deflection $y(x)$ must satisfy the following two conditions at the embedded end $x = 0$:

- $y(0) = 0$ since there is no deflection, and
- $y'(0) = 0$ since the deflection curve is tangent to the x-axis (in other words, the slope of the deflection curve is zero at this point).

At $x = L$ the free end conditions are

- $y''(L) = 0$ since the bending moment is zero, and
- $y'''(L) = 0$ since the shear force is zero.

The function $F(x) = dM/dx = EI\, d^3y/dx^3$ is called the shear force. If an end of a beam is **simply supported** (also called **pin supported**, **fulcrum supported**, and **hinged**), then we must have $y = 0$ and $y'' = 0$ at that end. The following table summarizes the boundary conditions that are associated with (4).

Ends of the Beam	Boundary Conditions
embedded	$y = 0,\ y' = 0$
free	$y'' = 0,\ y''' = 0$
simply supported	$y = 0,\ y'' = 0$

EXAMPLE 1 Embedded Beam

A beam of length L is embedded at both ends. Find the deflection of the beam if a constant load w_0 is uniformly distributed along its length, that is, $w(x) = w_0$, $0 < x < L$.

Solution From the preceding discussion, we see that the deflection $y(x)$ satisfies

$$EI\frac{d^4y}{dx^4} = w_0.$$

Because the beam is embedded at both its left ($x = 0$) and its right end ($x = L$), there is no vertical deflection and the line of deflection is horizontal at these points, and so the boundary conditions are

$$y(0) = 0, \quad y'(0) = 0, \qquad y(L) = 0, \quad y'(L) = 0.$$

We can solve the nonhomogeneous differential equation in the usual manner (find y_c by observing that $m = 0$ is a root of multiplicity four of the auxiliary equation $m^4 = 0$ and then find a particular solution y_p by undetermined coefficients), or we can simply integrate the equation $d^4y/dx^4 = w_0/EI$ four times in succession. Either way, we find the general solution of the equation to be

$$y(x) = c_1 + c_2x + c_3x^2 + c_4x^3 + \frac{w_0}{24EI}x^4.$$

Now the conditions $y(0) = 0$ and $y'(0) = 0$ give in turn $c_1 = 0$ and $c_2 = 0$. Whereas the remaining conditions $y(L) = 0$ and $y'(L) = 0$ applied to $y(x) = c_3x^2 + c_4x^3 + \dfrac{w_0}{24EI}x^4$ yield the equations

$$c_3L^2 + c_4L^3 + \frac{w_0}{24EI}L^4 = 0$$

$$2c_3L + 3c_4L^2 + \frac{w_0}{6EI}L^3 = 0.$$

Solving this system gives $c_3 = w_0L^2/24EI$ and $c_4 = -w_0L/12EI$. Thus the deflection is

$$y(x) = \frac{w_0L^2}{24EI}x^2 - \frac{w_0L}{12EI}x^3 + \frac{w_0}{24EI}x^4 = \frac{w_0}{24EI}x^2(x - L)^2.$$

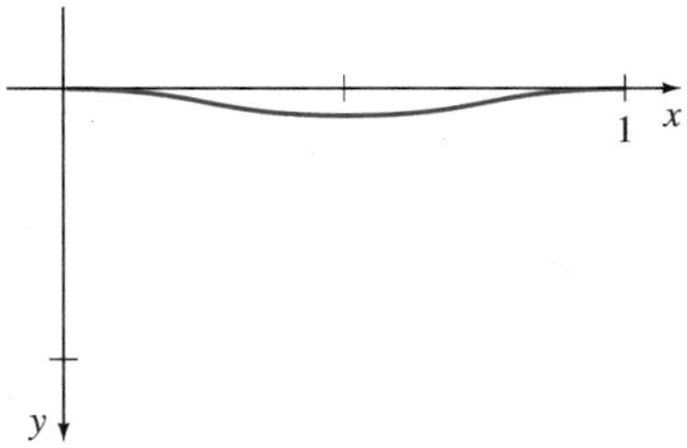

FIGURE 3.32

By choosing $w_0 = 24EI$, and $L = 1$, we obtain the graph of the deflection curve in Figure 3.32.

Eigenvalues and Eigenfunctions

Many applied problems demand that we solve a two-point boundary-value problem involving a linear differential equation that contains a parameter λ. We seek the values of λ for which the boundary-value problem has nontrivial solutions.

EXAMPLE 2 Nontrivial Solutions of BVP

Solve the boundary-value problem

$$y'' + \lambda y = 0, \qquad y(0) = 0, \quad y(L) = 0.$$

Solution We consider three cases: $\lambda = 0$, $\lambda < 0$, and $\lambda > 0$.

CASE I For $\lambda = 0$ the solution of $y'' = 0$ is $y = c_1 x + c_2$. The conditions $y(0) = 0$ and $y(L) = 0$ imply, in turn, $c_2 = 0$ and $c_1 = 0$. Hence for $\lambda = 0$ the only solution of the boundary-value problem is the trivial solution $y = 0$.

CASE II For $\lambda < 0$ we have $y = c_1 \cosh \sqrt{-\lambda}x + c_2 \sinh \sqrt{-\lambda}x$.* Again, $y(0) = 0$ gives $c_1 = 0$, and so $y = c_2 \sinh \sqrt{-\lambda}x$. The second condition $y(L) = 0$ dictates that $c_2 \sinh \sqrt{-\lambda}L = 0$. Since $\sinh \sqrt{-\lambda}L \neq 0$, we must have $c_2 = 0$. Thus $y = 0$.

CASE III For $\lambda > 0$ the general solution of $y'' + \lambda y = 0$ is $y = c_1 \cos \sqrt{\lambda}x + c_2 \sin \sqrt{\lambda}x$. As before, $y(0) = 0$ yields $c_1 = 0$, but $y(L) = 0$ implies

$$c_2 \sin \sqrt{\lambda}L = 0.$$

If $c_2 = 0$, then necessarily $y = 0$. However, if $c_2 \neq 0$, then $\sin \sqrt{\lambda}L = 0$. The last condition implies that the argument of the sine function must be an integer multiple of π:

$$\sqrt{\lambda}L = n\pi \quad \text{or} \quad \lambda = \frac{n^2\pi^2}{L^2}, \qquad n = 1, 2, 3, \ldots.$$

Therefore for any real nonzero c_2, $y = c_2 \sin(n\pi x/L)$ is a solution of the problem for each n. Since the differential equation is homogeneous, we may, if desired, not write c_2. In other words, for a given number in the sequence

$$\frac{\pi^2}{L^2}, \frac{4\pi^2}{L^2}, \frac{9\pi^2}{L^2}, \ldots,$$

* $\sqrt{-\lambda}$ looks a little strange, but bear in mind that $\lambda < 0$ is equivalent to $-\lambda > 0$.

the *corresponding* function in the sequence

$$\sin \frac{\pi}{L} x, \sin \frac{2\pi}{L} x, \sin \frac{3\pi}{L} x, \ldots$$

is a nontrivial solution of the original problem.

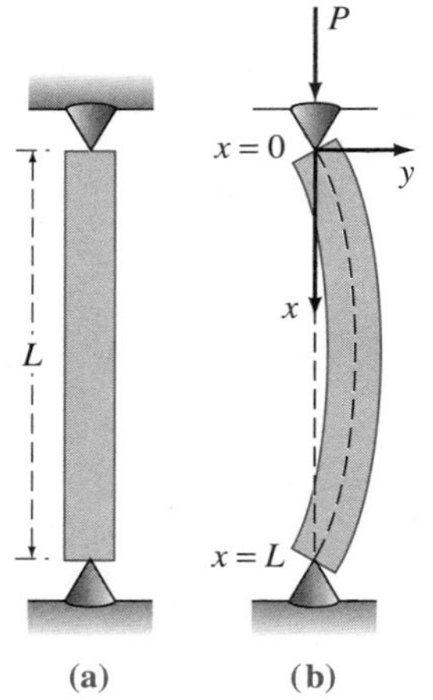

FIGURE 3.33

The numbers $\lambda = n^2\pi^2/L^2$, $n = 1, 2, 3, \ldots$, for which the boundary-value problem in Example 2 has a nontrivial solution are known as **characteristic values** or, more commonly, **eigenvalues**. The solutions depending on these values of λ, $y = c_2 \sin(n\pi x/L)$, or simply $y = \sin(n\pi x/L)$, are called **characteristic functions**, or **eigenfunctions**.

Buckling of a Thin Vertical Column

In the eighteenth century Leonhard Euler was one of the first mathematicians to study an eigenvalue problem in analyzing how a thin elastic column buckles under a compressive axial force.

Consider a long slender vertical column of uniform cross section and length L. Let $y(x)$ denote the deflection of the column when a constant vertical compressive force, or load, P is applied to its top, as shown in Figure 3.33. By comparing bending moments at any point along the column we obtain

$$EI\frac{d^2y}{dx^2} = -Py \quad \text{or} \quad EI\frac{d^2y}{dx^2} + Py = 0, \tag{5}$$

where E is Young's modulus of elasticity and I is the moment of inertia of a cross section about a vertical line through its centroid.

EXAMPLE 3 An Eigenvalue Problem

Find the deflection of a thin vertical homogeneous column of length L subjected to a constant axial load P if the column is hinged at both ends.

Solution The boundary-value problem to be solved is

$$EI\frac{d^2y}{dx^2} + Py = 0, \qquad y(0) = 0, \quad y(L) = 0.$$

First note that $y = 0$ is a perfectly good solution of this problem. This solution has a simple intuitive interpretation: If the load P is not great enough, there is no deflection. The question then is this: For what values of P will the column bend? In mathematical terms: For what values of P does the given boundary-value problem possess *nontrivial* solutions?

By writing $\lambda = P/EI$ we see that

$$y'' + \lambda y = 0, \qquad y(0) = 0, \quad y(L) = 0$$

is identical to the problem in Example 2. From case III of that discussion we see that the deflections are $y(x) = c_2 \sin(n\pi x/L)$ corresponding to the eigenvalues $\lambda_n = P_n/EI = n^2\pi^2/L^2$, $n = 1, 2, 3, \ldots$. Physically this means that the column will buckle or deflect only when the compressive force is one of the values $P_n = n^2\pi^2 EI/L^2$, $n = 1, 2, 3, \ldots$. These different forces are called **critical loads**. The deflection curve corresponding to the smallest critical load $P_1 = \pi^2 EI/L^2$, called the **Euler load**, is $y_1(x) = c_2 \sin(\pi x/L)$ and is known as the **first buckling mode**. ●

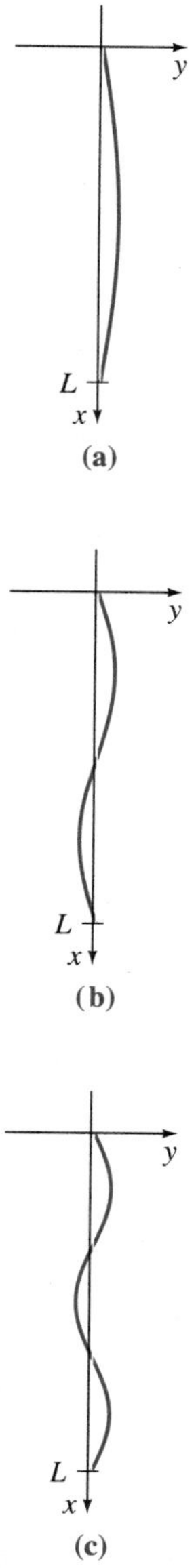

FIGURE 3.34

The deflection curves in Example 3 corresponding to $n = 1$, $n = 2$, and $n = 3$ are shown in Figure 3.34. Note that if the original column has some sort of physical restraint put on it at $x = L/2$, then the smallest critical load would be $P_2 = 4\pi^2 EI/L^2$ and the deflection curve would be as shown in Figure 3.34(b). If restraints are put on the column at $x = L/3$ and at $x = 2L/3$, then the column will not buckle until the critical load $P_3 = 9\pi^2 EI/L^2$ is applied and the deflection curve would be as shown in Figure 3.34(c). Where should physical restraints be placed on the column if we want the Euler load to be P_4?

Numerical Solutions

There are ways of approximating the solution of a boundary-value problem. In Problem 31 we indicate how to replace the linear differential equation in

$$y'' + P(x)y' + Q(x)y = f(x), \qquad y(a) = \alpha, \quad y(b) = \beta \tag{6}$$

by a **finite difference equation**. The latter equation enables us to approximate the value of a solution $y(x)$ of (6) at selected points $x_1, x_2, \ldots, x_{n-1}$ within the interval $[a, b]$. The method results in $n - 1$ algebraic equations in $n - 1$ unknowns $y_1, y_2, \ldots, y_{n-1}$, where $y_1 \approx y(x_1)$ and so on. For reasonable values of n, say 4, 6, or 10, systems of three, five, and nine equations are readily solved using a CAS. See *Computer Lab Experiment* 1 for this section.

Another way of approximating a solution of a boundary-value problem $y'' = f(x, y, y')$, $y(a) = \alpha$, $y(b) = \beta$ is called the **shooting method**. The starting point in this method is the replacement of the BVP by an IVP:

$$y'' = f(x, y, y'), \qquad y(a) = \alpha, \quad y'(a) = m_1. \tag{7}$$

The number m_1 in (7) is simply a guess for the unknown slope of the solution curve at the known point $(a, y(a))$. We then apply one of the step-by-step numerical techniques, such as the improved Euler method or the Runge-Kutta method, to the second-order equation in (7) (see Section 3.10) to find an approximation β_1 for the value of $y(b)$. If β_1 agrees with the given value $y(b) = \beta$ to some preassigned tolerance we stop, otherwise we repeat the calculations, starting with a different guess $y'(a) = m_2$ to obtain a second approximation β_2 for $y(b)$. This method can be continued in a trial-and-error manner, or the subsequent slopes $m_3, m_4, \ldots$ can be adjusted in some systematic way; linear interpolation is particularly successful when the differential equation in (7) is linear. The procedure is analogous to shooting (the "aim" is the choice of the initial slope) at a target until the bullseye $y(b)$ is hit. See *Computer Lab Experiment* 2 for this section.

Of course, underlying the use of these numerical methods is the assumption, which we know is not always warranted, that a solution of the boundary-value problem exists.

EXERCISES 3.8

Answers to odd-numbered problems begin on page AN-15.

In Problems 1–4 the beam is of length L and w_0 is a constant.

1. (a) Solve (4) when the beam is embedded at its left end and free at its right end and $w(x) = w_0$, $0 < x < L$.
   (b) Use a computer graphing program to obtain the graph of the deflection curve of the beam when $w_0 = 24EI$ and $L = 1$.
2. (a) Solve (4) when the beam is simply supported at both ends and $w(x) = w_0$, $0 < x < L$.
   (b) Use a computer graphing program to obtain the graph of the deflection curve of the beam when $w_0 = 24EI$ and $L = 1$.
3. (a) Solve (4) embedded at its left end and simply supported at its right end, and $w(x) = w_0$, $0 < x < L$.
   (b) Use a computer graphing program to obtain the graph of the deflection curve of the beam when $w_0 = 48EI$ and $L = 1$.
4. (a) Solve (4) when the beam is embedded at its left end and simply supported at its right end, and $w(x) = w_0 \sin(\pi x/L)$, $0 < x < L$.
   (b) Use a computer graphing program to obtain the graph of the deflection curve of the beam when $w_0 = 2\pi^3 EI$ and $L = 1$.
5. (a) Find the maximum deflection of the cantilever beam in Problem 1.
   (b) How does the maximum deflection of a beam that is half as long compare with the value in part (a)?
6. (a) Find the maximum deflection of the simply supported beam in Problem 2.
   (b) How does the maximum deflection of the simply supported beam compare with the value of maximum deflection of the embedded beam in Example 1?
7. A cantilever beam of length L is embedded at its right end and a horizontal tensile force of P pounds is applied to its free left end. When the origin is taken at its free end, as shown in Figure 3.35, the deflection $y(x)$ of the beam can be shown to satisfy the differential equation

$$EIy'' = Py - w(x)\frac{x}{2}.$$

Find the deflection of the cantilever beam if $w(x) = w_0 x$, $0 < x < L$, and $y(0) = 0$, $y'(L) = 0$.

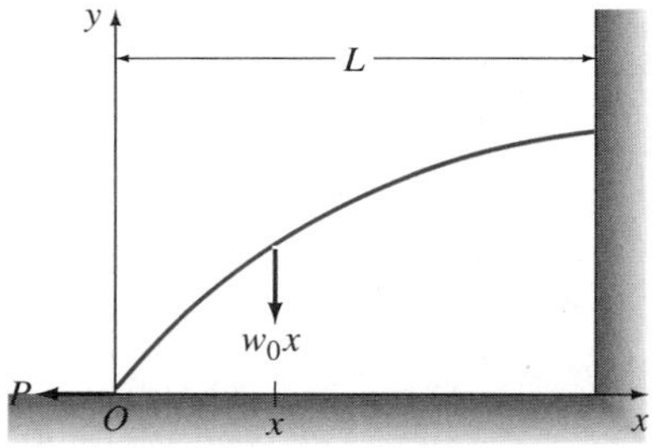

FIGURE 3.35

8. When a compressive instead of a tensile force is applied at the free end of the beam in Problem 7, the differential equation of the deflection is

$$EIy'' = -Py - w(x)\frac{x}{2}.$$

Solve this equation if $w(x) = w_0 x$, $0 < x < L$, and $y(0) = 0$, $y'(L) = 0$.

In Problems 9–22 find the eigenvalues and eigenfunctions for the given boundary-value problem.

9. $y'' + \lambda y = 0$, $y(0) = 0$, $y(\pi) = 0$
10. $y'' + \lambda y = 0$, $y(0) = 0$, $y(\pi/4) = 0$
11. $y'' + \lambda y = 0$, $y'(0) = 0$, $y(L) = 0$
12. $y'' + \lambda y = 0$, $y(0) = 0$, $y'(\pi/2) = 0$

13. $y'' + \lambda y = 0, \quad y'(0) = 0, \quad y'(\pi) = 0$

14. $y'' + \lambda y = 0, \quad y(-\pi) = 0, \quad y(\pi) = 0$

15. $y'' + 2y' + (\lambda + 1)y = 0, \quad y(0) = 0, \quad y(5) = 0$

16. $y'' + (\lambda + 1)y = 0, \quad y'(0) = 0, \quad y'(1) = 0$

17. $y'' + \lambda^2 y = 0, \quad y(0) = 0, \quad y(L) = 0$

18. $y'' + \lambda^2 y = 0, \quad y(0) = 0, \quad y'(3\pi) = 0$

19. $x^2y'' + xy' + \lambda y = 0, \quad y(1) = 0, \quad y(e^{\pi}) = 0$

20. $x^2y'' + xy' + \lambda y = 0, \quad y'(e^{-1}) = 0, \quad y(1) = 0$

21. $x^2y'' + xy' + \lambda y = 0, \quad y'(1) = 0, \quad y'(e^2) = 0$

22. $x^2y'' + 2xy' + \lambda y = 0, \quad y(1) = 0, \quad y(e^2) = 0$

23. **(a)** Show that the eigenfunctions of the boundary-value problem $y'' + \lambda y = 0$, $y(0) = 0$, $y(1) + y'(1) = 0$ are $y = \sin\sqrt{\lambda_n}x$ where the eigenvalues λ_n of the problem are $\lambda_n = x_n^2$, where x_n, $n = 1, 2, 3, \ldots$ are the consecutive *positive* roots of the equation $\tan\sqrt{\lambda} = -\sqrt{\lambda}$.

(b) Use a computer graphing program to convince yourself that the equation $\tan x = -x$ has an infinite number of roots. Explain why the negative roots of the equation can be ignored. Explain why $\lambda = 0$ is not an eigenvalue, even though $x = 0$ is an obvious root of the equation.

(c) Use a numerical procedure and a computer to approximate the first four eigenvalues λ_1, λ_2, λ_3, and λ_4.

24. ***PROJECT PROBLEM*** The point in this problem is to find the mathematical model for the shape of a rotating string or jump rope twirling in a synchronous manner, as shown in Figure 3.36(a).

Suppose the string has length L, constant linear density ρ (mass per unit length), is stretched along the x-axis and is fixed at $x = 0$ and at $x = L$. Suppose the string is then rotated about that axis at a constant angular speed ω. Let $y(x)$ denote the shape of the string or the deflection curve away from its initial position. See Figure 3.36(b).

(a) If the magnitude T of the tension $\mathbf{T}$ acting tangential to the string is constant along the string then use Figure 3.36(c) to express the net vertical force F acting on the string on the interval $[x, x + \Delta x]$ in terms of θ_1 and θ_2.

(b) If the deflection is small on the interval explain why it makes sense that $\sin\theta_1 \approx \tan\theta_1$ and $\sin\theta_2 \approx \tan\theta_2$.

(c) Use the results in part (b) to express the net force F in part (a) in terms of $y'(x)$ and $y'(x + \Delta x)$.

(d) The net force F in part (a) is also given by Newton's second law $F = ma$, where m denotes the mass of the string on the interval and a is the centripetal acceleration of that portion of the string. Show that $F \approx -(\rho\Delta x)y\omega^2$.

(e) Use the preceding parts of this problem to conclude that $T\dfrac{d^2y}{dx^2} + \rho\omega^2 y = 0$.

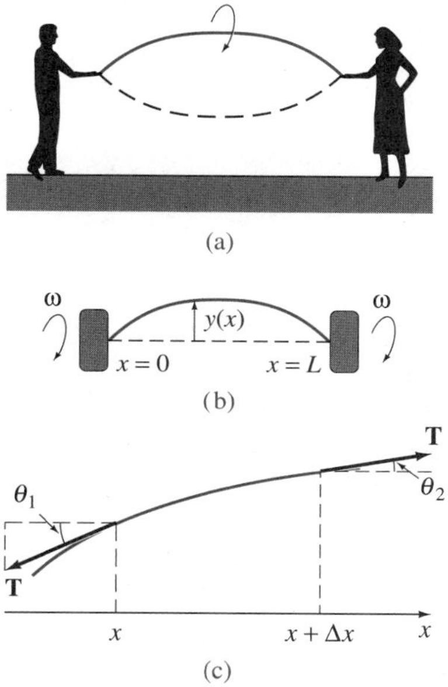

FIGURE 3.36

25. Consider the boundary-value problem introduced in Problem 24:

$$T\frac{d^2y}{dx^2} + \rho\omega^2 y = 0, \qquad y(0) = 0, \quad y(L) = 0.$$

Find the critical speeds ω_n and corresponding deflection curves of the problem.

26. The differential equation for the displacement $y(x)$ of a rotating string in which the tension T is not constant is given by

$$\frac{d}{dx}\left[T(x)\frac{dy}{dx}\right] + \rho\omega^2 y = 0.$$

Suppose $1 < x < e$ and that the tension is given by $T(x) = x^2$.

(a) If $y(1) = 0$, $y(e) = 0$, and $\rho\omega^2 > 0.25$, show that the critical speeds of angular rotation are $\omega_n = \frac{1}{2}\sqrt{(4n^2\pi^2 + 1)/\rho}$ and the corresponding deflection curves are $y_n(x) = c_2x^{-1/2}\sin(n\pi\ln x)$, $n = 1, 2, 3, \ldots$.

(b) Use a computer graphing program to graph the deflection curves on the interval $[1, e]$ for $n = 1, 2, 3$. Choose $c_2 = 1$.

27. Consider two concentric spheres of radius $r = a$ and $r = b$, $a < b$, as shown in Figure 3.37. The temperature $u(r)$ in the region between the spheres is determined from the boundary-value problem

$$r\frac{d^2u}{dr^2} + 2\frac{du}{dr} = 0, \qquad u(a) = u_0, \quad u(b) = u_1,$$

where u_0 and u_1 are constants. Solve for $u(r)$.

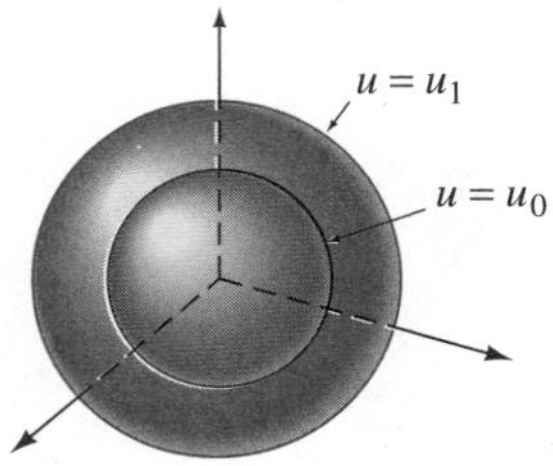

FIGURE 3.37

28. The temperature $u(r)$ in the circular ring shown in Figure 3.38 is determined from the boundary-value problem

$$r\frac{d^2u}{dr^2} + \frac{du}{dr} = 0, \qquad u(a) = u_0, \quad u(b) = u_1,$$

where u_0 and u_1 are constants. Show that

$$u(r) = \frac{u_0 \ln(r/b) - u_1 \ln(r/a)}{\ln(a/b)}.$$

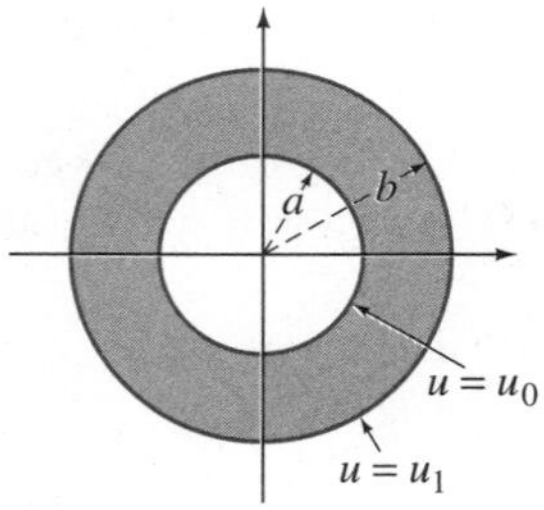

FIGURE 3.38

29. *PROJECT PROBLEM* A uniform solid sphere of radius $r = 1$ at a constant initial temperature throughout is dropped into a large container of fluid that is also kept at a constant, but higher, temperature for all time. See Figure 3.39. To find the temperature $u(r, t)$ inside the sphere for time $t > 0$, it is necessary to solve the following boundary-value problem

$$r\frac{d^2R}{dr^2} + 2\frac{dR}{dr} + \lambda^2 rR = 0, \qquad 0 < r < 1,$$

$$R'(1) = -hR(1), \quad h > 0.$$

This problem determines the radial component of the temperature $u(r, t)$.

(a) Solve the differential equation by means of the substitution $R(r) = y(r)/r$.

(b) Do some extra reading to supply a physical interpretation for the given boundary condition.

(c) Find the first three positive eigenvalues when $h = 0.2$. First supply a physically plausible but implicit condition that can be used to simplify the general solution of the differential equation.

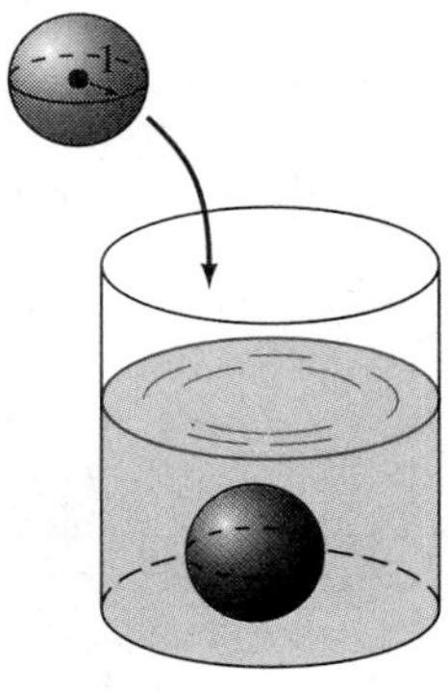

FIGURE 3.39

30. *PROJECT PROBLEM* The critical loads of thin columns depend on the end conditions of the column. The value of the Euler load P_1 in Example 3 was derived under the assumption that the column was hinged at both ends.

Suppose a thin vertical homogeneous column is embedded at its base ($x = 0$) and free at its top ($x = L$) and that an axial load P is applied to its free end. Either this load causes a small deflection δ, as shown in Figure 3.40, or it does not cause such a deflection. In either case the differential equation for the deflection $y(x)$ is

$$EI\frac{d^2y}{dx^2} + Py = P\delta.$$

(a) What is the predicted deflection when $\delta = 0$?

(b) When $\delta \neq 0$, show that the Euler load for this column is one-fourth of the Euler load for the hinged column in Example 3.

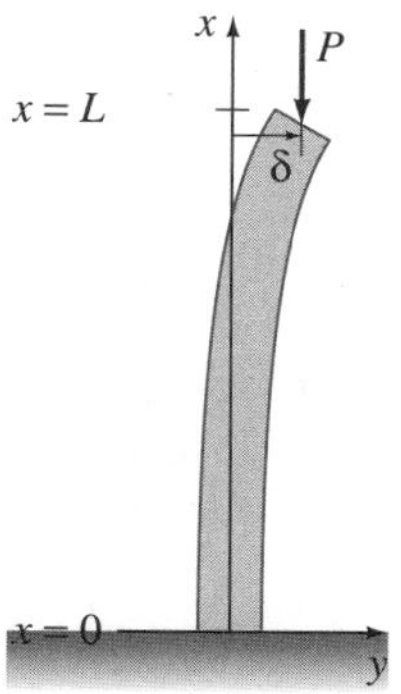

FIGURE 3.40

31. *WRITING PROJECT* The point of this problem is to learn how to approximate solutions of boundary-value problems.

(a) Do some outside reading and write a report on **finite differences**. In your report define **forward difference**, **backward difference**, and **central differences**. Show how the first derivative $y'(x)$ can be approximated by **difference quotients** involving forward, backward, and central differences. Show how the second derivative $y''(x)$ can be approximated by a difference quotient involving a central difference.

(b) Consider the boundary-value problem

$$y'' + P(x)y' + Q(x)y = f(x), \qquad y(a) = \alpha, \quad y(b) = \beta,$$

and suppose that $a = x_0 < x_1 < x_2 < \cdots < x_{n-1} < x_n = b$ is a regular partition of the interval $[a, b]$. For $h = (b - a)/n$, the points $x_i = a + ih$, $i = 1, 2, \ldots, n - 1$ are called interior mesh points of the interval. Let $y_i = y(x_i)$, $P_i = P(x_i)$, $Q_i = Q(x_i)$, $f_i = f(x_i)$, $y_0 = y(a) = \alpha$, and $y_n = y(b) = \beta$. Use the information in part (a) to show how the differential equation can be replaced by the **finite difference equation**

$$\left(1 + \frac{h}{2}P_i\right)y_{i+1} + (-2 + h^2 Q_i)y_i + \left(1 - \frac{h}{2}P_i\right)y_{i-1} = h^2 f_i.$$

For an application of this formula see the *Computer Lab Experiment* 1 for Section 3.8.

32. *WRITING PROJECT* Particular solutions of linear **partial differential equations** are often obtained by a method known as **separation of variables**. This method reduces a PDE to two or more ODEs. Write a report illustrating separation of variables using three classical partial differential equations: the one-dimensional **heat equation**, the one-dimensional **wave equation**, and **Laplace's equation** in two dimensions. Describe typical boundary-value problems involving these equations. Show how boundary conditions associated with the PDE can, in some circumstances, translate into boundary conditions for ODEs. Show how eigenvalues and eigenfunctions arise naturally in this context.

33. *WRITING PROJECT* One of the most important problems, which furnishes sort of a bridge between ordinary differential equations and aspects of applied partial differential equations, is called the **Sturm-Liouville problem**. Write a report on this problem, that includes the properties of the solutions of the Sturm-Liouville problem. Also define and illustrate the concepts of the **self-adjoint form** of a differential equation and **orthogonal set of functions**. If directed by your instructor, illustrate how this theory is useful in finding a particular solution of a boundary-value problem associated with one of the classic partial differential equations mentioned in Problem 32.

3.9 SYSTEMS OF DIFFERENTIAL EQUATIONS

Simultaneous ordinary differential equations involve two or more equations that contain derivatives of two or more unknown functions of a single independent variable. If x, y, and z are functions of the variable t, then

$$4\frac{d^2x}{dt^2} = -5x + y \qquad\qquad x' - 3x + y' + z' = 5$$

$$\text{and} \quad x' - y' + 2z' = t^2$$

$$2\frac{d^2y}{dt^2} = 3x - y \qquad\qquad x + y' - 6z' = t - 1$$

are two examples of systems of simultaneous differential equations.

Solution of a System

A **solution** of a system of differential equations is a set of differentiable functions $x = f(t)$, $y = g(t)$, $z = h(t)$, and so on, that satisfies each equation of the system on some interval I.

Systematic Elimination

The first technique we consider for solving systems is based on the fundamental principle of **systematic algebraic elimination** of variables. We shall see that the analogue of multiplying an algebraic equation by a constant is operating on a differential equation with some combination of derivatives. In calculus, the symbol D^n is often used to denote the nth-order derivative of a function $D^n y = d^n y/dt^n$. Hence a linear differential equation

$$a_n y^{(n)} + a_{n-1} y^{(n-1)} + \cdots + a_1 y' + a_0 y = g(t),$$

where the a_i, $i = 0, 1, \ldots, n$ are constants, can be written as

$$a_n D^n y + a_{n-1} D^{n-1} y + \cdots + a_1 Dy + a_0 y = g(t)$$

or

$$(a_n D^n + a_{n-1} D^{n-1} + \cdots + a_1 D + a_0)y = g(t).$$

The expression $a_n D^n + a_{n-1} D^{n-1} + \cdots + a_1 D + a_0$ is called a **linear nth-order differential operator** and is often abbreviated as $P(D)$. Since $P(D)$ is a polynomial in the symbol D, we may be able to factor it into differential operators of lower order. Moreover, the factors of $P(D)$ commute. For example, the linear second-order differential operator $D^2 + 5D + 6$ can be written as $(D + 2)(D + 3)$ or $(D + 3)(D + 2)$.

EXAMPLE 1 System Written in Operator Notation

Write the system of differential equations

$$x'' + 2x' + y'' = x + 3y + \sin t$$
$$x' + y' = -4x + 2y + e^{-t}$$

in operator notation.

Solution Rewrite the given system as

$$\begin{aligned} x'' + 2x' - x + y'' - 3y &= \sin t \\ x' + 4x + y' - 2y &= e^{-t} \end{aligned} \quad \text{so that} \quad \begin{aligned} (D^2 + 2D - 1)x + (D^2 - 3)y &= \sin t \\ (D + 4)x + (D - 2)y &= e^{-t}. \end{aligned}$$

Method of Solution

Consider the simple system of linear first-order equations

$$\begin{aligned} Dy &= 2x \\ Dx &= 3y \end{aligned} \qquad \textbf{(1)}$$

or equivalently,

$$\begin{aligned} 2x - Dy &= 0 \\ Dx - 3y &= 0. \end{aligned} \tag{2}$$

Operating on the first equation in (2) by D while multiplying the second by 2 and then subtracting, eliminates x from the system. It follows that

$$-D^2y + 6y = 0 \quad \text{or} \quad D^2y - 6y = 0.$$

Since the roots of the auxiliary equation are $m_1 = \sqrt{6}$ and $m_2 = -\sqrt{6}$, we obtain

$$y(t) = c_1e^{\sqrt{6}t} + c_2e^{-\sqrt{6}t}. \tag{3}$$

Multiplying the first equation by -3 while operating on the second by D and then adding give the differential equation for x, $D^2x - 6x = 0$. It follows immediately that

$$x(t) = c_3e^{\sqrt{6}t} + c_4e^{-\sqrt{6}t}. \tag{4}$$

Now (3) and (4) do not satisfy the system (1) for every choice of c_1, c_2, c_3, and c_4. Substituting $x(t)$ and $y(t)$ into the first equation of the original system (1) gives, after we simplify,

$$(\sqrt{6}c_1 - 2c_3)e^{\sqrt{6}t} + (-\sqrt{6}c_2 - 2c_4)e^{-\sqrt{6}t} = 0.$$

Since the latter expression is to be zero for all values of t, we must have

$$\sqrt{6}c_1 - 2c_3 = 0 \quad \text{and} \quad -\sqrt{6}c_2 - 2c_4 = 0$$

or

$$c_3 = \frac{\sqrt{6}}{2}c_1, \quad c_4 = -\frac{\sqrt{6}}{2}c_2. \tag{5}$$

Hence we conclude that a solution of the system must be

$$\begin{aligned} x(t) &= \frac{\sqrt{6}}{2}c_1e^{\sqrt{6}t} - \frac{\sqrt{6}}{2}c_2e^{-\sqrt{6}t} \\ y(t) &= c_1e^{\sqrt{6}t} + c_2e^{-\sqrt{6}t}. \end{aligned}$$

You are urged to substitute (3) and (4) into the second equation of (1) and verify that the same relationship (5) holds between the constants.

EXAMPLE 2 Solution by Elimination

Solve

$$\begin{aligned} Dx + (D + 2)y &= 0 \\ (D - 3)x - \quad 2y &= 0. \end{aligned} \tag{6}$$

Solution Operating on the first equation by $D - 3$ and on the second by D and subtracting eliminate x from the system. It follows that the differential equation for y is

$$[(D - 3)(D + 2) + 2D]y = 0 \quad \text{or} \quad (D^2 + D - 6)y = 0.$$

Since the characteristic equation of this last differential equation is $m^2 + m - 6 = (m - 2)(m + 3) = 0$, we obtain the solution

$$y(t) = c_1e^{2t} + c_2e^{-3t}. \tag{7}$$

Eliminating y in a similar manner yields $(D^2 + D - 6)x = 0$, from which we find

$$x(t) = c_3e^{2t} + c_4e^{-3t}. \tag{8}$$

As we noted in the foregoing discussion, a solution of (6) does not contain four independent constants since the system itself puts a constraint on the actual number that can be chosen arbitrarily. Substituting (7) and (8) into the first equation of (6) gives

$$(4c_1 + 2c_3)e^{2t} + (-c_2 - 3c_4)e^{-3t} = 0.$$

From $4c_1 + 2c_3 = 0$ and $-c_2 - 3c_4 = 0$ we get $c_3 = -2c_1$ and $c_4 = -\frac{1}{3}c_2$. Accordingly, a solution of the system is

$$x(t) = -2c_1e^{2t} - \frac{1}{3}c_2e^{-3t}$$

$$y(t) = c_1e^{2t} + c_2e^{-3t}.$$

Since we could just as easily solve for c_3 and c_4 in terms of c_1 and c_2, the solution in Example 2 can be written in the alternative form

$$x(t) = c_3e^{2t} + c_4e^{-3t}$$

$$y(t) = -\frac{1}{2}c_3e^{2t} - 3c_4e^{-3t}.$$

Also, it sometimes pays to keep one's eyes open when solving systems. Had we solved for x first, then y could be found, along with the relationship between the constants by using the last equation in (6). You should verify that substituting $x(t)$ in $y = \frac{1}{2}(Dx - 3x)$ yields $y = -\frac{1}{2}c_3e^{2t} - 3c_4e^{-3t}$.

EXAMPLE 3 Solution by Elimination

Solve

$$\begin{aligned} x' - 4x + y'' &= t^2 \\ x' + x + y' &= 0. \end{aligned} \tag{9}$$

Solution First we write the system in differential operator notation:

$$\begin{aligned} (D - 4)x + D^2y &= t^2 \\ (D + 1)x + Dy &= 0. \end{aligned} \tag{10}$$

Then, by eliminating x, we obtain

$$[(D + 1)D^2 - (D - 4)D]y = (D + 1)t^2 - (D - 4)0$$

or

$$(D^3 + 4D)y = t^2 + 2t.$$

Since the roots of the auxiliary equation $m(m^2 + 4) = 0$ are $m_1 = 0$, $m_2 = 2i$, and $m_3 = -2i$, the complementary function is

$$y_c = c_1 + c_2 \cos 2t + c_3 \sin 2t.$$

To determine the particular solution y_p, we use undetermined coefficients by assuming $y_p = At^3 + Bt^2 + Ct$. Therefore

$$y_p' = 3At^2 + 2Bt + C, \quad y_p'' = 6At + 2B, \quad y_p''' = 6A,$$

$$y_p''' + 4y_p' = 12At^2 + 8Bt + 6A + 4C = t^2 + 2t.$$

The last equality implies

$$12A = 1, \quad 8B = 2, \quad 6A + 4C = 0,$$

and hence $A = 1/12$, $B = 1/4$, $C = -1/8$. Thus

$$y = y_c + y_p = c_1 + c_2 \cos 2t + c_3 \sin 2t + \frac{1}{12}t^3 + \frac{1}{4}t^2 - \frac{1}{8}t. \tag{11}$$

Eliminating y from the system (10) leads to

$$[(D - 4) - D(D + 1)]x = t^2 \quad \text{or} \quad (D^2 + 4)x = -t^2.$$

It should be obvious that

$$x_c = c_4 \cos 2t + c_5 \sin 2t$$

and that undetermined coefficients can be applied to obtain a particular solution of the form $x_p = At^2 + Bt + C$. In this case the usual differentiations and algebra yield $x_p = -\frac{1}{4}t^2 + \frac{1}{8}$ and so

$$x = x_c + x_p = c_4 \cos 2t + c_5 \sin 2t - \frac{1}{4}t^2 + \frac{1}{8}. \tag{12}$$

Now c_4 and c_5 can be expressed in terms of c_2 and c_3 by substituting (11) and (12) into either equation of (9). By using the second equation, we find, after combining terms,

$$(c_5 - 2c_4 - 2c_2) \sin 2t + (2c_5 + c_4 + 2c_3) \cos 2t = 0$$

so that $\quad c_5 - 2c_4 - 2c_2 = 0 \quad \text{and} \quad 2c_5 + c_4 + 2c_3 = 0.$

Solving for c_4 and c_5 in terms of c_2 and c_3 gives

$$c_4 = -\frac{1}{5}(4c_2 + 2c_3) \quad \text{and} \quad c_5 = \frac{1}{5}(2c_2 - 4c_3).$$

Finally, a solution of (9) is found to be

$$x(t) = -\frac{1}{5}(4c_2 + 2c_3) \cos 2t + \frac{1}{5}(2c_2 - 4c_3) \sin 2t - \frac{1}{4}t^2 + \frac{1}{8}$$

$$y(t) = c_1 + c_2 \cos 2t + c_3 \sin 2t + \frac{1}{12}t^3 + \frac{1}{4}t^2 - \frac{1}{8}t.$$

Use of Determinants

Symbolically, if L_1, L_2, L_3, and L_4 denote linear differential operators with constant coefficients, then a system of linear differential equations in two variables x and y can be written as

$$\begin{aligned} L_1x + L_2y &= g_1(t) \\ L_3x + L_4y &= g_2(t). \end{aligned} \tag{13}$$

Eliminating variables, as we would for algebraic equations, leads to

$$(L_1L_4 - L_2L_3)x = f_1(t) \quad \text{and} \quad (L_1L_4 - L_2L_3)y = f_2(t), \tag{14}$$

where

$$f_1(t) = L_4g_1(t) - L_2g_2(t) \quad \text{and} \quad f_2(t) = L_1g_2(t) - L_3g_1(t).$$

Formally the results in (14) can be written in terms of determinants similar to those used in Cramer's rule:

$$\begin{vmatrix} L_1 & L_2 \\ L_3 & L_4 \end{vmatrix} x = \begin{vmatrix} g_1 & L_2 \\ g_2 & L_4 \end{vmatrix} \quad \text{and} \quad \begin{vmatrix} L_1 & L_2 \\ L_3 & L_4 \end{vmatrix} y = \begin{vmatrix} L_1 & g_1 \\ L_3 & g_2 \end{vmatrix}. \tag{15}$$

The left-hand determinant in each equation in (15) can be expanded in the usual algebraic sense, with the result then operating on the functions $x(t)$ and $y(t)$. However, some care should be exercised in the expansion of the right-hand determinants in (15). We must expand these determinants in the sense of the internal differential operators actually operating upon the functions $g_1(t)$ and $g_2(t)$.

If
$$\begin{vmatrix} L_1 & L_2 \\ L_3 & L_4 \end{vmatrix} \neq 0$$

in (15) and is a differential operator of order n, then:

- The system (13) can be decoupled into two nth-order differential equations in x and y.
- The characteristic equation and hence the complementary function of each of these differential equations are the same.
- Since x and y both contain n constants, there is a total of $2n$ constants appearing.
- The total number of *independent* constants in the solution of the system is n.

If
$$\begin{vmatrix} L_1 & L_2 \\ L_3 & L_4 \end{vmatrix} = 0$$

in (13), then the system may have a solution containing any number of independent constants or may have no solution at all. Similar remarks hold for systems larger than indicated in (13).

EXAMPLE 4 Solution by Determinants

Solve

$$\begin{aligned} x' &= 3x - y - 12 \\ y' &= x + y + 4e^t. \end{aligned} \tag{16}$$

Solution Write the system in terms of differential operators:

$$\begin{aligned} (D - 3)x + \qquad y &= -12 \\ -x + (D - 1)y &= 4e^t \end{aligned}$$

and then use determinants:

$$\begin{vmatrix} D-3 & 1 \\ -1 & D-1 \end{vmatrix} x = \begin{vmatrix} -12 & 1 \\ 4e^t & D-1 \end{vmatrix}$$

$$\begin{vmatrix} D-3 & 1 \\ -1 & D-1 \end{vmatrix} y = \begin{vmatrix} D-3 & -12 \\ -1 & 4e^t \end{vmatrix}.$$

After expanding, we find that

$$(D - 2)^2x = 12 - 4e^t \quad \text{and} \quad (D - 2)^2y = -12 - 8e^t.$$

By the usual methods it follows that

$$x = x_c + x_p = c_1e^{2t} + c_2te^{2t} + 3 - 4e^t \tag{17}$$

$$y = y_c + y_p = c_3e^{2t} + c_4te^{2t} - 3 - 8e^t. \tag{18}$$

Substituting (17) and (18) into the second equation of (16) gives

$$(c_3 - c_1 + c_4)e^{2t} + (c_4 - c_2)te^{2t} = 0,$$

which then implies $c_4 = c_2$ and $c_3 = c_1 - c_4 = c_1 - c_2$. Thus a solution of (16) is

$$\begin{aligned} x(t) &= c_1e^{2t} + c_2te^{2t} + 3 - 4e^t \\ y(t) &= (c_1 - c_2)e^{2t} + c_2te^{2t} - 3 - 8e^t. \end{aligned}$$

●

EXERCISES 3.9

Answers to odd-numbered problems begin on page AN-16.

In Problems 1–22 solve, if possible, the given system of differential equations by either systematic elimination or determinants.

1. $\dfrac{dx}{dt} = 2x - y$
   $\dfrac{dy}{dt} = x$

2. $\dfrac{dx}{dt} = 4x + 7y$
   $\dfrac{dy}{dt} = x - 2y$

3. $\dfrac{dx}{dt} = -y + t$
   $\dfrac{dy}{dt} = x - t$

4. $\dfrac{dx}{dt} - 4y = 1$
   $x + \dfrac{dy}{dt} = 2$

5. $(D^2 + 5)x - \qquad 2y = 0$
   $-2x + (D^2 + 2)y = 0$

6. $(D+1)x + (D-1)y = 2$
$3x + (D+2)y = -1$

7. $\dfrac{d^2x}{dt^2} = 4y + e^t$
$\dfrac{d^2y}{dt^2} = 4x - e^t$

8. $\dfrac{d^2x}{dt^2} + \dfrac{dy}{dt} = -5x$
$\dfrac{dx}{dt} + \dfrac{dy}{dt} = -x + 4y$

9. $Dx + D^2y = e^{3t}$
$(D+1)x + (D-1)y = 4e^{3t}$

10. $D^2x - Dy = t$
$(D+3)x + (D+3)y = 2$

11. $(D^2-1)x - y = 0$
$(D-1)x + Dy = 0$

12. $(2D^2 - D - 1)x - (2D+1)y = 1$
$(D-1)x + Dy = -1$

13. $2\dfrac{dx}{dt} - 5x + \dfrac{dy}{dt} = e^t$
$\dfrac{dx}{dt} - x + \dfrac{dy}{dt} = 5e^t$

14. $\dfrac{dx}{dt} + \dfrac{dy}{dt} = e^t$
$-\dfrac{d^2x}{dt^2} + \dfrac{dx}{dt} + x + y = 0$

15. $(D-1)x + (D^2+1)y = 1$
$(D^2-1)x + (D+1)y = 2$

16. $D^2x - 2(D^2 + D)y = \sin t$
$x + Dy = 0$

17. $Dx = y$
$Dy = z$
$Dz = x$

18. $Dx + z = e^t$
$(D-1)x + Dy + Dz = 0$
$x + 2y + Dz = e^t$

19. $\dfrac{dx}{dt} - 6y = 0$
$x - \dfrac{dy}{dt} + z = 0$
$x + y - \dfrac{dz}{dt} = 0$

20. $\dfrac{dx}{dt} = -x + z$
$\dfrac{dy}{dt} = -y + z$
$\dfrac{dz}{dt} = -x + y$

21. $2Dx + (D-1)y = t$
$Dx + Dy = t^2$

22. $Dx - 2Dy = t^2$
$(D+1)x - 2(D+1)y = 1$

In Problems 23 and 24 solve the given system subject to the indicated initial conditions.

23. $\dfrac{dx}{dt} = -5x - y$
$\dfrac{dy}{dt} = 4x - y$
$x(1) = 0,\ y(1) = 1$

24. $\dfrac{dx}{dt} = y - 1$
$\dfrac{dy}{dt} = -3x + 2y$
$x(0) = 0,\ y(0) = 0$

3.10 NUMERICAL SOLUTIONS

Second-Order Initial-Value Problems

In Sections 2.6–2.8 we focused on numerical techniques that could be applied to obtain an approximation to the solution of a first-order initial-value problem $y' = f(x, y)$, $y(x_0) = y_0$. To approximate the solution of a second-order initial-value problem

$$y'' = f(x, y, y'), \qquad y(x_0) = y_0, \quad y'(x_0) = y_1, \tag{1}$$

we reduce the differential equation to a system of two first-order equations. When we let $y' = u$, the initial-value problem in (1) becomes

$$\begin{aligned} y' &= u \\ u' &= f(x, y, u) \end{aligned} \tag{2}$$
$$y(x_0) = y_0, \quad u(x_0) = y_1.$$

We can now solve this system numerically by adapting the techniques discussed in Sections 2.6–2.8 to the system. We do this by simply applying a particular method to each equation in the system. For example, **Euler's method** applied to the system (2) would be

$$\begin{aligned} y_{n+1} &= y_n + hu_n \\ u_{n+1} &= u_n + hf(x_n, y_n, u_n). \end{aligned} \tag{3}$$

EXAMPLE 1 Euler's Method

Use Euler's method to obtain the approximate value of $y(0.2)$, where $y(x)$ is the solution of the initial-value problem

$$y'' + xy' + y = 0, \qquad y(0) = 1, \quad y'(0) = 2.$$

Solution In terms of the substitution $y' = u$, the equation is equivalent to the system

$$\begin{aligned} y' &= u \\ u' &= -xu - y. \end{aligned}$$

Thus from (3) we obtain

$$\begin{aligned} y_{n+1} &= y_n + hu_n \\ u_{n+1} &= u_n + h[-x_n u_n - y_n]. \end{aligned}$$

Using the step size $h = 0.1$ and $y_0 = 1$, $u_0 = 2$, we find

$$\begin{aligned} y_1 &= y_0 + (0.1)u_0 = 1 + (0.1)2 = 1.2 \\ u_1 &= u_0 + (0.1)[-x_0u_0 - y_0] = 2 + (0.1)[-(0)(2) - 1] = 1.9 \\ y_2 &= y_1 + (0.1)u_1 = 1.2 + (0.1)(1.9) = 1.39 \\ u_2 &= u_1 + (0.1)[-x_1u_1 - y_1] = 1.9 + (0.1)[-(0.1)(1.9) - 1.2] = 1.761. \end{aligned}$$

In other words, $y(0.2) \approx 1.39$ and $y'(0.2) \approx 1.761$. ●

In general, we can reduce every nth-order differential equation $y^{(n)} = f(x, y, y', \ldots, y^{(n-1)})$ to a system of n first-order equations using the substitutions $y = u_1$, $y' = u_2$, $y'' = u_3, \ldots, y^{(n-1)} = u_n$.

Systems Reduced to First-Order Systems

Using a procedure similar to that just discussed, we can often reduce a system of higher-order differential equations to a system of first-order equations by first solving for the highest order derivative of each dependent variable and then making appropriate substitutions for the lower order derivatives.

EXAMPLE 2 A System Rewritten as a First-Order System

Write
$$x'' - x' + 5x + 2y'' = e^t$$
$$-2x + y'' + 2y = 3t^2$$
as a system of first-order differential equations.

Solution Write the system as
$$x'' + 2y'' = e^t - 5x + x'$$
$$y'' = 3t^2 + 2x - 2y$$
and then eliminate y'' by multiplying the second equation by 2 and subtracting. This gives
$$x'' = -9x + 4y + x' + e^t - 6t^2.$$
Since the second equation of the system already expresses the highest-order derivative of y in terms of the remaining functions, we are now in a position to introduce new variables. If we let $x' = u$ and $y' = v$, the expressions for x'' and y'' become, respectively,
$$u' = x'' = -9x + 4y + u + e^t - 6t^2$$
$$v' = y'' = 2x - 2y + 3t^2.$$
The original system can then be written in the form
$$x' = u$$
$$y' = v$$
$$u' = -9x + 4y + u + e^t - 6t^2$$
$$v' = 2x - 2y + 3t^2.$$

The reductions illustrated in Example 2 may not always be possible to carry out. See Problem 13 in Exercises 3.10.

Numerical Solution of a System

The solution of a system of the form
$$\frac{dx_1}{dt} = f_1(t, x_1, x_2, \ldots, x_n)$$
$$\frac{dx_2}{dt} = f_2(t, x_1, x_2, \ldots, x_n)$$
$$\vdots$$
$$\frac{dx_n}{dt} = f_n(t, x_1, x_2, \ldots, x_n)$$

can be approximated by a version of the Euler, Runge-Kutta, or Adams-Bashforth/Adams-Moulton methods adapted to the system. For example, the **fourth-order Runge-Kutta method** applied to the system

$$x' = f(t, x, y)$$
$$y' = g(t, x, y) \tag{4}$$
$$x(t_0) = x_0, \quad y(t_0) = y_0$$

looks like this:

$$x_{n+1} = x_n + \frac{1}{6}(m_1 + 2m_2 + 2m_3 + m_4)$$
$$y_{n+1} = y_n + \frac{1}{6}(k_1 + 2k_2 + 2k_3 + k_4), \tag{5}$$

where

$$m_1 = hf(t_n, x_n, y_n) \qquad k_1 = hg(t_n, x_n, y_n)$$
$$m_2 = hf\left(t_n + \frac{1}{2}h, x_n + \frac{1}{2}m_1, y_n + \frac{1}{2}k_1\right) \qquad k_2 = hg\left(t_n + \frac{1}{2}h, x_n + \frac{1}{2}m_1, y_n + \frac{1}{2}k_1\right)$$
$$m_3 = hf\left(t_n + \frac{1}{2}h, x_n + \frac{1}{2}m_2, y_n + \frac{1}{2}k_2\right) \qquad k_3 = hg\left(t_n + \frac{1}{2}h, x_n + \frac{1}{2}m_2, y_n + \frac{1}{2}k_2\right) \tag{6}$$
$$m_4 = hf(t_n + h, x_n + m_3, y_n + k_3) \qquad k_4 = hg(t_n + h, x_n + m_3, y_n + k_3).$$

EXAMPLE 3 Runge-Kutta Method

Consider the initial-value problem

$$x' = 2x + 4y$$
$$y' = -x + 6y$$
$$x(0) = -1, \quad y(0) = 6.$$

Use the fourth-order Runge-Kutta method to approximate $x(0.6)$ and $y(0.6)$. Compare the results for $h = 0.2$ and $h = 0.1$.

Solution We illustrate the computations of x_1 and y_1 with the step size $h = 0.2$. With the identifications $f(t, x, y) = 2x + 4y$, $g(t, x, y) = -x + 6y$, $t_0 = 0$, $x_0 = -1$, and $y_0 = 6$, we see from (6) that

$$m_1 = hf(t_0, x_0, y_0) = 0.2f(0, -1, 6) = 0.2[2(-1) + 4(6)] = 4.4000$$
$$k_1 = hg(t_0, x_0, y_0) = 0.2g(0, -1, 6) = 0.2[-1(-1) + 6(6)] = 7.4000$$
$$m_2 = hf\left(t_0 + \frac{1}{2}h, x_0 + \frac{1}{2}m_1, y_0 + \frac{1}{2}k_1\right) = 0.2f(0.1, 1.2, 9.7) = 8.2400$$

$$k_2 = hg\left(t_0 + \frac{1}{2}h, x_0 + \frac{1}{2}m_1, y_0 + \frac{1}{2}k_1\right) = 0.2g(0.1, 1.2, 9.7) = 11.4000$$

$$m_3 = hf\left(t_0 + \frac{1}{2}h, x_0 + \frac{1}{2}m_2, y_0 + \frac{1}{2}k_2\right) = 0.2f(0.1, 3.12, 11.7) = 10.6080$$

$$k_3 = hg\left(t_0 + \frac{1}{2}h, x_0 + \frac{1}{2}m_2, y_0 + \frac{1}{2}k_2\right) = 0.2g(0.1, 3.12, 11.7) = 13.4160$$

$$m_4 = hf(t_0 + h, x_0 + m_3, y_0 + k_3) = 0.2f(0.2, 8, 20.216) = 19.3760$$

$$k_4 = hg(t_0 + h, x_0 + m_3, y_0 + k_3) = 0.2g(0.2, 8, 20.216) = 21.3776.$$

Therefore, from (5) we get

$$x_1 = x_0 + \frac{1}{6}(m_1 + 2m_2 + 2m_3 + m_4)$$

$$= -1 + \frac{1}{6}(4.4 + 2(8.24) + 2(10.608) + 19.3760) = 9.2453$$

$$y_1 = y_0 + \frac{1}{6}(k_1 + 2k_2 + 2k_3 + k_4)$$

$$= 6 + \frac{1}{6}(7.4 + 2(11.4) + 2(13.416) + 21.3776) = 19.0683,$$

where, as usual, the computed values are rounded to four decimal places. These numbers give us the approximations $x_1 \approx x(0.2)$ and $y_1 \approx y(0.2)$. The subsequent values, obtained with the aid of a computer, are summarized in Tables 3.2 and 3.3. ●

TABLE 3.2 Runge-Kutta Method with $h = 0.2$

m_1	m_2	m_3	m_4	k_1	k_2	k_3	k_4	t_n	x_n	y_n
								0.00	−1.0000	6.0000
4.4000	8.2400	10.6080	19.3760	7.4000	11.4000	13.4160	21.3776	0.20	9.2453	19.0683
18.9527	31.1564	37.8870	63.6848	21.0329	31.7573	36.9716	57.8214	0.40	46.0327	55.1203
62.5093	97.7863	116.0063	187.3669	56.9378	84.8495	98.0688	151.4191	0.60	158.9430	150.8192

TABLE 3.3 Runge-Kutta Method with $h = 0.1$

m_1	m_2	m_3	m_4	k_1	k_2	k_3	k_4	t_n	x_n	y_n
								0.00	−1.0000	6.0000
2.2000	3.1600	3.4560	4.8720	3.7000	4.7000	4.9520	6.3256	0.10	2.3840	10.8883
4.8321	6.5742	7.0778	9.5870	6.2946	7.9413	8.3482	10.5957	0.20	9.3379	19.1332
9.5208	12.5821	13.4258	17.7609	10.5461	13.2339	13.8872	17.5358	0.30	22.5541	32.8539
17.6524	22.9090	24.3055	31.6554	17.4569	21.8114	22.8549	28.7393	0.40	46.5103	55.4420
31.4788	40.3496	42.6387	54.9202	28.6141	35.6245	37.2840	46.7207	0.50	88.5729	93.3006
54.6348	69.4029	73.1247	93.4107	46.5231	57.7482	60.3774	75.4370	0.60	160.7563	152.0025

You should verify that the solution of the initial-value problem in Example 3 is given by $x(t) = (26t - 1)e^{4t}$, $y(t) = (13t + 6)e^{4t}$. From these equations we see that the exact values are $x(0.6) = 160.9384$ and $y(0.6) = 152.1198$.

In conclusion, we state **Euler's method** for the general system (4) as

$$x_{n+1} = x_n + hf(t_n, x_n, y_n)$$

$$y_{n+1} = y_n + hg(t_n, x_n, y_n).$$

EXERCISES 3.10

Answers to odd-numbered problems begin on page AN-16.

1. Use the Euler method to approximate $y(0.2)$, where $y(x)$ is the solution of the initial-value problem

$$y'' - 4y' + 4y = 0, \qquad y(0) = -2, \quad y'(0) = 1.$$

Use $h = 0.1$. Find the exact solution of the problem and compare the exact value of $y(0.2)$ with y_2.

2. Use the Euler method to approximate $y(1.2)$, where $y(x)$ is the solution of the initial-value problem

$$x^2y'' - 2xy' + 2y = 0, \qquad y(1) = 4, \quad y'(1) = 9,$$

where $x > 0$. Use $h = 0.1$. Find the exact solution of the problem and compare the exact value of $y(1.2)$ with y_2.

3. Repeat Problem 1 using the Runge-Kutta method with $h = 0.2$ and $h = 0.1$.

4. Repeat Problem 2 using the Runge-Kutta method with $h = 0.2$ and $h = 0.1$.

5. Use the Runge-Kutta method to obtain the approximate value of $y(0.2)$, where $y(x)$ is a solution of the initial-value problem

$$y'' - 2y' + 2y = e^t \cos t, \qquad y(0) = 1, \quad y'(0) = 2.$$

Use $h = 0.2$ and $h = 0.1$.

6. When $E = 100$ volts, $R = 10$ ohms, and $L = 1$ henry, the system of differential equations for the currents $i_1(t)$ and $i_3(t)$ in the electrical network given in Figure 3.41 is

$$\frac{di_1}{dt} = -20i_1 + 10i_3 + 100$$

$$\frac{di_3}{dt} = 10i_1 - 20i_3,$$

where $i_1(0) = 0$ and $i_3(0) = 0$. Use the Runge-Kutta method to approximate $i_1(t)$ and $i_3(t)$ at $t = 0.1, 0.2, 0.3, 0.4$, and 0.5. Use $h = 0.1$.

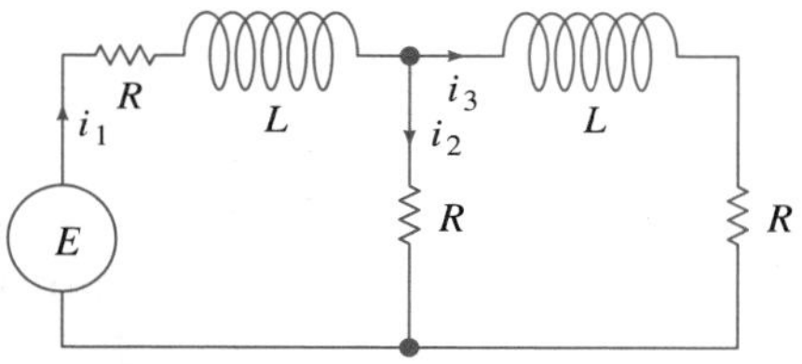

FIGURE 3.41

In Problems 7–12 use the Runge-Kutta method to approximate $x(0.2)$ and $y(0.2)$. Compare the results for $h = 0.2$ and $h = 0.1$.

7. $x' = 2x - y$
$y' = x$
$x(0) = 6,\ y(0) = 2$

8. $x' = x + 2y$
$y' = 4x + 3y$
$x(0) = 1,\ y(0) = 1$

9. $x' = -y + t$
$y' = x - t$
$x(0) = -3,\ y(0) = 5$

10. $x' = 6x + y + 6t$
$y' = 4x + 3y - 10t + 4$
$x(0) = 0.5,\ y(0) = 0.2$

11. $x' + 4x - y' = 7t$
$x' + y' - 2y = 3t$
$x(0) = 1,\ y(0) = -2$

12. $x'' + y' = 4t$
$-x'' + y' + y = 6t^2 + 10$
$x(0) = 3,\ y(0) = -1$

13. A system of linear equations is said to have **linear normal form** if it can be expressed in the form

$$\begin{aligned}\frac{dx_1}{dt} &= a_{11}(t)x_1 + a_{12}(t)x_2 + \cdots + a_{1n}(t)x_n + f_1(t)\\ \frac{dx_2}{dt} &= a_{21}(t)x_1 + a_{22}(t)x_2 + \cdots + a_{2n}(t)x_n + f_2(t)\\ &\ \ \vdots \qquad\qquad \vdots \qquad\qquad\qquad\qquad \vdots\\ \frac{dx_n}{dt} &= a_{n1}(t)x_1 + a_{n2}(t)x_2 + \cdots + a_{nn}(t)x_n + f_n(t),\end{aligned} \tag{7}$$

where the functions $a_{ij}(t)$ and $f_i(t)$ are continuous on a common interval. Show that the following systems cannot be expressed in linear normal form.

(a) $x' + x + y' + y = 0$
$2x' + 2y' + y = 0$

(b) $2x' + x - 2y' = 4$
$x' - y' = e^t$

14. Consider the system

$$a_1x' - b_1x + a_2y' - b_2y = 0$$

$$a_3x' - b_3x + a_4y' - b_4y = 0,$$

where the a_i are nonzero constants. Determine a condition on the a_i such that the system cannot be expressed in the form (7).

3.11 NONLINEAR HIGHER-ORDER EQUATIONS

Introduction

There are several significant differences between linear and nonlinear differential equations. We saw in Section 3.1 that homogeneous linear equations of order two or higher have the property that a linear combination of solutions is also a solution (Theorem 3.2). Nonlinear equations do not possess this property of superposability. In Chapter 2 we saw that we could solve a few nonlinear first-order differential equations by recognizing them as separable, homogeneous, exact, or perhaps Bernoulli equations. Even though the solutions of these equations were in the form of a one-parameter family, this family did not, as a rule, represent the general solution of the differential equation. On the other hand, by paying attention to certain continuity conditions, we obtained general solutions of linear first-order equations. Stated another way, nonlinear first-order differential equations can possess singular solutions whereas linear equations cannot.

But the major difference between linear and nonlinear equations of order two or higher lies in the realm of solvability. Given a linear equation, there is a chance that we can find some form of a solution that we can look at. On the other hand, nonlinear higher-order differential equations virtually defy solution. This does not mean a nonlinear higher-order differential equation has no solution, but rather there are no general methods whereby either an explicit or implicit solution can be found. Although this sounds disheartening, there are still things that can be done. We can always analyze a nonlinear equation quantitatively (approximate a solution using a numerical procedure, graph a solution using an ODE solver) or qualitatively (examine its critical points, sketch its phase portrait).

Nonlinear Springs

In the discussion on spring/mass systems in Section 3.8, we examined mathematical models in which the damping imparted to the motion was proportional to the instantaneous velocity dx/dt, and the restoring force of a spring was given by the linear function $F(x) = kx$. But these were simply assumptions; in more realistic situations, damping could be proportional to some power of the instantaneous velocity dx/dt, and the restoring force F of a spring could be proportional to, say, the cube of the displacement x of the mass beyond its equilibrium position or even be a linear combination of powers of the displacement.

The autonomous second-order differential equation

$$m\frac{d^2x}{dt^2} + F(x) = 0 \qquad (1)$$

represents a general model for the free, undamped vibrations of a mass m attached to a spring. When $F(x) = kx$, the model is **linear**, whereas for $F(x) = kx^3$ or $F(x) = kx + k_1x^3$ the model is **nonlinear**.

Note that both $F(x) = kx^3$ and $F(x) = kx + k_1x^3$ are odd functions of x. To see why a polynomial function containing only odd powers of x provides a reasonable model for the restoring force, let us express F as power series centered at the equilibrium position $x = 0$:

$$F(x) = c_0 + c_1x + c_2x^2 + c_3x^3 + \cdots.$$

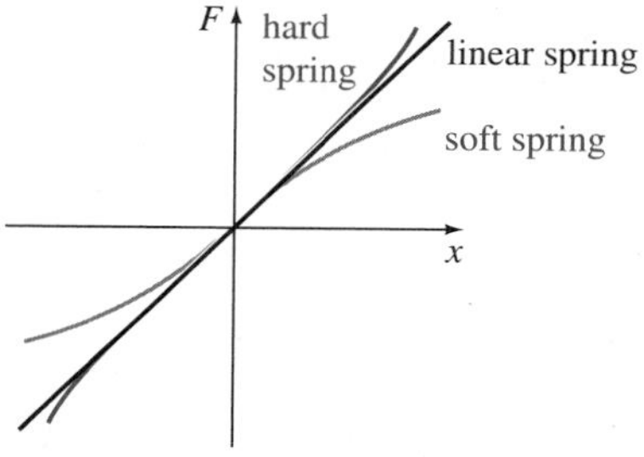

FIGURE 3.42

When the displacements x are small, the values of x^n are neglible for n sufficiently large. If we truncate the power series with, say, the fourth term, then

$$F(x) = c_0 + c_1x + c_2x^2 + c_3x^3.$$

In order that the force at $x > 0$ ($F(x) = c_0 + c_1x + c_2x^2 + c_3x^3$) and the force at $-x < 0$ ($F(-x) = c_0 - c_1x + c_2x^2 - c_3x^3$) have the same magnitude but act in opposite directions, we must have $F(-x) = -F(x)$. Since this means that F is an odd function, we must have $c_0 = 0$ and $c_2 = 0$, so $F(x) = c_1x + c_3x^3$. Had we used only the first two terms in the series, the same argument would yield $F(x) = c_1x$. For discussion purposes, we write $c_1 = k$ and $c_2 = k_1$. A restoring force with mixed powers, such as $F(x) = kx + k_1x^2$ and the corresponding vibrations, are said to be unsymmetrical.

Hard and Soft Springs

Let us take a closer look at the equation in (1) in the case where the restoring force is given by $F(x) = kx + k_1x^3$, $k > 0$. The spring is said to be **hard** if $k_1 > 0$ and **soft** if $k_1 < 0$. Graphs of three types of restoring forces are illustrated in Figure 3.42. With the aid of an ODE solver, the next example illustrates two special cases of the autonomous differential equation

$$m\frac{d^2x}{dt^2} + kx + k_1x^3 = 0, \tag{2}$$

where $m > 0$, $k > 0$.

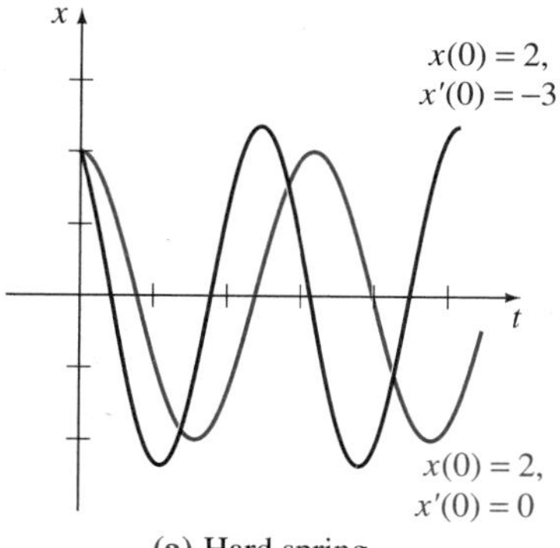

(a) Hard spring

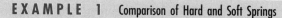

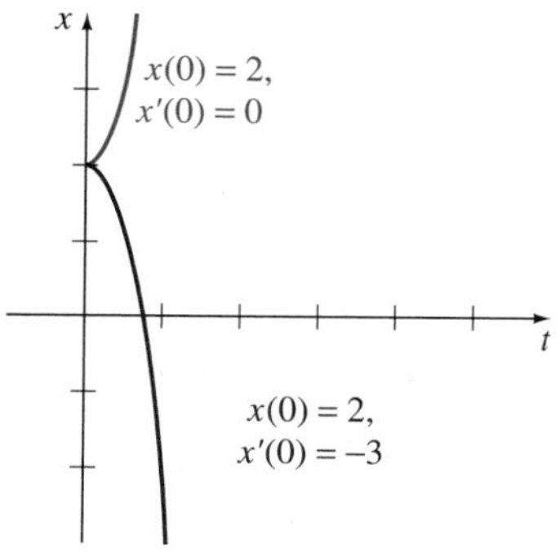

(b) Soft spring

FIGURE 3.43

EXAMPLE 1 Comparison of Hard and Soft Springs

The following differential equations are special cases of (2)

$$\frac{d^2x}{dt^2} + x + x^3 = 0 \tag{3}$$

$$\frac{d^2x}{dt^2} + x - x^3 = 0 \tag{4}$$

and are models of a hard spring and soft spring, respectively. Figure 3.43(a) shows two solutions of (3), and Figure 3.43(b) shows two solutions of (4). The curves shown in black are solutions satisfying the initial conditions $x(0) = 2$, $x'(0) = -3$; the two curves drawn in color are solutions satisfying $x(0) = 2$, $x'(0) = 0$. These solution curves certainly suggest that the motion of a mass on the hard spring is oscillatory, whereas motion of a mass on the soft spring is not oscillatory. But we must be careful about drawing conclusions based on a couple of solution curves. A more complete picture of the nature of the solutions of both of these equations can be obtained from their phase portraits. ●

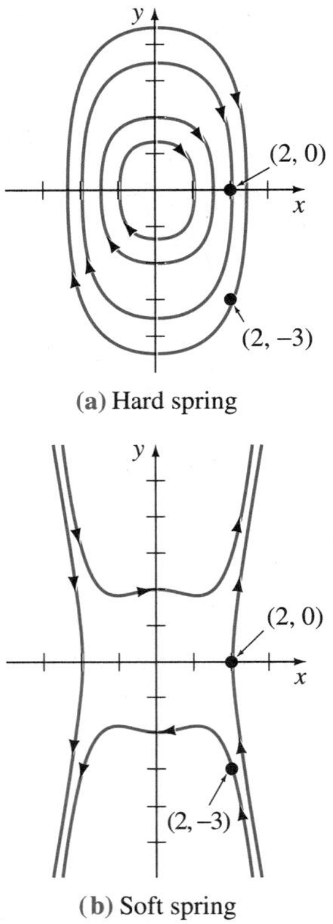

FIGURE 3.44

Phase Portraits

We have seen in earlier sections that a phase portrait of an autonomous second-order differential equation can be obtained by expressing the equation as an autonomous system of two first-order equations and then using computer software to plot trajectories through specified points (x_0, y_0). If $dx/dt = y$, then the systems corresponding to (3) and (4) are, respectively,

$$\frac{dx}{dt} = y$$
$$\frac{dy}{dt} = -x - x^3 \qquad (5)$$

and

$$\frac{dx}{dt} = y$$
$$\frac{dy}{dt} = -x + x^3 \qquad (6)$$

Recall from Section 3.3 that a critical point of an autonomous second-order differential equation is a critical point of its equivalent autonomous system. A critical point of an autonomous system is a point where both dx/dt and dy/dt are zero. For example, in (5) $dx/dt = 0$ at $y = 0$ and $dy/dt = 0$ at the solution of $-x - x^3 = 0$. Because 0 is the only real solution of the last equation, we conclude that $x = 0$; $y = 0$ is a constant or equilibrium solution of this system. In other words, (0, 0) is the only critical point of (5) and consequently the only critical point of equation (3). On the other hand, examination of (6) shows that (0, 0), (1, 0), and $(-1, 0)$ are three critical points of (4).

Using the systems (5) and (6), we obtain, in turn, the phase portraits in Figure 3.44(a) and Figure 3.44(b). The closed trajectories in Figure 3.44(a) indicate what we had intuitively expected, that each corresponding solution $x(t)$ is periodic. The solution curves for the hard spring given in Figure 3.43(a) correspond to the two trajectories in Figure 3.44(a) that pass through the indicated points (2, 0) and $(2, -3)$. The solution curves for the soft spring in Figure 3.43(b) correspond to the two trajectories passing through (2, 0) and $(2, -3)$ in Figure 3.44(b) and, of course, verify the unbounded motion of the spring/mass system for the stipulated initial conditions. However, Figure 3.44(b) is not an informative phase portrait of (6) since *it is important to examine trajectories close to the critical points*. Figure 3.45(a) shows trajectories of (6) obtained by choosing initial points close to the three critical points. Observe that we have discovered that there do indeed exist periodic solutions of (4), provided that the magnitudes

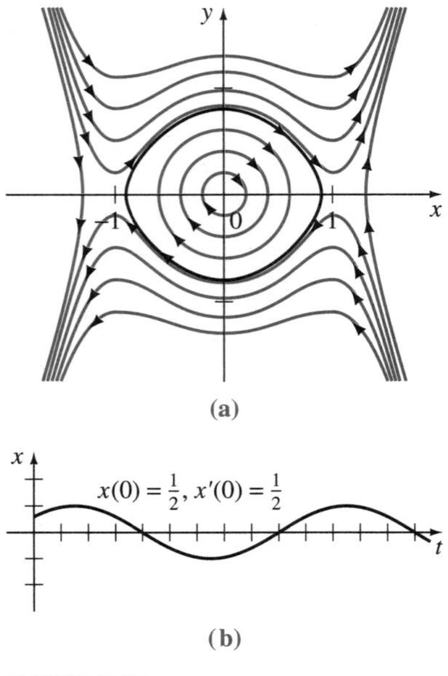

FIGURE 3.45

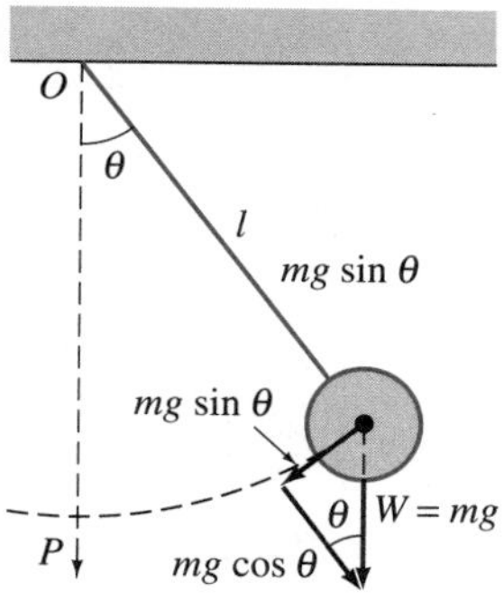

FIGURE 3.46

of the initial displacement and initial velocity are not too large. Figure 3.45(b) illustrates the periodic solution of (4) subject to the initial conditions $x(0) = \frac{1}{2}$, $x'(0) = \frac{1}{2}$ and corresponds to the closed trajectory shown in black in Figure 3.45(a).

Nonlinear Pendulum

Any object that swings back and forth is called a **physical pendulum**. The **simple pendulum** is a special case of the physical pendulum and consists of a rod of length l to which a mass m is attached at one end.

In describing the motion of a simple pendulum in a vertical plane, we make the simplifying assumptions that the mass of the rod is negligible and that no external damping or driving forces act on the system. The displacement angle θ of the pendulum, measured from the vertical, as shown in Figure 3.46, is considered positive when measured to the right of OP and negative when to the left of OP. Now recall that the arc s of a circle of radius l is related to the central angle θ by the formulas $s = l\theta$. Hence angular acceleration is

$$a = \frac{d^2s}{dt^2} = l\frac{d^2\theta}{dt^2}.$$

From Newton's second law we then have

$$F = ma = ml\frac{d^2\theta}{dt^2}.$$

From Figure 3.46 we see that the tangential component of the force due to the weight W is $mg \sin \theta$. We equate the two different formulations of the tangential force to obtain $ml\, d^2\theta/dt^2 = -mg \sin \theta$ or

$$\frac{d^2\theta}{dt^2} + \frac{g}{l}\sin \theta = 0. \tag{7}$$

Linearization

Because of the presence of $\sin \theta$, equation (7) is nonlinear. In the study of nonlinear higher-order differential equations, we often try to simplify a problem by replacing nonlinear terms by certain approximations. For example, the Maclaurin series for $\sin \theta$ is given by

$$\sin \theta = \theta - \frac{\theta^3}{3!} + \frac{\theta^5}{5!} - \cdots,$$

and so if we use the approximation $\sin \theta \approx \theta - \theta^3/6$, equation (7) becomes $d^2\theta/dt^2 + (g/l)\theta - (g/6l)\theta^3 = 0$. Observe that this last equation is the same as the nonlinear equation (2), with $m = 1$, $k = g/l$, and $k_1 = -g/6l$. However, if we assume that the displacements θ are small enough to justify using the replacement $\sin \theta \approx \theta$, then (7) becomes

$$\frac{d^2\theta}{dt^2} + \frac{g}{l}\theta = 0. \tag{8}$$

If we set $\omega^2 = g/l$, (8) has the same structure as the differential equation (2) of Section 3.7, governing the free undamped vibrations of a linear spring/mass system. Equation (8) is called a **linearization** of equation (7). Since the general solution of (8) is $\theta(t) = c_1 \cos \omega t + c_2 \sin \omega t$, this linearization suggests that, for initial conditions amenable to small oscillations, the motion of the pendulum described by (7) will be periodic.

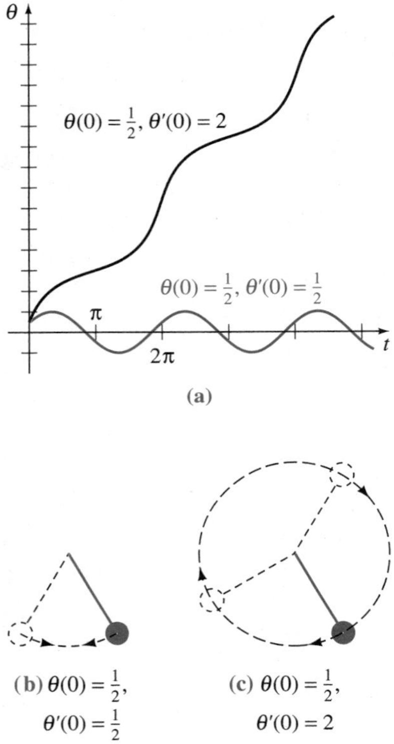

FIGURE 3.47

EXAMPLE 2 Nonlinear Pendulum

The graphs in Figure 3.47(a) were obtained with the aid of an ODE solver and represent solution curves of (7) when $\omega^2 = 1$. The colored curve depicts the solution of (7) that satisfies the initial conditions $\theta(0) = 1/2$, $\theta'(0) = 1/2$, whereas the black curve is the solution of (7) that satisfies $\theta(0) = 1/2$, $\theta'(0) = 2$. The colored curve represents a periodic solution—the pendulum oscillating back and forth, as shown in Figure 3.47(b), with an apparent amplitude $A < 1$. The black curve shows that θ increases without bound as time increases—the pendulum, starting from the same initial displacement, is given an initial velocity of magnitude great enough to send it over the top; in other words, the pendulum is whirling about its pivot, as shown in Figure 3.47(c). In the absence of damping, the motion in each case continues indefinitely. ●

Phase Portrait

If we denote $x = \theta(t)$ and $\omega^2 = g/l$, then (7) is equivalent to the autonomous system of first-order equations

$$\frac{dx}{dt} = y$$
$$\frac{dx}{dt} = -\omega^2 \sin x. \tag{9}$$

Solving $y = 0$ and $\sin x = 0$ gives the critical points $(0, 0)$, $(\pi, 0)$, $(-\pi, 0)$, ... or $(n\pi, 0)$, $n = 0, \pm 1, \pm 2, \ldots$. Using (9) with $\omega^2 = 1$, we obtain the phase portrait in Figure 3.48. The closed trajectories in Figure 3.48 correspond to periodic solutions of (8), whereas the wavy lines above and below the closed curves correspond to solutions that represent the whirling pendulum.

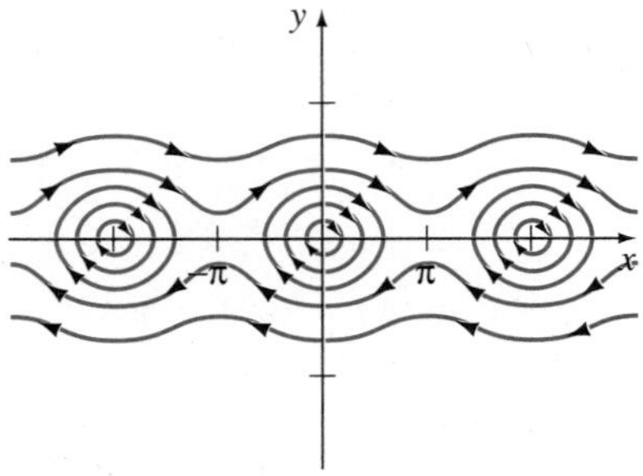

FIGURE 3.48

Stability

As in previous discussions, critical points of autonomous nonlinear second-order differential equations are classified as stable, asymptotically stable, or unstable. From the phase portrait for the hard spring defined by (3) (Figure 3.44(a)), it should be apparent that any solution starting at an initial point (x_0, y_0) near the critical point $(0, 0)$ always stays close to the critical point. Thus $(0, 0)$ is stable. Moreover, because the behavior of the simple closed curves representing periodic solutions is similar to the ellipses in Figure 3.7(e), we also say that $(0, 0)$ is a center. From the phase portrait, Figure 3.45(a), for the soft spring defined by (4), we conclude that $(0, 0)$ is a center but that $(-1, 0)$ and $(1, 0)$ are unstable

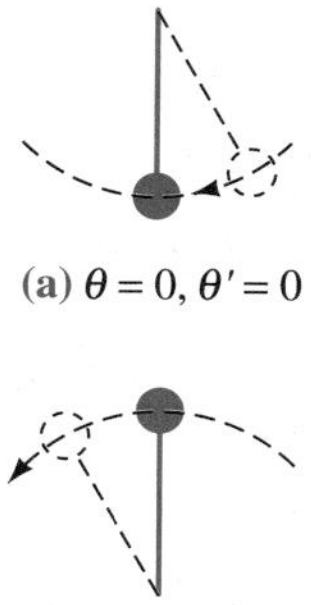

(a) $\theta = 0, \theta' = 0$

(b) $\theta = \pi, \theta' = 0$

FIGURE 3.49

critical points. Because the behavior of the trajectories near these latter two points in Figure 3.45(a) resembles Figure 3.5(e), we further say that $(-1, 0)$ and $(1, 0)$ are saddle points. In other words, a critical point (x_1, y_1) of an autonomous nonlinear second-order differential equation is assigned one of the names: node, saddle point, spiral point, or center, if the behavior of the trajectories near (x_1, y_1) resembles the behavior of the trajectories near $(0, 0)$ in Figures 3.5, 3.6, and 3.7. Keep in mind that a center is always a stable critical point and that a saddle point is always an unstable critical point. We note in passing that, unlike Figure 3.7(e), the closed curves surrounding a center of a nonlinear differential equation, as in Figures 3.44(a) and 3.45(a), need not be ellipses.

In the case of the nonlinear pendulum, the equilibrium solutions $\theta(t) = 0$ and $\theta(t) = \pi$ correspond, respectively, to the pendulum in its lowest position hanging vertically downward, as shown in Figure 3.49(a), and the pendulum in its highest position pointing vertically upward, as in Figure 3.49(b). Our physical intuition tells us an undamped pendulum that starts *near* $(0, 0)$, in other words that starts with an initial displacement and velocity of reasonably small magnitude, will exhibit periodic oscillations about $\theta = 0$, of reasonably small amplitude. This is, of course, the nature of a stable critical point. Similarly we can argue that $(\pi, 0)$ is an unstable critical point, because when the pendulum is at rest at its highest position, the slightest displacement or the slightest push will result in it falling rapidly away from that position. These conjectures are verified by the portrait in Figure 3.48; $(0, 0)$ along with $(2k\pi, 0)$, $k = \pm 1, \pm 2, \ldots$ are centers; $(\pi, 0)$ along with $((2k + 1)\pi, 0)$, $k = \pm 1, \pm 2, \ldots$ are saddle points.

EXERCISES 3.11

Answers to odd-numbered problems begin on page AN-16.

In Problems 1–4 the given differential equation is a model of an undamped spring/mass system in which the restoring force $F(x)$ in (1) is nonlinear.

(a) For each equation use an ODE solver to obtain the solution curves satisfying the given initial conditions. If the solution is periodic, use the solution curve to estimate the period T of oscillations.

(b) Find the critical points of each equation.

(c) Obtain a phase portrait of each equation and identify the trajectories corresponding to the solution curves obtained in part (a). Use the phase portrait to classify the critical points as stable or unstable. If possible, determine whether the descriptive names: node, saddle point, spiral point, or center, are applicable to the critical points.

1. $\dfrac{d^2x}{dt^2} + x^3 = 0$

$x(0) = 1, x'(0) = 1;$

$x(0) = \frac{1}{2}, x'(0) = -1$

2. $\dfrac{d^2x}{dt^2} + 16x - 4x^3 = 0$

$x(0) = 1, x'(0) = 1;$

$x(0) = -2, x'(0) = 2$

3. $\dfrac{d^2x}{dt^2} + 2x - x^2 = 0$

$x(0) = 1, x'(0) = 1;$

$x(0) = \frac{3}{2}, x'(0) = -1$

4. $\dfrac{d^2x}{dt^2} + xe^{0.01x} = 0$

$x(0) = 1, x'(0) = 1;$

$x(0) = 3, x'(0) = -1$

5. (a) In Problem 3, suppose that the mass is released from the initial position $x(0) = 1$ with an initial velocity $x'(0) = x_1$. Use a phase portrait of the equation to estimate the smallest value of $|x_1|$ at which the motion of the mass is nonperiodic.

(b) In Problem 3, suppose that the mass is released from the initial position $x(0) = x_0$ with an initial velocity $x'(0) = 1$. Use a phase portrait of the equation to estimate an interval $a \le x_0 \le b$ for which the motion is oscillatory.

6. In Problem 3, suppose that the mass is released from rest from the initial position $x(0) = 1$. Since the restoring force $F(x) = 2x - x^2$ is asymmetric, it makes sense to assume that the mass oscillates between $x = 1$ and some value $x = a < 0$, where $a \neq -1$.

(a) Use an ODE solver to find the solution curve and use this curve to estimate the extreme displacements of the mass from the line $x = 0$.

(b) Find a Cartesian equation of the trajectory corresponding to these initial conditions and use this trajectory to obtain the exact values of the extreme displacements of the mass from the line $x = 0$.

7. Find a linearization of the differential equation in Problem 4.

8. Consider the model of an undamped nonlinear spring/mass system given by

$$\frac{d^2x}{dt^2} + 8x - 6x^3 + x^5 = 0.$$

Discuss in detail the nature of the oscillations of the system by examining critical points, a phase portrait, and solution curves obtained from an ODE solver.

In Problems 9 and 10 the given differential equation is a model of a damped nonlinear spring/mass system.

(a) Predict the behavior of each system as $t \to \infty$.
(b) For each equation use an ODE solver to obtain the solution curves satisfying the given initial conditions.
(c) Find the critical points of each equation.
(d) Obtain a phase portrait of each equation and identify the trajectories corresponding to the solution curves obtained in part (b). Use the phase portrait to classify the critical points as stable or unstable. If possible, determine whether the descriptive names: node, saddle point, spiral point, or center, are applicable to the critical points.

9. $\dfrac{d^2x}{dt^2} + \dfrac{dx}{dt} + x + x^3 = 0$

$x(0) = -3,\ x'(0) = 4;$

$x(0) = 0,\ x'(0) = -8$

10. $\dfrac{d^2x}{dt^2} + \dfrac{dx}{dt} + x - x^3 = 0$

$x(0) = 0,\ x'(0) = \frac{3}{2};$

$x(0) = -1,\ x'(0) = 1$

11. In Section 3.7 we saw that an undriven spring/mass system with linear damping was described by the differential equation $x'' + 2\lambda x' + \omega^2 x = 0$, $\lambda > 0$. Such a system is overdamped if $\lambda^2 - \omega^2 > 0$ and underdamped if $\lambda^2 - \omega^2 < 0$. See pages 152 and 153. A model of an undriven spring/mass system in which damping is nonlinear is given by

$$x'' + 2\lambda x'|x'| + \omega^2 x = 0.$$

In this case the damping is proportional to the square of the velocity. Use a computer graphing program to investigate this nonlinear model by means of phase plane analysis for selected values of λ and ω to determine whether there exist overdamped and underdamped systems.

12. The model $mx'' + kx + k_1x^3 = F_0 \cos \omega t$ of an undamped, periodically driven spring/mass system is called **Duffing's differential equation**. (Note that the equation is not autonomous.)

(a) Consider the initial-value problem

$$x'' + x + k_1x^3 = 5 \cos t, \qquad x(0) = 1, \quad x'(0) = 0.$$

Use an ODE solver to investigate the behavior of the system for values of $k_1 > 0$ ranging from $k_1 = 0.01$ to $k_1 = 100$.

(b) Find values of $k_1 < 0$ for which the system in part (a) is oscillatory.

(c) Consider the initial-value problem

$$x'' + x + k_1x^3 = \cos \tfrac{3}{2}t, \qquad x(0) = 0, \quad x'(0) = 0.$$

Find the values for $k_1 < 0$ for which the system is oscillatory.

13. (a) Show that a Cartesian equation of the phase-plane trajectories of equation (2) is given by $\frac{1}{2}my^2 + \frac{1}{2}kx^2 + \frac{1}{4}k_1x^4 = E$, where E represents the constant energy of the system.

(b) In the case of a soft spring, where $m = 1$, $k = 1$, and $k_1 = -1$, show that

$$y = \pm\sqrt{2E - x^2 + \tfrac{1}{2}x^4}.$$

(c) Use the equation in part (b) and a computer graphing program to obtain the phase portrait of equation (4). First choose values of E satisfying $0 < E < \frac{1}{4}$ and then values satisfying $E > \frac{1}{4}$. Finally, plot the curve corresponding to $E = \frac{1}{4}$. This last curve is called a **separatrix** because it separates the phase plane into regions corresponding to the different behaviors of the spring/mass system.

14. Consider the model of the damped nonlinear pendulum

$$\frac{d^2\theta}{dt^2} + 2\lambda\frac{d\theta}{dt} + \omega^2 \sin \theta = 0.$$

(a) Find the critical points of the equation.

(b) Obtain the phase portrait of the equation in part (a) in the underdamped case where $\lambda^2 - \omega^2 < 0$ by selecting $\lambda = \frac{1}{2}$, $\omega^2 = 1$. Use the phase portrait to classify the critical points as stable or unstable. If possible, determine whether the descriptive names: node, saddle point, spiral point, or center, are applicable to the critical points.

(c) Repeat part (b) in the overdamped case where $\lambda^2 - \omega^2 > 0$ by selecting $\lambda = 2$, $\omega^2 = 1$.

15. *PROJECT PROBLEM* Consider a spring/mass system in which the mass m slides over a dry horizontal surface whose coefficient of kinetic friction is μ. If the force of kinetic friction has the constant magni-

tude $f_k = \mu mg$ and acts, as shown in Figure 3.50, opposite to the direction of motion, then the motion is described by the piecewise defined differential equation

$$mx'' = -kx - f_k \operatorname{sgn}(x') \quad \text{or} \quad mx'' + f_k \operatorname{sgn}(x') + kx = 0,$$

where
$$\operatorname{sgn}(x') = \begin{cases} 1, & x' > 0 \\ -1, & x' < 0. \end{cases}$$

Although the differential equation consists of two linear equations, the problem is considered nonlinear since the motion cannot be represented by a single equation.

(a) Find the Cartesian equations for the phase-plane trajectories in the special case $x'' + \operatorname{sgn}(x') + x = 0$. Find the phase portrait of the differential equation.

(b) Use the results in part (a) to trace the motion of the mass, using the appropriate trajectories in the phase plane if it is released from rest from the initial position $x(0) = 9$. Indicate each point where the mass comes to rest and give the position of the mass at these points. Find the position where the motion of the mass stops completely.

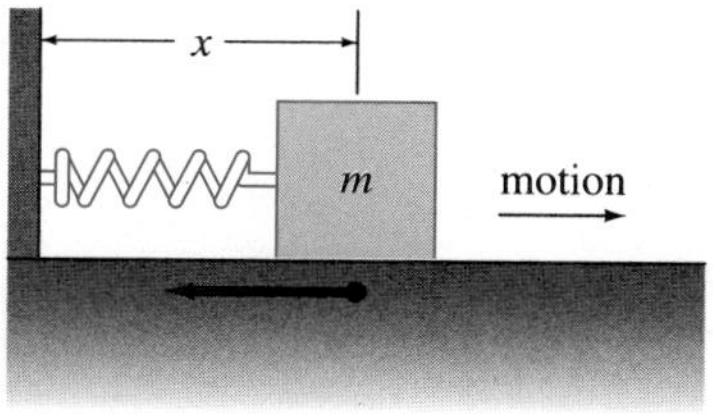

FIGURE 3.50

16. **PROJECT PROBLEM** From (8) we can see that the period of small oscillations of a linear simple pendulum is given by $T = 2\pi\sqrt{l/g}$. Note that this period does not depend on the initial angular displacement θ_0. Now consider the initial-value problem for the nonlinear pendulum

$$\frac{d^2\theta}{dt^2} + \frac{g}{l}\sin\theta = 0, \qquad \theta(0) = \theta_0, \quad \theta'(0) = 0.$$

(a) Show that by multiplying this differential equation by $2d\theta/dt$ we obtain the first-order equation

$$\left(\frac{d\theta}{dt}\right)^2 = \frac{2g}{l}(\cos\theta - \cos\theta_0).$$

(b) Use the differential equation in part (a) to show that the period of large oscillations depends on the initial angular displacement θ_0 and is given by

$$T_{\theta_0} = 2\sqrt{\frac{2l}{g}}\int_0^{\theta_0} \frac{d\theta}{\sqrt{\cos\theta - \cos\theta_0}}.$$

(c) Use the trigonometric identities

$$\cos\theta = 1 - 2\sin^2\frac{\theta}{2} \quad \text{and} \quad \cos\theta_0 = 1 - 2\sin^2\frac{\theta_0}{2}$$

to show that the integral in part (b) can be written as

$$T_{\theta_0} = 2\sqrt{\frac{l}{g}}\int_0^{\theta_0} \frac{d\theta}{\sqrt{\sin^2(\theta_0/2) - \sin^2(\theta/2)}}.$$

(d) Define $k = \sin\dfrac{\theta_0}{2}$ and $\sin\dfrac{\theta}{2} = k\sin\phi$ to show that

$$T_{\theta_0} = \frac{2T}{\pi}K(k),$$

where

$$K(k) = \int_0^{\pi/2} \frac{d\phi}{\sqrt{1 - k^2\sin\phi}}$$

is called the **complete elliptic integral of the first kind** and T is the period of the linear simple pendulum.

(e) Use the result from part (d) to find k for $\theta_0 = 60°$ and $\theta_0 = 2°$. Use tables or a CAS to evaluate the elliptic integral for $\theta_0 = 60°$ and $\theta_0 = 2°$. Compare the periods of oscillation $T_{60°}$ and $T_{2°}$ of the nonlinear pendulum with the period T of the linear simple pendulum.

(f) Use the difference $(T_{2°} - T)/T$ to compare the time disparity between the nonlinear and linear pendulums over a time interval of one year (365 days).

CHAPTER 4

SYSTEMS OF FIRST-ORDER DIFFERENTIAL EQUATIONS

4.1 PRELIMINARY THEORY: LINEAR SYSTEMS

To the Student Matrix notation and properties are used extensively throughout this chapter. You should review Appendix I if you are unfamiliar with these concepts.

In Section 3.9 of Chapter 3 we dealt with systems of differential equations of the form

$$\begin{aligned} P_{11}(D)x_1 + P_{12}(D)x_2 + \cdots + P_{1n}(D)x_n &= b_1(t) \\ P_{21}(D)x_1 + P_{22}(D)x_2 + \cdots + P_{2n}(D)x_n &= b_2(t) \\ \vdots \qquad\qquad\qquad &\quad \vdots \\ P_{n1}(D)x_1 + P_{n2}(D)x_2 + \cdots + P_{nn}(D)x_n &= b_n(t), \end{aligned} \tag{1}$$

where the P_{ij} were polynomials of various degrees in the differential operator D. In this chapter we confine our study to systems of first-order differential equations

$$\begin{aligned} \frac{dx_1}{dt} &= g_1(t, x_1, x_2, \ldots, x_n) \\ \frac{dx_2}{dt} &= g_2(t, x_1, x_2, \ldots, x_n) \\ \vdots \quad &\qquad \vdots \\ \frac{dx_n}{dt} &= g_n(t, x_1, x_2, \ldots, x_n). \end{aligned} \tag{2}$$

System (2) of n first-order equations is called an ***n*th-order system**. When the right-hand function of each equation in (2) is free of the independent variable t the system is said to be **autonomous**. We have in several instances expressed an autonomous second-order differential equation as an autonomous second-order system. In general, *any* nth-order differential equation $y^{(n)} = g(x, y, y', \ldots, y^{(n-1)})$ can be expressed as an nth-order system.

Linear Systems

In this and the next two sections, we shall be interested in an important special case of (2), namely, **systems of linear first-order equations**:

$$\begin{aligned} \frac{dx_1}{dt} &= a_{11}(t)x_1 + a_{12}(t)x_2 + \cdots + a_{1n}(t)x_n + f_1(t) \\ \frac{dx_2}{dt} &= a_{21}(t)x_1 + a_{22}(t)x_2 + \cdots + a_{2n}(t)x_n + f_2(t) \\ \vdots \quad &\qquad \vdots \\ \frac{dx_n}{dt} &= a_{n1}(t)x_1 + a_{n2}(t)x_2 + \cdots + a_{nn}(t)x_n + f_n(t). \end{aligned} \tag{3}$$

We also refer to a system of the form (3) as an ***n*th-order linear system** or simply as a **linear system**. We assume that the coefficients a_{ij} and the functions f_i are continuous on a common interval I. When $f_i(t) = 0$, $i = 1, 2, \ldots, n$, the linear system is said to be **homogeneous**; otherwise it is **nonhomogeneous**.

Matrix Form of a Linear System

If $\mathbf{X}$, $\mathbf{A}(t)$, and $\mathbf{F}(t)$ denote the respective matrices

$$\mathbf{X} = \begin{pmatrix} x_1(t) \\ x_2(t) \\ \vdots \\ x_n(t) \end{pmatrix}, \quad \mathbf{A}(t) = \begin{pmatrix} a_{11}(t) & a_{12}(t) & \cdots & a_{1n}(t) \\ a_{21}(t) & a_{22}(t) & \cdots & a_{2n}(t) \\ \vdots & & & \vdots \\ a_{n1}(t) & a_{n2}(t) & \cdots & a_{nn}(t) \end{pmatrix}, \quad \mathbf{F}(t) = \begin{pmatrix} f_1(t) \\ f_2(t) \\ \vdots \\ f_n(t) \end{pmatrix},$$

then the system of linear first-order differential equations (3) can be written as

$$\frac{d}{dt}\begin{pmatrix} x_1 \\ x_2 \\ \vdots \\ x_n \end{pmatrix} = \begin{pmatrix} a_{11}(t) & a_{12}(t) & \cdots & a_{1n}(t) \\ a_{21}(t) & a_{22}(t) & \cdots & a_{2n}(t) \\ \vdots & & & \vdots \\ a_{n1}(t) & a_{n2}(t) & \cdots & a_{nn}(t) \end{pmatrix}\begin{pmatrix} x_1 \\ x_2 \\ \vdots \\ x_n \end{pmatrix} + \begin{pmatrix} f_1(t) \\ f_2(t) \\ \vdots \\ f_n(t) \end{pmatrix}$$

or simply

$$\mathbf{X}' = \mathbf{AX} + \mathbf{F}. \tag{4}$$

If the system is homogeneous, its matrix form is then

$$\mathbf{X}' = \mathbf{AX}. \tag{5}$$

EXAMPLE 1 Systems Written in Matrix Notation

(a) If $\mathbf{X} = \begin{pmatrix} x \\ y \end{pmatrix}$ then the matrix form of the homogeneous system

$$\frac{dx}{dt} = 3x + 4y$$
$$\frac{dy}{dt} = 5x - 7y$$

is $\mathbf{X}' = \begin{pmatrix} 3 & 4 \\ 5 & -7 \end{pmatrix}\mathbf{X}$.

(b) If $\mathbf{X} = \begin{pmatrix} x \\ y \\ z \end{pmatrix}$ then the matrix form of the nonhomogeneous system

$$\frac{dx}{dt} = 6x + y + z + t$$
$$\frac{dy}{dt} = 8x + 7y - z + 10t$$
$$\frac{dz}{dt} = 2x + 9y - z + 6t$$

is $\mathbf{X}' = \begin{pmatrix} 6 & 1 & 1 \\ 8 & 7 & -1 \\ 2 & 9 & -1 \end{pmatrix}\mathbf{X} + \begin{pmatrix} t \\ 10t \\ 6t \end{pmatrix}$.

The system in part (a) of Example 1 is an autonomous system.

DEFINITION 4.1 Solution Vector

A **solution vector** on an interval I is any column matrix

$$\mathbf{X} = \begin{pmatrix} x_1(t) \\ x_2(t) \\ \vdots \\ x_n(t) \end{pmatrix}$$

whose entries are differentiable functions satisfying the system (4) on the interval.

EXAMPLE 2 Verification of Solutions

Verify that on the interval $(-\infty, \infty)$

$$\mathbf{X}_1 = \begin{pmatrix} 1 \\ -1 \end{pmatrix} e^{-2t} = \begin{pmatrix} e^{-2t} \\ -e^{-2t} \end{pmatrix} \quad \text{and} \quad \mathbf{X}_2 = \begin{pmatrix} 3 \\ 5 \end{pmatrix} e^{6t} = \begin{pmatrix} 3e^{6t} \\ 5e^{6t} \end{pmatrix}$$

are solutions of

$$\mathbf{X}' = \begin{pmatrix} 1 & 3 \\ 5 & 3 \end{pmatrix} \mathbf{X}. \tag{6}$$

Solution From $\mathbf{X}_1' = \begin{pmatrix} -2e^{-2t} \\ 2e^{-2t} \end{pmatrix}$ and $\mathbf{X}_2' = \begin{pmatrix} 18e^{6t} \\ 30e^{6t} \end{pmatrix}$ we see

$$\mathbf{A}\mathbf{X}_1 = \begin{pmatrix} 1 & 3 \\ 5 & 3 \end{pmatrix} \begin{pmatrix} e^{-2t} \\ -e^{-2t} \end{pmatrix} = \begin{pmatrix} e^{-2t} - 3e^{-2t} \\ 5e^{-2t} - 3e^{-2t} \end{pmatrix} = \begin{pmatrix} -2e^{-2t} \\ 2e^{-2t} \end{pmatrix} = \mathbf{X}_1'$$

and

$$\mathbf{A}\mathbf{X}_2 = \begin{pmatrix} 1 & 3 \\ 5 & 3 \end{pmatrix} \begin{pmatrix} 3e^{6t} \\ 5e^{6t} \end{pmatrix} = \begin{pmatrix} 3e^{6t} + 15e^{6t} \\ 15e^{6t} + 15e^{6t} \end{pmatrix} = \begin{pmatrix} 18e^{6t} \\ 30e^{6t} \end{pmatrix} = \mathbf{X}_2'.$$

●

Much of the theory of systems of n linear first-order differential equations is similar to that of linear nth-order differential equations.

Initial-Value Problem

Let t_0 denote a point on an interval I and

$$\mathbf{X}(t_0) = \begin{pmatrix} x_1(t_0) \\ x_2(t_0) \\ \vdots \\ x_n(t_0) \end{pmatrix} \quad \text{and} \quad \mathbf{X}_0 = \begin{pmatrix} \gamma_1 \\ \gamma_2 \\ \vdots \\ \gamma_n \end{pmatrix},$$

where the γ_i, $i = 1, 2, \ldots, n$, are given constants. Then the problem

$$\begin{aligned} &\textit{Solve:} \quad \mathbf{X}' = \mathbf{A}(t)\mathbf{X} + \mathbf{F}(t) \\ &\textit{Subject to:} \quad \mathbf{X}(t_0) = \mathbf{X}_0 \end{aligned} \tag{7}$$

is an **initial-value problem** on the interval.

THEOREM 4.1 Existence of a Unique Solution

Let the entries of the matrices $\mathbf{A}(t)$ and $\mathbf{F}(t)$ be functions continuous on a common interval I that contains the point t_0. Then there exists a unique solution of the initial-value problem (7) on the interval.

Homogeneous Systems

In the next several definitions and theorems we are concerned only with homogeneous systems. Without stating it, we shall always assume that the a_{ij} and the f_i are continuous functions of t on some common interval I.

Superposition Principle

The following result is a **superposition principle** for solutions of linear systems.

THEOREM 4.2 Superposition Principle

Let $\mathbf{X}_1, \mathbf{X}_2, \ldots, \mathbf{X}_k$ be a set of solution vectors of the homogeneous system (5) on an interval I. Then the linear combination

$$\mathbf{X} = c_1\mathbf{X}_1 + c_2\mathbf{X}_2 + \cdots + c_k\mathbf{X}_k,$$

where the c_i, $i = 1, 2, \ldots, k$, are arbitrary constants, is also a solution on the interval.

It follows from Theorem 4.2 that a constant multiple of any solution vector of a homogeneous system of linear first-order differential equations is also a solution.

EXAMPLE 3 Using the Superposition Principle

You should practice by verifying that the two vectors

$$\mathbf{X}_1 = \begin{pmatrix} \cos t \\ -\frac{1}{2}\cos t + \frac{1}{2}\sin t \\ -\cos t - \sin t \end{pmatrix} \quad \text{and} \quad \mathbf{X}_2 = \begin{pmatrix} 0 \\ e^t \\ 0 \end{pmatrix}$$

are solutions of the system

$$\mathbf{X}' = \begin{pmatrix} 1 & 0 & 1 \\ 1 & 1 & 0 \\ -2 & 0 & -1 \end{pmatrix}\mathbf{X}. \tag{8}$$

By the superposition principle the linear combination

$$\mathbf{X} = c_1\mathbf{X}_1 + c_2\mathbf{X}_2 = c_1\begin{pmatrix} \cos t \\ -\frac{1}{2}\cos t + \frac{1}{2}\sin t \\ -\cos t - \sin t \end{pmatrix} + c_2\begin{pmatrix} 0 \\ e^t \\ 0 \end{pmatrix}$$

is yet another solution of the system.

Linear Dependence and Linear Independence

We are primarily interested in linearly independent solutions of the homogeneous system (5).

DEFINITION 4.2 Linear Dependence/Independence

Let $\mathbf{X}_1, \mathbf{X}_2, \ldots, \mathbf{X}_k$ be a set of solution vectors of the homogeneous system (5) on an interval I. We say that the set is **linearly dependent** on the interval if there exist constants $c_1, c_2, \ldots, c_k$, not all zero, such that

$$c_1\mathbf{X}_1 + c_2\mathbf{X}_2 + \cdots + c_k\mathbf{X}_k = \mathbf{0}$$

for every t in the interval. If the set of vectors is not linearly dependent on the interval, it is said to be **linearly independent**.

The case when $k = 2$ should be clear; two solution vectors $\mathbf{X}_1$ and $\mathbf{X}_2$ are linearly dependent if one is a constant multiple of the other, and conversely. For $k > 2$ a set of solution vectors is linearly dependent if we can express at least one solution vector as a linear combination of the remaining vectors.

Wronskian

As in our earlier consideration of the theory of a single ordinary differential equation, we can introduce the concept of the **Wronskian** determinant as a test for linear independence. We state the following theorem without proof.

THEOREM 4.3 Criterion for Linearly Independent Solutions

Let $$\mathbf{X}_1 = \begin{pmatrix} x_{11} \\ x_{21} \\ \vdots \\ x_{n1} \end{pmatrix}, \quad \mathbf{X}_2 = \begin{pmatrix} x_{12} \\ x_{22} \\ \vdots \\ x_{n2} \end{pmatrix}, \ldots, \quad \mathbf{X}_n = \begin{pmatrix} x_{1n} \\ x_{2n} \\ \vdots \\ x_{nn} \end{pmatrix}$$

be n solution vectors of the homogeneous system (5) on an interval I. Then the set of solution vectors is linearly independent on I if and only if the **Wronskian**

$$W(\mathbf{X}_1, \mathbf{X}_2, \ldots, \mathbf{X}_n) = \begin{vmatrix} x_{11} & x_{12} & \cdots & x_{1n} \\ x_{21} & x_{22} & \cdots & x_{2n} \\ \vdots & \vdots & & \vdots \\ x_{n1} & x_{n2} & \cdots & x_{nn} \end{vmatrix} \neq 0 \tag{9}$$

for every t in the interval.

In fact it can be shown that if $\mathbf{X}_1, \mathbf{X}_2, \ldots, \mathbf{X}_n$ are solution vectors of (5), then for every t in I either $W(\mathbf{X}_1, \mathbf{X}_2, \ldots, \mathbf{X}_n) \neq 0$ or $W(\mathbf{X}_1, \mathbf{X}_2, \ldots, \mathbf{X}_n) = 0$. Thus if we can show that $W \neq 0$ for some t_0 in I, then $W \neq 0$ for every t, and hence the solutions are linearly independent on the interval.

Notice that, unlike our previous definition of the Wronskian, the determinant (9) does not involve differentiation.

EXAMPLE 4 Linearly Independent Solutions

In Example 2 we saw that $\mathbf{X}_1 = \begin{pmatrix} 1 \\ -1 \end{pmatrix} e^{-2t}$ and $\mathbf{X}_2 = \begin{pmatrix} 3 \\ 5 \end{pmatrix} e^{6t}$ are solutions of system (6). Clearly, $\mathbf{X}_1$ and $\mathbf{X}_2$ are linearly independent on the interval $(-\infty, \infty)$ since neither vector is a constant multiple of the other. In addition, we have

$$W(\mathbf{X}_1, \mathbf{X}_2) = \begin{vmatrix} e^{-2t} & 3e^{6t} \\ -e^{-2t} & 5e^{6t} \end{vmatrix} = 8e^{4t} \neq 0$$

for all real values of t. ●

DEFINITION 4.3 Fundamental Set of Solutions

Any set $\mathbf{X}_1, \mathbf{X}_2, \ldots, \mathbf{X}_n$ of n linearly independent solution vectors of the homogeneous system (5) on an interval I is said to be a **fundamental set of solutions** on the interval.

THEOREM 4.4 Existence of a Fundamental Set

There exists a fundamental set of solutions for the homogeneous system (5) on an interval I.

THEOREM 4.5 General Solution—Homogeneous Systems

Let $\mathbf{X}_1, \mathbf{X}_2, \ldots, \mathbf{X}_n$ be a fundamental set of solutions of the homogeneous system (5) on an interval I. Then the **general solution** of the system on the interval is

$$\mathbf{X} = c_1\mathbf{X}_1 + c_2\mathbf{X}_2 + \cdots + c_n\mathbf{X}_n,$$

where the c_i, $i = 1, 2, \ldots, n$, are arbitrary constants.

EXAMPLE 5 General Solution of System (6)

From Example 2 we know that $\mathbf{X}_1 = \begin{pmatrix} 1 \\ -1 \end{pmatrix} e^{-2t}$ and $\mathbf{X}_2 = \begin{pmatrix} 3 \\ 5 \end{pmatrix} e^{6t}$ are linearly independent solutions of (6) on $(-\infty, \infty)$. Hence $\mathbf{X}_1$ and $\mathbf{X}_2$ form a fundamental set of solutions on the interval. The general solution of the system on the interval is then

$$\mathbf{X} = c_1\mathbf{X}_1 + c_2\mathbf{X}_2 = c_1\begin{pmatrix} 1 \\ -1 \end{pmatrix} e^{-2t} + c_2\begin{pmatrix} 3 \\ 5 \end{pmatrix} e^{6t}. \tag{10}$$

EXAMPLE 6 General Solution of System (8)

The vectors

$$\mathbf{X}_1 = \begin{pmatrix} \cos t \\ -\frac{1}{2}\cos t + \frac{1}{2}\sin t \\ -\cos t - \sin t \end{pmatrix}, \quad \mathbf{X}_2 = \begin{pmatrix} 0 \\ 1 \\ 0 \end{pmatrix} e^t, \quad \mathbf{X}_3 = \begin{pmatrix} \sin t \\ -\frac{1}{2}\sin t - \frac{1}{2}\cos t \\ -\sin t + \cos t \end{pmatrix}$$

are solutions of the system (8) in Example 3 (see Problem 16 in Exercises 4.1). Now

$$W(\mathbf{X}_1, \mathbf{X}_2, \mathbf{X}_3) = \begin{vmatrix} \cos t & 0 & \sin t \\ -\frac{1}{2}\cos t + \frac{1}{2}\sin t & e^t & -\frac{1}{2}\sin t - \frac{1}{2}\cos t \\ -\cos t - \sin t & 0 & -\sin t + \cos t \end{vmatrix} = e^t \neq 0$$

for all real values of t. We conclude that $\mathbf{X}_1$, $\mathbf{X}_2$, and $\mathbf{X}_3$ form a fundamental set of solutions on $(-\infty, \infty)$. Thus the general solution of the system on the interval is the linear combination $\mathbf{X} = c_1\mathbf{X}_1 + c_2\mathbf{X}_2 + c_3\mathbf{X}_3$, that is,

$$\mathbf{X} = c_1\begin{pmatrix} \cos t \\ -\frac{1}{2}\cos t + \frac{1}{2}\sin t \\ -\cos t - \sin t \end{pmatrix} + c_2\begin{pmatrix} 0 \\ 1 \\ 0 \end{pmatrix} e^t + c_3\begin{pmatrix} \sin t \\ -\frac{1}{2}\sin t - \frac{1}{2}\cos t \\ -\sin t + \cos t \end{pmatrix}.$$

Nonhomogeneous Systems

For nonhomogeneous systems a **particular solution** $\mathbf{X}_p$ on an interval I is any vector, free of arbitrary parameters, whose entries are functions that satisfy the system (4).

THEOREM 4.6 General Solution—Nonhomogeneous Systems

Let $\mathbf{X}_p$ be a given solution of the nonhomogeneous system (4) on an interval I, and let

$$\mathbf{X}_c = c_1\mathbf{X}_1 + c_2\mathbf{X}_2 + \cdots + c_n\mathbf{X}_n$$

denote the general solution on the same interval of the corresponding homogeneous system (5). Then the **general solution** of the nonhomogeneous system on the interval is

$$\mathbf{X} = \mathbf{X}_c + \mathbf{X}_p.$$

The general solution $\mathbf{X}_c$ of the homogeneous system (5) is called the **complementary function** of the nonhomogeneous system (4).

EXAMPLE 7 General Solution—Nonhomogeneous System

The vector $\mathbf{X}_p = \begin{pmatrix} 3t-4 \\ -5t+6 \end{pmatrix}$ is a particular solution of the nonhomogeneous system

$$\mathbf{X}' = \begin{pmatrix} 1 & 3 \\ 5 & 3 \end{pmatrix}\mathbf{X} + \begin{pmatrix} 12t-11 \\ -3 \end{pmatrix} \tag{11}$$

on the interval $(-\infty, \infty)$. (Verify this.) The complementary function of (11) on the same interval, or the general solution of $\mathbf{X}' = \begin{pmatrix} 1 & 3 \\ 5 & 3 \end{pmatrix}\mathbf{X}$, was seen in (10) of Example 5 to be $\mathbf{X}_c = c_1\begin{pmatrix} 1 \\ -1 \end{pmatrix}e^{-2t} + c_2\begin{pmatrix} 3 \\ 5 \end{pmatrix}e^{6t}$. Hence by Theorem 4.6,

$$\mathbf{X} = \mathbf{X}_c + \mathbf{X}_p = c_1\begin{pmatrix} 1 \\ -1 \end{pmatrix}e^{-2t} + c_2\begin{pmatrix} 3 \\ 5 \end{pmatrix}e^{6t} + \begin{pmatrix} 3t-4 \\ -5t+6 \end{pmatrix}$$

is the general solution of (11) on $(-\infty, \infty)$. ●

EXERCISES 4.1

Answers to odd-numbered problems begin on page AN-18.

In Problems 1–6 write the given system in matrix form.

1. $\dfrac{dx}{dt} = 3x - 5y$

$\dfrac{dy}{dt} = 4x + 8y$

2. $\dfrac{dx}{dt} = 4x - 7y$

$\dfrac{dy}{dt} = 5x$

3. $\dfrac{dx}{dt} = -3x + 4y - 9z$

$\dfrac{dy}{dt} = 6x - y$

$\dfrac{dz}{dt} = 10x + 4y + 3z$

4. $\dfrac{dx}{dt} = x - y$

$\dfrac{dy}{dt} = x + 2z$

$\dfrac{dz}{dt} = -x + z$

5. $\dfrac{dx}{dt} = x - y + z + t - 1$

$\dfrac{dy}{dt} = 2x + y - z - 3t^2$

$\dfrac{dz}{dt} = x + y + z + t^2 - t + 2$

6. $\dfrac{dx}{dt} = -3x + 4y + e^{-t}\sin 2t$

$\dfrac{dy}{dt} = 5x + 9y + 4e^{-t}\cos 2t$

In Problems 7–10 write the given system without the use of matrices.

7. $\mathbf{X}' = \begin{pmatrix} 4 & 2 \\ -1 & 3 \end{pmatrix}\mathbf{X} + \begin{pmatrix} 1 \\ -1 \end{pmatrix}e^{t}$

8. $\mathbf{X}' = \begin{pmatrix} 7 & 5 & -9 \\ 4 & 1 & 1 \\ 0 & -2 & 3 \end{pmatrix}\mathbf{X} + \begin{pmatrix} 0 \\ 2 \\ 1 \end{pmatrix}e^{5t} - \begin{pmatrix} 8 \\ 0 \\ 3 \end{pmatrix}e^{-2t}$

9. $\dfrac{d}{dt}\begin{pmatrix} x \\ y \\ z \end{pmatrix} = \begin{pmatrix} 1 & -1 & 2 \\ 3 & -4 & 1 \\ -2 & 5 & 6 \end{pmatrix}\begin{pmatrix} x \\ y \\ z \end{pmatrix} + \begin{pmatrix} 1 \\ 2 \\ 2 \end{pmatrix}e^{-t} - \begin{pmatrix} 3 \\ -1 \\ 1 \end{pmatrix}t$

10. $\dfrac{d}{dt}\begin{pmatrix} x \\ y \end{pmatrix} = \begin{pmatrix} 3 & -7 \\ 1 & 1 \end{pmatrix}\begin{pmatrix} x \\ y \end{pmatrix} + \begin{pmatrix} 4 \\ 8 \end{pmatrix}\sin t + \begin{pmatrix} t-4 \\ 2t+1 \end{pmatrix} e^{4t}$

In Problems 11–16 verify that the vector X is a solution of the given system.

11. $\dfrac{dx}{dt} = 3x - 4y$

$\dfrac{dy}{dt} = 4x - 7y; \quad \mathbf{X} = \begin{pmatrix} 1 \\ 2 \end{pmatrix} e^{-5t}$

12. $\dfrac{dx}{dt} = -2x + 5y$

$\dfrac{dy}{dt} = -2x + 4y; \quad \mathbf{X} = \begin{pmatrix} 5\cos t \\ 3\cos t - \sin t \end{pmatrix} e^{t}$

13. $\mathbf{X}' = \begin{pmatrix} -1 & \frac{1}{4} \\ 1 & -1 \end{pmatrix}\mathbf{X}; \quad \mathbf{X} = \begin{pmatrix} -1 \\ 2 \end{pmatrix} e^{-3t/2}$

14. $\mathbf{X}' = \begin{pmatrix} 2 & 1 \\ -1 & 0 \end{pmatrix}\mathbf{X}; \quad \mathbf{X} = \begin{pmatrix} 1 \\ 3 \end{pmatrix} e^{t} + \begin{pmatrix} 4 \\ -4 \end{pmatrix} te^{t}$

15. $\mathbf{X}' = \begin{pmatrix} 1 & 2 & 1 \\ 6 & -1 & 0 \\ -1 & -2 & -1 \end{pmatrix}\mathbf{X}; \quad \mathbf{X} = \begin{pmatrix} 1 \\ 6 \\ -13 \end{pmatrix}$

16. $\mathbf{X}' = \begin{pmatrix} 1 & 0 & 1 \\ 1 & 1 & 0 \\ -2 & 0 & -1 \end{pmatrix}\mathbf{X}; \quad \mathbf{X} = \begin{pmatrix} \sin t \\ -\frac{1}{2}\sin t - \frac{1}{2}\cos t \\ -\sin t + \cos t \end{pmatrix}$

In Problems 17–20 the given vectors are solutions of a system X′ = AX. Determine whether the vectors form a fundamental set on (−∞, ∞).

17. $\mathbf{X}_1 = \begin{pmatrix} 1 \\ 1 \end{pmatrix} e^{-2t}, \mathbf{X}_2 = \begin{pmatrix} 1 \\ -1 \end{pmatrix} e^{-6t}$

18. $\mathbf{X}_1 = \begin{pmatrix} 1 \\ -1 \end{pmatrix} e^{t}, \mathbf{X}_2 = \begin{pmatrix} 2 \\ 6 \end{pmatrix} e^{t} + \begin{pmatrix} 8 \\ -8 \end{pmatrix} te^{t}$

19. $\mathbf{X}_1 = \begin{pmatrix} 1 \\ -2 \\ 4 \end{pmatrix} + t\begin{pmatrix} 1 \\ 2 \\ 2 \end{pmatrix}, \mathbf{X}_2 = \begin{pmatrix} 1 \\ -2 \\ 4 \end{pmatrix},$

$\mathbf{X}_3 = \begin{pmatrix} 3 \\ -6 \\ 12 \end{pmatrix} + t\begin{pmatrix} 2 \\ 4 \\ 4 \end{pmatrix}$

20. $\mathbf{X}_1 = \begin{pmatrix} 1 \\ 6 \\ -13 \end{pmatrix}, \mathbf{X}_2 = \begin{pmatrix} 1 \\ -2 \\ -1 \end{pmatrix} e^{-4t}, \mathbf{X}_3 = \begin{pmatrix} 2 \\ 3 \\ -2 \end{pmatrix} e^{3t}$

In Problems 21–24 verify that the vector $\mathbf{X}_p$ is a particular solution of the given system.

21. $\dfrac{dx}{dt} = x + 4y + 2t - 7$

$\dfrac{dy}{dt} = 3x + 2y - 4t - 18; \quad \mathbf{X}_p = \begin{pmatrix} 2 \\ -1 \end{pmatrix} t + \begin{pmatrix} 5 \\ 1 \end{pmatrix}$

22. $\mathbf{X}' = \begin{pmatrix} 2 & 1 \\ 1 & -1 \end{pmatrix}\mathbf{X} + \begin{pmatrix} -5 \\ 2 \end{pmatrix}; \quad \mathbf{X}_p = \begin{pmatrix} 1 \\ 3 \end{pmatrix}$

23. $\mathbf{X}' = \begin{pmatrix} 2 & 1 \\ 3 & 4 \end{pmatrix}\mathbf{X} - \begin{pmatrix} 1 \\ 7 \end{pmatrix} e^{t}; \quad \mathbf{X}_p = \begin{pmatrix} 1 \\ 1 \end{pmatrix} e^{t} + \begin{pmatrix} 1 \\ -1 \end{pmatrix} te^{t}$

24. $\mathbf{X}' = \begin{pmatrix} 1 & 2 & 3 \\ -4 & 2 & 0 \\ -6 & 1 & 0 \end{pmatrix}\mathbf{X} + \begin{pmatrix} -1 \\ 4 \\ 3 \end{pmatrix}\sin 3t;$

$\mathbf{X}_p = \begin{pmatrix} \sin 3t \\ 0 \\ \cos 3t \end{pmatrix}$

25. Prove that the general solution of

$$\mathbf{X}' = \begin{pmatrix} 0 & 6 & 0 \\ 1 & 0 & 1 \\ 1 & 1 & 0 \end{pmatrix}\mathbf{X}$$

on the interval $(-\infty, \infty)$ is

$$\mathbf{X} = c_1\begin{pmatrix} 6 \\ -1 \\ -5 \end{pmatrix} e^{-t} + c_2\begin{pmatrix} -3 \\ 1 \\ 1 \end{pmatrix} e^{-2t} + c_3\begin{pmatrix} 2 \\ 1 \\ 1 \end{pmatrix} e^{3t}.$$

26. Prove that the general solution of

$$\mathbf{X}' = \begin{pmatrix} -1 & -1 \\ -1 & 1 \end{pmatrix}\mathbf{X} + \begin{pmatrix} 1 \\ 1 \end{pmatrix} t^2 + \begin{pmatrix} 4 \\ -6 \end{pmatrix} t + \begin{pmatrix} -1 \\ 5 \end{pmatrix}$$

on the interval $(-\infty, \infty)$ is

$$\mathbf{X} = c_1\begin{pmatrix} 1 \\ -1-\sqrt{2} \end{pmatrix} e^{\sqrt{2}t} + c_2\begin{pmatrix} 1 \\ -1+\sqrt{2} \end{pmatrix} e^{-\sqrt{2}t}$$

$$+ \begin{pmatrix} 1 \\ 0 \end{pmatrix} t^2 + \begin{pmatrix} -2 \\ 4 \end{pmatrix} t + \begin{pmatrix} 1 \\ 0 \end{pmatrix}.$$

4.2 Homogeneous Linear Systems

4.2.1 Distinct Real Eigenvalues

We saw in Example 5 of Section 4.1 that the general solution of the homogeneous system $\mathbf{X}' = \begin{pmatrix} 1 & 3 \\ 5 & 3 \end{pmatrix}\mathbf{X}$ is $\mathbf{X} = c_1\begin{pmatrix} 1 \\ -1 \end{pmatrix}e^{-2t} + c_2\begin{pmatrix} 3 \\ 5 \end{pmatrix}e^{6t}$. Since both solution vectors have the form $\mathbf{X}_i = \begin{pmatrix} k_1 \\ k_2 \end{pmatrix}e^{\lambda_i t}$, $i = 1, 2$, where k_1 and k_2 are constants, we are prompted to ask whether we can always find a solution of the form

$$\mathbf{X} = \begin{pmatrix} k_1 \\ k_2 \\ \vdots \\ k_n \end{pmatrix} e^{\lambda t} = \mathbf{K}e^{\lambda t} \tag{1}$$

for the general homogeneous linear first-order system

$$\mathbf{X}' = \mathbf{AX}, \tag{2}$$

where $\mathbf{A}$ is an $n \times n$ matrix of constants.

Eigenvalues and Eigenvectors

If (1) is to be a solution vector of (2), then $\mathbf{X}' = \mathbf{K}\lambda e^{\lambda t}$ so that the system becomes

$$\mathbf{K}\lambda e^{\lambda t} = \mathbf{AK}e^{\lambda t}.$$

After dividing out $e^{\lambda t}$ and rearranging, we obtain $\mathbf{AK} = \lambda\mathbf{K}$ or

$$(\mathbf{A} - \lambda\mathbf{I})\mathbf{K} = \mathbf{0}. \tag{3}$$

Equation (3) is equivalent to the simultaneous algebraic equations

$$\begin{aligned}
(a_{11} - \lambda)k_1 + \quad a_{12}k_2 + \cdots + \quad a_{1n}k_n &= 0 \\
a_{21}k_1 + (a_{22} - \lambda)k_2 + \cdots + \quad a_{2n}k_n &= 0 \\
\vdots \qquad\qquad\qquad & \vdots \\
a_{n1}k_1 + \quad a_{n2}k_2 + \cdots + (a_{nn} - \lambda)k_n &= 0.
\end{aligned}$$

Thus to find a nontrivial solution $\mathbf{X}$ of (2) we must first find a nontrivial solution of the foregoing system; in other words, we must find a nontrivial vector $\mathbf{K}$ that satisfies (3). But in order for (3) to have nontrivial solutions we must have

$$\det(\mathbf{A} - \lambda\mathbf{I}) = 0.$$

The last equation is the **characteristic equation** of the matrix $\mathbf{A}$. In other words, $\mathbf{X} = \mathbf{K}e^{\lambda t}$ will be a solution of the system of differential equations (2) if and only if λ is an **eigenvalue** of $\mathbf{A}$ and $\mathbf{K}$ is an **eigenvector** corresponding to λ.

When the $n \times n$ matrix $\mathbf{A}$ possesses n distinct real eigenvalues $\lambda_1, \lambda_2, \ldots, \lambda_n$, then a set of n linearly independent eigenvectors $\mathbf{K}_1, \mathbf{K}_2, \ldots, \mathbf{K}_n$ can always be found and

$$\mathbf{X}_1 = \mathbf{K}_1 e^{\lambda_1 t},\ \mathbf{X}_2 = \mathbf{K}_2 e^{\lambda_2 t}, \ldots,\ \mathbf{X}_n = \mathbf{K}_n e^{\lambda_n t}$$

is a fundamental set of solutions of (2) on $(-\infty, \infty)$.

THEOREM 4.7 General Solution—Homogeneous Systems

Let $\lambda_1, \lambda_2, \ldots, \lambda_n$ be n distinct real eigenvalues of the coefficient matrix $\mathbf{A}$ of the homogeneous system (2) and let $\mathbf{K}_1, \mathbf{K}_2, \ldots, \mathbf{K}_n$ be the corresponding eigenvectors. Then the **general solution** of (2) on the interval $(-\infty, \infty)$ is given by

$$\mathbf{X} = c_1\mathbf{K}_1 e^{\lambda_1 t} + c_2\mathbf{K}_2 e^{\lambda_2 t} + \cdots + c_n\mathbf{K}_n e^{\lambda_n t}.$$

EXAMPLE 1 Distinct Eigenvalues

Solve

$$\begin{aligned}\frac{dx}{dt} &= 2x + 3y \\ \frac{dy}{dt} &= 2x + y.\end{aligned} \tag{4}$$

Solution We first find the eigenvalues and eigenvectors of the matrix of coefficients.

From the characteristic equation

$$\det(\mathbf{A} - \lambda\mathbf{I}) = \begin{vmatrix} 2-\lambda & 3 \\ 2 & 1-\lambda \end{vmatrix} = \lambda^2 - 3\lambda - 4 = (\lambda + 1)(\lambda - 4) = 0$$

we see that the eigenvalues are $\lambda_1 = -1$ and $\lambda_2 = 4$.

Now for $\lambda_1 = -1$, (3) is equivalent to

$$\begin{aligned}3k_1 + 3k_2 &= 0 \\ 2k_1 + 2k_2 &= 0.\end{aligned}$$

Thus $k_1 = -k_2$. When $k_2 = -1$ the related eigenvector is

$$\mathbf{K}_1 = \begin{pmatrix} 1 \\ -1 \end{pmatrix}.$$

For $\lambda_2 = 4$ we have

$$\begin{aligned}-2k_1 + 3k_2 &= 0 \\ 2k_1 - 3k_2 &= 0\end{aligned}$$

so that $k_1 = 3k_2/2$, and therefore with $k_2 = 2$, the corresponding eigenvector is

$$\mathbf{K}_2 = \begin{pmatrix} 3 \\ 2 \end{pmatrix}.$$

Since the matrix of coefficients **A** is a 2×2 matrix and since we have found two linearly independent solutions of (4),

$$\mathbf{X}_1 = \begin{pmatrix} 1 \\ -1 \end{pmatrix} e^{-t} \quad \text{and} \quad \mathbf{X}_2 = \begin{pmatrix} 3 \\ 2 \end{pmatrix} e^{4t},$$

we conclude that the general solution of the system is

$$\mathbf{X} = c_1\mathbf{X}_1 + c_2\mathbf{X}_2 = c_1 \begin{pmatrix} 1 \\ -1 \end{pmatrix} e^{-t} + c_2 \begin{pmatrix} 3 \\ 2 \end{pmatrix} e^{4t}. \tag{5}$$

For the sake of review, you should keep firmly in mind that a solution of a system of first-order differential equations, when written in terms of matrices, is simply an alternative to the method that we employed in Section 3.10, namely, listing the individual functions and the relationships between the constants. If we add the vectors on the right side in (5) and equate the entries with the corresponding entries in the vector on the left, we obtain the more familiar statement

$$\begin{aligned} x(t) &= c_1 e^{-t} + 3c_2 e^{4t} \\ y(t) &= -c_1 e^{-t} + 2c_2 e^{4t}. \end{aligned}$$

EXAMPLE 2 Distinct Eigenvalues

Solve

$$\begin{aligned} \frac{dx}{dt} &= -4x + y + z \\ \frac{dy}{dt} &= x + 5y - z \\ \frac{dz}{dt} &= y - 3z. \end{aligned} \tag{6}$$

Solution Using the cofactors of the third row, we find

$$\det(\mathbf{A} - \lambda\mathbf{I}) = \begin{vmatrix} -4-\lambda & 1 & 1 \\ 1 & 5-\lambda & -1 \\ 0 & 1 & -3-\lambda \end{vmatrix} = -(\lambda + 3)(\lambda + 4)(\lambda - 5) = 0$$

and so the eigenvalues are $\lambda_1 = -3$, $\lambda_2 = -4$, $\lambda_3 = 5$.

For $\lambda_1 = -3$, Gauss-Jordan elimination gives

$$(\mathbf{A} + 3\mathbf{I}|\mathbf{0}) = \left(\begin{array}{ccc|c} -1 & 1 & 1 & 0 \\ 1 & 8 & -1 & 0 \\ 0 & 1 & 0 & 0 \end{array}\right) \xrightarrow{\text{row operations}} \left(\begin{array}{ccc|c} 1 & 0 & -1 & 0 \\ 0 & 1 & 0 & 0 \\ 0 & 0 & 0 & 0 \end{array}\right).$$

Therefore $k_1 = k_3$ and $k_2 = 0$. The choice $k_3 = 1$ gives an eigenvector and corresponding solution vector

$$\mathbf{K}_1 = \begin{pmatrix} 1 \\ 0 \\ 1 \end{pmatrix}, \quad \mathbf{X}_1 = \begin{pmatrix} 1 \\ 0 \\ 1 \end{pmatrix} e^{-3t}. \tag{7}$$

Similarly, for $\lambda_2 = -4$,

$$(\mathbf{A} + 4\mathbf{I}|\mathbf{0}) = \left(\begin{array}{ccc|c} 0 & 1 & 1 & 0 \\ 1 & 9 & -1 & 0 \\ 0 & 1 & 1 & 0 \end{array}\right) \xrightarrow{\text{row operations}} \left(\begin{array}{ccc|c} 1 & 0 & -10 & 0 \\ 0 & 1 & 1 & 0 \\ 0 & 0 & 0 & 0 \end{array}\right)$$

implies $k_1 = 10k_3$ and $k_2 = -k_3$. Choosing $k_3 = 1$, we get a second eigenvector and solution vector

$$\mathbf{K}_2 = \begin{pmatrix} 10 \\ -1 \\ 1 \end{pmatrix}, \quad \mathbf{X}_2 = \begin{pmatrix} 10 \\ -1 \\ 1 \end{pmatrix} e^{-4t}. \tag{8}$$

Finally, when $\lambda_3 = 5$, the augmented matrices

$$(\mathbf{A} - 5\mathbf{I}|\mathbf{0}) = \left(\begin{array}{ccc|c} -9 & 1 & 1 & 0 \\ 1 & 0 & -1 & 0 \\ 0 & 1 & -8 & 0 \end{array}\right) \xrightarrow{\text{row operations}} \left(\begin{array}{ccc|c} 1 & 0 & -1 & 0 \\ 0 & 1 & -8 & 0 \\ 0 & 0 & 0 & 0 \end{array}\right)$$

yield

$$\mathbf{K}_3 = \begin{pmatrix} 1 \\ 8 \\ 1 \end{pmatrix}, \quad \mathbf{X}_3 = \begin{pmatrix} 1 \\ 8 \\ 1 \end{pmatrix} e^{5t}. \tag{9}$$

The general solution of (6) is a linear combination of the solution vectors in (7), (8), and (9):

$$\mathbf{X} = c_1 \begin{pmatrix} 1 \\ 0 \\ 1 \end{pmatrix} e^{-3t} + c_2 \begin{pmatrix} 10 \\ -1 \\ 1 \end{pmatrix} e^{-4t} + c_3 \begin{pmatrix} 1 \\ 8 \\ 1 \end{pmatrix} e^{5t}.$$

●

4.2.2 Repeated Eigenvalues

Of course, not all of the n eigenvalues $\lambda_1, \lambda_2, \ldots, \lambda_n$ of an $n \times n$ matrix $\mathbf{A}$ may be distinct, that is, some of the eigenvalues may be repeated. For example, the characteristic equation of the coefficient matrix in the system

$$\mathbf{X}' = \begin{pmatrix} 3 & -18 \\ 2 & -9 \end{pmatrix} \mathbf{X} \tag{10}$$

is readily shown to be $(\lambda + 3)^2 = 0$, and therefore $\lambda_1 = \lambda_2 = -3$ is a root of *multiplicity two*. For this value we find the single eigenvector

$$\mathbf{K}_1 = \begin{pmatrix} 3 \\ 1 \end{pmatrix}, \quad \text{so} \quad \mathbf{X}_1 = \begin{pmatrix} 3 \\ 1 \end{pmatrix} e^{-3t} \tag{11}$$

is one solution of (10). But since we are obviously interested in forming the general solution of the system, we need to pursue the question of finding a second solution.

In general, if m is a positive integer and $(\lambda - \lambda_1)^m$ is a factor of the characteristic equation, while $(\lambda - \lambda_1)^{m+1}$ is not a factor, then λ_1 is said to be an **eigenvalue of multiplicity *m***. We distinguish two possibilities:

(i) For some $n \times n$ matrices $\mathbf{A}$ it may be possible to find m linearly independent eigenvectors $\mathbf{K}_1, \mathbf{K}_2, \ldots, \mathbf{K}_m$ corresponding to an eigenvalue λ_1 of multiplicity $m \leq n$. In this case the general solution of the system contains the linear combination

$$c_1\mathbf{K}_1 e^{\lambda_1 t} + c_2\mathbf{K}_2 e^{\lambda_1 t} + \cdots + c_m\mathbf{K}_m e^{\lambda_1 t}.$$

(ii) If there is only one eigenvector corresponding to the eigenvalue λ_1 of multiplicity m, then m linearly independent solutions of the form

$$\begin{aligned}
\mathbf{X}_1 &= \mathbf{K}_{11} e^{\lambda_1 t} \\
\mathbf{X}_2 &= \mathbf{K}_{21} t e^{\lambda_1 t} + \mathbf{K}_{22} e^{\lambda_1 t} \\
&\ \vdots \\
\mathbf{X}_m &= \mathbf{K}_{m1} \frac{t^{m-1}}{(m-1)!} e^{\lambda_1 t} + \mathbf{K}_{m2} \frac{t^{m-2}}{(m-2)!} e^{\lambda_1 t} + \cdots + \mathbf{K}_{mm} e^{\lambda_1 t},
\end{aligned}$$

where $\mathbf{K}_{ij}$ are column vectors, can always be found.

Eigenvalue of Multiplicity Two

We begin by considering eigenvalues of multiplicity two. In the first example we illustrate a matrix for which we can find two distinct eigenvectors corresponding to a double eigenvalue.

EXAMPLE 3 Repeated Eigenvalues

Solve

$$\mathbf{X}' = \begin{pmatrix} 1 & -2 & 2 \\ -2 & 1 & -2 \\ 2 & -2 & 1 \end{pmatrix} \mathbf{X}.$$

Solution Expanding the determinant in the characteristic equation

$$\det(\mathbf{A} - \lambda\mathbf{I}) = \begin{vmatrix} 1-\lambda & -2 & 2 \\ -2 & 1-\lambda & -2 \\ 2 & -2 & 1-\lambda \end{vmatrix} = 0$$

yields $-(\lambda + 1)^2(\lambda - 5) = 0$. We see that $\lambda_1 = \lambda_2 = -1$ and $\lambda_3 = 5$.

For $\lambda_1 = -1$, Gauss-Jordan elimination gives immediately

$$(\mathbf{A} + \mathbf{I}|\mathbf{0}) = \left(\begin{array}{ccc|c} 2 & -2 & 2 & 0 \\ -2 & 2 & -2 & 0 \\ 2 & -2 & 2 & 0 \end{array}\right) \xrightarrow{\text{row operations}} \left(\begin{array}{ccc|c} 1 & -1 & 1 & 0 \\ 0 & 0 & 0 & 0 \\ 0 & 0 & 0 & 0 \end{array}\right).$$

The first row of the last matrix means $k_1 - k_2 + k_3 = 0$ or $k_1 = k_2 - k_3$. The choices $k_2 = 1$, $k_3 = 0$ and $k_2 = 1$, $k_3 = 1$, yield, in turn, $k_1 = 1$ and $k_1 = 0$. Thus two eigenvectors corresponding to $\lambda_1 = -1$ are

$$\mathbf{K}_1 = \begin{pmatrix} 1 \\ 1 \\ 0 \end{pmatrix} \quad \text{and} \quad \mathbf{K}_2 = \begin{pmatrix} 0 \\ 1 \\ 1 \end{pmatrix}.$$

Since neither eigenvector is a constant multiple of the other, we have found, corresponding to the same eigenvalue, two linearly independent solutions

$$\mathbf{X}_1 = \begin{pmatrix} 1 \\ 1 \\ 0 \end{pmatrix} e^{-t} \quad \text{and} \quad \mathbf{X}_2 = \begin{pmatrix} 0 \\ 1 \\ 1 \end{pmatrix} e^{-t}.$$

Last, for $\lambda_3 = 5$, the reduction

$$(\mathbf{A} - 5\mathbf{I}|\mathbf{0}) = \left(\begin{array}{ccc|c} -4 & -2 & 2 & 0 \\ -2 & -4 & -2 & 0 \\ 2 & -2 & -4 & 0 \end{array}\right) \xrightarrow{\text{row operations}} \left(\begin{array}{ccc|c} 1 & 0 & -1 & 0 \\ 0 & 1 & 1 & 0 \\ 0 & 0 & 0 & 0 \end{array}\right)$$

implies $k_1 = k_3$ and $k_2 = -k_3$. Picking $k_3 = 1$ gives $k_1 = 1$, $k_2 = -1$, and thus a third eigenvector is

$$\mathbf{K}_3 = \begin{pmatrix} 1 \\ -1 \\ 1 \end{pmatrix}.$$

We conclude that the general solution of the system is

$$\mathbf{X} = c_1 \begin{pmatrix} 1 \\ 1 \\ 0 \end{pmatrix} e^{-t} + c_2 \begin{pmatrix} 0 \\ 1 \\ 1 \end{pmatrix} e^{-t} + c_3 \begin{pmatrix} 1 \\ -1 \\ 1 \end{pmatrix} e^{5t}.$$

Second Solution

Now suppose that λ_1 is an eigenvalue of multiplicity two and that there is only one eigenvector associated with this value. A second solution can be found of the form

$$\mathbf{X}_2 = \mathbf{K}te^{\lambda_1 t} + \mathbf{P}e^{\lambda_1 t}, \tag{12}$$

where

$$\mathbf{K} = \begin{pmatrix} k_1 \\ k_2 \\ \vdots \\ k_n \end{pmatrix} \quad \text{and} \quad \mathbf{P} = \begin{pmatrix} p_1 \\ p_2 \\ \vdots \\ p_n \end{pmatrix}.$$

To see this we substitute (12) into the system $\mathbf{X}' = \mathbf{AX}$ and simplify:

$$(\mathbf{AK} - \lambda_1\mathbf{K})te^{\lambda_1 t} + (\mathbf{AP} - \lambda_1\mathbf{P} - \mathbf{K})e^{\lambda_1 t} = \mathbf{0}.$$

Since this last equation is to hold for all values of t, we must have

$$(\mathbf{A} - \lambda_1\mathbf{I})\mathbf{K} = \mathbf{0} \tag{13}$$

and

$$(\mathbf{A} - \lambda_1\mathbf{I})\mathbf{P} = \mathbf{K}. \tag{14}$$

Equation (13) simply states that $\mathbf{K}$ must be an eigenvector of $\mathbf{A}$ associated with λ_1. By solving (13), we find one solution $\mathbf{X}_1 = \mathbf{K}e^{\lambda_1 t}$. To find the second solution $\mathbf{X}_2$ we need only solve the additional system (14) for the vector $\mathbf{P}$.

EXAMPLE 4 Repeated Eigenvalues

Find the general solution of the system given in (10).

Solution From (11) we know that $\lambda_1 = -3$ and that one solution is $\mathbf{X}_1 = \begin{pmatrix} 3 \\ 1 \end{pmatrix} e^{-3t}$. Identifying $\mathbf{K} = \begin{pmatrix} 3 \\ 1 \end{pmatrix}$ and $\mathbf{P} = \begin{pmatrix} p_1 \\ p_2 \end{pmatrix}$, we find from (14) that we must now solve

$$(\mathbf{A} + 3\mathbf{I})\mathbf{P} = \mathbf{K} \quad \text{or} \quad \begin{aligned} 6p_1 - 18p_2 &= 3 \\ 2p_1 - 6p_2 &= 1 \end{aligned}.$$

Since this system is obviously equivalent to one equation, we have an infinite number of choices for p_1 and p_2. For example, by choosing $p_1 = 1$, we find $p_2 = 1/6$. However, for simplicity, we shall choose $p_1 = 1/2$ so that $p_2 = 0$. Hence $\mathbf{P} = \begin{pmatrix} \frac{1}{2} \\ 0 \end{pmatrix}$. Thus from (12) we find

$$\mathbf{X}_2 = \begin{pmatrix} 3 \\ 1 \end{pmatrix} te^{-3t} + \begin{pmatrix} \frac{1}{2} \\ 0 \end{pmatrix} e^{-3t}.$$

The general solution of (10) is then

$$\mathbf{X} = c_1 \begin{pmatrix} 3 \\ 1 \end{pmatrix} e^{-3t} + c_2 \left[\begin{pmatrix} 3 \\ 1 \end{pmatrix} te^{-3t} + \begin{pmatrix} \frac{1}{2} \\ 0 \end{pmatrix} e^{-3t} \right].$$

Eigenvalues of Multiplicity Three

When a matrix **A** has only one eigenvector associated with an eigenvalue λ_1 of multiplicity three, we can find a second solution of form (12) and a third solution of the form

$$\mathbf{X}_3 = \mathbf{K}\frac{t^2}{2}e^{\lambda_1 t} + \mathbf{P}te^{\lambda_1 t} + \mathbf{Q}e^{\lambda_1 t}, \tag{15}$$

where $\mathbf{K} = \begin{pmatrix} k_1 \\ k_2 \\ \vdots \\ k_n \end{pmatrix}$, $\mathbf{P} = \begin{pmatrix} p_1 \\ p_2 \\ \vdots \\ p_n \end{pmatrix}$, and $\mathbf{Q} = \begin{pmatrix} q_1 \\ q_2 \\ \vdots \\ q_n \end{pmatrix}$.

By substituting (15) into the system $\mathbf{X}' = \mathbf{AX}$, we find the column vectors **K**, **P**, and **Q** must satisfy

$$(\mathbf{A} - \lambda_1\mathbf{I})\mathbf{K} = \mathbf{0} \tag{16}$$

$$(\mathbf{A} - \lambda_1\mathbf{I})\mathbf{P} = \mathbf{K} \tag{17}$$

and

$$(\mathbf{A} - \lambda_1\mathbf{I})\mathbf{Q} = \mathbf{P}. \tag{18}$$

Of course the solutions of (16) and (17) can be used in forming the solutions $\mathbf{X}_1$ and $\mathbf{X}_2$.

EXAMPLE 5 Repeated Eigenvalues

Solve
$$\mathbf{X}' = \begin{pmatrix} 2 & 1 & 6 \\ 0 & 2 & 5 \\ 0 & 0 & 2 \end{pmatrix}\mathbf{X}.$$

Solution The characteristic equation $(\lambda - 2)^3 = 0$ shows that $\lambda_1 = 2$ is an eigenvalue of multiplicity three. We next solve the three systems

$$(\mathbf{A} - 2\mathbf{I})\mathbf{K} = \mathbf{0}, \quad (\mathbf{A} - 2\mathbf{I})\mathbf{P} = \mathbf{K}, \quad \text{and} \quad (\mathbf{A} - 2\mathbf{I})\mathbf{Q} = \mathbf{P}$$

in succession and find that

$$\mathbf{K} = \begin{pmatrix} 1 \\ 0 \\ 0 \end{pmatrix}, \quad \mathbf{P} = \begin{pmatrix} 0 \\ 1 \\ 0 \end{pmatrix}, \quad \text{and} \quad \mathbf{Q} = \begin{pmatrix} 0 \\ -\frac{6}{5} \\ \frac{1}{5} \end{pmatrix}.$$

Using (12) and (15), the general solution of the system is

$$\mathbf{X} = c_1\begin{pmatrix} 1 \\ 0 \\ 0 \end{pmatrix}e^{2t} + c_2\left[\begin{pmatrix} 1 \\ 0 \\ 0 \end{pmatrix}te^{2t} + \begin{pmatrix} 0 \\ 1 \\ 0 \end{pmatrix}e^{2t}\right] + c_3\left[\begin{pmatrix} 1 \\ 0 \\ 0 \end{pmatrix}\frac{t^2}{2}e^{2t} + \begin{pmatrix} 0 \\ 1 \\ 0 \end{pmatrix}te^{2t} + \begin{pmatrix} 0 \\ -\frac{6}{5} \\ \frac{1}{5} \end{pmatrix}e^{2t}\right].$$ ●

4.2.3 Complex Eigenvalues

If $\lambda_1 = \alpha + i\beta$ and $\lambda_2 = \alpha - i\beta$, $i^2 = -1$, are complex eigenvalues of the coefficient matrix $\mathbf{A}$, we can then certainly expect their corresponding eigenvectors to also have complex entries.*

For example, the characteristic equation of the system

$$\begin{aligned}\frac{dx}{dt} &= 6x - y\\ \frac{dy}{dt} &= 5x + 4y\end{aligned} \tag{19}$$

is $$\det(\mathbf{A} - \lambda\mathbf{I}) = \begin{vmatrix} 6-\lambda & -1 \\ 5 & 4-\lambda \end{vmatrix} = \lambda^2 - 10\lambda + 29 = 0.$$

From the quadratic formula we find $\lambda_1 = 5 + 2i$, $\lambda_2 = 5 - 2i$.

Now for $\lambda_1 = 5 + 2i$ we must solve

$$\begin{aligned}(1-2i)k_1 - \quad\quad k_2 &= 0\\ 5k_1 - (1+2i)k_2 &= 0.\end{aligned}$$

Since $k_2 = (1 - 2i)k_1$,† the choice $k_1 = 1$ gives the following eigenvector and a solution vector

$$\mathbf{K}_1 = \begin{pmatrix} 1 \\ 1-2i \end{pmatrix}, \quad \mathbf{X}_1 = \begin{pmatrix} 1 \\ 1-2i \end{pmatrix} e^{(5+2i)t}.$$

In like manner, for $\lambda_2 = 5 - 2i$ we find

$$\mathbf{K}_2 = \begin{pmatrix} 1 \\ 1+2i \end{pmatrix}, \quad \mathbf{X}_2 = \begin{pmatrix} 1 \\ 1+2i \end{pmatrix} e^{(5-2i)t}.$$

We can verify by means of the Wronskian that these solution vectors are linearly independent and so the general solution of (19) is

$$\mathbf{X} = c_1 \begin{pmatrix} 1 \\ 1-2i \end{pmatrix} e^{(5+2i)t} + c_2 \begin{pmatrix} 1 \\ 1+2i \end{pmatrix} e^{(5-2i)t}. \tag{20}$$

Note that the entries in $\mathbf{K}_2$ corresponding to λ_2 are the conjugates of the entries in $\mathbf{K}_1$ corresponding to λ_1. The conjugate of λ_1 is, of course, λ_2. We write this as $\lambda_2 = \bar{\lambda}_1$ and $\mathbf{K}_2 = \bar{\mathbf{K}}_1$. We have illustrated the following general result.

---

* When the characteristic equation has real coefficients, complex eigenvalues always appear in conjugate pairs.

† Note that the second equation is simply $(1 + 2i)$ times the first.

THEOREM 4.8 Solutions Corresponding to a Complex Eigenvalue

Let **A** be the coefficient matrix having real entries of the homogeneous system (2), and let $\mathbf{K}_1$ be an eigenvector corresponding to the complex eigenvalue $\lambda_1 = \alpha + i\beta$, α and β real. Then

$$\mathbf{K}_1 e^{\lambda_1 t} \quad \text{and} \quad \overline{\mathbf{K}}_1 e^{\bar{\lambda}_1 t}$$

are solutions of (2).

It is desirable and relatively easy to rewrite a solution such as (20) in terms of real functions. To this end we first use Euler's formula to write

$$e^{(5+2i)t} = e^{5t}e^{2ti} = e^{5t}(\cos 2t + i \sin 2t)$$

$$e^{(5-2i)t} = e^{5t}e^{-2ti} = e^{5t}(\cos 2t - i \sin 2t).$$

Then, after we multiply complex numbers, collect terms, and replace $c_1 + c_2$ by C_1 and $(c_1 - c_2)i$ by C_2, (20) becomes

$$\mathbf{X} = C_1\mathbf{X}_1 + C_2\mathbf{X}_2 \tag{21}$$

where

$$\mathbf{X}_1 = \left[\begin{pmatrix} 1 \\ 1 \end{pmatrix} \cos 2t - \begin{pmatrix} 0 \\ -2 \end{pmatrix} \sin 2t\right] e^{5t}$$

$$\mathbf{X}_2 = \left[\begin{pmatrix} 0 \\ -2 \end{pmatrix} \cos 2t + \begin{pmatrix} 1 \\ 1 \end{pmatrix} \sin 2t\right] e^{5t}$$

It is now important to realize that the two vectors $\mathbf{X}_1$ and $\mathbf{X}_2$ in (21) are themselves linearly independent *real* solutions of the original system. Consequently, we are justified in ignoring the relationship between C_1, C_2 and c_1, c_2 and regard C_1 and C_2 as completely arbitrary and real. In other words, the linear combination (21) is an alternative general solution of (19).

The foregoing process can be generalized. Let $\mathbf{K}_1$ be an eigenvector of the coefficient matrix **A** (with real entries) corresponding to the complex eigenvalue $\lambda_1 = \alpha + i\beta$. Then the two solution vectors in Theorem 4.8 can be written as

$$\mathbf{K}_1 e^{\lambda_1 t} = \mathbf{K}_1 e^{\alpha t} e^{i\beta t} = \mathbf{K}_1 e^{\alpha t}(\cos \beta t + i \sin \beta t)$$

$$\overline{\mathbf{K}}_1 e^{\bar{\lambda}_1 t} = \overline{\mathbf{K}}_1 e^{\alpha t} e^{-i\beta t} = \overline{\mathbf{K}}_1 e^{\alpha t}(\cos \beta t - i \sin \beta t).$$

By the superposition principle, Theorem 4.2, the following vectors are also solutions

$$\mathbf{X}_1 = \frac{1}{2}(\mathbf{K}_1 e^{\lambda_1 t} + \overline{\mathbf{K}}_1 e^{\bar{\lambda}_1 t}) = \frac{1}{2}(\mathbf{K}_1 + \overline{\mathbf{K}}_1)e^{\alpha t} \cos \beta t - \frac{i}{2}(-\mathbf{K}_1 + \overline{\mathbf{K}}_1)e^{\alpha t} \sin \beta t$$

$$\mathbf{X}_2 = \frac{i}{2}(-\mathbf{K}_1 e^{\lambda_1 t} + \overline{\mathbf{K}}_1 e^{\bar{\lambda}_1 t}) = \frac{i}{2}(-\mathbf{K}_1 + \overline{\mathbf{K}}_1)e^{\alpha t} \cos \beta t + \frac{1}{2}(\mathbf{K}_1 + \overline{\mathbf{K}}_1)e^{\alpha t} \sin \beta t.$$

For *any* complex number $z = a + ib$, both $\frac{1}{2}(z + \bar{z}) = a$ and $\frac{i}{2}(-z + \bar{z}) = b$ are *real* numbers. Therefore, the entries in the column vectors $\frac{1}{2}(\mathbf{K}_1 + \overline{\mathbf{K}}_1)$ and

$\frac{i}{2}(-\mathbf{K}_1 + \overline{\mathbf{K}}_1)$ are real numbers. By defining

$$\mathbf{B}_1 = \frac{1}{2}(\mathbf{K}_1 + \overline{\mathbf{K}}_1) \quad \text{and} \quad \mathbf{B}_2 = \frac{i}{2}(-\mathbf{K}_1 + \overline{\mathbf{K}}_1), \tag{22}$$

we are led to the following theorem.

THEOREM 4.9 Real Solutions Corresponding to a Complex Eigenvalue

Let $\lambda_1 = \alpha + i\beta$ be a complex eigenvalue of the coefficient matrix $\mathbf{A}$ in the homogeneous system (2) and let $\mathbf{B}_1$ and $\mathbf{B}_2$ denote the column vectors defined in (22). Then

$$\begin{aligned} \mathbf{X}_1 &= [\mathbf{B}_1 \cos \beta t - \mathbf{B}_2 \sin \beta t]e^{\alpha t} \\ \mathbf{X}_2 &= [\mathbf{B}_2 \cos \beta t + \mathbf{B}_1 \sin \beta t]e^{\alpha t} \end{aligned} \tag{23}$$

are linearly independent solutions of (2) on $(-\infty, \infty)$.

The matrices $\mathbf{B}_1$ and $\mathbf{B}_2$ in (22) are often denoted by

$$\mathbf{B}_1 = \operatorname{Re}(\mathbf{K}_1) \quad \text{and} \quad \mathbf{B}_2 = \operatorname{Im}(\mathbf{K}_1) \tag{24}$$

since these vectors are, in turn, the *real* and *imaginary* parts of the eigenvector $\mathbf{K}_1$. For example, (21) follows from (23) with

$$\mathbf{K}_1 = \begin{pmatrix} 1 \\ 1 - 2i \end{pmatrix} = \begin{pmatrix} 1 \\ 1 \end{pmatrix} + i\begin{pmatrix} 0 \\ -2 \end{pmatrix}$$

$$\mathbf{B}_1 = \operatorname{Re}(\mathbf{K}_1) = \begin{pmatrix} 1 \\ 1 \end{pmatrix} \quad \text{and} \quad \mathbf{B}_2 = \operatorname{Im}(\mathbf{K}_1) = \begin{pmatrix} 0 \\ -2 \end{pmatrix}.$$

EXAMPLE 6 Complex Eigenvalues

Solve
$$\mathbf{X}' = \begin{pmatrix} 2 & 8 \\ -1 & -2 \end{pmatrix}\mathbf{X}.$$

Solution First we obtain the eigenvalues from

$$\det(\mathbf{A} - \lambda\mathbf{I}) = \begin{vmatrix} 2 - \lambda & 8 \\ -1 & -2 - \lambda \end{vmatrix} = \lambda^2 + 4 = 0.$$

Thus the eigenvalues are $\lambda_1 = 2i$ and $\lambda_2 = \overline{\lambda}_1 = -2i$. For λ_1 the system

$$\begin{aligned} (2 - 2i)k_1 + \qquad\quad 8k_2 &= 0 \\ -k_1 + (-2 - 2i)k_2 &= 0 \end{aligned}$$

gives $k_1 = -(2 + 2i)k_2$. By choosing $k_2 = -1$, we get

$$\mathbf{K}_1 = \begin{pmatrix} 2 + 2i \\ -1 \end{pmatrix} = \begin{pmatrix} 2 \\ -1 \end{pmatrix} + i\begin{pmatrix} 2 \\ 0 \end{pmatrix}.$$

Now from (24) we form

$$\mathbf{B}_1 = \operatorname{Re}(\mathbf{K}_1) = \begin{pmatrix} 2 \\ -1 \end{pmatrix} \quad \text{and} \quad \mathbf{B}_2 = \operatorname{Im}(\mathbf{K}_1) = \begin{pmatrix} 2 \\ 0 \end{pmatrix}.$$

Since $\alpha = 0$, it follows from (23) that the general solution of the system is

$$\begin{aligned} \mathbf{X} &= c_1\left[\begin{pmatrix} 2 \\ -1 \end{pmatrix}\cos 2t - \begin{pmatrix} 2 \\ 0 \end{pmatrix}\sin 2t\right] + c_2\left[\begin{pmatrix} 2 \\ 0 \end{pmatrix}\cos 2t + \begin{pmatrix} 2 \\ -1 \end{pmatrix}\sin 2t\right] \\ &= c_1\begin{pmatrix} 2\cos 2t - 2\sin 2t \\ -\cos 2t \end{pmatrix} + c_2\begin{pmatrix} 2\cos 2t + 2\sin 2t \\ -\sin 2t \end{pmatrix}. \end{aligned}$$

EXERCISES 4.2

Answers to odd-numbered problems begin on page AN-19.

4.2.1

In Problems 1–12 find the general solution of the given system.

1. $\dfrac{dx}{dt} = x + 2y$, $\dfrac{dy}{dt} = 4x + 3y$

2. $\dfrac{dx}{dt} = 2y$, $\dfrac{dy}{dt} = 8x$

3. $\dfrac{dx}{dt} = -4x + 2y$, $\dfrac{dy}{dt} = -\dfrac{5}{2}x + 2y$

4. $\dfrac{dx}{dt} = \dfrac{1}{2}x + 9y$, $\dfrac{dy}{dt} = \dfrac{1}{2}x + 2y$

5. $\mathbf{X}' = \begin{pmatrix} 10 & -5 \\ 8 & -12 \end{pmatrix}\mathbf{X}$

6. $\mathbf{X}' = \begin{pmatrix} -6 & 2 \\ -3 & 1 \end{pmatrix}\mathbf{X}$

7. $\dfrac{dx}{dt} = x + y - z$, $\dfrac{dy}{dt} = 2y$, $\dfrac{dz}{dt} = y - z$

8. $\dfrac{dx}{dt} = 2x - 7y$, $\dfrac{dy}{dt} = 5x + 10y + 4z$, $\dfrac{dz}{dt} = 5y + 2z$

9. $\mathbf{X}' = \begin{pmatrix} -1 & 1 & 0 \\ 1 & 2 & 1 \\ 0 & 3 & -1 \end{pmatrix}\mathbf{X}$

10. $\mathbf{X}' = \begin{pmatrix} 1 & 0 & 1 \\ 0 & 1 & 0 \\ 1 & 0 & 1 \end{pmatrix}\mathbf{X}$

11. $\mathbf{X}' = \begin{pmatrix} -1 & -1 & 0 \\ \frac{3}{4} & -\frac{3}{2} & 3 \\ \frac{1}{8} & \frac{1}{4} & -\frac{1}{2} \end{pmatrix}\mathbf{X}$

12. $\mathbf{X}' = \begin{pmatrix} -1 & 4 & 2 \\ 4 & -1 & -2 \\ 0 & 0 & 6 \end{pmatrix}\mathbf{X}$

In Problems 13 and 14 solve the given system subject to the indicated initial condition.

13. $\mathbf{X}' = \begin{pmatrix} \frac{1}{2} & 0 \\ 1 & -\frac{1}{2} \end{pmatrix}\mathbf{X}$, $\mathbf{X}(0) = \begin{pmatrix} 3 \\ 5 \end{pmatrix}$

14. $\mathbf{X}' = \begin{pmatrix} 1 & 1 & 4 \\ 0 & 2 & 0 \\ 1 & 1 & 1 \end{pmatrix}\mathbf{X}$, $\mathbf{X}(0) = \begin{pmatrix} 1 \\ 3 \\ 0 \end{pmatrix}$

4.2.2

In Problems 15–24 find the general solution of the given system.

15. $\dfrac{dx}{dt} = 3x - y$, $\dfrac{dy}{dt} = 9x - 3y$

16. $\dfrac{dx}{dt} = -6x + 5y$, $\dfrac{dy}{dt} = -5x + 4y$

17. $\dfrac{dx}{dt} = -x + 3y$

$\dfrac{dy}{dt} = -3x + 5y$

18. $\dfrac{dx}{dt} = 12x - 9y$

$\dfrac{dy}{dt} = 4x$

19. $\dfrac{dx}{dt} = 3x - y - z$

$\dfrac{dy}{dt} = x + y - z$

$\dfrac{dz}{dt} = x - y + z$

20. $\dfrac{dx}{dt} = 3x + 2y + 4z$

$\dfrac{dy}{dt} = 2x + 2z$

$\dfrac{dz}{dt} = 4x + 2y + 3z$

21. $\mathbf{X}' = \begin{pmatrix} 5 & -4 & 0 \\ 1 & 0 & 2 \\ 0 & 2 & 5 \end{pmatrix}\mathbf{X}$

22. $\mathbf{X}' = \begin{pmatrix} 1 & 0 & 0 \\ 0 & 3 & 1 \\ 0 & -1 & 1 \end{pmatrix}\mathbf{X}$

23. $\mathbf{X}' = \begin{pmatrix} 1 & 0 & 0 \\ 2 & 2 & -1 \\ 0 & 1 & 0 \end{pmatrix}\mathbf{X}$

24. $\mathbf{X}' = \begin{pmatrix} 4 & 1 & 0 \\ 0 & 4 & 1 \\ 0 & 0 & 4 \end{pmatrix}\mathbf{X}$

In Problems 25 and 26 solve the given system subject to the indicated initial condition.

25. $\mathbf{X}' = \begin{pmatrix} 2 & 4 \\ -1 & 6 \end{pmatrix}\mathbf{X}, \quad \mathbf{X}(0) = \begin{pmatrix} -1 \\ 6 \end{pmatrix}$

26. $\mathbf{X}' = \begin{pmatrix} 0 & 0 & 1 \\ 0 & 1 & 0 \\ 1 & 0 & 0 \end{pmatrix}\mathbf{X}, \quad \mathbf{X}(0) = \begin{pmatrix} 1 \\ 2 \\ 5 \end{pmatrix}$

4.2.3

In Problems 27–38 find the general solution of the given system.

27. $\dfrac{dx}{dt} = 6x - y$

$\dfrac{dy}{dt} = 5x + 2y$

28. $\dfrac{dx}{dt} = x + y$

$\dfrac{dy}{dt} = -2x - y$

29. $\dfrac{dx}{dt} = 5x + y$

$\dfrac{dy}{dt} = -2x + 3y$

30. $\dfrac{dx}{dt} = 4x + 5y$

$\dfrac{dy}{dt} = -2x + 6y$

31. $\mathbf{X}' = \begin{pmatrix} 4 & -5 \\ 5 & -4 \end{pmatrix}\mathbf{X}$

32. $\mathbf{X}' = \begin{pmatrix} 1 & -8 \\ 1 & -3 \end{pmatrix}\mathbf{X}$

33. $\dfrac{dx}{dt} = z$

$\dfrac{dy}{dt} = -z$

$\dfrac{dz}{dt} = y$

34. $\dfrac{dx}{dt} = 2x + y + 2z$

$\dfrac{dy}{dt} = 3x + 6z$

$\dfrac{dz}{dt} = -4x - 3z$

35. $\mathbf{X}' = \begin{pmatrix} 1 & -1 & 2 \\ -1 & 1 & 0 \\ -1 & 0 & 1 \end{pmatrix}\mathbf{X}$

36. $\mathbf{X}' = \begin{pmatrix} 4 & 0 & 1 \\ 0 & 6 & 0 \\ -4 & 0 & 4 \end{pmatrix}\mathbf{X}$

37. $\mathbf{X}' = \begin{pmatrix} 2 & 5 & 1 \\ -5 & -6 & 4 \\ 0 & 0 & 2 \end{pmatrix}\mathbf{X}$

38. $\mathbf{X}' = \begin{pmatrix} 2 & 4 & 4 \\ -1 & -2 & 0 \\ -1 & 0 & -2 \end{pmatrix}\mathbf{X}$

In Problems 39 and 40 solve the given system subject to the indicated initial condition.

39. $\mathbf{X}' = \begin{pmatrix} 1 & -12 & -14 \\ 1 & 2 & -3 \\ 1 & 1 & -2 \end{pmatrix}\mathbf{X}, \quad \mathbf{X}(0) = \begin{pmatrix} 4 \\ 6 \\ -7 \end{pmatrix}$

40. $\mathbf{X}' = \begin{pmatrix} 6 & -1 \\ 5 & 4 \end{pmatrix}\mathbf{X}, \quad \mathbf{X}(0) = \begin{pmatrix} -2 \\ 8 \end{pmatrix}$

4.3 VARIATION OF PARAMETERS

Before developing a matrix version of variation of parameters for nonhomogeneous linear systems $\mathbf{X}' = \mathbf{AX} + \mathbf{F}$, we need to examine a special matrix that is formed out of the solution vectors of the corresponding homogeneous system $\mathbf{X}' = \mathbf{AX}$.

A Fundamental Matrix

If $\mathbf{X}_1, \mathbf{X}_2, \ldots, \mathbf{X}_n$ is a fundamental set of solutions for the homogeneous system $\mathbf{X}' = \mathbf{AX}$ on an interval I, then its general solution on the interval is

$$\mathbf{X} = c_1\mathbf{X}_1 + c_2\mathbf{X}_2 + \cdots + c_n\mathbf{X}_n$$

$$= c_1\begin{pmatrix} x_{11} \\ x_{21} \\ \vdots \\ x_{n1} \end{pmatrix} + c_2\begin{pmatrix} x_{12} \\ x_{22} \\ \vdots \\ x_{n2} \end{pmatrix} + \cdots + c_n\begin{pmatrix} x_{1n} \\ x_{2n} \\ \vdots \\ x_{nn} \end{pmatrix} = \begin{pmatrix} c_1x_{11} + c_2x_{12} + \cdots + c_nx_{1n} \\ c_1x_{21} + c_2x_{22} + \cdots + c_nx_{2n} \\ \vdots \\ c_1x_{n1} + c_2x_{n2} + \cdots + c_nx_{nn} \end{pmatrix}. \quad (1)$$

The last matrix in (1) is recognized as the product of an $n \times n$ matrix with an $n \times 1$ matrix. In other words, the general solution (1) can be written as

$$\mathbf{X} = \mathbf{\Phi}(t)\mathbf{C} \quad \mathbf{(2)}$$

where $\mathbf{C}$ is an $n \times 1$ column vector of arbitrary constants, and the $n \times n$ matrix, whose columns consist of the entries of the solution vectors of system $\mathbf{X}' = \mathbf{AX}$,

$$\mathbf{\Phi}(t) = \begin{pmatrix} x_{11} & x_{12} & \cdots & x_{1n} \\ x_{21} & x_{22} & \cdots & x_{2n} \\ \vdots & & & \vdots \\ x_{n1} & x_{n2} & \cdots & x_{nn} \end{pmatrix}$$

is called a **fundamental matrix** of the system on the interval.

In the discussion that follows we need to use two properties of a fundamental matrix:

- A fundamental matrix $\mathbf{\Phi}(t)$ is nonsingular.
- If $\mathbf{\Phi}(t)$ is a fundamental matrix of the system $\mathbf{X}' = \mathbf{AX}$, then

$$\mathbf{\Phi}'(t) = \mathbf{A\Phi}(t). \quad \mathbf{(3)}$$

A reexamination of (9) of Theorem 4.3 shows that det $\mathbf{\Phi}(t)$ is the same as the Wronskian $W(\mathbf{X}_1, \mathbf{X}_2, \ldots, \mathbf{X}_n)$. Hence the linear independence of the columns of $\mathbf{\Phi}(t)$ on the interval I guarantees that det $\mathbf{\Phi}(t) \neq 0$ for every t in the interval. Since $\mathbf{\Phi}(t)$ is nonsingular, the multiplicative inverse $\mathbf{\Phi}^{-1}(t)$ exists for every t in the interval. The result given in (3) follows immediately from the fact that every column of $\mathbf{\Phi}(t)$ is a solution vector of $\mathbf{X}' = \mathbf{AX}$.

Variation of Parameters

Analogous to the procedure in Section 3.5, we ask whether it is possible to replace the matrix of constants $\mathbf{C}$ in (2) by a column matrix of functions

$$\mathbf{U}(t) = \begin{pmatrix} u_1(t) \\ u_2(t) \\ \vdots \\ u_n(t) \end{pmatrix} \quad \text{so that} \quad \mathbf{X}_p = \mathbf{\Phi}(t)\mathbf{U}(t) \quad \mathbf{(4)}$$

is a particular solution of the nonhomogeneous system

$$\mathbf{X}' = \mathbf{AX} + \mathbf{F}(t). \tag{5}$$

By the product rule, the derivative of the last expression in (4) is

$$\mathbf{X}_p' = \mathbf{\Phi}(t)\mathbf{U}'(t) + \mathbf{\Phi}'(t)\mathbf{U}(t). \tag{6}$$

Note that the order of the products in (6) is very important. Since $\mathbf{U}(t)$ is a column matrix, the products $\mathbf{U}'(t)\mathbf{\Phi}(t)$ and $\mathbf{U}(t)\mathbf{\Phi}'(t)$ are not defined. Substituting (4) and (6) into (5) gives

$$\mathbf{\Phi}(t)\mathbf{U}'(t) + \mathbf{\Phi}'(t)\mathbf{U}(t) = \mathbf{A}\boldsymbol{\phi}(t)\mathbf{U}(t) + \mathbf{F}(t). \tag{7}$$

Now using (3) to replace $\mathbf{\Phi}'(t)$, (7) becomes

$$\mathbf{\Phi}(t)\mathbf{U}'(t) + \mathbf{A\Phi}(t)\mathbf{U}(t) = \mathbf{A}\boldsymbol{\phi}(t)\mathbf{U}(t) + \mathbf{F}(t)$$

or

$$\mathbf{\Phi}(t)\mathbf{U}'(t) = \mathbf{F}(t). \tag{8}$$

Multiplying both sides of equation (8) by $\mathbf{\Phi}^{-1}(t)$ gives

$$\mathbf{U}'(t) = \mathbf{\Phi}^{-1}(t)\,\mathbf{F}(t) \quad \text{and so} \quad \mathbf{U}(t) = \int \mathbf{\Phi}^{-1}(t)\,\mathbf{F}(t)\,dt.$$

Since $\mathbf{X}_p = \mathbf{\Phi}(t)\mathbf{U}(t)$ we conclude that a particular solution of (5) is

$$\mathbf{X}_p = \mathbf{\Phi}(t)\int \mathbf{\Phi}^{-1}(t)\mathbf{F}(t)\,dt. \tag{9}$$

To calculate the indefinite integral of the column matrix $\mathbf{\Phi}^{-1}(t)\mathbf{F}(t)$ in (9) we integrate each entry. Thus the general solution of the system (5) is $\mathbf{X} = \mathbf{X}_c + \mathbf{X}_p$ or

$$\mathbf{X} = \mathbf{\Phi}(t)\mathbf{C} + \mathbf{\Phi}(t)\int \mathbf{\Phi}^{-1}(t)\mathbf{F}(t)\,dt. \tag{10}$$

EXAMPLE 1 Variation of Parameters

Find the general solution of the nonhomogeneous system

$$\mathbf{X}' = \begin{pmatrix} -3 & 1 \\ 2 & -4 \end{pmatrix}\mathbf{X} + \begin{pmatrix} 3t \\ e^{-t} \end{pmatrix} \tag{11}$$

on the interval $(-\infty, \infty)$.

Solution We first solve the homogeneous system

$$\mathbf{X}' = \begin{pmatrix} -3 & 1 \\ 2 & -4 \end{pmatrix}\mathbf{X}. \tag{12}$$

The characteristic equation of the coefficient matrix is

$$\det(\mathbf{A} - \lambda\mathbf{I}) = \begin{vmatrix} -3-\lambda & 1 \\ 2 & -4-\lambda \end{vmatrix} = (\lambda + 2)(\lambda + 5) = 0,$$

so the eigenvalues are $\lambda_1 = -2$ and $\lambda_2 = -5$. By the usual method we find that the eigenvectors corresponding to λ_1 and λ_2 are, respectively,

$$\begin{pmatrix} 1 \\ 1 \end{pmatrix} \quad \text{and} \quad \begin{pmatrix} 1 \\ -2 \end{pmatrix}.$$

The solution vectors of the system (12) are then

$$\mathbf{X}_1 = \begin{pmatrix} 1 \\ 1 \end{pmatrix} e^{-2t} = \begin{pmatrix} e^{-2t} \\ e^{-2t} \end{pmatrix} \quad \text{and} \quad \mathbf{X}_2 = \begin{pmatrix} 1 \\ -2 \end{pmatrix} e^{-5t} = \begin{pmatrix} e^{-5t} \\ -2e^{-5t} \end{pmatrix}.$$

The entries in $\mathbf{X}_1$ form the first column of $\mathbf{\Phi}(t)$, and the entries in $\mathbf{X}_2$ form the second column of $\mathbf{\Phi}(t)$. Hence

$$\mathbf{\Phi}(t) = \begin{pmatrix} e^{-2t} & e^{-5t} \\ e^{-2t} & -2e^{-5t} \end{pmatrix} \quad \text{and} \quad \mathbf{\Phi}^{-1}(t) = \begin{pmatrix} \frac{2}{3}e^{2t} & \frac{1}{3}e^{2t} \\ \frac{1}{3}e^{5t} & -\frac{1}{3}e^{5t} \end{pmatrix}.$$

From (9) we obtain

$$\begin{aligned}
\mathbf{X}_p = \mathbf{\Phi}(t) \int \mathbf{\Phi}^{-1}(t)\mathbf{F}(t)\,dt &= \begin{pmatrix} e^{-2t} & e^{-5t} \\ e^{-2t} & -2e^{-5t} \end{pmatrix} \int \begin{pmatrix} \frac{2}{3}e^{2t} & \frac{1}{3}e^{2t} \\ \frac{1}{3}e^{5t} & -\frac{1}{3}e^{5t} \end{pmatrix} \begin{pmatrix} 3t \\ e^{-t} \end{pmatrix} dt \\
&= \begin{pmatrix} e^{-2t} & e^{-5t} \\ e^{-2t} & -2e^{-5t} \end{pmatrix} \int \begin{pmatrix} 2te^{2t} + \frac{1}{3}e^{t} \\ te^{5t} - \frac{1}{3}e^{4t} \end{pmatrix} dt \\
&= \begin{pmatrix} e^{-2t} & e^{-5t} \\ e^{-2t} & -2e^{-5t} \end{pmatrix} \begin{pmatrix} te^{2t} - \frac{1}{2}e^{2t} + \frac{1}{3}e^{t} \\ \frac{1}{5}te^{5t} - \frac{1}{25}e^{5t} - \frac{1}{12}e^{4t} \end{pmatrix} \\
&= \begin{pmatrix} \frac{6}{5}t - \frac{27}{50} + \frac{1}{4}e^{-t} \\ \frac{3}{5}t - \frac{21}{50} + \frac{1}{2}e^{-t} \end{pmatrix}.
\end{aligned}$$

Hence from (10) the general solution of (11) on the interval is

$$\begin{aligned}
\mathbf{X} &= \begin{pmatrix} e^{-2t} & e^{-5t} \\ e^{-2t} & -2e^{-5t} \end{pmatrix} \begin{pmatrix} c_1 \\ c_1 \end{pmatrix} + \begin{pmatrix} \frac{6}{5}t - \frac{27}{50} + \frac{1}{4}e^{-t} \\ \frac{3}{5}t - \frac{21}{50} + \frac{1}{2}e^{-t} \end{pmatrix} \\
&= c_1 \begin{pmatrix} 1 \\ 1 \end{pmatrix} e^{-2t} + c_2 \begin{pmatrix} 1 \\ -2 \end{pmatrix} e^{-5t} + \begin{pmatrix} \frac{6}{5} \\ \frac{3}{5} \end{pmatrix} t - \begin{pmatrix} \frac{27}{50} \\ \frac{21}{50} \end{pmatrix} + \begin{pmatrix} \frac{1}{4} \\ \frac{1}{2} \end{pmatrix} e^{-t}.
\end{aligned}$$

Initial-Value Problem

The general solution of (5) on an interval can be written in the alternative manner

$$\mathbf{X} = \mathbf{\Phi}(t)\mathbf{C} + \mathbf{\Phi}(t) \int_{t_0}^{t} \mathbf{\Phi}^{-1}(s)\mathbf{F}(s)\,ds, \tag{13}$$

where t and t_0 are points in the interval. This last form is useful in solving (5), subject to an initial condition $\mathbf{X}(t_0) = \mathbf{X}_0$. Substituting $t = t_0$ in (13) yields

$\mathbf{X}_0 = \mathbf{\Phi}(t_0)\,\mathbf{C}$, from which we get $\mathbf{C} = \mathbf{\Phi}^{-1}(t_0)\,\mathbf{X}_0$. Substituting this last result in (13) gives the following solution of the initial-value problem:

$$\mathbf{X} = \mathbf{\Phi}(t)\mathbf{\Phi}^{-1}(t_0)\mathbf{X}_0 + \mathbf{\Phi}(t)\int_{t_0}^{t} \mathbf{\Phi}^{-1}(s)\mathbf{F}(s)\,ds. \tag{14}$$

EXERCISES 4.3

Answers to odd-numbered problems begin on page AN-19.

In Problems 1–20 use variation of parameters to solve the given system.

1. $\dfrac{dx}{dt} = 3x - 3y + 4$
$\dfrac{dy}{dt} = 2x - 2y - 1$

2. $\dfrac{dx}{dt} = 2x - y$
$\dfrac{dy}{dt} = 3x - 2y + 4t$

3. $\mathbf{X}' = \begin{pmatrix} 3 & -5 \\ \frac{3}{4} & -1 \end{pmatrix}\mathbf{X} + \begin{pmatrix} 1 \\ -1 \end{pmatrix} e^{t/2}$

4. $\mathbf{X}' = \begin{pmatrix} 2 & -1 \\ 4 & 2 \end{pmatrix}\mathbf{X} + \begin{pmatrix} \sin 2t \\ 2\cos 2t \end{pmatrix} e^{2t}$

5. $\mathbf{X}' = \begin{pmatrix} 0 & 2 \\ -1 & 3 \end{pmatrix}\mathbf{X} + \begin{pmatrix} 1 \\ -1 \end{pmatrix} e^{t}$

6. $\mathbf{X}' = \begin{pmatrix} 0 & 2 \\ -1 & 3 \end{pmatrix}\mathbf{X} + \begin{pmatrix} 2 \\ e^{-3t} \end{pmatrix}$

7. $\mathbf{X}' = \begin{pmatrix} 1 & 8 \\ 1 & -1 \end{pmatrix}\mathbf{X} + \begin{pmatrix} 12 \\ 12 \end{pmatrix} t$

8. $\mathbf{X}' = \begin{pmatrix} 1 & 8 \\ 1 & -1 \end{pmatrix}\mathbf{X} + \begin{pmatrix} e^{-t} \\ te^{t} \end{pmatrix}$

9. $\mathbf{X}' = \begin{pmatrix} 3 & 2 \\ -2 & -1 \end{pmatrix}\mathbf{X} + \begin{pmatrix} 2e^{-t} \\ e^{-t} \end{pmatrix}$

10. $\mathbf{X}' = \begin{pmatrix} 3 & 2 \\ -2 & -1 \end{pmatrix}\mathbf{X} + \begin{pmatrix} 1 \\ 1 \end{pmatrix}$

11. $\mathbf{X}' = \begin{pmatrix} 0 & -1 \\ 1 & 0 \end{pmatrix}\mathbf{X} + \begin{pmatrix} \sec t \\ 0 \end{pmatrix}$

12. $\mathbf{X}' = \begin{pmatrix} 1 & -1 \\ 1 & 1 \end{pmatrix}\mathbf{X} + \begin{pmatrix} 3 \\ 3 \end{pmatrix} e^{t}$

13. $\mathbf{X}' = \begin{pmatrix} 1 & -1 \\ 1 & 1 \end{pmatrix}\mathbf{X} + \begin{pmatrix} \cos t \\ \sin t \end{pmatrix} e^{t}$

14. $\mathbf{X}' = \begin{pmatrix} 2 & -2 \\ 8 & -6 \end{pmatrix}\mathbf{X} + \begin{pmatrix} 1 \\ 3 \end{pmatrix} \dfrac{e^{-2t}}{t}$

15. $\mathbf{X}' = \begin{pmatrix} 0 & 1 \\ -1 & 0 \end{pmatrix}\mathbf{X} + \begin{pmatrix} 0 \\ \sec t \tan t \end{pmatrix}$

16. $\mathbf{X}' = \begin{pmatrix} 0 & 1 \\ -1 & 0 \end{pmatrix}\mathbf{X} + \begin{pmatrix} 1 \\ \cot t \end{pmatrix}$

17. $\mathbf{X}' = \begin{pmatrix} 1 & 2 \\ -\frac{1}{2} & 1 \end{pmatrix}\mathbf{X} + \begin{pmatrix} \csc t \\ \sec t \end{pmatrix} e^{t}$

18. $\mathbf{X}' = \begin{pmatrix} 1 & -2 \\ 1 & -1 \end{pmatrix}\mathbf{X} + \begin{pmatrix} \tan t \\ 1 \end{pmatrix}$

19. $\mathbf{X}' = \begin{pmatrix} 1 & 1 & 0 \\ 1 & 1 & 0 \\ 0 & 0 & 3 \end{pmatrix}\mathbf{X} + \begin{pmatrix} e^{t} \\ e^{2t} \\ te^{3t} \end{pmatrix}$

20. $\mathbf{X}' = \begin{pmatrix} 3 & -1 & -1 \\ 1 & 1 & -1 \\ 1 & -1 & 1 \end{pmatrix}\mathbf{X} + \begin{pmatrix} 0 \\ t \\ 2e^{t} \end{pmatrix}$

In Problems 21 and 22 use (14) to solve the given system, subject to the indicated initial condition.

21. $\mathbf{X}' = \begin{pmatrix} 3 & -1 \\ -1 & 3 \end{pmatrix}\mathbf{X} + \begin{pmatrix} 4e^{2t} \\ 4e^{4t} \end{pmatrix}, \quad \mathbf{X}(0) = \begin{pmatrix} 1 \\ 1 \end{pmatrix}$

22. $\mathbf{X}' = \begin{pmatrix} 1 & -1 \\ 1 & -1 \end{pmatrix}\mathbf{X} + \begin{pmatrix} 1/t \\ 1/t \end{pmatrix}, \quad \mathbf{X}(1) = \begin{pmatrix} 2 \\ -1 \end{pmatrix}$

23. The system of differential equations for the currents $i_1(t)$ and $i_2(t)$ in the electrical network shown in Figure 4.1 is

$$\frac{d}{dt}\begin{pmatrix} i_1 \\ i_2 \end{pmatrix} = \begin{pmatrix} -(R_1 + R_2)/L_2 & R_2/L_2 \\ R_2/L_1 & -R_2/L_1 \end{pmatrix}\begin{pmatrix} i_1 \\ i_2 \end{pmatrix} + \begin{pmatrix} E/L_2 \\ 0 \end{pmatrix}.$$

Solve the system if $R_1 = 8$ ohms, $R_2 = 3$ ohms, $L_1 = 1$ henry, $L_2 = 1$ henry, $E(t) = 100 \sin t$ volts, $i_1(0) = 0$, and $i_2(0) = 0$.

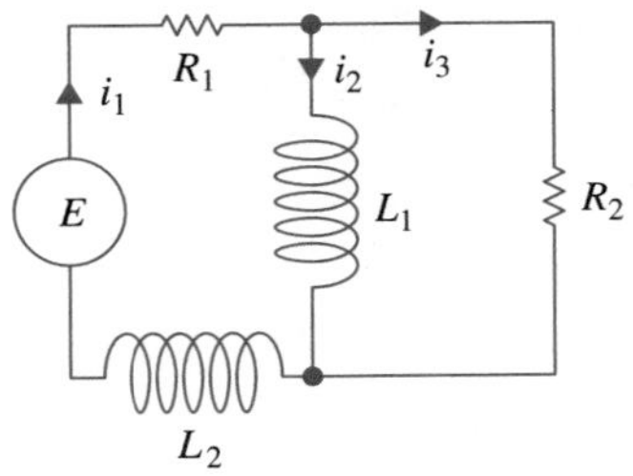

FIGURE 4.1

24. *WRITING PROJECT* Write a short report on how the **method of undetermined coefficients** can be applied to a nonhomogeneous linear system $\mathbf{X}' = \mathbf{AX} + \mathbf{F}(t)$. Discuss the types of functions that can be entries in the matrix $\mathbf{F}(t)$ and any difficulties that one may encounter when using this method.

25. *WRITING PROJECT* The general solution of the simple linear first-order differential equation $x' = ax$, a a constant, is an exponential function $x = ce^{at}$. Similarly, the general solution of the linear system $\mathbf{X}' = \mathbf{AX}$ can be expressed in terms of a **matrix exponential** $e^{\mathbf{A}t}$.

(a) Write a short report on the definition of $e^{\mathbf{A}t}$ and the various ways of computing a matrix exponential. Show how the general solutions of the linear systems $\mathbf{X}' = \mathbf{AX}$ and $\mathbf{X}' = \mathbf{AX} + \mathbf{F}(t)$ can be expressed in terms of a matrix exponential.

(b) Use a matrix exponential to solve the system

$$\mathbf{X}' = \begin{pmatrix} 4 & 2 \\ 3 & 3 \end{pmatrix}\mathbf{X}.$$

(c) Use a matrix exponential to solve the system

$$\mathbf{X}' = \begin{pmatrix} 4 & 2 \\ 3 & 3 \end{pmatrix}\mathbf{X} + \begin{pmatrix} t \\ e^{4t} \end{pmatrix}.$$

4.4 STABILITY

Introduction

The importance of systems of first-order differential equations cannot be overemphasized. We have seen that it is useful at times to express a single differential equation as a system of differential equations. To analyze a differential equation of order $n \geq 2$ quantitatively, we had to apply a specific numerical procedure to each equation in a system of n first-order equations. Likewise, to investigate the qualitative aspects of an autonomous second-order differential equation, we found the critical points of the equation by finding the critical points of an equivalent autonomous system of two first-order equations. We also used this system to find a phase portrait of the equation.

In the discussion that follows we shall be concerned with the qualitative aspects of autonomous systems of two first-order equations

$$\begin{aligned} \frac{dx}{dt} &= f(x, y) \\ \frac{dy}{dt} &= g(x, y). \end{aligned} \tag{1}$$

Recall at the outset that the word *autonomous* indicates that the independent variable t does not appear explicitly in the functions f and g. Also, a **critical point** of (1) is a point (x_1, y_1) for which we have $f(x_1, y_1) = 0$ and $g(x_1, y_1) = 0$. We shall assume hereafter that critical points are isolated and that the functions f and g in the general system (1) are continuous and possess continuous first partial derivatives in a neighborhood of an isolated critical point (x_1, y_1).

A system such as (1) is often referred to as a **plane autonomous system**, since a solution $(x(t), y(t))$ of the system can be interpreted as a parameterized curve or **trajectory** in the xy-plane or **phase plane**. We shall use the words *solution* and *trajectory* interchangeably.

4.4.1 Linear Systems

The linear second-order differential $ax'' + bx' + cx = 0$, $c \neq 0$ is equivalent to the linear system of two equations

$$\begin{aligned} \frac{dx}{dt} &= y \\ \frac{dy}{dt} &= -\frac{c}{a}x - \frac{b}{a}y. \end{aligned} \tag{2}$$

System (2) is, in turn, a special case of the general autonomous linear system of two first-order equations

$$\begin{aligned} \frac{dx}{dt} &= ax + by \\ \frac{dy}{dt} &= cx + dy. \end{aligned} \tag{3}$$

Of course, (3) is a just special case of (1) when f and g are the linear functions $f(x, y) = ax + by$ and $g(x, y) = cx + dy$.

We have already seen that (0, 0) was the only critical point of the equation $ax'' + bx' + cx = 0$, $c \neq 0$, since $x = 0$ and $y = 0$ are the only values that make the right-hand members of system (2) simultaneously zero. Similarly, inspection of (3) reveals that (0, 0) is the only critical point of the system provided that $ad - bc \neq 0$. We assume here that the coefficients a, b, c, and d in (3) are real numbers.

Classification of a Critical Point by Type and Stability

For a linear system (3), the origin (0, 0) can be one of four **types** of critical points—node, saddle point, spiral point, or center—and has one **stability** designation—asymptotically stable, stable (but not asymptotically stable), or unstable. Because (2) is just a special form of (3), the discussion on stability in Section 3.3 is subsumed in the current discussion. Consequently, a phase portrait of a linear system, except in one instance, is basically one of those in the gallery of phase portraits given in Figures 3.5, 3.6, and 3.7.

Our earlier considerations of stability in Sections 2.1 and 3.3 were purposely kept at an intuitive level. With this prior insight, the precise meaning of this concept, as set forth in the following definition, should make sense. We refer to the general system (1) so that the definition applies to linear systems as well as nonlinear systems considered in the next subsection.

DEFINITION 4.4 Stability of an Autonomous System

Let $(x(t), y(t))$ be a solution of the autonomous system (1) that satisfies $x(t_0) = x_0$, $y(t_0) = y_0$. Then a critical point (x_1, y_1) of the system is said to be

(i) **stable** if for every $\varepsilon > 0$ there exists a $\delta > 0$ such that

$$[x(t_0) - x_1]^2 + [y(t_0) - y_1]^2 < \delta^2 \text{ implies } [x(t) - x_1]^2 + [y(t) - y_1]^2 < \varepsilon^2$$

for all $t > t_0$;

(ii) **asymptotically stable** if it is stable and a $\delta_1 > 0$ can be chosen so that

$$[x(t_0) - x_1]^2 + [y(t_0) - y_1]^2 < \delta_1^2 \text{ implies } (x(t), y(t)) \to (x_1, y_1)$$

as $t \to \infty$; or

(iii) **unstable** if it is not stable.

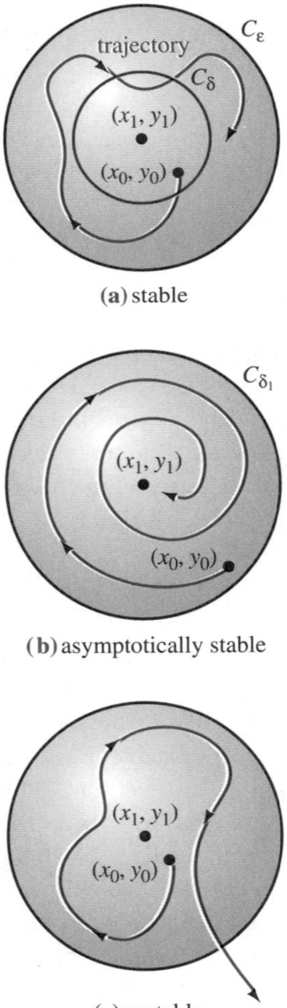

(a) stable

(b) asymptotically stable

(c) unstable

FIGURE 4.2

An inequality such as $[x(t_0) - x_1]^2 + [y(t_0) - y_1]^2 < \delta^2$ indicates that the point $(x(t_0), y(t_0))$ is in the interior of a circle of radius δ centered at (x_1, y_1). Interpreted geometrically, a critical point (x_1, y_1) is stable provided that, for every circle C_ε we draw of radius ε centered at (x_1, y_1), we can find another circle C_δ of radius δ centered at (x_1, y_1) such that any trajectory starting at a point in C_δ never leaves C_ε. See Figure 4.2(a). If all trajectories that start sufficiently close, that is, within some circle C_{δ_1} centered at a stable critical point (x_1, y_1), also approach (x_1, y_1) as $t \to \infty$, then (x_1, y_1) is asymptotically stable. See Figure 4.2(b). A critical point is unstable if there exist trajectories that move away, or recede, from (x_1, y_1) regardless of how close we choose an initial point (x_0, y_0) to (x_1, y_1). See Figure 4.2(c).

More Words

As we pointed out in Section 3.3, the study of stability is larded with terminology. The definition of asymptotic stability just given is, in some texts, called **strict stability**. A stable critical point that is not strictly stable is said to be **neutrally stable**. Since asymptotic stability is often referred to as a strong kind of stability, neutral stability is naturally also called **weak stability**.

Three Cases

When (3) is written as $\mathbf{X}' = \mathbf{AX}$, the characteristic equation of the coefficient matrix $\mathbf{A} = \begin{pmatrix} a & b \\ c & d \end{pmatrix}$ is the quadratic equation

$$\lambda^2 - (a + d)\lambda + ad - bc = 0, \tag{4}$$

where $\det \mathbf{A} = ad - bc \neq 0$. The roots λ_1 and λ_2 of (4) are the eigenvalues of $\mathbf{A}$. As in Section 4.2, we have the following three major cases.

CASE I Distinct real eigenvalues: $\lambda_1 \neq \lambda_2$.

CASE II Repeated real eigenvalues: $\lambda_1 = \lambda_2$.

CASE III Conjugate complex eigenvalues: $\lambda_1 = \alpha + i\beta$, $\lambda_2 = \alpha - i\beta$.

Under each of these cases we can further distinguish subcases. These various subcases, except in one instance, are analogous to those listed in the table on page 116 in our discussion of the autonomous equation $ax'' + bx' + cx = 0$.

CASE I General solution $$\mathbf{X} = c_1\mathbf{K}_1e^{\lambda_1 t} + c_2\mathbf{K}_2e^{\lambda_2 t} \tag{5}$$

(i) When both eigenvalues are negative, $\lambda_1 < \lambda_2 < 0$, then for any values of c_1 and c_2, $\mathbf{X} \to \mathbf{0}$ as $t \to \infty$. In other words, points (that is, solutions) $(x(t), y(t))$ on every trajectory approach, or are attracted to, the critical point (0, 0):

$$\lim_{t\to\infty} x(t) = 0 \quad \text{and} \quad \lim_{t\to\infty} y(t) = 0.$$

To understand this a little better, let us suppose for the sake of discussion that $\mathbf{K}_1 = \begin{pmatrix} k_{11} \\ k_{21} \end{pmatrix}$ and $\mathbf{K}_2 = \begin{pmatrix} k_{12} \\ k_{22} \end{pmatrix}$. When $c_2 = 0$, then $\mathbf{X} = c_1\mathbf{K}_1e^{\lambda_1 t}$ is the same as

$$x = c_1k_{11}e^{\lambda_1 t}, \quad y = c_1k_{21}e^{\lambda_1 t}. \tag{6}$$

Eliminating the parameter t from these two equations shows that $\mathbf{X} \to \mathbf{0}$ along a half-line trajectory defined by $y = (k_{21}/k_{11})x$. The last equation defines only a half-line, since $e^{\lambda_1 t}$ is positive for all t and therefore x in (6) is either always positive or always negative. Moreover, there are *two* half-line trajectories with slope k_{21}/k_{11}, since we can choose $c_1 > 0$ or $c_1 < 0$ in (6). If we interpret the eigenvector $\mathbf{K}_1$ as a point (k_{11}, k_{21}), then we have shown that $\mathbf{X} \to \mathbf{0}$ along a *half-line determined by the eigenvector*, because the slope of a line through (0, 0) and (k_{11}, k_{21}) is also k_{21}/k_{11}. The second half-line is determined by $-\mathbf{K}_1$. When $c_1 = 0$, a similar argument shows that $\mathbf{X} \to \mathbf{0}$ along the two half-line trajectories defined by $y = (k_{22}/k_{12})x$, that is, half-lines determined by the eigenvectors $\mathbf{K}_2$ and $-\mathbf{K}_2$. Finally, when $c_1 \neq 0$ and $c_2 \neq 0$, it follows from the quotient

$$\frac{y}{x} = \frac{c_1k_{21}e^{\lambda_1 t} + c_2k_{22}e^{\lambda_2 t}}{c_1k_{11}e^{\lambda_1 t} + c_2k_{12}e^{\lambda_2 t}} = \frac{c_1k_{21}e^{(\lambda_1 - \lambda_2)t} + c_2k_{22}}{c_1k_{11}e^{(\lambda_1 - \lambda_2)t} + c_2k_{12}}$$

and $\lambda_1 - \lambda_2 < 0$, that $y/x \to k_{22}/k_{12}$ as $t \to \infty$. We can interpret this result to mean that points on all other curved paths or trajectories approach (0, 0) as $t \to \infty$ *tangentially*, either to the half-line determined by the eigenvector $\mathbf{K}_2$ or to the half-line determined by $-\mathbf{K}_2$. The point (0, 0) is **asymptotically stable** and is called a **node**.

(ii) When both eigenvalues are positive, $\lambda_1 > \lambda_2 > 0$, the situation described in (i) is now reversed; that is, points $(x(t), y(t))$ on every trajectory move away or are repelled from (0, 0) as $t \to \infty$. Every solution $\mathbf{X}$ of the system becomes unbounded as $t \to \infty$. By repeating the argument of choosing $c_2 = 0$, $c_1 \neq 0$ and then $c_1 = 0$, $c_2 \neq 0$, we see that there are again four half-line trajectories determined by the eigenvectors. The critical point (0, 0) is now an **unstable node**.

(iii) Suppose the eigenvalues have opposite signs, $\lambda_1 < 0 < \lambda_2$. If $c_2 = 0$ then, since $\lambda_1 < 0$, $\mathbf{X} = c_1\mathbf{K}_1e^{\lambda_1 t} \to \mathbf{0}$ along the half-line trajectories determined by $\mathbf{K}_1$ and $-\mathbf{K}_1$. If $c_1 = 0$ then, since $\lambda_2 > 0$, $\mathbf{X} = c_2\mathbf{K}_2e^{\lambda_2 t}$ becomes unbounded on the half-line trajectories determined by $\mathbf{K}_2$ and $-\mathbf{K}_2$. If $c_1 \neq 0$ and $c_2 \neq 0$, then the trajectories are curved paths. Observe from (5) that as $t \to \infty$, $e^{\lambda_1 t} \to 0$, and so $\mathbf{X} \approx c_2\mathbf{K}_2e^{\lambda_2 t}$. This means $\mathbf{X}$ becomes unbounded but does so in a manner that is asymptotic to the half-line trajectories determined by $\mathbf{K}_2$ and $-\mathbf{K}_2$. On the other hand, if we let $t \to -\infty$, then $e^{\lambda_2 t} \to 0$ in (5) and so $\mathbf{X} \approx c_1\mathbf{K}_1e^{\lambda_1 t}$ shows $\mathbf{X}$ is asymptotic to the half-line trajectories determined by $\mathbf{K}_1$ and $-\mathbf{K}_1$. Roughly, this means that the curved trajectories must "start" (that is, as $t \to -\infty$) near one set of half-lines and then "finish" (that is, as $t \to \infty$) near the other set of half-lines. The critical point (0, 0) is a **saddle point**.

Recall from the discussion in Section 3.3 that a saddle point is always unstable and that a node is characterized by the fact that points $(x(t), y(t))$ on *every* trajectory are either attracted to it or repelled from it as $t \to \infty$. For a linear system, a saddle point is characterized by the existence of *two* half-line trajectories on which points $(x(t), y(t))$ approach it as $t \to \infty$, whereas points $(x(t), y(t))$ on every other trajectory recede from it as $t \to \infty$.

Each of the three foregoing subcases is illustrated in the next example.

EXAMPLE 1 Distinct Eigenvalues/Phase Portraits

Classify the critical point (0, 0) and obtain the phase portrait of each system:

$$\textbf{(a)}\ \frac{dx}{dt} = -2x + 3y \qquad \textbf{(b)}\ \frac{dx}{dt} = -x + 2y \qquad \textbf{(c)}\ \frac{dx}{dt} = 2x + 3y$$

$$\frac{dy}{dt} = x - 4y \qquad \frac{dy}{dt} = -7x + 8y \qquad \frac{dy}{dt} = 2x + y.$$

Solution

(a) The eigenvalues of the matrix $\begin{pmatrix} -2 & 3 \\ 1 & -4 \end{pmatrix}$ and the corresponding eigenvectors are found to be

$$\lambda_1 = -5, \quad \mathbf{K}_1 = \begin{pmatrix} -1 \\ 1 \end{pmatrix}, \qquad \lambda_2 = -1, \quad \mathbf{K}_2 = \begin{pmatrix} 3 \\ 1 \end{pmatrix}.$$

Since the eigenvalues are both negative, the critical point (0, 0) is an asymptotically stable node. With the aid of computer software we obtain the phase portrait of the system given in Figure 4.3(a).

(b) The eigenvalues of the matrix $\begin{pmatrix} -1 & 2 \\ -7 & 8 \end{pmatrix}$ and the corresponding eigenvectors are found to be

$$\lambda_1 = 6, \quad \mathbf{K}_1 = \begin{pmatrix} 2 \\ 7 \end{pmatrix}, \qquad \lambda_2 = 1, \quad \mathbf{K}_2 = \begin{pmatrix} 1 \\ 1 \end{pmatrix}.$$

Since the eigenvalues are both positive, (0, 0) is an unstable node. A phase portrait of the system is given in Figure 4.3(b).

(c) The eigenvalues of the matrix $\begin{pmatrix} 2 & 3 \\ 2 & 1 \end{pmatrix}$ and the corresponding eigenvectors are found to be

$$\lambda_1 = -1, \quad \mathbf{K}_1 = \begin{pmatrix} 1 \\ -1 \end{pmatrix}, \qquad \lambda_2 = 4, \quad \mathbf{K}_2 = \begin{pmatrix} 3 \\ 2 \end{pmatrix}.$$

Because the eigenvalues have opposite signs, (0, 0) is a saddle point. A phase portrait of the system is given in Figure 4.3(c). ●

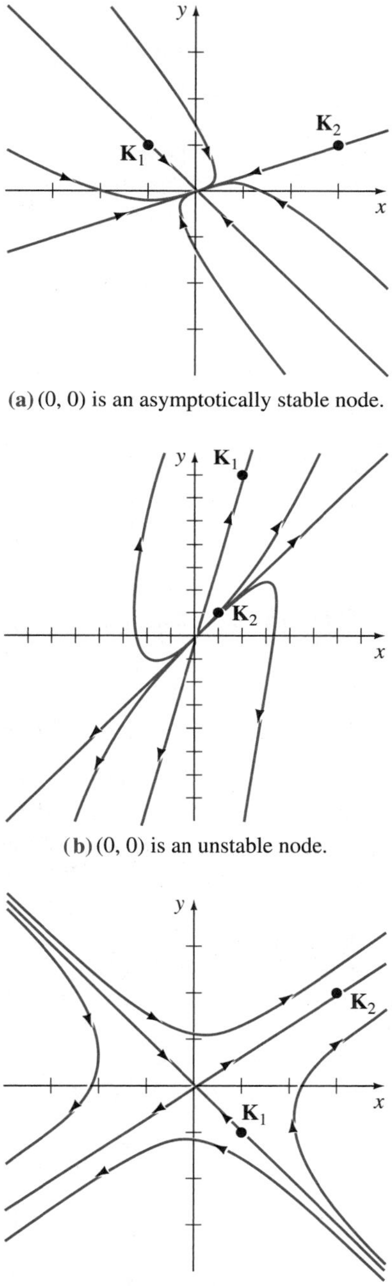

(a) (0, 0) is an asymptotically stable node.

(b) (0, 0) is an unstable node.

(c) (0, 0) is a saddle point.

FIGURE 4.3

CASE II General solution $\quad \mathbf{X} = c_1\mathbf{K}_1e^{\lambda_1 t} + c_2(\mathbf{K}_1te^{\lambda_1 t} + \mathbf{P}e^{\lambda_1 t}) \quad$ **(7)**

(i) For the moment let us rule out the possibility that $a = d \neq 0$ and $b = c = 0$ in the linear system (3). For all other choices of a, b, c, and d, in (3) that lead to a repeated eigenvalue $\lambda_1 = \lambda_2$, the form of the general solution of the system is as given in (7). Recall that $\mathbf{P}$ is a solution of $(\mathbf{A} - \lambda\mathbf{I})\mathbf{K}_1 = \mathbf{P}$. When $c_2 = 0$, $\mathbf{X} = c_1\mathbf{K}_1e^{\lambda_1 t}$, and so, as in the discussion of Case I, the trajectories are two half-lines determined by $\mathbf{K}_1$ and $-\mathbf{K}_1$. When $c_2 \neq 0$ the trajectories are curved paths. If $\mathbf{K}_1 = \begin{pmatrix} k_1 \\ k_2 \end{pmatrix}$ and $\mathbf{P} = \begin{pmatrix} p_1 \\ p_2 \end{pmatrix}$, then we see from the quotient

$$\frac{y}{x} = \frac{c_1k_2e^{\lambda_1 t} + c_2(k_2te^{\lambda_1 t} + p_2e^{\lambda_1 t})}{c_1k_1e^{\lambda_1 t} + c_2(k_1te^{\lambda_1 t} + p_1e^{\lambda_1 t})} = \frac{c_1k_2/t + c_2(k_2 + p_2/t)}{c_1k_1/t + c_2(k_1 + p_1/t)} \qquad \textbf{(8)}$$

that $y/x \to k_2/k_1$ as $t \to \infty$. If $\lambda_1 < 0$, then $e^{\lambda_1 t} \to 0$ and $te^{\lambda_1 t} \to 0$ as $t \to \infty$ and so $\mathbf{X} \to \mathbf{0}$. In particular, the quotient shows that points on the curved trajectories approach (0, 0) tangentially to the half-lines determined by $\mathbf{K}_1$ and $-\mathbf{K}_1$. Thus for $\lambda_1 < 0$, the critical point (0, 0) is **asymptotically stable**. Because the points on the trajectories approach the origin either on or tangent to two half-lines, rather than four half-lines as in (i) and (ii) of Case I, (0, 0) is called a **degenerate node**.

(ii) If $\lambda_1 > 0$ then $e^{\lambda_1 t}$ and $te^{\lambda_1 t}$ become unbounded as $t \to \infty$, so points on all the trajectories move away from (0, 0). Notice that the quotient (8) also shows that $y/x \to k_2/k_1$ as $t \to -\infty$. This means that the points "start" moving away from (0, 0) either on or near the two half-lines. Therefore, for $\lambda_1 > 0$, (0, 0) is an **unstable degenerate node**.

(iii) This subcase has no analog on page 116. It is the one instance in which a linear system (3) can have a repeated eigenvalue but the general solution of the system is *not* of the form given in (7). If we let $a = d \neq 0$ and $b = c = 0$, then (3) is

$$\frac{dx}{dt} = ax$$
$$\frac{dy}{dt} = ay. \tag{9}$$

Since the differential equations are independent of each other the system is actually *uncoupled*. Nevertheless, if we write (9) in the matrix form $\mathbf{X}' = \begin{pmatrix} a & 0 \\ 0 & a \end{pmatrix}\mathbf{X}$, we see that the characteristic equation is $(\lambda - a)^2 = 0$ so $\lambda = a$ is an eigenvalue of multiplicity two. It is worth verifying that *any* vector $\begin{pmatrix} c_1 \\ c_2 \end{pmatrix}$ is an eigenvector corresponding to $\lambda = a$. Hence the general solution of (9) is given by $\mathbf{X} = \begin{pmatrix} c_1 \\ c_2 \end{pmatrix} e^{at} = c_1 \begin{pmatrix} 1 \\ 0 \end{pmatrix} e^{at} + c_2 \begin{pmatrix} 0 \\ 1 \end{pmatrix} e^{at}$. Notice that this solution is not of the form given in (7). By eliminating t from $x = c_1 e^{at}$ and $y = c_2 e^{at}$ we see that the trajectories of the system are half-lines defined by $c_1 y = c_2 x$. But since c_1 and c_2 can be chosen arbitrarily, there are an infinite number of half-line trajectories. If $\lambda = a < 0$, then $\mathbf{X} \to \mathbf{0}$ as $t \to \infty$ and the critical point (0, 0) is **asymptotically stable**. If $\lambda = a > 0$, then $\mathbf{X}$ becomes unbounded as $t \to \infty$ and (0, 0) is an **unstable** critical point. The critical point (0, 0) in this special case is a **degenerate node** that is sometimes referred to as a **star node**.

EXAMPLE 2 Repeated Eigenvalues/Phase Portraits

Classify the critical point (0, 0) and obtain the phase portrait of each system:

(a) $\dfrac{dx}{dt} = 3x - 18y$ **(b)** $\dfrac{dx}{dt} = x$

$\dfrac{dy}{dt} = 2x - 9y$ $\dfrac{dy}{dt} = y.$

Solution

(a) The repeated eigenvalue of this system is $\lambda_1 = \lambda_2 = -3 < 0$, and so from the foregoing discussion, (0, 0) is an asymptotically stable

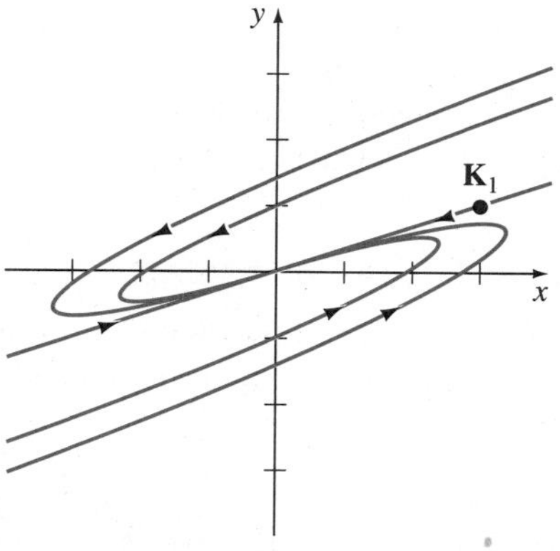

(a) (0, 0) is an asymptotically stable degenerate node.

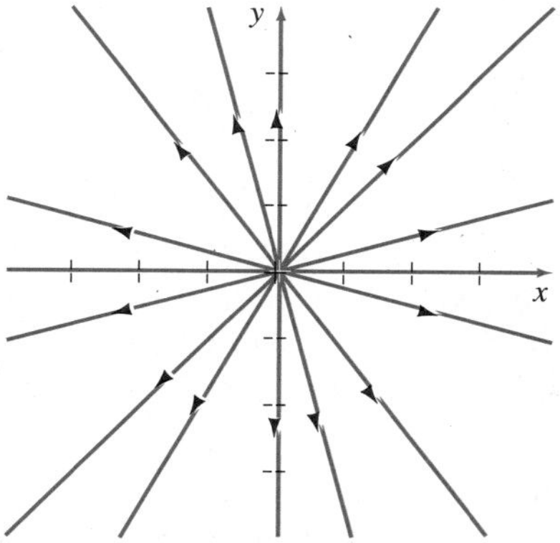

(b) (0, 0) is an unstable degenerate node.

FIGURE 4.4

degenerate node. The eigenvector indicated in the phase portrait in Figure 4.4(a) is $\mathbf{K}_1 = \begin{pmatrix} 3 \\ 1 \end{pmatrix}$.

(b) By treating these two independent equations as a system we find $\lambda_1 = \lambda_2 = 1$. A computer is not necessary to obtain the phase portrait of the system, it is simply half-lines with every possible slope. The arrows on the trajectories in Figure 4.4(b) point away from (0, 0); since the eigenvalue is positive the origin is an unstable degenerate star node. ●

CASE III General solution

$$\mathbf{X} = c_1[\mathbf{B}_1 \cos \beta t - \mathbf{B}_2 \sin \beta t]e^{\alpha t} + c_2[\mathbf{B}_2 \cos \beta t + \mathbf{B}_1 \sin \beta t]e^{\alpha t}, \quad \textbf{(10)}$$

where $\mathbf{B}_1 = \text{Re}(\mathbf{K}_1)$, $\mathbf{B}_2 = \text{Im}(\mathbf{K}_1)$, and $\mathbf{K}_1$ is an eigenvector corresponding to $\lambda_1 = \alpha + i\beta$.

Before considering the three subcases of Case III, let us examine the special linear system

$$\begin{aligned} \frac{dx}{dt} &= \alpha x - \beta y \\ \frac{dy}{dt} &= \beta x + \alpha y. \end{aligned} \quad \textbf{(11)}$$

System (11) should be considered as the archetype of every linear system with complex eigenvalues; you should verify that its coefficient matrix has the eigenvalues $\lambda_1 = \alpha + i\beta$, $\lambda_2 = \alpha - i\beta$. It is now helpful to introduce polar coordinates. When we substitute $x = r \cos \theta$, $y = r \sin \theta$, where r and θ are functions of t, (11) becomes

$$\begin{aligned} (-r \sin \theta)\theta' + (\cos \theta)r' &= \alpha r \cos \theta - \beta r \sin \theta \\ (r \cos \theta)\theta' + (\sin \theta)r' &= \beta r \cos \theta + \alpha r \sin \theta. \end{aligned}$$

Solving this last system (say, by Cramer's rule) for the symbols θ' and r', we find, after simplification, the first-order differential equations

$$r' = \alpha r \quad \text{and} \quad \theta' = \beta. \quad \textbf{(12)}$$

Dividing the first equation in (12) by the second gives $dr/d\theta = (\alpha/\beta)r$. Solving yields

$$r = c_1 e^{(\alpha/\beta)\theta}. \quad \textbf{(13)}$$

For $\alpha \neq 0$, $c_1 \neq 0$, (13) is a polar equation of a logarithmic spiral.

(i) If $\alpha < 0$, $\beta \neq 0$, we see in (10) that $\mathbf{X} \to \mathbf{0}$ since $e^{\alpha t} \to 0$ as $t \to \infty$. Points on each trajectory approach (0, 0), not from a specific direction as in the case of a node, but by winding around (0, 0) an infinite number of times, getting closer and closer as $t \to \infty$. The critical point (0, 0) is **asymptotically stable** and is called **spiral point**.

(ii) If $\alpha > 0$, $\beta \neq 0$, points spiral away from (0, 0) as $t \to \infty$. The critical point (0, 0) is an **unstable spiral point**.

(iii) If $\alpha = 0$, $\beta \neq 0$, the eigenvalues of the system (2) are pure imaginary. Equation (13) shows that when $\alpha = 0$, r is constant, which is a polar equation of a family of circles. Bear in mind that the last remark only pertains to the special system (12). In general, when $\alpha = 0$ the solutions in (10) have the form

$$x = c_{11} \cos \beta t + c_{12} \sin \beta t, \quad y = c_{21} \cos \beta t + c_{22} \sin \beta t,$$

which are parametric equations of a family of ellipses with center at the origin. The critical point (0, 0) is **stable** (but not asymptotically stable) and is naturally called a **center**.

EXAMPLE 3 Complex Eigenvalues/Phase Portraits

Classify the critical point (0, 0) and obtain the phase portrait of each system:

(a) $$\frac{dx}{dt} = x + 2y \qquad \frac{dy}{dt} = -2x + y$$

(b) $$\frac{dx}{dt} = 2x + 8y \qquad \frac{dy}{dt} = -x - 2y.$$

Solution

(a) The eigenvalues of this system are the complex numbers $\lambda_1 = 1 + 2i$, $\lambda_2 = 1 - 2i$. Because the real part of $\lambda_1 (\alpha = 1)$ is positive we know that (0, 0) is an unstable spiral point. Several of the spiral trajectories are shown in Figure 4.5(a).

(b) The eigenvalues of the system are the pure imaginary numbers $\lambda_1 = 2i$, $\lambda_2 = -2i$. Hence from the above discussion we know that (0, 0) is a center. The phase portrait of the system consists of the family of ellipses shown in Figure 4.5(b). ●

For convenience we summarize the main results of the discussion in the form of a theorem.

THEOREM 4.10 Stability of Linear Systems

Let λ_1 and λ_2 be the eigenvalues of the coefficient matrix **A** of system (3). Then the critical point (0, 0) of the system is

(i) asymptotically stable if λ_1 and λ_2 are real and both negative, or if λ_1 and λ_2 are complex with negative real parts;

(ii) stable (but not asymptotically stable) if λ_1 and λ_2 are pure imaginary numbers; and

(iii) unstable if λ_1 and λ_2 are real and both positive, if λ_1 and λ_2 are real and have opposite signs, or if λ_1 and λ_2 are complex with positive real parts.

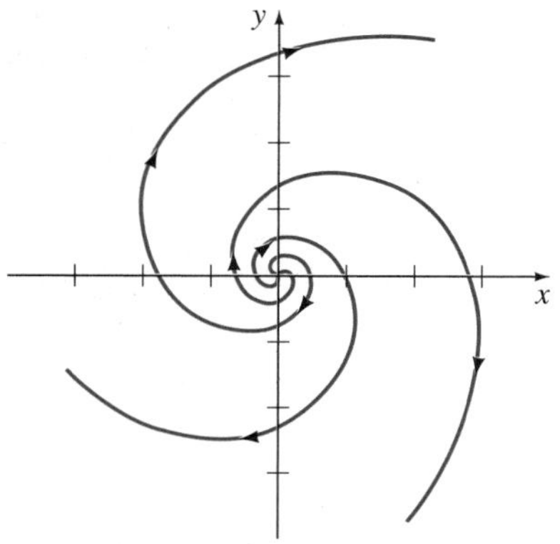

(a) (0, 0) is an unstable spiral point.

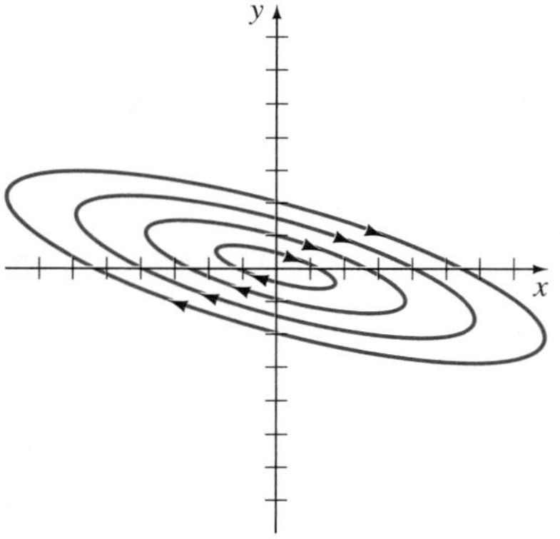

(b) (0, 0) is a center.

FIGURE 4.5

4.4.2 Nonlinear Systems

We shall be concerned only with nonlinear systems from this point on. For example

$$\frac{dx}{dt} = \sin y \qquad \frac{dx}{dt} = 4.5x - 0.9xy$$
$$\text{and} \tag{14}$$
$$\frac{dy}{dt} = \cos x \qquad \frac{dy}{dt} = -0.16y + 0.08xy$$

are nonlinear systems.

We dealt briefly with special nonlinear systems in Section 3.11, where we expressed autonomous nonlinear second-order differential equations as equivalent autonomous nonlinear systems. Of course one of the big differences between linear differential equations and nonlinear differential equations also carries over to systems of differential equations: We are always able to find an explicit solution of a linear system of the form given in (3), but it is rarely possible to find, in the sense of displaying, any solution of a nonlinear system.

Critical Points

In contrast with the linear systems that we have just studied, a nonlinear system can have more than one critical point or no critical points at all. As usual, we shall consider systems that possess only isolated critical points. Note for example, (0, 0) and (2, 5) are isolated critical points of the second system in (14). As in the linear case, critical points for autonomous nonlinear systems are classified according to type (node, saddle point, spiral point, center) and stability (stable, asymptotically stable, unstable) *when it is possible to do so.*

When discussing an isolated critical point (x_1, y_1) we can, without any loss of generality, assume that $x_1 = 0$, $y_1 = 0$, since a change of variables can always transform a point to the origin. We will illustrate this later on.

Global and Local Stability

Recall that a critical point is stable if all solutions that start close to the critical point stay close to that point for all $t > t_0$. Stability for a linear system of the form (3) is a **global** stability. If the critical point (0, 0) is, say, asymptotically stable, closeness of an initial starting point (x_0, y_0) to the critical point really does not matter, all solutions or points $(x(t), y(t))$ must approach (0, 0) as $t \to \infty$. We see this global stability in Figures 4.3(a) and 4.5(b). Put another way, the entire phase portrait of a linear system is characterized by the behavior of a few trajectories near the critical point.

On the other hand, stability for nonlinear systems is often a **local** phenomenon. For example, in the case of the soft spring discussed in Section 3.11, we found that there exist periodic solutions of the nonlinear model $x'' + x - x^3 = 0$ by discovering, in Figure 3.45(a), that (0, 0) is a stable center only in a neighborhood of the origin. It might be worthwhile to go back and verify that points

$(x(t), y(t))$ on the trajectory whose initial point is (0.5, 0.6) move away from the critical point, but the trajectory whose initial point is (0.5, 0.5) is a closed path surrounding (0, 0).

Classifying a critical point of a nonlinear system is not as straightforward as it is for a linear system. But in this brief introduction to nonlinear systems we are primarily going to consider those systems for which the behavior of the trajectories near a critical point is comparable to or is *approximated* by trajectories of associated linear systems near the point.

Linearization

If f is a function of a single variable x and differentiable at x_1, then an equation of the tangent line to the graph at $(x_1, f(x_1))$ is $y = f(x_1) + f'(x_1)(x - x_1)$. The tangent line, which we write as $L(x) = f(x_1) + f'(x_1)(x - x_1)$, is said to be a **linearization** of f at x_1. When x is close to x_1 and the tangent line is close to the graph of f, we can approximate the functional value $f(x)$ using the corresponding y-coordinate on the tangent line. We say that $f(x) \approx L(x)$ is a **local linear approximation** of $f(x)$. Similarly, if f is a function of two variables x and y that has continuous first partial derivatives in a neighborhood of (x_1, y_1), then using an equation of the tangent plane at $(x_1, y_1, f(x_1, y_1))$, the linearization of f at (x_1, y_1) is

$$L(x, y) = f(x_1, y_1) + f_x(x_1, y_1)(x - x_1) + f_y(x_1, y_1)(y - y_1), \tag{15}$$

where f_x and f_y denote the first partial derivatives. Thus when (x, y) is close to (x_1, y_1),

$$f(x, y) \approx L(x, y) \tag{16}$$

is a local linear approximation of $f(x, y)$.

Now suppose that (0, 0) is an isolated critical point of the system in (1). If we use (15) on both f and g, then, since $f(0, 0) = 0$ and $g(0, 0) = 0$, equation (16) gives

$$\begin{aligned} f(x, y) &\approx f_x(0, 0)x + f_y(0, 0)y \\ g(x, y) &\approx g_x(0, 0)x + g_y(0, 0)y \end{aligned} \tag{17}$$

whenever (x, y) is close to (0, 0). By replacing the right-hand members in (1) by the right-hand members in (17), we obtain

$$\begin{aligned} \frac{dx}{dt} &= ax + by \\ \frac{dy}{dt} &= cx + dy, \end{aligned} \tag{18}$$

where $a = f_x(0, 0)$, $b = f_y(0, 0)$, $c = g_x(0, 0)$, and $d = g_y(0, 0)$. As before, we assume that the determinant of the coefficients is not zero; that is,

$$\begin{vmatrix} f_x(0, 0) & f_y(0, 0) \\ g_x(0, 0) & g_y(0, 0) \end{vmatrix} = ad - bc \neq 0. \tag{19}$$

The system given in (18) is said to be a **linearization** of the system (1) at (0, 0). If a nonlinear system (1) yields a linearization defined by (18) with an isolated critical point, we say that (1) is a **nearly linear system**.

The next theorem asserts that in some instances we can determine the type and stability of a critical point of a nonlinear system from the type and stability of the linearization of the system at that point.

THEOREM 4.11 Stability of Nonlinear Systems

The critical point (0, 0) of a nonlinear system (1) is of the same general type (node, saddle point, spiral point) and stability (stable, asymptotically stable, unstable) as the critical point (0, 0) of the linearized system (18). There are two possible exceptions:

(i) If the eigenvalues λ_1 and λ_2 of the linearized system (18) are equal, then the critical point (0, 0) of the nearly linear system (1) may either be a node or a spiral point and is asymptotically stable if $\lambda_1 = \lambda_2 < 0$ and unstable if $\lambda_1 = \lambda_2 > 0$.

(ii) If the eigenvalues λ_1 and λ_2 of the linearized system (18) are pure imaginary, then the critical point (0, 0) of the nonlinear system (1) may either be a center or a spiral point. The stability of the spiral point is not determined by the stability of (0, 0) of the linearized system.

Note that in (a) of Theorem 4.11 we can determine the *stability* but not the *type* of critical point of the nonlinear system from the linearized system. Part (b) of the theorem is inconclusive; we can neither determine the stability nor the type of critical point of the nonlinear system from the fact that the linearized system has a center.

EXAMPLE 4 Linearization of a Nonlinear System

Observe that (0, 0) is a critical point of the nonlinear system

$$\frac{dx}{dt} = 3x + y - \sin y$$
$$\frac{dy}{dt} = 2x - 5y + 2x\cos x. \tag{20}$$

With

$$f(x, y) = 3x + y - \sin y, \quad g(x, y) = 2x - 5y + 2x\cos x$$

$$f_x(x, y) = 3, \quad f_y(x, y) = 1 - \cos y, \quad g_x(x, y) = 2 - 2x\sin x + 2\cos x, \quad g_y(x, y) = -5$$

$$a = f_x(0, 0) = 3, \quad b = f_y(0, 0) = 0, \quad c = g_x(0, 0) = 4, \quad d = g_y(0, 0) = -5$$

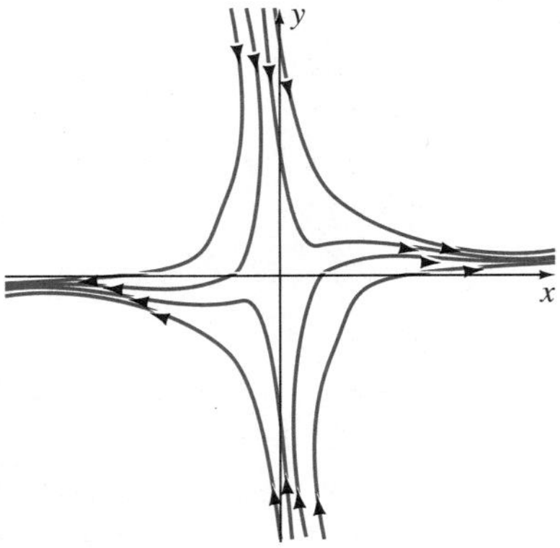

(a) nonlinear system

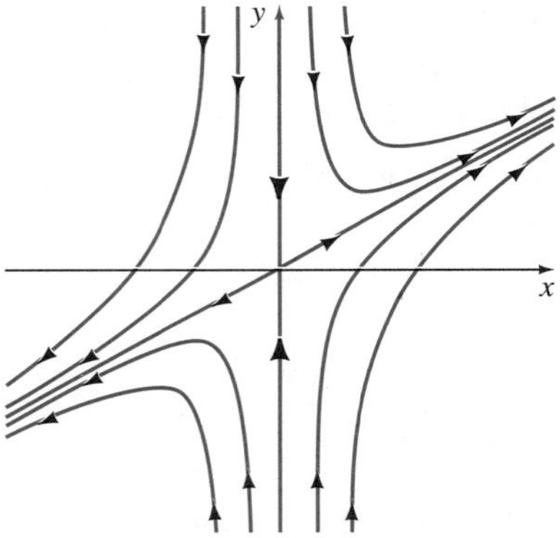

(b) linear system

FIGURE 4.6

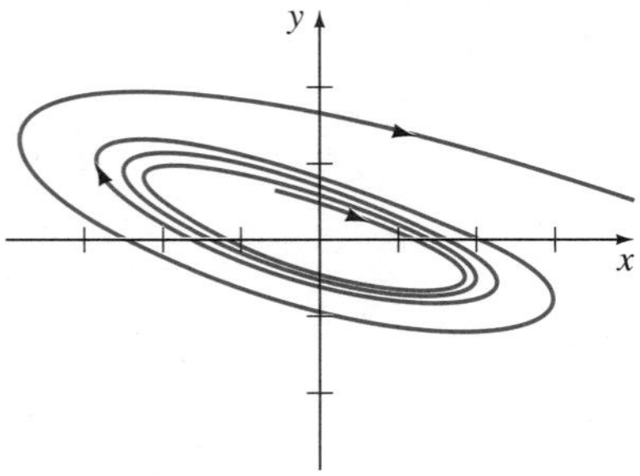

FIGURE 4.7

we find from (18) that a linearization of (20) is

$$\frac{dx}{dt} = 3x$$
$$\frac{dy}{dt} = 4x - 5y. \tag{21}$$

In the usual manner we find that the eigenvalues of (21) are $\lambda_1 = -3$ and $\lambda_2 = 5$. Since these eigenvalues are real and distinct, but with opposite signs, it follows from Case I(iii) and Theorem 4.10 that (0, 0) is an (unstable) saddle point. Continuing, it follows from Theorem 4.11 that (0, 0) is likewise a saddle point for the nonlinear system (20).

With computer software we get the phase portraits in Figure 4.6. Notice that the trajectories near the origin behave in a similar manner. ●

EXAMPLE 5 Linearization of a Nonlinear System

Consider the nonlinear system

$$\frac{dx}{dt} = 2x + 8y$$
$$\frac{dy}{dt} = -x - 2y + \tfrac{1}{2}y^3.$$

Proceeding as in Example 4 we find that a linearization of the system at (0, 0) is

$$\frac{dx}{dt} = 2x + 8y$$
$$\frac{dy}{dt} = -x - 2y.$$

We saw in part (b) of Example 3 that this linear system has imaginary eigenvalues $\lambda_1 = 2i$ and $\lambda_2 = -2i$, and so (0, 0) is a center. Although we can draw no conclusion regarding the nature of the critical point (0, 0) of the nonlinear system from Theorem 4.11, it *appears* that (0, 0) is an unstable spiral point in Figure 4.7. ●

Critical Point Not (0, 0)

Suppose that (x_1, y_1) is a critical point of (1) and suppose that this point is not at the origin. We can always use a change of variables in (1) obtained from a translation of axes defined by the equations

$$X = x - x_1, \quad Y = y - y_1 \tag{22}$$

so that the point (x_1, y_1) in the xy-plane is transformed into the origin $X = 0$, $Y = 0$ in the XY-plane.

EXAMPLE 6 Critical Point Not at Origin

Find and classify the critical points of the system:

$$\frac{dx}{dt} = x + y^3 - 12$$

$$\frac{dy}{dt} = x - y^2.$$

Solution Using the second equation of the algebraic system

$$x + y^3 - 12 = 0$$

$$x - y^2 = 0$$

to substitute into the first, we get $y^3 + y^2 - 12 = 0$. By inspection we see that $y = 2$ is a root, and so by dividing the left member of the equation by $y - 2$ we get

$$y^3 + y^2 - 12 = (y - 2)(y^2 + 3y + 6).$$

The roots of the second quadratic factor are complex so that the only real critical point of the system is (4, 2).

Now we use (22) in the form $x = X + 4$, $y = Y + 2$ and substitute these equations into the original system:

$$\frac{dX}{dt} = X + Y^3 + 6Y^2 + 12Y$$

$$\frac{dY}{dt} = X - 4Y^2 - 4Y.$$

Notice (0, 0) is a critical point of the last system. The linearization of the last system at (0, 0) is

$$\frac{dX}{dt} = X + 12Y$$

$$\frac{dY}{dt} = X - 4Y.$$

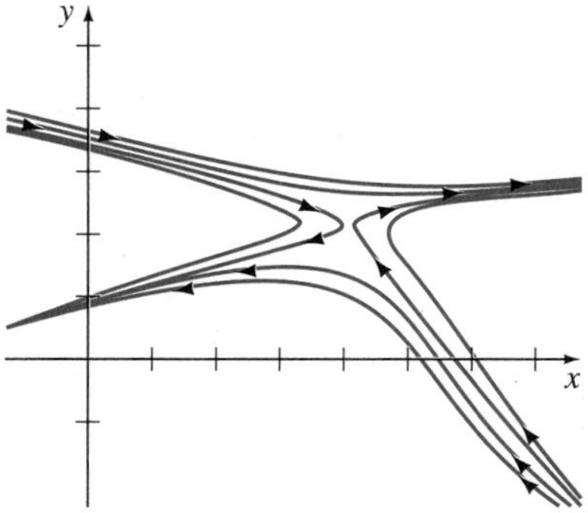

FIGURE 4.8

The eigenvalues of the linear system are real and distinct but with opposite signs ($\lambda_1 \approx -5.8$, $\lambda_2 \approx 2.8$). We conclude that (0, 0) is a saddle point of the linear system and consequently (4, 2) is a saddle point of the original nonlinear system. A phase portrait of the nonlinear system near the point (4, 2) is given in Figure 4.8. ●

Critical Points That Cannot Be Classified by Type

You should not get the impression from the foregoing discussion that linearization and phase portraits will tell us everything we want to know about a nonlinear

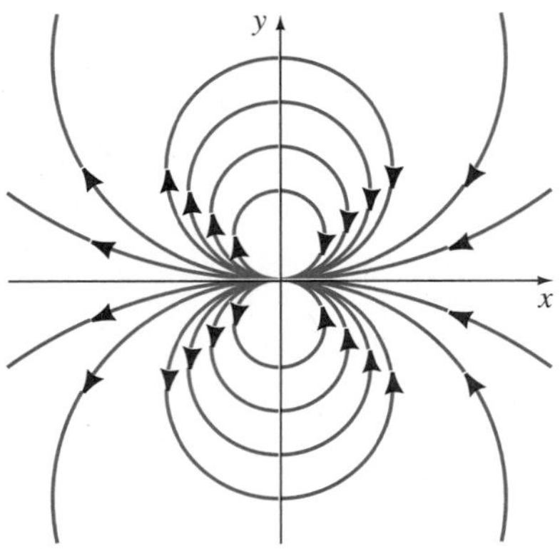

FIGURE 4.9

system. Linearization has at least two shortcomings. As mentioned earlier we can draw no conclusion about the nature of a critical point of a nonlinear system from Theorem 4.11 when (0, 0) is a center of the linearized system, and, unfortunately, not all nonlinear systems are nearly linear. For example, (0, 0) is an isolated critical point of the nonlinear system

$$\frac{dx}{dt} = x^2 - y^2$$
$$\frac{dy}{dt} = 2xy. \qquad (23)$$

Since all first partials of $f(x, y) = x^2 - y^2$ and $g(x, y) = 2xy$ are zero at (0, 0) the linearization of this system is $dx/dt = 0$, $dy/dt = 0$. We can extract no useful information from this trivial linear system. With computer software we can, however, obtain the phase portrait of the nonlinear system shown in Figure 4.9. By dividing the second equation of (23) and solving the homogeneous equation $dy/dx = 2xy/(x^2 - y^2)$ we get a Cartesian equation that appears to be a family of circles passing through (0, 0). See Problem 41. The trajectories are circles but with one point missing. Points $(x(t), y(t))$ on the trajectories "start" arbitrarily close to the origin, loop around, and "finish" again arbitrarily close to the origin. The trajectories do not pass through the origin. For this system (0, 0) is obviously not one of the four types: node, saddle point, spiral point, and center. This last example illustrates a fact of life in the study of nonlinear systems: The local behavior of trajectories near the isolated critical points of a nonlinear system *that is not nearly linear* may differ wildly from the behavior of trajectories of linear systems near the isolated critical point (0, 0). We simply may not be able to assign to a critical point of a nonlinear system a classification according to type; for general nonlinear systems there are an unlimited variety of types of critical points.

Using the tools at hand, namely, linearization and phase portraits, it may also not be possible to classify a critical point according to stability. A phase portrait can be so complicated or so subtle that the stability of a critical point is not obvious. For example, the phase portrait in Figure 4.9 clearly indicates that points $(x(t), y(t))$ on the trajectories approach the critical point (0, 0) for increasing values of t. Isn't this sufficient to say that (0, 0) is asymptotically stable? The answer is no. It must be remembered that to be asymptotically stable a critical point (x_1, y_1) must first be stable and then $(x(t), y(t)) \to (x_1, y_1)$ as $t \to \infty$ on all trajectories. In this example it is easy to show using a parameterization that the trajectories approach the origin for a *finite* value of t. Moreover, and this is worth thinking about, the critical point (0, 0) for (23) actually is unstable. See Problem 41. We note in passing that there exist examples of nonlinear systems for which (x_1, y_1) is a critical point and $(x(t), y(t)) \to (x_1, y_1)$ as $t \to \infty$ on all trajectories and yet (x_1, y_1) is unstable.

We often have to appeal to an advanced technique to answer the question of stability.

A Predator-Prey Model

Suppose two different species of animals interact within the same environment or ecosystem and suppose further that one species eats only vegetation and the second eats only the first species. In other words, one species is a predator and the other is a prey. For example, wolves hunt grass eating caribou, sharks devour little fish, and the snowy owl pursues an arctic rodent called the lemming. For the sake of discussion, let us imagine that the predators are foxes and the prey are rabbits.

Let $x(t)$ and $y(t)$ denote, respectively, the rabbit and fox populations at any time t. If there were no foxes, then one might expect that the rabbits would, with an added assumption of unlimited food supply, grow at a rate that is proportional to the number of rabbits present at any time:

$$\frac{dx}{dt} = ax, \quad a > 0. \tag{24}$$

However, when foxes are present in the environment it seems reasonable that the number of encounters or interactions between these two species per unit time is jointly proportional to their populations x and y, that is, proportional to the product xy. Thus a model for the rabbit population is (24) reduced by the rate bxy, $b > 0$, at which the rabbits are eaten during these encounters:

$$\frac{dx}{dt} = ax - bxy. \tag{25}$$

On the other hand were there no rabbits then the foxes, lacking an adequate food supply, would decline in number according to

$$\frac{dy}{dt} = -cy, \quad c > 0. \tag{26}$$

But when rabbits are present, there is a supply of food and so foxes are added to the system at a rate dxy, $d > 0$. Adding this last rate to (26) gives a model for the fox population

$$\frac{dy}{dt} = -cy + dxy. \tag{27}$$

Equations (25) and (27) constitute an autonomous system of nonlinear differential equations

$$\begin{aligned} \frac{dx}{dt} &= \quad ax - bxy = x(a - by) \\ \frac{dy}{dt} &= -cy + dxy = y(-c + dx), \end{aligned} \tag{28}$$

where a, b, c, and d are positive constants. This famous system of equations is known as the **Lotka-Volterra predator-prey model**.

Except for the trivial solution $x(t) = 0$, $y(t) = 0$ (which means the extinction of both species), the nonlinear system (28) cannot be solved in terms of elementary

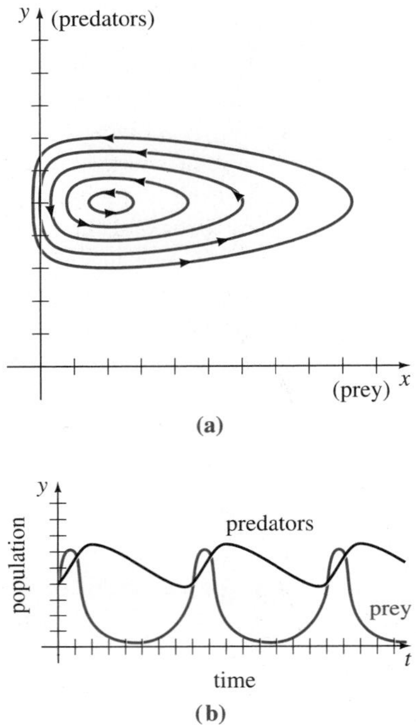

FIGURE 4.10

functions. From the equations $x(a - by) = 0$, $y(-c + dx) = 0$ we see that the critical points of the system are $(0, 0)$ and $(c/d, a/b)$. Linearization of (28) at $(0, 0)$ demonstrates that this critical point is a saddle point, whereas linearization at $(c/d, a/b)$ leads to a center which, unfortunately, is the inconclusive case (b) in Theorem 4.11. Nonetheless by dividing the second equation in (28) by the first, and solving the differential equation $dy/dx = (-cy + dxy)/(ax - bxy)$, it can be proved from the Cartesian equation of the trajectories that $(c/d, a/b)$ is indeed a center (see Problem 42 in Exercises 4.4).

EXAMPLE 7 Predator-Prey Model

Suppose

$$\frac{dx}{dt} = 4.5x - 0.9xy$$

$$\frac{dy}{dt} = -0.16y + 0.08xy$$

represents a predator-prey model. Since we are dealing with populations, we have $x(t) \geq 0$, $y(t) \geq 0$ and so the phase portrait of the system in Figure 4.10(a) shows trajectories confined to the first quadrant. The closed trajectories in a neighborhood of $(2, 5)$ affirms the fact that this critical point is a center. In other words, the model predicts that both populations x and y are periodic. You should think about the cyclic nature of the predicted populations in Figure 4.10(a): As the number of prey decreases the predator population eventually decreases due to a diminished food supply, but attendant to a decrease in the number of predators the number of prey increase, and this in turn gives rise to an increased number of predators which ultimately brings about a decrease in the number of prey. Figure 4.10(b) shows typical population curves of the predators and prey for this model. ●

Competition Models

Now suppose that two different species of animals occupy the same ecosystem, not as predator and prey but rather as competitors for the same resources (such as food, living space) in the system. In the absence of the other, let us assume that the rate at which each population grows is given, in turn, by

$$\frac{dx}{dt} = ax \quad \text{and} \quad \frac{dy}{dt} = cy. \tag{29}$$

Since the two species compete another assumption might be that each of these rates are diminished simply by the influence, or existence, of the other population. Thus a model for the two populations is given by the linear system

$$\begin{aligned} \frac{dx}{dt} &= ax - by \\ \frac{dy}{dt} &= cy - dx, \end{aligned} \tag{30}$$

where a, b, c, and d are positive constants. On the other hand, we might assume, as we did in (25), that each growth rate in (29) should be reduced by a rate proportional to the number of interactions between the two species:

$$\frac{dx}{dt} = ax - bxy$$
$$\frac{dy}{dt} = cy - dxy. \tag{31}$$

This nonlinear system is, on inspection, similar to the Lotka-Volterra predator-prey model. Lastly, it might be more realistic to replace the rates in (29) which indicate that the population of each species, in isolation, grows exponentially with rates that indicate each population grows logistically (that is, over a long time the population is bounded),

$$\frac{dx}{dt} = a_1x - b_1x^2 \quad \text{and} \quad \frac{dy}{dt} = a_2y - b_2y^2. \tag{32}$$

When these new rates are decreased by rates proportional to the number of interactions we obtain another nonlinear model,

$$\frac{dx}{dt} = a_1x - b_1x^2 - c_1xy = x(a_1 - b_1x - c_1y)$$
$$\frac{dy}{dt} = a_2y - b_2y^2 - c_2xy = y(a_2 - b_2y - c_2x), \tag{33}$$

where all coefficients are positive. The linear system (30) and the nonlinear systems (31) and (33) are, of course, called **competition models**.

EXERCISES 4.4

Answers to odd-numbered problems begin on page AN-20.

4.4.1

In Problems 1–8 classify the critical point (0, 0) for the given autonomous linear system. Use a computer graphing program to obtain a phase portrait of each system.

1. $\dfrac{dx}{dt} = x + y$, $\dfrac{dy}{dt} = y$

2. $\dfrac{dx}{dt} = x - y$, $\dfrac{dy}{dt} = 9x - y$

3. $\dfrac{dx}{dt} = -4x + 2y$, $\dfrac{dy}{dt} = 2x - 4y$

4. $\dfrac{dx}{dt} = 2y$, $\dfrac{dy}{dt} = 5x + 7y$

5. $\dfrac{dx}{dt} = -3x + 6y$, $\dfrac{dy}{dt} = -2x + 3y$

6. $\dfrac{dx}{dt} = -x + y$, $\dfrac{dy}{dt} = -3x - 3y$

7. $\dfrac{dx}{dt} = -2x + 5y$, $\dfrac{dy}{dt} = -2x + 4y$

8. $\dfrac{dx}{dt} = 7x + \frac{3}{2}y$, $\dfrac{dy}{dt} = -4x - 2y$

In Problems 9–12 find the critical point (x_1, y_1) of the given autonomous linear system. Use the change of variables $x = X + x_1$, $y = Y + y_1$ as an aid in classifying the critical point. Use a computer graphing program to obtain a phase portrait of each system.

9. $$\frac{dx}{dt} = 2x + 3y - 6$$
$$\frac{dy}{dt} = -x - 2y + 5$$

10. $$\frac{dx}{dt} = -5x + 9y + 13$$
$$\frac{dy}{dt} = -x - 11y - 23$$

11. $$\frac{dx}{dt} = 0.1x - 0.2y + 0.35$$
$$\frac{dy}{dt} = 0.1x + 0.1y - 0.25$$

12. $$\frac{dx}{dt} = 3x - 2y - 1$$
$$\frac{dy}{dt} = 5x - 3y - 2$$

13. Determine whether it is possible to find values of c for the linear system

$$\frac{dx}{dt} = x + y$$
$$\frac{dy}{dt} = cx + y$$

so that (0, 0) is **(a)** a spiral point, **(b)** an unstable node, **(c)** an asymptotically stable node, **(d)** a degenerate node, and **(e)** a saddle point.

14. Determine values of c so that (0, 0) is a center for the linear system

$$\frac{dx}{dt} = 2x - cy$$
$$\frac{dy}{dt} = cx - 2y.$$

15. Determine values of a so that (0, 0) is an asymptotically stable spiral point for the linear system

$$\frac{dx}{dt} = ax - 4y$$
$$\frac{dy}{dt} = x + y.$$

16. **(a)** Find the critical points of the linear system

$$\frac{dx}{dt} = 3x + y$$
$$\frac{dy}{dt} = 3x + y.$$

(b) Find a Cartesian equation of the trajectories.
(c) Without the aid of a computer, sketch a phase portrait of the system. Indicate by means of arrowheads whether points $(x(t), y(t))$ on a trajectory move toward or away from a critical point as $t \to \infty$.

17. The characteristic equation (4) is

$$\lambda^2 - (a + d)\lambda + ad - bc = 0.$$

(a) Explain why $a + d = \lambda_1 + \lambda_2$ and $ad - bc = \lambda_1\lambda_2$.
(b) If (3) is written $\mathbf{X}' = \mathbf{AX}$, explain why (0, 0) is a saddle point if and only if det $\mathbf{A} < 0$. Can you draw any conclusion about (0, 0) if det $\mathbf{A} > 0$?

18. *PROJECT PROBLEM* If we let $p = a + d$ and $q = ad - bc$ in the characteristic equation given in Problem 17, then by the quadratic formula, the eigenvalues λ_1, λ_2 are given by $\dfrac{p \pm \sqrt{p^2 - 4q}}{2}$.

(a) Fill in each cell in the following table with one of the symbols: > 0, < 0, $= 0$, —, to denote whether the given expression (that is, $p^2 - 4q$, p, or q) is positive, negative, equal to zero, or can be any value, in describing the critical point under the various cases.
(b) Consider the pq-plane as shown in Figure 4.11. Use the table in part (a) to fill in the cells with a name such as *stable nodes*, *centers*, and so forth for each of the eight indicated lines, curves, and regions.

CASE I $p^2 - 4q$ ☐			*CASE II* $p^2 - 4q$ ☐			*CASE III* $p^2 - 4q$ ☐		
Critical point (0, 0)	p	q	**Critical point** (0, 0)	p	q	**Critical point** (0, 0)	p	q
stable node	☐	☐	stable degenerate node	☐	☐	stable spiral point	☐	☐
unstable node	☐	☐	unstable degenerate node	☐	☐	unstable spiral point	☐	☐
saddle point	☐	☐				center	☐	☐

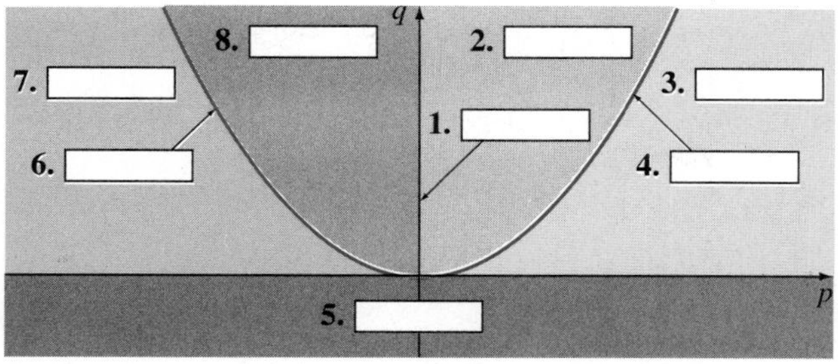

FIGURE 4.11

4.4.2

In Problems 19–26, (0, 0) is an isolated critical point of the given system. Use linearization at (0, 0) to classify the critical point, if possible, as to type and stability. If Theorem 4.11 is not applicable, use a computer graphing program to obtain a phase portrait of the system. Use the phase portrait as an aid in classifying the point.

19. $\frac{dx}{dt} = 2e^{2x} + \frac{1}{2}y - 2$, $\frac{dy}{dt} = y^2 + 4y + \sin 2x$

20. $\frac{dx}{dt} = \sin(-x + y)$, $\frac{dy}{dt} = e^{x-2y} + \sin 5y$

21. $\frac{dx}{dt} = 3x - 6y - xy$, $\frac{dy}{dt} = x - y + x^2y$

22. $\frac{dx}{dt} = -y - x^3$, $\frac{dy}{dt} = x - y^3$

23. $\frac{dx}{dt} = \sin x + 3\sin y$, $\frac{dy}{dt} = 2x\cos x - 2y + y^2$

24. $\frac{dx}{dt} = -x - 4y + xy^3$, $\frac{dy}{dt} = x - 6y - x^2 + y^2$

25. $\frac{dx}{dt} = x(2 + y)^2$, $\frac{dy}{dt} = e^{x+4y} - 1$

26. $\frac{dx}{dt} = -x + 3y - \sin x$, $\frac{dy}{dt} = -x + 2ye^y$

In Problems 27–30 find all critical points of the given system. Use linearization as illustrated in Example 6 to classify all the critical points as to type and stability. Obtain a phase portrait of the system to verify your results or as an aid in classifying a critical point if Theorem 4.11 is not applicable.

27. $\frac{dx}{dt} = y + x^2 - 2$, $\frac{dy}{dt} = y - x^2$

28. $\frac{dx}{dt} = y - x^2$, $\frac{dy}{dt} = xy - 2x$

29. $\frac{dx}{dt} = (x - 1)(3 - x - y)$, $\frac{dy}{dt} = (y - 1)(2x - 3)$

30. $\frac{dx}{dt} = -5x - 4y + x^2 + 14$, $\frac{dy}{dt} = 2x - 5y + 4$

In Problems 31–34 find the critical points of the given system. Use a computer graphing program to obtain a phase portrait of each system. Use the phase portrait to classify the critical points, if possible, as to type and stability.

31. $\frac{dx}{dt} = \sin y$, $\frac{dy}{dt} = \cos x$

32. $\frac{dx}{dt} = 10 - y^2 + x$, $\frac{dy}{dt} = 2x + 2$

33. $\frac{dx}{dt} = -2xy$, $\frac{dy}{dt} = x^2 - y^3$

34. $\frac{dx}{dt} = -2x + y + 2x^2 - xy$, $\frac{dy}{dt} = (y - 2)^2$

35. **(a)** Show that the two nonlinear systems

$$\frac{dx}{dt} = y \qquad \frac{dx}{dt} = y$$
$$\text{and}$$
$$\frac{dy}{dt} = -x - y^2 \qquad \frac{dy}{dt} = -x - y^3$$

possess the same linearized system.

(b) Classify the critical point (0, 0) of the linearized system found in part (a).

(c) Classify, if possible, the critical point (0, 0) of each of the nonlinear systems. Obtain a phase portrait of each system.

36. A model of the nonlinear pendulum with a constant driving function is $d^2\theta/dt^2 + \sin\theta = \frac{1}{2}$.

(a) Let $x = \theta(t)$ and express the differential equation as an equivalent autonomous system.

(b) Show that $(\pi/6, 0)$ is a critical point of the equation. Find all other critical points.

(c) Verify that $(\pi/6, 0)$ cannot be classified using linearization.

(d) Obtain a phase portrait of the differential equation and use it as an aid in classifying the critical points.

(e) Use the phase portrait to determine whether the differential equation possesses a periodic solution satisfying $\theta(0) = 1$, $\theta'(0) = 1$. Satisfying $\theta(0) = 1$, $\theta'(0) = 1.1$. Use an ODE solver to obtain the solution curve corresponding to each set of initial conditions.

37. Investigate the nonlinear differential equation $d^2\theta/dt^2 + \sin\theta = 1$ to determine whether it possesses nonconstant periodic solutions.

38. In 1927 the Dutch electrical engineer Balthasar van der Pol showed that certain properties of electrical circuits containing vacuum tubes could be deduced from the nonlinear differential equation

$$\frac{d^2x}{dt^2} + \mu(x^2 - 1)\frac{dx}{dt} + x = 0,$$

where μ is a positive constant. Show that (0, 0) is an unstable critical point of van der Pol's equation for $\mu > 0$.

In Problems 39 and 40, (0, 0) is an isolated critical point of the given nonlinear system. Use polar coordinates, $x = r\cos\theta$, $y = r\sin\theta$, where r and θ are functions of t, in the system. Classify the critical point at the origin by finding a polar equation of the trajectories.

39. $\dfrac{dx}{dt} = -y + xy^2 + x^3$

$\dfrac{dy}{dt} = x + x^2y + y^3$

40. $\dfrac{dx}{dt} = -y - x\sqrt{x^2 + y^2}$

$\dfrac{dy}{dt} = x - y\sqrt{x^2 + y^2}$

41. **(a)** Show that an equation of the trajectory passing through an initial point (x_0, y_0) for the system given in (23) is $cy = x^2 + y^2$, where $c = (x_0^2 + y_0^2)/y_0$, $y_0 \neq 0$.

(b) Use the result in part (a) to prove that (0, 0) is an unstable critical point of the system.

42. ***PROJECT PROBLEM***

(a) Show that a Cartesian equation for the trajectories of the Lotka-Volterra equations (28) is $(x^c e^{-dx})(y^a e^{-by}) = C$, where C is a constant.

(b) Use a CAS with a contour plot feature to convince yourself that the Cartesian equation in part (a) defines a family of closed trajectories for $x > 0$, $y > 0$.

(c) There is an algebraic way of proving that $(x^c e^{-dx})(y^a e^{-by}) = C$ defines a family of closed curves for $x > 0$, $y > 0$. Consider the two functions $f(x) = x^c e^{-dx}$ and $g(y) = y^a e^{-by}$. Prove that f has a maximum M_f for $x > 0$ and g has a maximum M_g for $y > 0$. Explain why $(x^c e^{-dx})(y^a e^{-by}) = C$ has no solution when $C > M_f M_g$. Explain why there is exactly one solution for $C = M_f M_g$. Now let $C = \mu M_g$, where $0 < \mu < M_f$. Explain why there are two solutions x_1 and x_2 for which $f(x) = \mu$. Explain why the equation $y^a e^{-by} = (\mu/x^c e^{-dx})M_g$ has no y solutions for $x < x_1$ and $x > x_2$, one y solution for $x = x_1$ and for $x = x_2$, and two y solutions for each x satisfying $x_1 < x < x_2$. Use figures when possible to illustrate your reasoning.

43. Consider the Lotka-Volterra predator-prey model defined by

$$\frac{dx}{dt} = 0.2x - 0.025xy$$

$$\frac{dy}{dt} = -0.1y + 0.02xy,$$

where the populations $x(t)$ (prey) and $y(t)$ (predators) are measured in the thousands.

(a) Find the critical point (x_1, y_1) in the first quadrant.

(b) Use a computer graphing program to obtain a phase portrait of the system. Find the trajectory passing through (6, 6) at $t = 0$.

(c) Approximate the period of $x(t)$ and $y(t)$ by solving the linearization of the system at (x_1, y_1) subject to $x(0) = 6$, $y(0) = 6$.

(d) Use an ODE solver with $x(0) = 6$, $y(0) = 6$ to approximate the time $t > 0$ when the two populations are first equal. Use the graphs to approximate the period of $x(t)$ and $y(t)$ and compare it with the result of part (c).

44. Each of the following systems is a competition model. Find all critical points. Analyze and interpret the model.

(a) $\dfrac{dx}{dt} = x(2 - 0.4x - 0.3y)$

$\dfrac{dy}{dt} = y(1 - 0.1y - 0.3x)$

(b) $\dfrac{dx}{dt} = x(1 - 0.1x - 0.05y)$

$\dfrac{dy}{dt} = y(1.7 - 0.1y - 0.15x)$

45. Each of the following systems is a predator-prey model. Find all critical points. Analyze and interpret the model.

(a) $\dfrac{dx}{dt} = x(6 - x - y)$

$\dfrac{dy}{dt} = y(-2 - y + x)$

(b) $\dfrac{dx}{dt} = x(2 - x - y)$

$\dfrac{dy}{dt} = y(-2 - y + x)$

46. A model for the spread of an epidemic throughout a population is given by

$$\frac{dx}{dt} = -axy$$

$$\frac{dy}{dt} = -by + axy,$$

where a and b are positive constants. Here $x(t)$ denotes the number of people in a community that are susceptible to a communicable disease but, as of time t, not yet infected with it, and $y(t)$ denotes the number of people who are infected with the disease.

(a) Determine the critical points of the system.

(b) Suppose that initial conditions are $x(0) = x_0 > 0$, $y(0) = y_0 > 0$. Explain why the model predicts an epidemic for $x_0 > a/b$ but no epidemic for $x_0 < a/b$. Verify your reasoning by obtaining a representative phase portrait of the system in the first quadrant.

47. *PROJECT PROBLEM* A bead of mass m is free to slide on a wire circular hoop of radius l. If the hoop rotates with a constant angular velocity ω about a vertical diameter, then the angle θ from the vertical to a radius drawn to the bead, as shown in Figure 4.12, satisfies the nonlinear differential equation

$$\frac{d^2\theta}{dt^2} + \sin\theta - \omega^2 \sin\theta\cos\theta = 0, \quad \omega \geq 0.$$

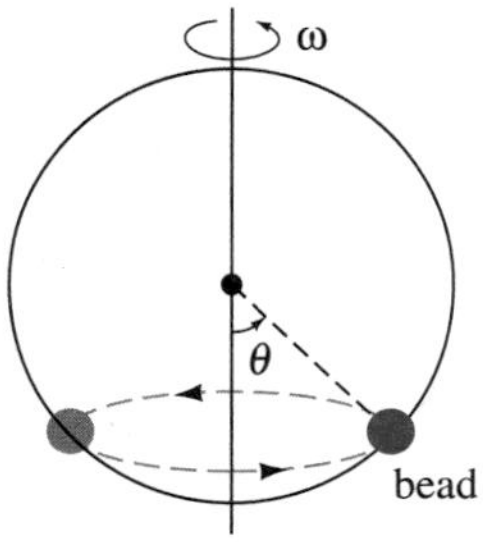

FIGURE 4.12

(a) Assume $-\pi \leq \theta \leq \pi$. Find all critical points of the differential equation. Consider the cases $0 < \omega < 1$, $\omega = 1$, and $\omega > 1$.

(b) Elaborate on the meaning of the word **bifurcation** as it applies to this differential equation.

48. *WRITING PROJECT* Write a short report on the concept of a **limit cycle**.

(a) Use a computer graphing program to illustrate a limit cycle for a nonlinear system with (0, 0) as a critical point. Discuss stability of the critical point and define stability of a limit cycle.

(b) Use a computer graphing program to illustrate a limit cycle for van der Pol's equation $\frac{d^2x}{dt^2} + \mu(x^2 - 1)\frac{dx}{dt} + x = 0$ for $\mu = 0.1, 1, 10$.

49. *WRITING PROJECT* Suppose that (0, 0) is a critical point of a nonlinear system. It was stated on page 238 that $(x(t), y(t)) \to (0, 0)$ as $t \to \infty$ on all trajectories of the system is not in itself sufficient to guarantee that (0, 0) is asymptotically stable; (0, 0) could still be unstable. Consult an advanced text to find an example of this and write up your findings in the form of a short report.

50. *WRITING PROJECT* Write a report relating the material discussed in this section with the **Lorenz equations**, **strange attractor**, and **chaos**.

CHAPTER 5

THE LAPLACE TRANSFORM

5.1 DEFINITION OF THE LAPLACE TRANSFORM

Introduction

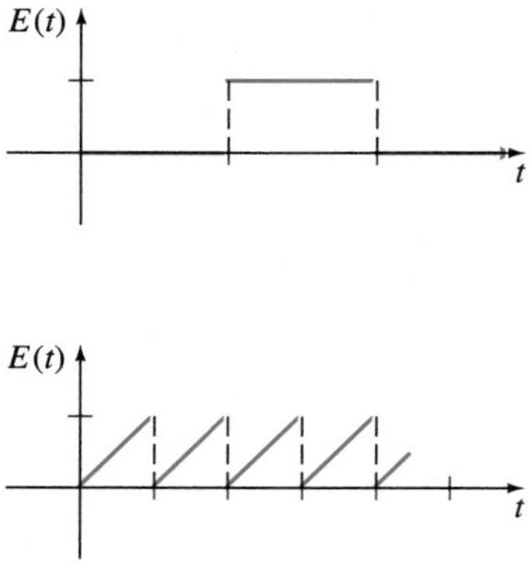

FIGURE 5.1

In the linear mathematical models for physical systems such as a spring/mass system or a series electrical circuit, the right-hand member of the differential equations

$$m\frac{d^2x}{dt^2} + \beta\frac{dx}{dt} + kx = f(t) \quad \text{or} \quad L\frac{d^2q}{dt^2} + R\frac{dq}{dt} + \frac{1}{C}q = E(t)$$

is a driving function and represents either an external force $f(t)$ or an impressed voltage $E(t)$. In Section 3.7 we solved problems in which the functions f and E were continuous. However, piecewise continuous driving functions are not uncommon. For example, the impressed voltage on a circuit could be as shown in Figure 5.1. Solving the differential equation of the circuit in this case is difficult but not impossible. The Laplace transform studied in this chapter is an invaluable aid in solving problems such as these.

Linearity Property

In elementary calculus you learned that differentiation and integration transform a function into another function. For example, the function $f(x) = x^2$ is transformed, in turn, into a linear function, a family of cubic polynomial functions, and a constant by the operations of differentiation, indefinite integration, and definite integration:

$$\frac{d}{dx}x^2 = 2x, \quad \int x^2\,dx = \frac{x^3}{3} + c, \quad \int_0^3 x^2\,dx = 9.$$

Moreover, these three operations possess the **linearity property**. This means that for any constants α and β,

$$\frac{d}{dx}[\alpha f(x) + \beta g(x)] = \alpha\frac{d}{dx}f(x) + \beta\frac{d}{dx}g(x)$$

$$\int [\alpha f(x) + \beta g(x)]\,dx = \alpha\int f(x)\,dx + \beta\int g(x)\,dx \qquad \textbf{(1)}$$

$$\int_a^b [\alpha f(x) + \beta g(x)]\,dx = \alpha\int_a^b f(x)\,dx + \beta\int_a^b g(x)\,dx,$$

provided each derivative and integral exists.

If $f(x, y)$ is a function of two variables, then a definite integral of f with respect to one of the variables leads to a function of the other variable. For example, by holding y constant, we see that $\int_1^2 2xy^2\,dx = 3y^2$. Similarly, a definite integral such as $\int_a^b K(s, t)f(t)\,dt$ transforms a function $f(t)$ into a function of the variable s. We are particularly interested in **integral transforms** of this last kind, where the interval of integration is the unbounded interval $[0, \infty)$.

Basic Definition

If $f(t)$ is defined for $t \geq 0$, then the improper integral $\int_0^\infty K(s, t)f(t)\,dt$ is defined as a limit:

$$\int_0^\infty K(s, t)f(t)\,dt = \lim_{b \to \infty} \int_0^b K(s, t)f(t)\,dt.$$

If the limit exists, the integral exists or is convergent; if the limit does not exist, the integral does not exist and is said to be divergent. The foregoing limit will, in general, exist for only certain values of the variable s. The choice $K(s, t) = e^{-st}$ gives us an especially important integral transform.

DEFINITION 5.1 Laplace Transform

Let f be a function defined for $t \geq 0$. Then the integral

$$\mathscr{L}\{f(t)\} = \int_0^\infty e^{-st} f(t)\,dt \tag{2}$$

is said to be the **Laplace transform** of f provided the integral converges.

When the defining integral (2) converges, the result is a function of s. In general discussion, we shall use a lowercase letter to denote the function being transformed and the corresponding capital letter to denote its Laplace transform; for example,

$$\mathscr{L}\{f(t)\} = F(s), \quad \mathscr{L}\{g(t)\} = G(s), \quad \mathscr{L}\{y(t)\} = Y(s).$$

EXAMPLE 1 Applying Definition 5.1

Evaluate $\mathscr{L}\{1\}$.

Solution

$$\mathscr{L}\{1\} = \int_0^\infty e^{-st}(1)\,dt = \lim_{b \to \infty} \int_0^b e^{-st}\,dt$$

$$= \lim_{b \to \infty} \frac{-e^{-st}}{s}\Bigg|_0^b = \lim_{b \to \infty} \frac{-e^{-sb} + 1}{s} = \frac{1}{s}$$

provided $s > 0$. In other words, when $s > 0$, the exponent $-sb$ is negative and $e^{-sb} \to 0$ as $b \to \infty$. When $s < 0$, the integral is divergent. ●

The use of the limit sign becomes somewhat tedious, so we shall adopt the notation $|_0^\infty$ as a shorthand to writing $\lim_{b \to \infty}(\;)|_0^b$. For example,

$$\mathscr{L}\{1\} = \int_0^\infty e^{-st}\,dt = \frac{-e^{-st}}{s}\Bigg|_0^\infty = \frac{1}{s}, \quad s > 0.$$

At the upper limit it is understood we mean $e^{-st} \to 0$ as $t \to \infty$ for $s > 0$.

$\mathscr{L}$ is a Linear Transform

For a sum of functions we can write

$$\int_0^\infty e^{-st}[\alpha f(t) + \beta g(t)]\, dt = \alpha \int_0^\infty e^{-st} f(t)\, dt + \beta \int_0^\infty e^{-st} g(t)\, dt,$$

whenever both integrals converge. Hence it follows that

$$\mathscr{L}\{\alpha f(t) + \beta g(t)\} = \alpha \mathscr{L}\{f(t)\} + \beta \mathscr{L}\{g(t)\} = \alpha F(s) + \beta G(s). \tag{3}$$

Because of the property given in (3), $\mathscr{L}$ is said to be a **linear transform**.

Sufficient Conditions for Existence of $\mathscr{L}\{f(t)\}$

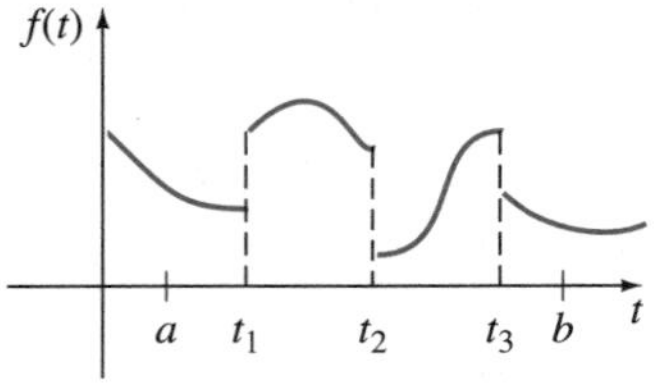

FIGURE 5.2

The integral that defines the Laplace transform does not have to converge. For example, neither $\mathscr{L}\{1/t\}$ nor $\mathscr{L}\{e^{t^2}\}$ exists. Sufficient conditions that guarantee the existence of $\mathscr{L}\{f(t)\}$ are that f be piecewise continuous on $[0, \infty)$ and that f be of exponential order for $t > T$. Recall that a function f is **piecewise continuous** on $[0, \infty)$ if, in any interval $0 \le a \le t \le b$, there are at most a finite number of points t_k, $k = 1, 2, \ldots, n$ $(t_{k-1} < t_k)$, at which f has finite discontinuities and is continuous on each open interval $t_{k-1} < t < t_k$. See Figure 5.2. The concept of **exponential order** is defined in the following manner.

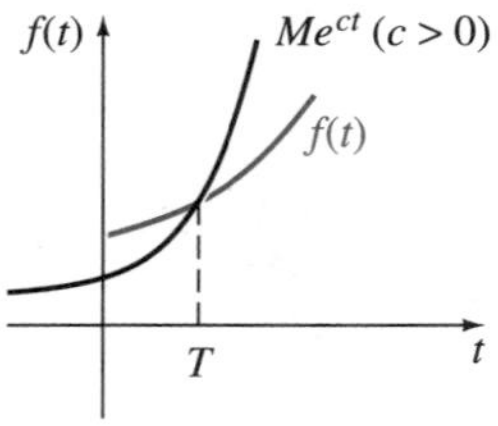

FIGURE 5.3

DEFINITION 5.2 Exponential Order

A function f is said to be of **exponential order** if there exist numbers c, $M > 0$, and $T > 0$ such that $|f(t)| \le Me^{ct}$ for $t > T$.

If, for example, f is an *increasing* function, then the condition $|f(t)| \le Me^{ct}$, $t > T$, simply states that the graph of f on the interval (T, ∞) does not grow faster than the graph of the exponential function Me^{ct}, where c is a positive constant. See Figure 5.3. The functions $f(t) = t$, $f(t) = e^{-t}$, and $f(t) = 2 \cos t$ are all of exponential order for $t > 0$ since we have, respectively,

$$|t| \le e^t, \quad |e^{-t}| \le e^t, \quad |2 \cos t| \le 2e^t.$$

A comparison of the graphs on the interval $[0, \infty)$ is given in Figure 5.4.

A function such as $f(t) = e^{t^2}$ is not of exponential order since, as shown in Figure 5.5 on page 251, its graph grows faster than any positive linear power of e for $t > c > 0$.

A positive integral power of t is always of exponential order since for $c > 0$,

$$|t^n| \le Me^{ct} \quad \text{or} \quad \left|\frac{t^n}{e^{ct}}\right| \le M \quad \text{for } t > T$$

is equivalent to showing that $\lim_{t\to\infty} t^n/e^{ct}$ is finite for $n = 1, 2, 3, \ldots$. The result follows by n applications of L'Hôpital's rule.

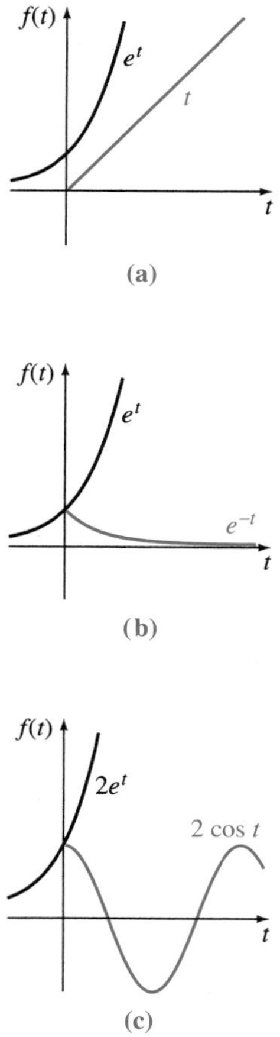

FIGURE 5.4

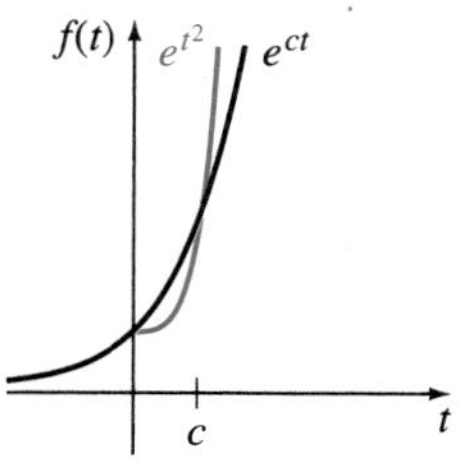

FIGURE 5.5

THEOREM 5.1 Sufficient Conditions for Existence

If $f(t)$ is piecewise continuous on the interval $[0, \infty)$ and of exponential order for $t > T$, then $\mathcal{L}\{f(t)\}$ exists for $s > c$.

Proof
$$\mathcal{L}\{f(t)\} = \int_0^T e^{-st}f(t)\,dt + \int_T^\infty e^{-st}f(t)\,dt = I_1 + I_2.$$

The integral I_1 exists because it can be written as a sum of integrals over intervals for which $e^{-st}f(t)$ is continuous. Now

$$|I_2| \le \int_T^\infty |e^{-st}f(t)|\,dt \le M\int_T^\infty e^{-st}e^{ct}\,dt$$

$$= M\int_T^\infty e^{-(s-c)t}\,dt = -M\frac{e^{-(s-c)t}}{s-c}\Bigg|_T^\infty = M\frac{e^{-(s-c)T}}{s-c}$$

for $s > c$. Since $\int_T^\infty Me^{-(s-c)t}\,dt$ converges, the integral $\int_T^\infty |e^{-st}f(t)|\,dt$ converges by the comparison test for improper integrals. This, in turn, implies that I_2 exists for $s > c$. The existence of I_1 and I_2 implies that $\mathcal{L}\{f(t)\} = \int_0^\infty e^{-st}f(t)\,dt$ exists for $s > c$. ○

Throughout this entire chapter we shall be concerned only with functions that are both piecewise continuous and of exponential order. We note, however, that these conditions are sufficient but not necessary for the existence of a Laplace transform. The function $f(t) = t^{-1/2}$ is not piecewise continuous on the interval $[0, \infty)$, but its Laplace transform exists. See Problems 41 and 42(f) in Exercises 5.1.

EXAMPLE 2 Applying Definition 5.1

Evaluate $\mathcal{L}\{t\}$.

Solution From Definition 5.1 we have $\mathcal{L}\{t\} = \int_0^\infty e^{-st}t\,dt$. Integrating by parts and using $\lim_{t\to\infty} te^{-st} = 0$, $s > 0$, along with the result of Example 1, we obtain

$$\mathcal{L}\{t\} = \frac{-te^{-st}}{s}\Bigg|_0^\infty + \frac{1}{s}\int_0^\infty e^{-st}\,dt = \frac{1}{s}\mathcal{L}\{1\} = \frac{1}{s}\left(\frac{1}{s}\right) = \frac{1}{s^2}.$$ ●

EXAMPLE 3 Applying Definition 5.1

Evaluate $\mathcal{L}\{e^{-3t}\}$.

Solution From Definition 5.1 we have

$$\mathcal{L}\{e^{-3t}\} = \int_0^\infty e^{-st}e^{-3t}\,dt = \int_0^\infty e^{-(s+3)t}\,dt$$

$$= \frac{-e^{-(s+3)t}}{s+3}\Bigg|_0^\infty$$

$$= \frac{1}{s+3}, \quad s > -3.$$

The result follows from the fact that $\lim_{t \to \infty} e^{-(s+3)t} = 0$ for $s + 3 > 0$ or $s > -3$.

EXAMPLE 4 Applying Definition 5.1

Evaluate $\mathscr{L}\{\sin 2t\}$.

Solution From Definition 5.1 and integration by parts we have

$$\mathscr{L}\{\sin 2t\} = \int_0^\infty e^{-st} \sin 2t \, dt = \frac{-e^{-st} \sin 2t}{s} \bigg|_0^\infty + \frac{2}{s} \int_0^\infty e^{-st} \cos 2t \, dt$$

$$= \frac{2}{s} \int_0^\infty e^{-st} \cos 2t \, dt, \quad s > 0 \qquad \text{Laplace transform of } \sin 2t \downarrow$$

$$= \frac{2}{s} \left[\frac{-e^{-st} \cos 2t}{s} \bigg|_0^\infty - \frac{2}{s} \int_0^\infty e^{-st} \sin 2t \, dt \right]$$

$$= \frac{2}{s^2} - \frac{4}{s^2} \mathscr{L}\{\sin 2t\}.$$

At this point we have an equation with $\mathscr{L}\{\sin 2t\}$ on both sides of the equality. Solving for that quantity yields the result

$$\mathscr{L}\{\sin 2t\} = \frac{2}{s^2 + 4}, \quad s > 0.$$

EXAMPLE 5 Using Linearity

Evaluate $\mathscr{L}\{3t - 5 \sin 2t\}$.

Solution From Examples 2 and 4 and the linearity property of the Laplace transform we can write

$$\mathscr{L}\{3t - 5 \sin 2t\} = 3\mathscr{L}\{t\} - 5\mathscr{L}\{\sin 2t\}$$

$$= 3 \cdot \frac{1}{s^2} - 5 \cdot \frac{2}{s^2 + 4}$$

$$= \frac{-7s^2 + 12}{s^2(s^2 + 4)}, \quad s > 0.$$

EXAMPLE 6 Applying Definition 5.1

Evaluate **(a)** $\mathscr{L}\{te^{-2t}\}$ and **(b)** $\mathscr{L}\{t^2e^{-2t}\}$.

Solution **(a)** From Definition 5.1 and integration by parts we have

$$\mathscr{L}\{te^{-2t}\} = \int_0^\infty e^{-st}(te^{-2t})\,dt = \int_0^\infty te^{-(s+2)t}\,dt$$

$$= \frac{-te^{-(s+2)t}}{s+2}\bigg|_0^\infty + \frac{1}{s+2}\int_0^\infty e^{-(s+2)t}\,dt$$

$$= \frac{-e^{-(s+2)t}}{(s+2)^2}\bigg|_0^\infty, \quad s > -2$$

$$= \frac{1}{(s+2)^2}, \qquad s > -2.$$

(b) Again, integration by parts gives

$$\mathscr{L}\{t^2e^{-2t}\} = \frac{-t^2e^{-(s+2)t}}{s+2}\bigg|_0^\infty + \frac{2}{s+2}\int_0^\infty te^{-(s+2)t}\,dt$$

$$= \frac{2}{s+2}\int_0^\infty e^{-st}(te^{-2t})\,dt, \quad s > -2$$

$$= \frac{2}{s+2}\mathscr{L}\{te^{-2t}\} = \frac{2}{s+2}\left[\frac{1}{(s+2)^2}\right] \quad \leftarrow \text{from part (a)}$$

$$= \frac{2}{(s+2)^3}, \quad s > -2.$$

EXAMPLE 7 Transform of a Piecewise Defined Function

Evaluate $\mathscr{L}\{f(t)\}$ for $f(t) = \begin{cases} 0, & 0 \le t < 3 \\ 2, & t \ge 3. \end{cases}$

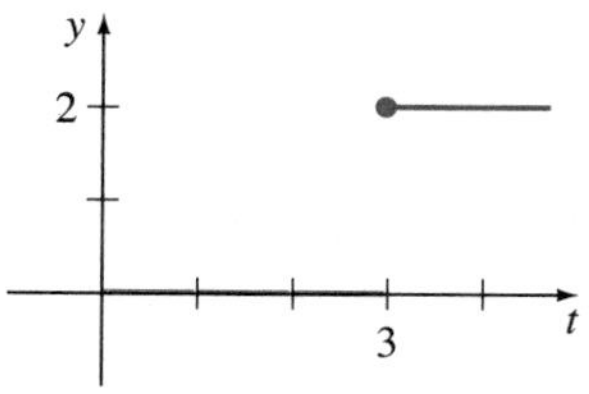

FIGURE 5.6

Solution This piecewise continuous function is shown in Figure 5.6. Since f is defined in two pieces, $\mathscr{L}\{f(t)\}$ is expressed as the sum of two integrals:

$$\mathscr{L}\{f(t)\} = \int_0^\infty e^{-st}f(t)\,dt = \int_0^3 e^{-st}(0)\,dt + \int_3^\infty e^{-st}(2)\,dt$$

$$= -\frac{2e^{-st}}{s}\bigg|_3^\infty$$

$$= \frac{2e^{-3s}}{s}, \quad s > 0.$$

We state the generalization of some of the preceding examples by means of the next theorem. From this point on we shall also refrain from stating any restrictions on s; it is understood that s is sufficiently restricted to guarantee the convergence of the appropriate Laplace transform.

THEOREM 5.2 Transforms of Some Basic Functions

(a) $\mathscr{L}\{1\} = \dfrac{1}{s}$ **(b)** $\mathscr{L}\{t^n\} = \dfrac{n!}{s^{n+1}}, \quad n = 1, 2, 3, \ldots$

(c) $\mathscr{L}\{e^{at}\} = \dfrac{1}{s-a}$ **(d)** $\mathscr{L}\{\sin kt\} = \dfrac{k}{s^2+k^2}$

(e) $\mathscr{L}\{\cos kt\} = \dfrac{s}{s^2+k^2}$ **(f)** $\mathscr{L}\{\sinh kt\} = \dfrac{k}{s^2-k^2}$

(g) $\mathscr{L}\{\cosh kt\} = \dfrac{s}{s^2-k^2}$

Part (b) of Theorem 5.2 can be justified in the following manner. Integration by parts yields

$$\mathscr{L}\{t^n\} = \int_0^\infty e^{-st}t^n\,dt = -\frac{1}{s}e^{-st}t^n\Big|_0^\infty + \frac{n}{s}\int_0^\infty e^{-st}t^{n-1}\,dt = \frac{n}{s}\int_0^\infty e^{-st}t^{n-1}\,dt$$

or

$$\mathscr{L}\{t^n\} = \frac{n}{s}\mathscr{L}\{t^{n-1}\}, \quad n = 1, 2, 3, \ldots.$$

Now $\mathscr{L}\{1\} = 1/s$, so it follows by iteration that

$$\mathscr{L}\{t\} = \frac{1}{s}\mathscr{L}\{1\} = \frac{1}{s^2}, \quad \mathscr{L}\{t^2\} = \frac{2}{s}\mathscr{L}\{t\} = \frac{2}{s^3}, \quad \mathscr{L}\{t^3\} = \frac{3}{s}\mathscr{L}\{t^2\} = \frac{3\cdot 2}{s^4} = \frac{3!}{s^4}$$

Although a rigorous proof requires mathematical induction, it seems reasonable to conclude from the foregoing results that in general

$$\mathscr{L}\{t^n\} = \frac{n}{s}\mathscr{L}\{t^{n-1}\} = \frac{n}{s}\left[\frac{(n-1)!}{s^n}\right] = \frac{n!}{s^{n+1}}.$$

The justifications of parts (f) and (g) of Theorem 5.2 are left to you. See Problems 33 and 34 in Exercises 5.1.

EXAMPLE 8 Trigonometric Identity and Linearity

Evaluate $\mathscr{L}\{\sin^2 t\}$.

Solution With the aid of a trigonometric identity, linearity, and parts (a) and (e) of Theorem 5.2, we obtain

$$\mathscr{L}\{\sin^2 t\} = \mathscr{L}\left\{\frac{1-\cos 2t}{2}\right\} = \frac{1}{2}\mathscr{L}\{1\} - \frac{1}{2}\mathscr{L}\{\cos 2t\}$$

$$= \frac{1}{2}\cdot\frac{1}{s} - \frac{1}{2}\cdot\frac{s}{s^2+4}$$

$$= \frac{2}{s(s^2+4)}.$$

EXERCISES 5.1

Answers to odd-numbered problems begin on page AN-23.

In Problems 1–18 use Definition 5.1 to find $\mathscr{L}\{f(t)\}$.

1. $f(t) = \begin{cases} -1, & 0 \le t < 1 \\ 1, & t \ge 1 \end{cases}$

2. $f(t) = \begin{cases} 4, & 0 \le t < 2 \\ 0, & t \ge 2 \end{cases}$

3. $f(t) = \begin{cases} t, & 0 \le t < 1 \\ 1, & t \ge 1 \end{cases}$

4. $f(t) = \begin{cases} 2t+1, & 0 \le t < 1 \\ 0, & t \ge 1 \end{cases}$

5. $f(t) = \begin{cases} \sin t, & 0 \le t < \pi \\ 0, & t \ge \pi \end{cases}$

6. $f(t) = \begin{cases} 0, & 0 \le t < \pi/2 \\ \cos t, & t \ge \pi/2 \end{cases}$

7.

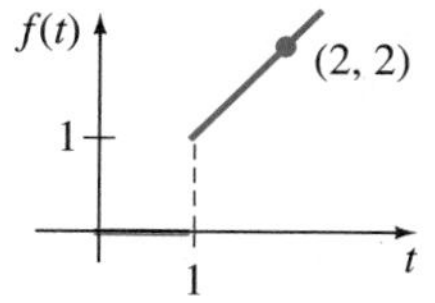

FIGURE 5.7

8.

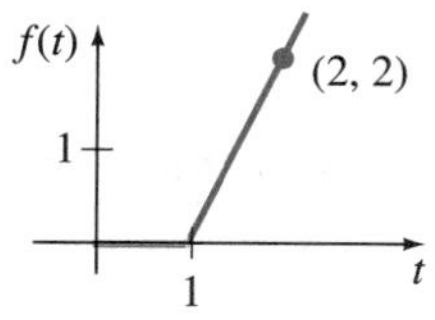

FIGURE 5.8

9.

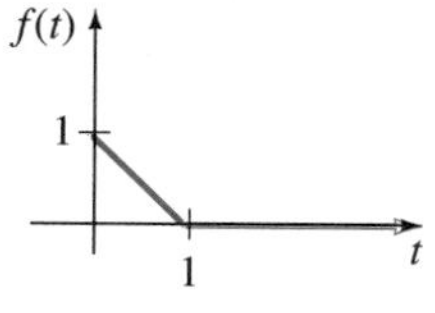

FIGURE 5.9

10.

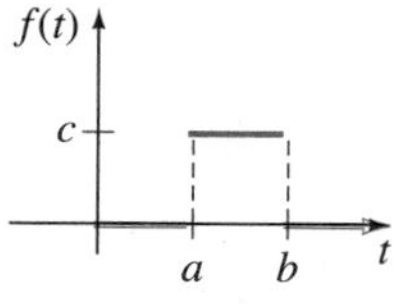

FIGURE 5.10

11. $f(t) = e^{t+7}$
12. $f(t) = e^{-2t-5}$
13. $f(t) = te^{4t}$
14. $f(t) = t^2e^{3t}$
15. $f(t) = e^{-t}\sin t$
16. $f(t) = e^{t}\cos t$
17. $f(t) = t\cos t$
18. $f(t) = t\sin t$

In Problems 19–38 use Theorem 5.2 to find $\mathscr{L}\{f(t)\}$.

19. $f(t) = 2t^4$
20. $f(t) = t^5$
21. $f(t) = 4t - 10$
22. $f(t) = 7t + 3$
23. $f(t) = t^2 + 6t - 3$
24. $f(t) = -4t^2 + 16t + 9$
25. $f(t) = (t+1)^3$
26. $f(t) = (2t-1)^3$
27. $f(t) = 1 + e^{4t}$
28. $f(t) = t^2 - e^{-9t} + 5$
29. $f(t) = (1 + e^{2t})^2$
30. $f(t) = (e^t - e^{-t})^2$
31. $f(t) = 4t^2 - 5\sin 3t$
32. $f(t) = \cos 5t + \sin 2t$
33. $f(t) = \sinh kt$
34. $f(t) = \cosh kt$
35. $f(t) = e^t \sinh t$
36. $f(t) = e^{-t}\cosh t$
37. $f(t) = \sin 2t \cos 2t$
38. $f(t) = \cos^2 t$

39. *WRITING PROJECT* Write a report on the life and contributions to mathematics and science of **Pierre Simon Marquis de Laplace**. Since it is generally conceded that Laplace did not ''invent'' the Laplace transform, explain in your report why his name is associated with it and which mathematician is generally thought to have invented the Laplace transform.

40. PROJECT PROBLEM A function is often defined by means of an integral. The **Bessel function** of order 0 (see Section 6.4), denoted by $J_0(t)$, can be defined by

$$J_0(t) = \frac{1}{\pi}\int_0^{\pi} \cos(t \sin x)\, dx.$$

(a) Make any necessary assumptions to show that

$$\mathscr{L}\{J_0(t)\} = \frac{s}{\pi}\int_0^{\pi} \frac{1}{s^2 + \sin^2 x}\, dx.$$

(b) Evaluate the integral in part (a). If necessary, use a CAS as an aid.

41. PROJECT PROBLEM Find a function f that is not of exponential order but yet possesses a Laplace transform. Prove your assertion.

42. PROJECT PROBLEM Every student of mathematics should know something about the **gamma function**. One definition of this function is given by the improper integral

$$\Gamma(\alpha) = \int_0^{\infty} t^{\alpha - 1} e^{-t}\, dt, \quad \alpha > 0.$$

(a) Show that $\Gamma(\alpha + 1) = \alpha\Gamma(\alpha)$.

(b) Show that $\Gamma(1) = 1$. Use the property in (a) and the last result as a basis for discussing why the gamma function is known as the ''generalized factorial function.''

(c) Show that $\Gamma(\frac{1}{2}) = \sqrt{\pi}$.

(d) Using Parts (a)–(c), give what you consider to be a reasonable definition of $(\frac{1}{2})!$

(e) Show that for $\alpha > -1$, $\mathscr{L}\{t^{\alpha}\} = \dfrac{\Gamma(\alpha + 1)}{s^{\alpha + 1}}$.

(f) Find the Laplace transforms of $f(t) = t^{-1/2}$, $f(t) = t^{1/2}$, and $f(t) = t^{3/2}$.

43. PROJECT PROBLEM

(a) Look up the definition of the equality of two complex numbers.

(b) Assume (or prove it at your instructor's direction) that the result in part (c) of Theorem 5.2 holds for the pure imaginary number $a = ikt$, where $i^2 = -1$ and k is real. Show how $\mathscr{L}\{e^{ikt}\}$ can be used to find $\mathscr{L}\{\cos kt\}$ and $\mathscr{L}\{\sin kt\}$.

44. PROJECT PROBLEM Show that the function $f(t) = 1/t^2$ does not possess a Laplace transform.

5.2 INVERSE TRANSFORM

In the preceding section we were concerned with the problem of transforming a function $f(t)$ into another function $F(s)$ by means of the integral $\int_0^{\infty} e^{-st} f(t)\, dt$. We denoted this symbolically by $\mathscr{L}\{f(t)\} = F(s)$. We now turn the problem around; namely, given $F(s)$, find the function $f(t)$ corresponding to this transform. We say $f(t)$ is the **inverse Laplace transform** of $F(s)$ and write

$$f(t) = \mathscr{L}^{-1}\{F(s)\}.$$

The analogue of Theorem 5.2 for the inverse transform is Theorem 5.3.

THEOREM 5.3 Some Inverse Transforms

(a) $1 = \mathscr{L}^{-1}\left\{\dfrac{1}{s}\right\}$

(b) $t^n = \mathscr{L}^{-1}\left\{\dfrac{n!}{s^{n+1}}\right\}, \quad n = 1, 2, 3, \ldots$

(c) $e^{at} = \mathscr{L}^{-1}\left\{\dfrac{1}{s - a}\right\}$

(d) $\sin kt = \mathscr{L}^{-1}\left\{\dfrac{k}{s^2 + k^2}\right\}$

(e) $\cos kt = \mathscr{L}^{-1}\left\{\dfrac{s}{s^2+k^2}\right\}$ **(f)** $\sinh kt = \mathscr{L}^{-1}\left\{\dfrac{k}{s^2-k^2}\right\}$

(g) $\cosh kt = \mathscr{L}^{-1}\left\{\dfrac{s}{s^2-k^2}\right\}$

$\mathscr{L}^{-1}$ is a Linear Transform

We assume that the inverse Laplace transform is itself a linear transform; that is, for constants α and β,

$$\mathscr{L}^{-1}\{\alpha F(s) + \beta G(s)\} = \alpha\mathscr{L}^{-1}\{F(s)\} + \beta\mathscr{L}^{-1}\{G(s)\},$$

where F and G are the transforms of some functions f and g.

The inverse Laplace transform of a function $F(s)$ may not be unique. It is possible that $\mathscr{L}\{f_1(t)\} = \mathscr{L}\{f_2(t)\}$ and yet $f_1 \neq f_2$, but for our purposes this is not anything to be concerned about. If f_1 and f_2 are piecewise continuous on $[0, \infty)$ and of exponential order for $t > 0$ and if $\mathscr{L}\{f_1(t)\} = \mathscr{L}\{f_2(t)\}$, then the functions f_1 and f_2 are *essentially* the same. See Problem 35 in Exercises 5.2. However, if f_1 and f_2 are continuous on $[0, \infty)$ and $\mathscr{L}\{f_1(t)\} = \mathscr{L}\{f_2(t)\}$, then $f_1 = f_2$ on the interval.

EXAMPLE 1 Applying Theorem 5.3

Evaluate $\mathscr{L}^{-1}\left\{\dfrac{1}{s^5}\right\}$.

Solution To match the form given in part (b) of Theorem 5.3, we identify $n = 4$ and then multiply and divide by 4!. It follows that

$$\mathscr{L}^{-1}\left\{\frac{1}{s^5}\right\} = \frac{1}{4!}\mathscr{L}^{-1}\left\{\frac{4!}{s^5}\right\} = \frac{1}{24}t^4.$$

EXAMPLE 2 Applying Theorem 5.3

Evaluate $\mathscr{L}^{-1}\left\{\dfrac{1}{s^2+64}\right\}$.

Solution Since $k^2 = 64$ we fix up the expression by multiplying and dividing by 8. From part (d) of Theorem 5.3,

$$\mathscr{L}^{-1}\left\{\frac{1}{s^2+64}\right\} = \frac{1}{8}\mathscr{L}^{-1}\left\{\frac{8}{s^2+64}\right\} = \frac{1}{8}\sin 8t.$$

EXAMPLE 3 Termwise Division and Linearity

Evaluate $\mathscr{L}^{-1}\left\{\dfrac{3s+5}{s^2+7}\right\}$.

Solution The given function of s can be written as two expressions by means of termwise division:

$$\frac{3s+5}{s^2+7} = \frac{3s}{s^2+7} + \frac{5}{s^2+7}.$$

From the linearity property of the inverse transform and parts (e) and (d) of Theorem 5.3, we then have

$$\mathscr{L}^{-1}\left\{\frac{3s+5}{s^2+7}\right\} = 3\mathscr{L}^{-1}\left\{\frac{s}{s^2+7}\right\} + \frac{5}{\sqrt{7}}\mathscr{L}^{-1}\left\{\frac{\sqrt{7}}{s^2+7}\right\}$$

$$= 3\cos\sqrt{7}t + \frac{5}{\sqrt{7}}\sin\sqrt{7}t.$$

Partial Fractions

Partial fractions play an important role in finding inverse Laplace transforms. As mentioned in Section 2.2, this fraction decomposition can be done quickly by means of a single command on some computer algebra systems. Indeed, some CASs have packages that implement Laplace transform and inverse Laplace transform commands. But for those of you without access to such software, in the next three examples we review the basic algebra in three cases of partial fraction decomposition. For example, the denominators of

$$\textbf{(i)}\ F(s) = \frac{1}{(s-1)(s+2)(s+4)} \quad \textbf{(ii)}\ F(s) = \frac{s+1}{s^2(s+2)^3} \quad \textbf{(iii)}\ F(s) = \frac{3s-2}{s^3(s^2+4)}$$

contain, respectively, distinct linear factors, repeated linear factors, and a quadratic expression with no real factors. You should consult a calculus text for a more complete review of this theory.

EXAMPLE 4 Partial Fractions and Linearity

Evaluate $\mathscr{L}^{-1}\left\{\dfrac{1}{(s-1)(s+2)(s+4)}\right\}$.

Solution There exist unique constants A, B, and C so that

$$\frac{1}{(s-1)(s+2)(s+4)} = \frac{A}{s-1} + \frac{B}{s+2} + \frac{C}{s+4}$$

$$= \frac{A(s+2)(s+4) + B(s-1)(s+4) + C(s-1)(s+2)}{(s-1)(s+2)(s+4)}.$$

Since the denominators are identical, the numerators are identical:

$$1 = A(s+2)(s+4) + B(s-1)(s+4) + C(s-1)(s+2).$$

By comparing coefficients of powers of s on both sides of the equality, we know that the last equation is equivalent to a system of three equations in the three unknowns A, B, and C. However, you might recall the following shortcut for determining these unknowns. If we set $s = 1$, $s = -2$, and $s = -4$, the zeros of the common denominator $(s-1)(s+2)(s+4)$, we obtain, in turn,

$$1 = A(3)(5), \qquad A = 1/15,$$

$$1 = B(-3)(2), \qquad B = -1/6,$$

$$1 = C(-5)(-2), \qquad C = 1/10.$$

Hence we can write

$$\frac{1}{(s-1)(s+2)(s+4)} = \frac{1/15}{s-1} - \frac{1/6}{s+2} + \frac{1/10}{s+4}$$

and thus, from part (c) of Theorem 5.3,

$$\mathscr{L}^{-1}\left\{\frac{1}{(s-1)(s+2)(s+4)}\right\} = \frac{1}{15}\mathscr{L}^{-1}\left\{\frac{1}{s-1}\right\} - \frac{1}{6}\mathscr{L}^{-1}\left\{\frac{1}{s+2}\right\} + \frac{1}{10}\mathscr{L}^{-1}\left\{\frac{1}{s+4}\right\}$$

$$= \frac{1}{15}e^{t} - \frac{1}{6}e^{-2t} + \frac{1}{10}e^{-4t}.$$

EXAMPLE 5 Partial Fractions and Linearity

Evaluate $\mathscr{L}^{-1}\left\{\dfrac{s+1}{s^2(s+2)^3}\right\}$.

Solution Assume

$$\frac{s+1}{s^2(s+2)^3} = \frac{A}{s} + \frac{B}{s^2} + \frac{C}{s+2} + \frac{D}{(s+2)^2} + \frac{E}{(s+2)^3}$$

so that

$$s + 1 = As(s+2)^3 + B(s+2)^3 + Cs^2(s+2)^2 + Ds^2(s+2) + Es^2.$$

Setting $s = 0$ and $s = -2$ gives $B = 1/8$ and $E = -1/4$, respectively. By equating the coefficients of s^4, s^3, and s, we obtain

$$0 = A + C, \quad 0 = 6A + B + 4C + D, \quad 1 = 8A + 12B,$$

from which it follows that $A = -1/16$, $C = 1/16$, and $D = 0$. Hence from parts (a), (b), and (c) of Theorem 5.3,

$$\mathcal{L}^{-1}\left\{\frac{s+1}{s^2(s+2)^3}\right\} = \mathcal{L}^{-1}\left\{-\frac{1/16}{s} + \frac{1/8}{s^2} + \frac{1/16}{s+2} - \frac{1/4}{(s+2)^3}\right\}$$

$$= -\frac{1}{16}\mathcal{L}^{-1}\left\{\frac{1}{s}\right\} + \frac{1}{8}\mathcal{L}^{-1}\left\{\frac{1}{s^2}\right\} + \frac{1}{16}\mathcal{L}^{-1}\left\{\frac{1}{s+2}\right\} - \frac{1}{8}\mathcal{L}^{-1}\left\{\frac{2}{(s+2)^3}\right\}$$

$$= -\frac{1}{16} + \frac{1}{8}t + \frac{1}{16}e^{-2t} - \frac{1}{8}t^2e^{-2t}.$$

Here we have also used $\mathcal{L}^{-1}\{2/(s+2)^3\} = t^2e^{-2t}$ from Example 6 of Section 5.1. ●

EXAMPLE 6 Partial Fractions and Linearity

Evaluate $\mathcal{L}^{-1}\left\{\dfrac{3s-2}{s^3(s^2+4)}\right\}$.

Solution Assume

$$\frac{3s-2}{s^3(s^2+4)} = \frac{A}{s} + \frac{B}{s^2} + \frac{C}{s^3} + \frac{Ds+E}{s^2+4}$$

so that

$$3s - 2 = As^2(s^2+4) + Bs(s^2+4) + C(s^2+4) + (Ds+E)s^3.$$

Setting $s = 0$ gives immediately $C = -1/2$. Now the coefficients of s^4, s^3, s^2, and s are, respectively,

$$0 = A + D, \quad 0 = B + E, \quad 0 = 4A + C, \quad 3 = 4B,$$

from which we obtain $B = 3/4$, $E = -3/4$, $A = 1/8$, and $D = -1/8$. Therefore from parts (a), (b), (e), and (d) of Theorem 5.3, we have

$$\mathcal{L}^{-1}\left\{\frac{3s-2}{s^3(s^2+4)}\right\} = \mathcal{L}^{-1}\left\{\frac{1/8}{s} + \frac{3/4}{s^2} - \frac{1/2}{s^3} + \frac{-s/8 - 3/4}{s^2+4}\right\}$$

$$= \frac{1}{8}\mathcal{L}^{-1}\left\{\frac{1}{s}\right\} + \frac{3}{4}\mathcal{L}^{-1}\left\{\frac{1}{s^2}\right\} - \frac{1}{4}\mathcal{L}^{-1}\left\{\frac{2}{s^3}\right\}$$

$$- \frac{1}{8}\mathcal{L}^{-1}\left\{\frac{s}{s^2+4}\right\} - \frac{3}{8}\mathcal{L}^{-1}\left\{\frac{2}{s^2+4}\right\}$$

$$= \frac{1}{8} + \frac{3}{4}t - \frac{1}{4}t^2 - \frac{1}{8}\cos 2t - \frac{3}{8}\sin 2t.$$ ●

As the next theorem indicates, not every arbitrary function of s is a Laplace transform of a piecewise continuous function of exponential order.

THEOREM 5.4 Behavior of $F(s)$ as $s \to \infty$

If $f(t)$ is piecewise continuous on $[0, \infty)$ and of exponential order for $t > T$, then $\lim_{s\to\infty} \mathscr{L}\{f(t)\} = 0$.

Proof Since $f(t)$ is piecewise continuous on $0 \le t \le T$, it is necessarily bounded on the interval. That is, $|f(t)| \le M_1 = M_1 e^{0t}$. Also, $|f(t)| \le M_2 e^{\gamma t}$ for $t > T$. If M denotes the maximum of $\{M_1, M_2\}$ and c denotes the maximum of $\{0, \gamma\}$, then

$$|\mathscr{L}\{f(t)\}| \le \int_0^\infty e^{-st}|f(t)|\,dt \le M\int_0^\infty e^{-st}\cdot e^{ct}\,dt = -M\frac{e^{-(s-c)t}}{s-c}\bigg|_0^\infty = \frac{M}{s-c}$$

for $s > c$. As $s \to \infty$, we have $|\mathscr{L}\{f(t)\}| \to 0$ and so $\mathscr{L}\{f(t)\} \to 0$. ○

In view of Theorem 5.4, we can say that $F_1(s) = 1$ and $F_2(s) = s/(s+1)$ are not the Laplace transforms of piecewise continuous functions of exponential order since $F_1(s) \not\to 0$ and $F_2(s) \not\to 0$ as $s \to \infty$. You should not conclude from this, for example, that $\mathscr{L}^{-1}\{F_1(s)\}$ does not exist. There are other kinds of functions.

EXERCISES 5.2

Answers to odd-numbered problems begin on page AN-24.

In Problems 1–34 use Theorem 5.3 to find the given inverse transform.

1. $\mathscr{L}^{-1}\left\{\frac{1}{s^3}\right\}$
2. $\mathscr{L}^{-1}\left\{\frac{1}{s^4}\right\}$
3. $\mathscr{L}^{-1}\left\{\frac{1}{s^2} - \frac{48}{s^5}\right\}$
4. $\mathscr{L}^{-1}\left\{\left(\frac{2}{s} - \frac{1}{s^3}\right)^2\right\}$
5. $\mathscr{L}^{-1}\left\{\frac{(s+1)^3}{s^4}\right\}$
6. $\mathscr{L}^{-1}\left\{\frac{(s+2)^2}{s^3}\right\}$
7. $\mathscr{L}^{-1}\left\{\frac{1}{s^2} - \frac{1}{s} + \frac{1}{s-2}\right\}$
8. $\mathscr{L}^{-1}\left\{\frac{4}{s} + \frac{6}{s^5} - \frac{1}{s+8}\right\}$
9. $\mathscr{L}^{-1}\left\{\frac{1}{4s+1}\right\}$
10. $\mathscr{L}^{-1}\left\{\frac{1}{5s-2}\right\}$
11. $\mathscr{L}^{-1}\left\{\frac{5}{s^2+49}\right\}$
12. $\mathscr{L}^{-1}\left\{\frac{10s}{s^2+16}\right\}$
13. $\mathscr{L}^{-1}\left\{\frac{4s}{4s^2+1}\right\}$
14. $\mathscr{L}^{-1}\left\{\frac{1}{4s^2+1}\right\}$
15. $\mathscr{L}^{-1}\left\{\frac{1}{s^2-16}\right\}$
16. $\mathscr{L}^{-1}\left\{\frac{10s}{s^2-25}\right\}$
17. $\mathscr{L}^{-1}\left\{\frac{2s-6}{s^2+9}\right\}$
18. $\mathscr{L}^{-1}\left\{\frac{s-1}{s^2+2}\right\}$
19. $\mathscr{L}^{-1}\left\{\frac{1}{s^2+3s}\right\}$
20. $\mathscr{L}^{-1}\left\{\frac{s+1}{s^2-4s}\right\}$
21. $\mathscr{L}^{-1}\left\{\frac{s}{s^2+2s-3}\right\}$
22. $\mathscr{L}^{-1}\left\{\frac{1}{s^2+s-20}\right\}$
23. $\mathscr{L}^{-1}\left\{\frac{0.9s}{(s-0.1)(s+0.2)}\right\}$
24. $\mathscr{L}^{-1}\left\{\frac{s-3}{(s-\sqrt{3})(s+\sqrt{3})}\right\}$
25. $\mathscr{L}^{-1}\left\{\frac{s}{(s-2)(s-3)(s-6)}\right\}$
26. $\mathscr{L}^{-1}\left\{\frac{s^2+1}{s(s-1)(s+1)(s-2)}\right\}$
27. $\mathscr{L}^{-1}\left\{\frac{2s+4}{(s-2)(s^2+4s+3)}\right\}$
28. $\mathscr{L}^{-1}\left\{\frac{s+1}{(s^2-4s)(s+5)}\right\}$
29. $\mathscr{L}^{-1}\left\{\frac{1}{s^2(s^2+4)}\right\}$
30. $\mathscr{L}^{-1}\left\{\frac{s-1}{s^2(s^2+1)}\right\}$
31. $\mathscr{L}^{-1}\left\{\frac{s}{(s^2+4)(s+2)}\right\}$
32. $\mathscr{L}^{-1}\left\{\frac{1}{s^4-9}\right\}$

33. $\mathscr{L}^{-1}\left\{\dfrac{1}{(s^2+1)(s^2+4)}\right\}$

34. $\mathscr{L}^{-1}\left\{\dfrac{6s+3}{(s^2+1)(s^2+4)}\right\}$

35. *WRITING PROJECT* Write a short report stating and elaborating on **Lerch's theorem**. Include examples in your report.

36. *WRITING PROJECT* The inverse Laplace transform $\mathscr{L}^{-1}\{F(s)\}$ is not just a symbol but is another integral, the evaluation of which demands the use of complex variables. Write a short report on the integral that defines the inverse Laplace transform. If possible, illustrate the evaluation of this integral by means of one or two examples. In your report elaborate on, and illustrate with other transforms, the general concept of **integral transform pairs**.

5.3 TRANSLATION THEOREMS AND DERIVATIVES OF A TRANSFORM

It is not convenient to use Definition 5.1 each time we wish to find the Laplace transform of a function $f(t)$. For example, the integration by parts involved in evaluating, say, $\mathscr{L}\{e^t t^2 \sin 3t\}$ is formidable to say the least. In the discussion that follows, we present several labor-saving theorems; these, in turn, enable us to build up a more extensive list of transforms without the necessity of using the definition of the Laplace transform. Indeed, we shall see that evaluating transforms such as $\mathscr{L}\{e^{4t}\cos 6t\}$, $\mathscr{L}\{t^3 \sin 2t\}$, and $\mathscr{L}\{t^{10}e^{-t}\}$ is fairly straightforward, provided we know $\mathscr{L}\{\cos 6t\}$, $\mathscr{L}\{\sin 2t\}$, and $\mathscr{L}\{t^{10}\}$, respectively. Though extensive tables can be constructed, and we have included a table in Appendix II, it is nonetheless a good idea to know the Laplace transforms of basic functions such as t^n, e^{at}, $\sin kt$, $\cos kt$, $\sinh kt$, and $\cosh kt$.

If we know $\mathscr{L}\{f(t)\} = F(s)$, we can compute the Laplace transform $\mathscr{L}\{e^{at}f(t)\}$ with no additional effort other than *translating*, or *shifting*, $F(s)$ to $F(s-a)$. This result is known as the **first translation theorem** or **first shifting theorem**.

THEOREM 5.5 First Translation Theorem

If $F(s) = \mathscr{L}\{f(t)\}$ and a is any real number, then

$$\mathscr{L}\{e^{at}f(t)\} = F(s-a).$$

Proof The proof is immediate, since by Definition 5.1,

$$\mathscr{L}\{e^{at}f(t)\} = \int_0^\infty e^{-st}e^{at}f(t)\,dt = \int_0^\infty e^{-(s-a)t}f(t)\,dt = F(s-a). \qquad \bigcirc$$

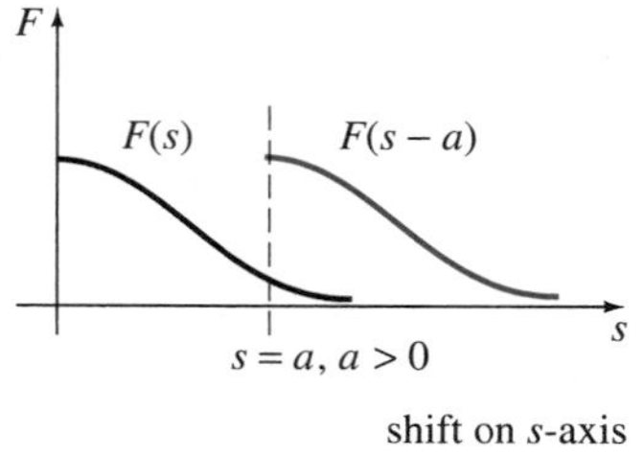

FIGURE 5.11

If we consider s a real variable, then the graph of $F(s-a)$ is the graph of $F(s)$ shifted on the s-axis by the amount $|a|$ units. If $a > 0$, the graph of $F(s)$ is shifted a units to the right, whereas if $a < 0$, the graph is shifted $|a|$ units to the left. See Figure 5.11.

For emphasis it is sometimes useful to use the symbolism

$$\mathscr{L}\{e^{at}f(t)\} = \mathscr{L}\{f(t)\}_{s \to s-a},$$

where $s \to s-a$ means that we replace s in $F(s)$ by $s-a$.

EXAMPLE 1 First Translation Theorem

Evaluate **(a)** $\mathscr{L}\{e^{5t}t^3\}$ and **(b)** $\mathscr{L}\{e^{-2t}\cos 4t\}$.

Solution The results follow from Theorem 5.5.

$$\text{(a)}\ \mathscr{L}\{e^{5t}t^3\} = \mathscr{L}\{t^3\}_{s\to s-5} = \frac{3!}{s^4}\bigg|_{s\to s-5} = \frac{6}{(s-5)^4}.$$

$$\text{(b)}\ \mathscr{L}\{e^{-2t}\cos 4t\} = \mathscr{L}\{\cos 4t\}_{s\to s+2} \quad \leftarrow a = -2 \text{ so } s - a = s - (-2) = s + 2$$

$$= \frac{s}{s^2+16}\bigg|_{s\to s+2} = \frac{s+2}{(s+2)^2+16}.$$

Inverse Form of the First Translation Theorem

If $f(t) = \mathscr{L}^{-1}\{F(s)\}$, the inverse form of Theorem 5.5 is

$$\mathscr{L}^{-1}\{F(s-a)\} = \mathscr{L}^{-1}\{F(s)|_{s\to s-a}\} = e^{at}f(t). \quad \textbf{(1)}$$

EXAMPLE 2 Completing the Square to Find $\mathscr{L}^{-1}$

Evaluate $\mathscr{L}^{-1}\left\{\dfrac{s}{s^2+6s+11}\right\}$.

Solution If $s^2 + 6s + 11$ had real factors, we would use partial fractions. Since this quadratic term does not factor, we complete the square.

$$\mathscr{L}^{-1}\left\{\frac{s}{s^2+6s+11}\right\} = \mathscr{L}^{-1}\left\{\frac{s}{(s+3)^2+2}\right\} \quad \leftarrow \text{completion of square}$$

$$= \mathscr{L}^{-1}\left\{\frac{s+3-3}{(s+3)^2+2}\right\} \quad \leftarrow \text{adding zero in the numerator}$$

$$= \mathscr{L}^{-1}\left\{\frac{s+3}{(s+3)^2+2} - \frac{3}{(s+3)^2+2}\right\} \quad \leftarrow \text{termwise division}$$

$$= \mathscr{L}^{-1}\left\{\frac{s+3}{(s+3)^2+2}\right\} - 3\mathscr{L}^{-1}\left\{\frac{1}{(s+3)^2+2}\right\} \quad \leftarrow \text{linearity of } \mathscr{L}^{-1}$$

$$= \mathscr{L}^{-1}\left\{\frac{s}{s^2+2}\bigg|_{s\to s+3}\right\} - \frac{3}{\sqrt{2}}\mathscr{L}^{-1}\left\{\frac{\sqrt{2}}{s^2+2}\bigg|_{s\to s+3}\right\}$$

$$= e^{-3t}\cos\sqrt{2}t - \frac{3}{\sqrt{2}}e^{-3t}\sin\sqrt{2}t. \quad \leftarrow \text{from (1) and Theorem 5.3}$$

EXAMPLE 3 Completing the Square and Linearity

Evaluate $\mathscr{L}^{-1}\left\{\dfrac{1}{(s-1)^3}+\dfrac{1}{s^2+2s-8}\right\}$.

Solution Completing the square in the second denominator and using linearity yield

$$\begin{aligned}\mathscr{L}^{-1}\left\{\frac{1}{(s-1)^3}+\frac{1}{s^2+2s-8}\right\} &= \mathscr{L}^{-1}\left\{\frac{1}{(s-1)^3}+\frac{1}{(s+1)^2-9}\right\}\\ &= \frac{1}{2!}\mathscr{L}^{-1}\left\{\frac{2!}{(s-1)^3}\right\}+\frac{1}{3}\mathscr{L}^{-1}\left\{\frac{3}{(s+1)^2-9}\right\}\\ &= \frac{1}{2!}\mathscr{L}^{-1}\left\{\left.\frac{2!}{s^3}\right|_{s\to s-1}\right\}+\frac{1}{3}\mathscr{L}^{-1}\left\{\left.\frac{3}{s^2-9}\right|_{s\to s+1}\right\}\\ &= \frac{1}{2}e^t t^2+\frac{1}{3}e^{-t}\sinh 3t.\end{aligned}$$

Unit Step Function

In engineering one frequently encounters functions that can be either "on" or "off." For example, an external force acting on a mechanical system or a voltage impressed on a circuit can be turned off after a period of time. It is thus convenient to define a special function called the **unit step function**.

DEFINITION 5.3 Unit Step Function

The function $\mathscr{U}(t-a)$ is defined to be

$$\mathscr{U}(t-a)=\begin{cases}0, & 0\le t<a\\ 1, & t\ge a.\end{cases}$$

Notice that we define $\mathscr{U}(t-a)$ only on the nonnegative t-axis since this is all that we are concerned with in the study of the Laplace transform. In a broader sense, $\mathscr{U}(t-a)=0$ for $t<a$.

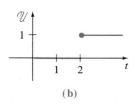

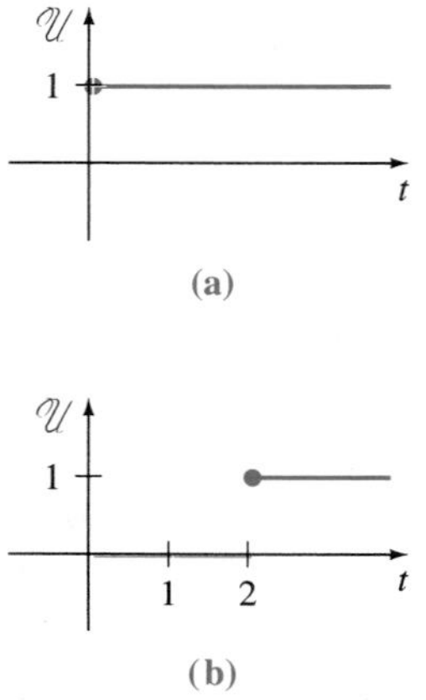

FIGURE 5.12

EXAMPLE 4 Graphs of Unit Step Functions

Graph **(a)** $\mathscr{U}(t)$ and **(b)** $\mathscr{U}(t-2)$.

Solution **(a)** $\mathscr{U}(t)=1,\quad t\ge 0$ **(b)** $\mathscr{U}(t-2)=\begin{cases}0, & 0\le t<2\\ 1, & t\ge 2.\end{cases}$

The respective graphs are given in Figure 5.12.

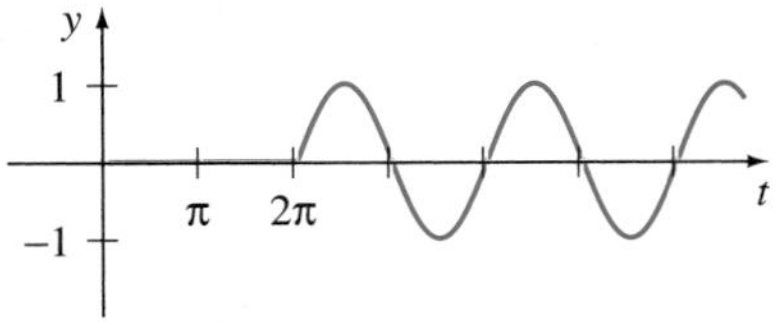

FIGURE 5.13

When multiplied by another function defined for $t \geq 0$, the unit step function "turns off" a portion of the graph of the function. For example, Figure 5.13 illustrates the graph of $\sin t$, $t \geq 0$, when multiplied by $\mathcal{U}(t - 2\pi)$:

$$f(t) = \sin t\,\mathcal{U}(t - 2\pi) = \begin{cases} 0, & 0 \leq t < 2\pi \\ \sin t, & t \geq 2\pi. \end{cases}$$

The unit step function can also be used to write piecewise-defined functions in a compact form. For instance, the piecewise-defined function

$$f(t) = \begin{cases} g(t), & 0 \leq t < a \\ h(t), & t \geq a \end{cases} \tag{2}$$

is the same as $f(t) = g(t) - g(t)\mathcal{U}(t - a) + h(t)\mathcal{U}(t - a)$. **(3)**

Similarly, a function of the type

$$f(t) = \begin{cases} 0, & 0 \leq t < a \\ g(t), & a \leq t < b \\ 0, & t \geq b \end{cases} \tag{4}$$

can be written $\quad f(t) = g(t)[\mathcal{U}(t - a) - \mathcal{U}(t - b)].$ **(5)**

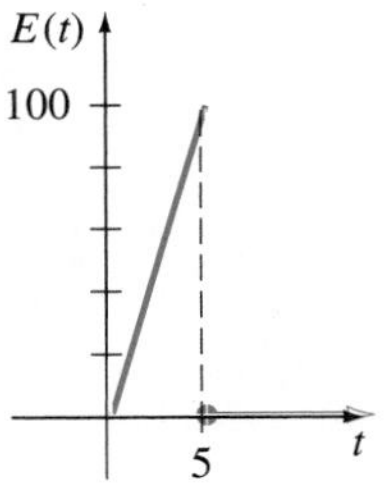

FIGURE 5.14

EXAMPLE 5 Function Expressed in Terms of a Unit Step Function

The voltage in a circuit is given by $E(t) = \begin{cases} 20t & 0 < t < 5 \\ 0, & t \geq 5 \end{cases}$. Graph $E(t)$. Express $E(t)$ in terms of unit step functions.

Solution The graph of this piecewise-defined function is given in Figure 5.14. Now from (2) and (3) with $g(t) = 20t$ and $h(t) = 0$, we get

$$E(t) = 20t - 20t\mathcal{U}(t - 5).$$

EXAMPLE 6 Comparison of Functions

Consider the function $y = f(t)$ defined by $f(t) = t^3$. Compare the graphs of **(a)** $f(t)$, $-\infty < t < \infty$, **(b)** $f(t)$, $t \geq 0$, **(c)** $f(t - 2)$, $t \geq 0$, **(d)** $f(t - 2)\ \mathcal{U}(t - 2)$, $t \geq 0$

Solution The respective graphs are given in Figure 5.15 on page 266.

In general, if $a > 0$, then the graph of $y = f(t - a)$ is the graph of $y = f(t)$, $t \geq 0$, shifted a units to the right on the t-axis. However, when $y = f(t - a)$ is multiplied by the unit step function $\mathcal{U}(t - a)$ in the manner illustrated in part (d) of Example 6, then the graph of the function

$$y = f(t - a)\mathcal{U}(t - a) \tag{6}$$

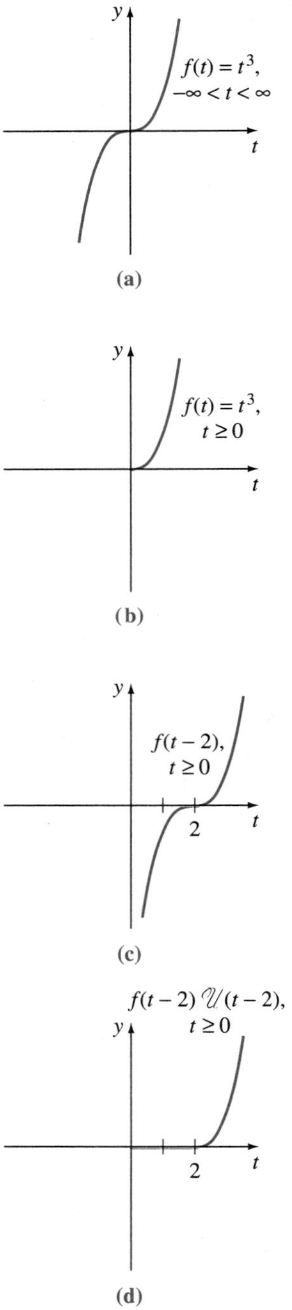

FIGURE 5.15

coincides with the graph of $y = f(t - a)$ for $t \geq a$ but is identically zero for $0 \leq t < a$. See Figure 5.16.

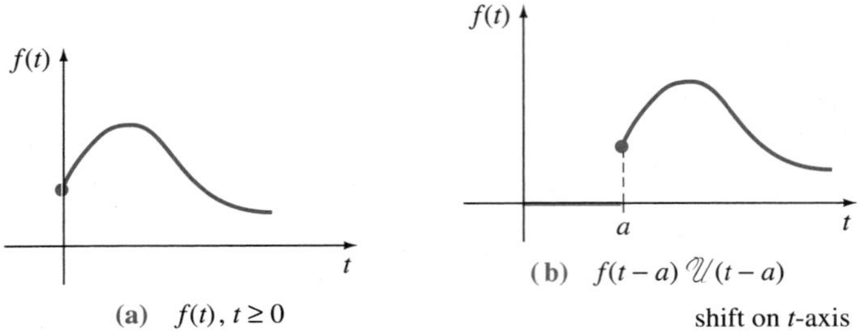

FIGURE 5.16

We saw in Theorem 5.5 that an exponential multiple of $f(t)$ results in a translation or shift of the transform $F(s)$ on the s-axis. In the next theorem we see that whenever $F(s)$ is multiplied by an appropriate exponential function, the inverse transform of this product is the shifted function given in (6). This result is called the **second translation theorem**, or **second shifting theorem**.

THEOREM 5.6 Second Translation Theorem

If $F(s) = \mathscr{L}\{f(t)\}$ and $a > 0$, then

$$\mathscr{L}\{f(t-a)\mathscr{U}(t-a)\} = e^{-as}F(s).$$

Proof We express $\int_0^\infty e^{-st}f(t-a)\mathscr{U}(t-a)\,dt$ as the sum of two integrals:

$$\mathscr{L}\{f(t-a)\mathscr{U}(t-a)\} = \int_0^a e^{-st}f(t-a)\underbrace{\mathscr{U}(t-a)}_{\substack{\text{zero for}\\ 0 \leq t < a}}\,dt + \int_a^\infty e^{-st}f(t-a)\underbrace{\mathscr{U}(t-a)}_{\substack{\text{one for}\\ t \geq a}}\,dt$$

$$= \int_a^\infty e^{-st}f(t-a)\,dt.$$

Now let $v = t - a$, $dv = dt$; then

$$\mathscr{L}\{f(t-a)\mathscr{U}(t-a)\} = \int_0^\infty e^{-s(v+a)}f(v)\,dv$$

$$= e^{-as}\int_0^\infty e^{-sv}f(v)\,dv = e^{-as}\mathscr{L}\{f(t)\}. \qquad \circ$$

EXAMPLE 7 Second Translation Theorem

Evaluate $\mathscr{L}\{(t-2)^3\mathscr{U}(t-2)\}$.

Solution With the identification $a = 2$, it follows from Theorem 5.6 that

$$\mathscr{L}\{(t-2)^3\mathscr{U}(t-2)\} = e^{-2s}\mathscr{L}\{t^3\} = e^{-2s}\frac{3!}{s^4} = \frac{6}{s^4}e^{-2s}. \qquad \bullet$$

We often wish to find the Laplace transform of just the unit step function. This can be found from either Definition 5.1 or Theorem 5.6. If we identify $f(t) = 1$ in Theorem 5.6, then $f(t - a) = 1$, $F(s) = \mathscr{L}\{1\} = 1/s$, and so

$$\mathscr{L}\{\mathscr{U}(t - a)\} = \frac{e^{-as}}{s}. \tag{7}$$

EXAMPLE 8 Function Expressed in Terms of Unit Step Functions

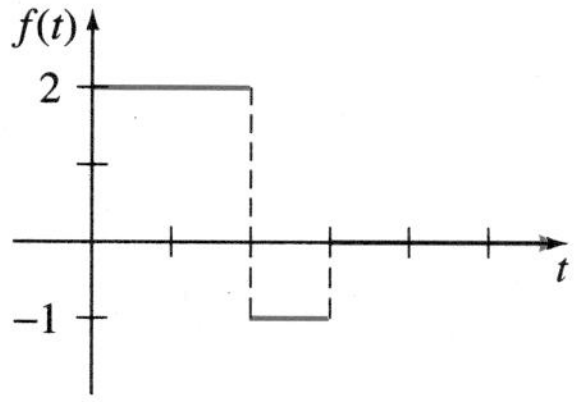

FIGURE 5.17

Find the Laplace transform of the function shown in Figure 5.17.

Solution With the aid of the unit step function, we can write

$$f(t) = 2 - 3\mathscr{U}(t - 2) + \mathscr{U}(t - 3).$$

Using linearity and the result in (7) it follows that

$$\mathscr{L}\{f(t)\} = \mathscr{L}\{2\} - 3\mathscr{L}\{\mathscr{U}(t - 2)\} + \mathscr{L}\{\mathscr{U}(t - 3)\}$$
$$= \frac{2}{s} - 3\frac{e^{-2s}}{s} + \frac{e^{-3s}}{s}.$$

EXAMPLE 9 Fixing Up a Function Before Transforming

Evaluate $\mathscr{L}\{\sin t\,\mathscr{U}(t - 2\pi)\}$.

Solution With $a = 2\pi$ we have from Theorem 5.6

$$\mathscr{L}\{\sin t\,\mathscr{U}(t - 2\pi)\} = \mathscr{L}\{\sin(t - 2\pi)\mathscr{U}(t - 2\pi)\} \quad \leftarrow \text{sin } t \text{ has period } 2\pi$$
$$= e^{-2\pi s}\mathscr{L}\{\sin t\}$$
$$= \frac{e^{-2\pi s}}{s^2 + 1}.$$

EXAMPLE 10 Fixing Up a Function Before Transforming

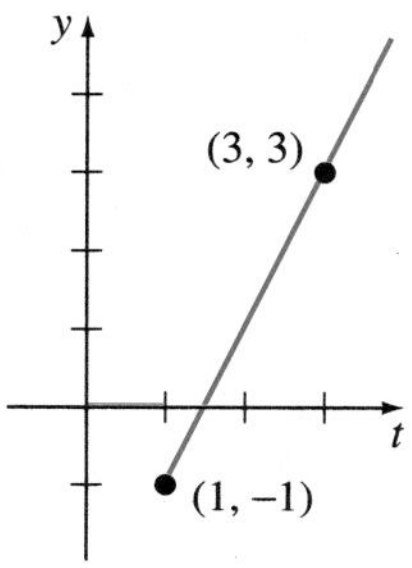

FIGURE 5.18

Find the Laplace transform of the function shown in Figure 5.18.

Solution An equation of the straight line is found to be $y = 2t - 3$. To "turn off" this graph on the interval $0 \le t < 1$ we form $(2t - 3)\mathscr{U}(t - 1)$. Now Theorem 5.6 is not immediately applicable in the evaluation of $\mathscr{L}\{(2t - 3)\,\mathscr{U}(t - 1)\}$, since the function being transformed lacks the precise form $f(t - a)\,\mathscr{U}(t - a)$. However, by rewriting $2t - 3$ in terms of $t - 1$,

$$2t - 3 = 2(t - 1) - 1,$$

we can identify $f(t - 1) = 2(t - 1) - 1$ and consequently $f(t) = 2t - 1$. We are now in a position to apply the second translation theorem:

$$\mathscr{L}\{(2t - 3)\mathscr{U}(t - 1)\} = \mathscr{L}\{(2(t - 1) - 1)\mathscr{U}(t - 1)\}$$
$$= e^{-s}\mathscr{L}\{2t - 1\}$$
$$= e^{-s}\left(\frac{2}{s^2} - \frac{1}{s}\right).$$

Inverse Form of the Second Translation Theorem

If $f(t) = \mathscr{L}^{-1}\{F(s)\}$, the inverse form of Theorem 5.6, $a > 0$, is

$$\mathscr{L}^{-1}\{e^{-as}F(s)\} = f(t-a)\mathscr{U}(t-a). \tag{8}$$

EXAMPLE 11 Inverse by Formula (8)

Evaluate $\mathscr{L}^{-1}\left\{\dfrac{e^{-\pi s/2}}{s^2+9}\right\}$.

Solution We identify $a = \dfrac{\pi}{2}$ and $f(t) = \mathscr{L}^{-1}\left\{\dfrac{1}{s^2+9}\right\} = \dfrac{1}{3}\sin 3t$. Thus from (8),

$$\begin{aligned}\mathscr{L}^{-1}\left\{\frac{e^{-\pi s/2}}{s^2+9}\right\} &= \frac{1}{3}\mathscr{L}^{-1}\left\{\frac{3}{s^2+9}\right\}\Bigg|_{t\to t-\pi/2}\mathscr{U}\left(t-\frac{\pi}{2}\right)\\ &= \frac{1}{3}\sin 3\left(t-\frac{\pi}{2}\right)\mathscr{U}\left(t-\frac{\pi}{2}\right)\\ &= \frac{1}{3}\cos 3t\,\mathscr{U}\left(t-\frac{\pi}{2}\right).\end{aligned}$$

●

If $F(s) = \mathscr{L}\{f(t)\}$ and if we assume that interchanging of differentiation and integration is possible, then

$$\frac{d}{ds}F(s) = \frac{d}{ds}\int_0^\infty e^{-st}f(t)\,dt = \int_0^\infty \frac{\partial}{\partial s}[e^{-st}f(t)]\,dt = -\int_0^\infty e^{-st}tf(t)\,dt = -\mathscr{L}\{tf(t)\},$$

that is,

$$\mathscr{L}\{tf(t)\} = -\frac{d}{ds}\mathscr{L}\{f(t)\}.$$

Similarly

$$\begin{aligned}\mathscr{L}\{t^2f(t)\} &= \mathscr{L}\{t\cdot tf(t)\} = -\frac{d}{ds}\mathscr{L}\{tf(t)\}\\ &= -\frac{d}{ds}\left(-\frac{d}{ds}\mathscr{L}\{f(t)\}\right) = \frac{d^2}{ds^2}\mathscr{L}\{f(t)\}.\end{aligned}$$

The preceding two cases suggest the general result for $\mathscr{L}\{t^nf(t)\}$.

THEOREM 5.7 Derivatives of Transforms

If $F(s) = \mathscr{L}\{f(t)\}$ and $n = 1, 2, 3, \ldots$, then

$$\mathscr{L}\{t^nf(t)\} = (-1)^n\frac{d^n}{ds^n}F(s).$$

EXAMPLE 12 Applying Theorem 5.7

Evaluate **(a)** $\mathscr{L}\{te^{3t}\}$, **(b)** $\mathscr{L}\{t \sin kt\}$, **(c)** $\mathscr{L}\{t^2 \sin kt\}$, and **(d)** $\mathscr{L}\{te^{-t} \cos t\}$.

Solution We make use of results (c), (d), and (e) of Theorem 5.2.

(a) Note in this first example we could also use the first translation theorem. To apply Theorem 5.7 we identify $n = 1$ and $f(t) = e^{3t}$:

$$\mathscr{L}\{te^{3t}\} = -\frac{d}{ds}\mathscr{L}\{e^{3t}\} = -\frac{d}{ds}\left(\frac{1}{s-3}\right) = \frac{1}{(s-3)^2}.$$

(b) $$\mathscr{L}\{t \sin kt\} = -\frac{d}{ds}\mathscr{L}\{\sin kt\} = -\frac{d}{ds}\left(\frac{k}{s^2+k^2}\right) = \frac{2ks}{(s^2+k^2)^2}.$$

(c) With $n = 2$ in Theorem 5.7 this transform can be written

$$\mathscr{L}\{t^2 \sin kt\} = \frac{d^2}{ds^2}\mathscr{L}\{\sin kt\}$$

and so by carrying out the two derivatives we obtain the result. Alternatively, we can make use of the result already obtained in part (b). Since $t^2 \sin kt = t(t \sin kt)$, we have

$$\begin{aligned}\mathscr{L}\{t^2 \sin kt\} &= -\frac{d}{ds}\mathscr{L}\{t \sin kt\}\\ &= -\frac{d}{ds}\left(\frac{2ks}{(s^2+k^2)^2}\right). \qquad \leftarrow \text{from part (b)}\end{aligned}$$

Differentiating and simplifying then give

$$\mathscr{L}\{t^2 \sin kt\} = \frac{6ks^2 - 2k^3}{(s^2+k^2)^3}.$$

(d) $$\begin{aligned}\mathscr{L}\{te^{-t} \cos t\} &= -\frac{d}{ds}\mathscr{L}\{e^{-t} \cos t\}\\ &= -\frac{d}{ds}\mathscr{L}\{\cos t\}_{s \to s+1} \qquad \leftarrow \text{first translation theorem}\\ &= -\frac{d}{ds}\left(\frac{s+1}{(s+1)^2+1}\right)\\ &= \frac{(s+1)^2 - 1}{[(s+1)^2+1]^2}.\end{aligned}$$ ●

EXERCISES 5.3

Answers to odd-numbered problems begin on page AN-24.

In Problems 1–44 find either $F(s)$ or $f(t)$ as indicated.

1. $\mathscr{L}\{te^{10t}\}$
2. $\mathscr{L}\{te^{-6t}\}$
3. $\mathscr{L}\{t^3e^{-2t}\}$
4. $\mathscr{L}\{t^{10}e^{-7t}\}$
5. $\mathscr{L}\{e^t \sin 3t\}$
6. $\mathscr{L}\{e^{-2t} \cos 4t\}$
7. $\mathscr{L}\{e^{5t} \sinh 3t\}$
8. $\mathscr{L}\left\{\dfrac{\cosh t}{e^t}\right\}$
9. $\mathscr{L}\{t(e^t + e^{2t})^2\}$
10. $\mathscr{L}\{e^{2t}(t - 1)^2\}$
11. $\mathscr{L}\{e^{-t} \sin^2 t\}$
12. $\mathscr{L}\{e^t \cos^2 3t\}$
13. $\mathscr{L}^{-1}\left\{\dfrac{1}{(s + 2)^3}\right\}$
14. $\mathscr{L}^{-1}\left\{\dfrac{1}{(s - 1)^4}\right\}$
15. $\mathscr{L}^{-1}\left\{\dfrac{1}{s^2 - 6s + 10}\right\}$
16. $\mathscr{L}^{-1}\left\{\dfrac{1}{s^2 + 2s + 5}\right\}$
17. $\mathscr{L}^{-1}\left\{\dfrac{s}{s^2 + 4s + 5}\right\}$
18. $\mathscr{L}^{-1}\left\{\dfrac{2s + 5}{s^2 + 6s + 34}\right\}$
19. $\mathscr{L}^{-1}\left\{\dfrac{s}{(s + 1)^2}\right\}$
20. $\mathscr{L}^{-1}\left\{\dfrac{5s}{(s - 2)^2}\right\}$
21. $\mathscr{L}^{-1}\left\{\dfrac{2s - 1}{s^2(s + 1)^3}\right\}$
22. $\mathscr{L}^{-1}\left\{\dfrac{(s + 1)^2}{(s + 2)^4}\right\}$
23. $\mathscr{L}\{(t - 1)\mathscr{U}(t - 1)\}$
24. $\mathscr{L}\{e^{2-t}\mathscr{U}(t - 2)\}$
25. $\mathscr{L}\{t\mathscr{U}(t - 2)\}$
26. $\mathscr{L}\{(3t + 1)\mathscr{U}(t - 3)\}$
27. $\mathscr{L}\{\cos 2t\,\mathscr{U}(t - \pi)\}$
28. $\mathscr{L}\left\{\sin t\,\mathscr{U}\left(t - \dfrac{\pi}{2}\right)\right\}$
29. $\mathscr{L}\{(t - 1)^3e^{t-1}\mathscr{U}(t - 1)\}$
30. $\mathscr{L}\{te^{t-5}\mathscr{U}(t - 5)\}$
31. $\mathscr{L}^{-1}\left\{\dfrac{e^{-2s}}{s^3}\right\}$
32. $\mathscr{L}^{-1}\left\{\dfrac{(1 + e^{-2s})^2}{s + 2}\right\}$
33. $\mathscr{L}^{-1}\left\{\dfrac{e^{-\pi s}}{s^2 + 1}\right\}$
34. $\mathscr{L}^{-1}\left\{\dfrac{se^{-\pi s/2}}{s^2 + 4}\right\}$
35. $\mathscr{L}^{-1}\left\{\dfrac{e^{-s}}{s(s + 1)}\right\}$
36. $\mathscr{L}^{-1}\left\{\dfrac{e^{-2s}}{s^2(s - 1)}\right\}$
37. $\mathscr{L}\{t \cos 2t\}$
38. $\mathscr{L}\{t \sinh 3t\}$
39. $\mathscr{L}\{t^2 \sinh t\}$
40. $\mathscr{L}\{t^2 \cos t\}$
41. $\mathscr{L}\{te^{2t} \sin 6t\}$
42. $\mathscr{L}\{te^{-3t} \cos 3t\}$
43. $\mathscr{L}^{-1}\left\{\dfrac{s}{(s^2 + 1)^2}\right\}$
44. $\mathscr{L}^{-1}\left\{\dfrac{s + 1}{(s^2 + 2s + 2)^2}\right\}$

In Problems 45–50 match the given graph with one of the functions in **(a)–(f)**. The graph of $f(t)$ is given in Figure 5.19.

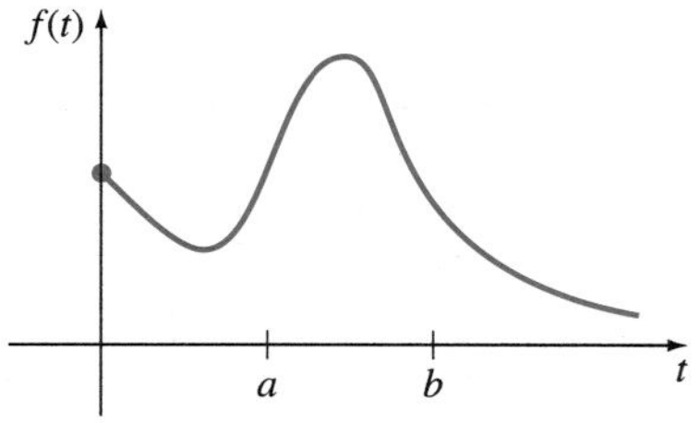

FIGURE 5.19

(a) $f(t) - f(t)\mathscr{U}(t - a)$
(b) $f(t - b)\mathscr{U}(t - b)$
(c) $f(t)\mathscr{U}(t - a)$
(d) $f(t) - f(t)\mathscr{U}(t - b)$
(e) $f(t)\mathscr{U}(t - a) - f(t)\mathscr{U}(t - b)$
(f) $f(t - a)\mathscr{U}(t - a) - f(t - a)\mathscr{U}(t - b)$

45.

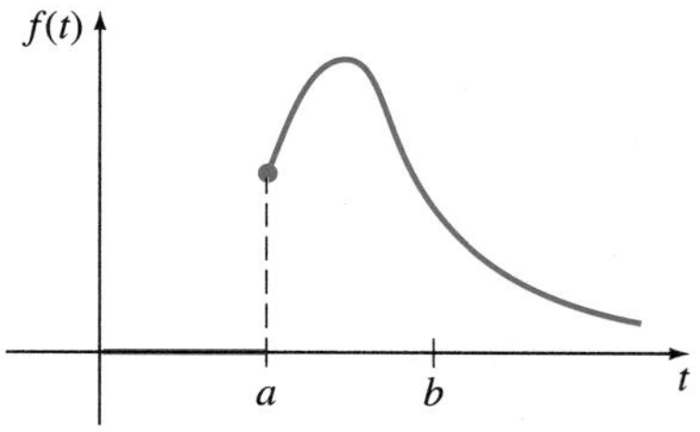

FIGURE 5.20

46.

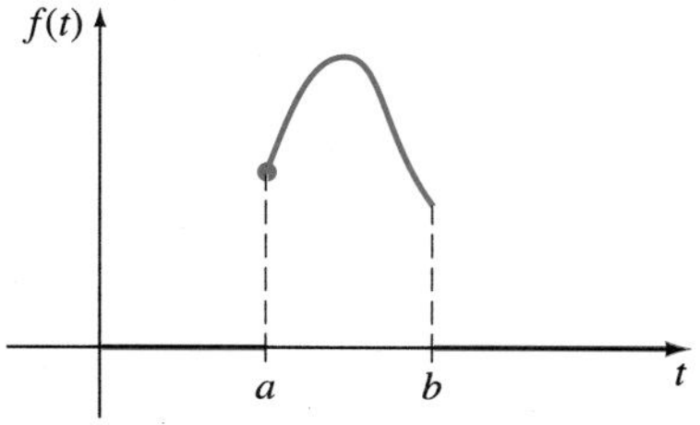

FIGURE 5.21

47.

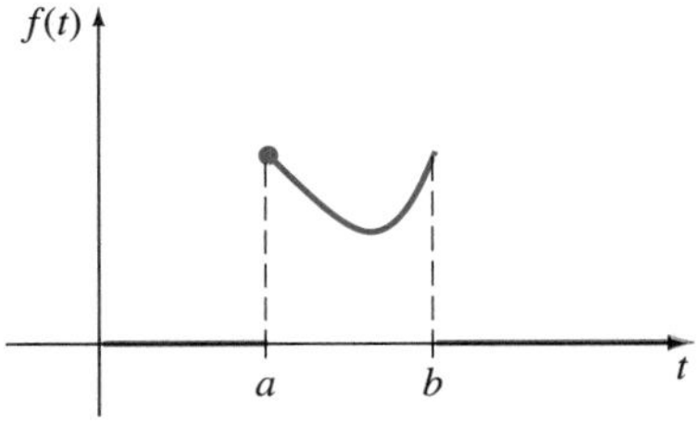

FIGURE 5.22

48.

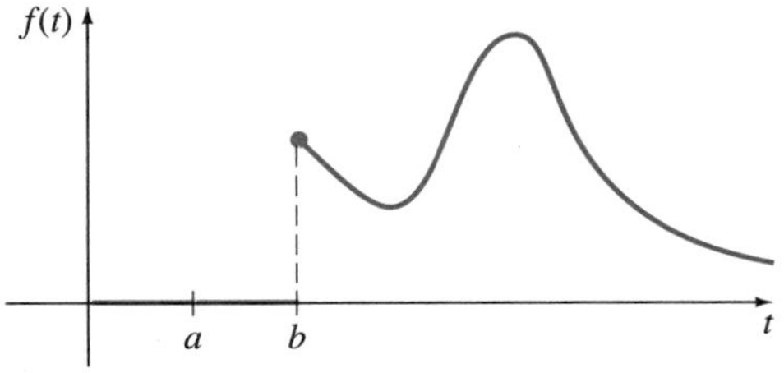

FIGURE 5.23

49.

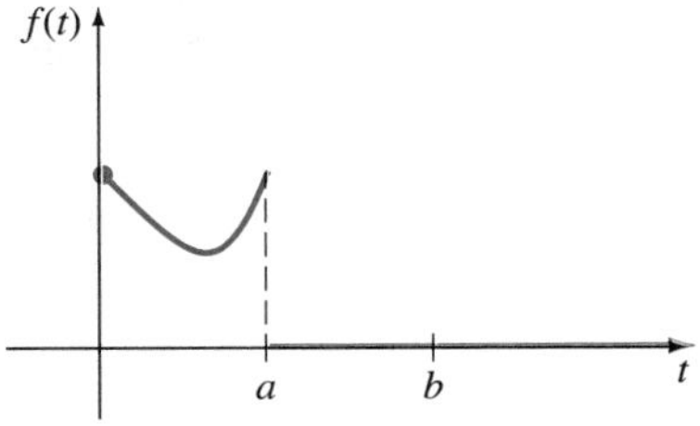

FIGURE 5.24

50.

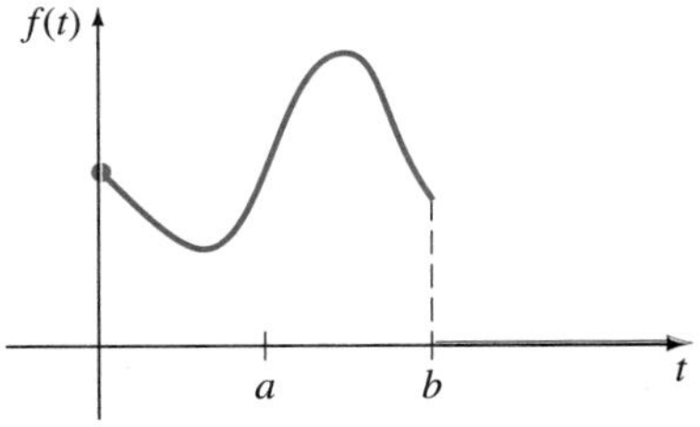

FIGURE 5.25

In Problems 51–58 write each function in terms of unit step functions. Find the Laplace transform of the given function.

51. $f(t) = \begin{cases} 2, & 0 \le t < 3 \\ -2, & t \ge 3 \end{cases}$

52. $f(t) = \begin{cases} 1, & 0 \le t < 4 \\ 0, & 4 \le t < 5 \\ 1, & t \ge 5 \end{cases}$

53. $f(t) = \begin{cases} 0, & 0 \le t < 1 \\ t^2, & t \ge 1 \end{cases}$

54. $f(t) = \begin{cases} 0, & 0 \le t < \dfrac{3\pi}{2} \\ \sin t, & t \ge \dfrac{3\pi}{2} \end{cases}$

55. $f(t) = \begin{cases} t & 0 \le t < 2 \\ 0, & t \ge 2 \end{cases}$

56. $f(t) = \begin{cases} \sin t, & 0 \le t < 2\pi \\ 0, & t \ge 2\pi \end{cases}$

57.

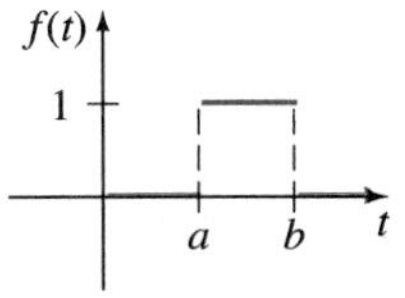

rectangular pulse

FIGURE 5.26

58.

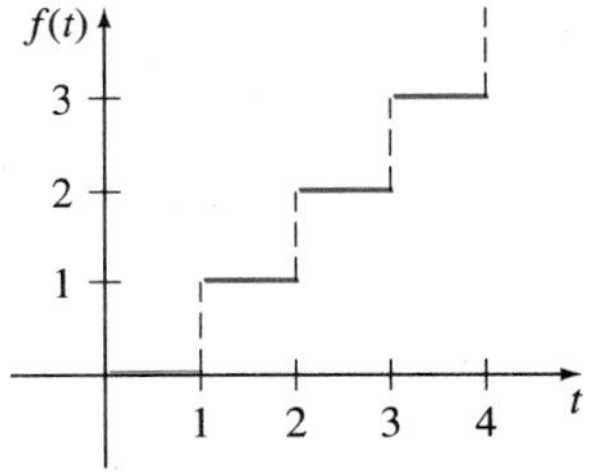

staircase function

FIGURE 5.27

In Problems 59 and 60 sketch the graph of the given function.

59. $f(t) = \mathscr{L}^{-1}\left\{\dfrac{1}{s^2} - \dfrac{e^{-s}}{s^2}\right\}$

60. $f(t) = \mathscr{L}^{-1}\left\{\dfrac{2}{s} - \dfrac{3e^{-s}}{s^2} + \dfrac{5e^{-2s}}{s^2}\right\}$

In Problems 61 and 62 use Theorem 5.7 in the form ($n = 1$),

$$f(t) = -\frac{1}{t}\mathscr{L}^{-1}\left\{\frac{d}{ds}F(s)\right\}$$

to evaluate the given inverse Laplace transform.

61. $\mathscr{L}^{-1}\left\{\ln\dfrac{s-3}{s+1}\right\}$

62. $\mathscr{L}^{-1}\left\{\ln\dfrac{s^2+1}{s^2+4}\right\}$

63. *PROJECT PROBLEM* The second translation theorem gives a means for transforming functions that have the precise form $f(t-a)\mathscr{U}(t-a)$,

$a > 0$. But often we wish to find the Laplace transform of a function of the form $f(t)\mathcal{U}(t - a)$ without bothering to "fix up" the function as we did in Example 10.

(a) Devise and prove a corollary to Theorem 5.6 that gives

$$\mathcal{L}\{f(t)\mathcal{U}(t - a)\},\ a > 0.$$

(b) Use your result in part (a) to rework Example 10.

(c) Use your result in part (a) to find $\mathcal{L}\{t\mathcal{U}(t - 2) + (t^2 + 2)\mathcal{U}(t - 5)\}$.

64. **PROJECT PROBLEM** The point of this problem is to find the Laplace transform of $f(t) = \ln t$, $t > 0$. Since f is discontinuous at $t = 0$, we define

$$\mathcal{L}\{f(t)\} = \lim_{\substack{b \to \infty \\ a \to 0}} \int_a^b e^{-st} \ln t\, dt.$$

(a) Find the definition of **Euler's constant** (sometimes called the **Euler-Mascheroni constant**) $\gamma = 0.57721566\ldots$. To solve part (b) you will also need to find a representation of γ as an improper integral.

(b) Show that $\mathcal{L}\{\ln t\} = -\dfrac{\gamma}{s} - \dfrac{\ln s}{s}$, $s > 0$.

5.4 TRANSFORMS OF DERIVATIVES, INTEGRALS, AND PERIODIC FUNCTIONS

Our goal is to use the Laplace transform to solve certain kinds of differential equations. To that end we need to evaluate quantities such as $\mathcal{L}\{dy/dt\}$ and $\mathcal{L}\{d^2y/dt^2\}$. For example, if f' is continuous for $t \geq 0$, then integration by parts gives

$$\mathcal{L}\{f'(t)\} = \int_0^\infty e^{-st} f'(t)\, dt = e^{-st} f(t)\Big|_0^\infty + s\int_0^\infty e^{-st} f(t)\, dt$$

$$= -f(0) + s\mathcal{L}\{f(t)\}$$

or

$$\mathcal{L}\{f'(t)\} = sF(s) - f(0). \tag{1}$$

Here we have assumed that $e^{-st}f(t) \to 0$ as $t \to \infty$. Similarly,

$$\mathcal{L}\{f''(t)\} = \int_0^\infty e^{-st} f''(t)\, dt = e^{-st} f'(t)\Big|_0^\infty + s\int_0^\infty e^{-st} f'(t)\, dt$$

$$= -f'(0) + s\mathcal{L}\{f'(t)\}$$

$$= s[sF(s) - f(0)] - f'(0)$$

or

$$\mathcal{L}\{f''(t)\} = s^2F(s) - sf(0) - f'(0). \tag{2}$$

The results in (1) and (2) are special cases of the next theorem, which gives the Laplace transform of the nth derivative of f. The proof is omitted.

THEOREM 5.8 Transform of a Derivative

If $f(t), f'(t), \ldots, f^{(n-1)}(t)$ are continuous on $[0, \infty)$ and are of exponential order and if $f^{(n)}(t)$ is piecewise continuous on $[0, \infty)$, then

$$\mathscr{L}\{f^{(n)}(t)\} = s^n F(s) - s^{n-1}f(0) - s^{n-2}f'(0) - \cdots - f^{(n-1)}(0),$$

where $F(s) = \mathscr{L}\{f(t)\}$.

EXAMPLE 1 Applying Theorem 5.8

Observe that the sum $kt \cos kt + \sin kt$ is the derivative of $t \sin kt$. Hence

$$\begin{aligned}
\mathscr{L}\{kt\cos kt + \sin kt\} &= \mathscr{L}\left\{\frac{d}{dt}(t\sin kt)\right\} \\
&= s\mathscr{L}\{t\sin kt\} \qquad \leftarrow \text{by (1)} \\
&= s\left(-\frac{d}{ds}\mathscr{L}\{\sin kt\}\right) \qquad \leftarrow \text{from Theorem 5.7} \\
&= s\left(\frac{2ks}{(s^2+k^2)^2}\right) = \frac{2ks^2}{(s^2+k^2)^2}.
\end{aligned}$$

Convolution

If functions f and g are piecewise continuous on $[0, \infty)$, then the **convolution** of f and g, denoted by $f * g$, is given by the integral

$$f * g = \int_0^t f(\tau)g(t-\tau)\,d\tau.$$

For example, the convolution of $f(t) = e^t$ and $g(t) = \sin t$ is

$$e^t * \sin t = \int_0^t e^\tau \sin(t-\tau)\,d\tau = \frac{1}{2}(-\sin t - \cos t + e^t). \tag{3}$$

It is left as an exercise to show that $\int_0^t f(\tau)g(t-\tau)\,d\tau = \int_0^t f(t-\tau)g(\tau)\,d\tau$; that is,

$$f * g = g * f.$$

See Problem 29 in Exercises 5.4. This means that the convolution of two functions is commutative.

It is possible to find the Laplace transform of the convolution of two functions without actually evaluating the integral as we did in (3). The result that follows is known as the **convolution theorem**.

THEOREM 5.9 Convolution Theorem

If $f(t)$ and $g(t)$ are piecewise continuous on $[0, \infty)$ and of exponential order, then

$$\mathscr{L}\{f * g\} = \mathscr{L}\{f(t)\}\mathscr{L}\{g(t)\} = F(s)G(s).$$

Proof Let
$$F(s) = \mathscr{L}\{f(t)\} = \int_0^\infty e^{-s\tau} f(\tau)\, d\tau$$

and
$$G(s) = \mathscr{L}\{g(t)\} = \int_0^\infty e^{-s\beta} g(\beta)\, d\beta.$$

Proceeding formally, we have

$$\begin{aligned} F(s)G(s) &= \left(\int_0^\infty e^{-s\tau} f(\tau)\, d\tau\right)\left(\int_0^\infty e^{-s\beta} g(\beta)\, d\beta\right) \\ &= \int_0^\infty \int_0^\infty e^{-s(\tau+\beta)} f(\tau) g(\beta)\, d\tau\, d\beta \\ &= \int_0^\infty f(\tau)\, d\tau \int_0^\infty e^{-s(\tau+\beta)} g(\beta)\, d\beta. \end{aligned}$$

Holding τ fixed, we let $t = \tau + \beta$, $dt = d\beta$ so that

$$F(s)G(s) = \int_0^\infty f(\tau)\, d\tau \int_\tau^\infty e^{-st} g(t - \tau)\, dt.$$

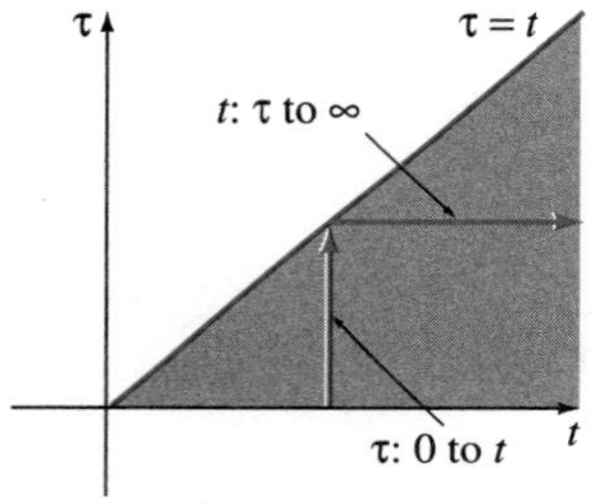

FIGURE 5.28

In the $t\tau$-plane we are integrating over the shaded region in Figure 5.28. Since f and g are piecewise continuous on $[0, \infty)$ and of exponential order, it is possible to interchange the order of integration:

$$F(s)G(s) = \int_0^\infty e^{-st}\, dt \int_0^t f(\tau) g(t - \tau)\, d\tau = \int_0^\infty e^{-st}\left\{\int_0^t f(\tau) g(t-\tau)\, d\tau\right\} dt = \mathscr{L}\{f * g\}. \quad \bigcirc$$

EXAMPLE 2 Transform of a Convolution

Evaluate $\mathscr{L}\left\{\int_0^t e^\tau \sin(t - \tau)\, d\tau\right\}$.

Solution With $f(t) = e^t$ and $g(t) = \sin t$, the convolution theorem states that the Laplace transform of the convolution of f and g is the product of their Laplace transforms:

$$\mathscr{L}\left\{\int_0^t e^\tau \sin(t - \tau)\, d\tau\right\} = \mathscr{L}\{e^t\} \cdot \mathscr{L}\{\sin t\} = \frac{1}{s-1} \cdot \frac{1}{s^2+1} = \frac{1}{(s-1)(s^2+1)}. \quad \bullet$$

Inverse Form of the Convolution Theorem

The convolution theorem is sometimes useful in finding the inverse Laplace transform of a product of two Laplace transforms. From Theorem 5.9 we have

$$\mathscr{L}^{-1}\{F(s)G(s)\} = f * g. \tag{4}$$

EXAMPLE 3 Inverse Transform as a Convolution

Evaluate $\mathscr{L}^{-1}\left\{\dfrac{1}{(s-1)(s+4)}\right\}$.

Solution Partial fractions could be used, but if we identify

$$F(s) = \frac{1}{s-1} \quad \text{and} \quad G(s) = \frac{1}{s+4},$$

then $\quad \mathscr{L}^{-1}\{F(s)\} = f(t) = e^t \quad \text{and} \quad \mathscr{L}^{-1}\{G(s)\} = g(t) = e^{-4t}.$

Hence from (4) we obtain

$$\begin{aligned}
\mathscr{L}^{-1}\left\{\frac{1}{(s-1)(s+4)}\right\} &= \int_0^t f(\tau)g(t-\tau)\,d\tau = \int_0^t e^{\tau}e^{-4(t-\tau)}\,d\tau \\
&= e^{-4t}\int_0^t e^{5\tau}\,d\tau \\
&= e^{-4t}\frac{1}{5}e^{5\tau}\Big|_0^t \\
&= \frac{1}{5}e^t - \frac{1}{5}e^{-4t}.
\end{aligned}$$

EXAMPLE 4 Inverse Transform as a Convolution

Evaluate $\mathscr{L}^{-1}\left\{\dfrac{1}{(s^2+k^2)^2}\right\}$.

Solution Let

$$F(s) = G(s) = \frac{1}{s^2+k^2}$$

so that

$$f(t) = g(t) = \frac{1}{k}\mathscr{L}^{-1}\left\{\frac{k}{s^2+k^2}\right\} = \frac{1}{k}\sin kt.$$

In this case (4) gives

$$\mathscr{L}^{-1}\left\{\frac{1}{(s^2+k^2)^2}\right\} = \frac{1}{k^2}\int_0^t \sin k\tau \sin k(t-\tau)\,d\tau. \tag{5}$$

Now recall from trigonometry that

$$\cos(A + B) = \cos A \cos B - \sin A \sin B$$

and $$\cos(A - B) = \cos A \cos B + \sin A \sin B.$$

Subtracting the first from the second gives the identity

$$\sin A \sin B = \frac{1}{2}[\cos(A - B) - \cos(A + B)].$$

If we set $A = k\tau$ and $B = k(t - \tau)$, we can carry out the integration in (5):

$$\begin{aligned}\mathscr{L}^{-1}\left\{\frac{1}{(s^2 + k^2)^2}\right\} &= \frac{1}{2k^2}\int_0^t [\cos k(2\tau - t) - \cos kt]\, d\tau \\ &= \frac{1}{2k^2}\left[\frac{1}{2k}\sin k(2\tau - t) - \tau\cos kt\right]_0^t \\ &= \frac{\sin kt - kt\cos kt}{2k^3}.\end{aligned}$$

Transform of an Integral

When $g(t) = 1$ and $\mathscr{L}\{g(t)\} = G(s) = 1/s$, the convolution theorem implies that the Laplace transform of the integral of f is

$$\mathscr{L}\left\{\int_0^t f(\tau)\, d\tau\right\} = \frac{F(s)}{s}. \tag{6}$$

The inverse form of (6),

$$\int_0^t f(\tau)\, d\tau = \mathscr{L}^{-1}\left\{\frac{F(s)}{s}\right\}, \tag{7}$$

can be used at times in lieu of partial fractions when s^n is a factor of the denominator and $f(t) = \mathscr{L}^{-1}\{F(s)\}$ is easy to integrate. For example, we know for $f(t) = \sin t$ that $F(s) = 1/(s^2 + 1)$, and so by (7),

$$\mathscr{L}^{-1}\left\{\frac{1}{s(s^2 + 1)}\right\} = \int_0^t \sin\tau\, d\tau = 1 - \cos t,$$

$$\mathscr{L}^{-1}\left\{\frac{1}{s^2(s^2 + 1)}\right\} = \int_0^t (1 - \cos\tau)\, d\tau = t - \sin t,$$

$$\mathscr{L}^{-1}\left\{\frac{1}{s^3(s^2 + 1)}\right\} = \int_0^t (\tau - \sin\tau)\, d\tau = \frac{1}{2}t^2 - 1 + \cos t,$$

and so on. We shall also use (6) in the next section on applications.

Transform of a Periodic Function

If a periodic function has period T, $T > 0$, then $f(t + T) = f(t)$. The Laplace transform of a **periodic function** can be obtained by an integration over one period.

THEOREM 5.10 Transform of a Periodic Function

If $f(t)$ is piecewise continuous on $[0, \infty)$, of exponential order, and periodic with period T, then

$$\mathscr{L}\{f(t)\} = \frac{1}{1 - e^{-sT}} \int_0^T e^{-st} f(t)\, dt. \tag{8}$$

Proof Write the Laplace transform as two integrals:

$$\mathscr{L}\{f(t)\} = \int_0^T e^{-st} f(t)\, dt + \int_T^\infty e^{-st} f(t)\, dt. \tag{9}$$

When we let $t = u + T$, the last integral in (9) becomes

$$\int_T^\infty e^{-st} f(t)\, dt = \int_0^\infty e^{-s(u+T)} f(u + T)\, du = e^{-sT} \int_0^\infty e^{-su} f(u)\, du = e^{-sT} \mathscr{L}\{f(t)\}.$$

Hence (9) is $$\mathscr{L}\{f(t)\} = \int_0^T e^{-st} f(t)\, dt + e^{-sT} \mathscr{L}\{f(t)\}.$$

Solving for $\mathscr{L}\{f(t)\}$ yields the result given in (8). ○

EXAMPLE 5 Transform of a Periodic Function

Find the Laplace transform of the periodic function shown in Figure 5.29.

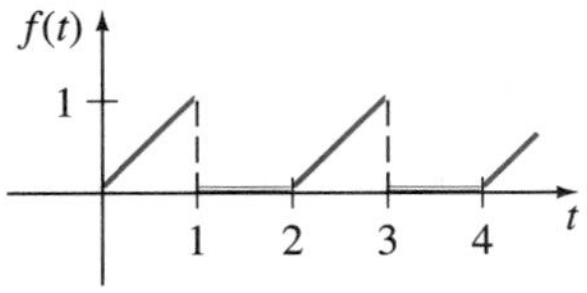

FIGURE 5.29

Solution On the interval $0 \le t < 2$ the function can be defined by

$$f(t) = \begin{cases} t, & 0 \le t < 1 \\ 0, & 1 \le t < 2 \end{cases}$$

and outside the interval by $f(t + 2) = f(t)$. Identifying $T = 2$, we use (8) and integration by parts to obtain

$$\mathscr{L}\{f(t)\} = \frac{1}{1 - e^{-2s}} \int_0^2 e^{-st} f(t)\, dt = \frac{1}{1 - e^{-2s}} \left[\int_0^1 e^{-st} t\, dt + \int_1^2 e^{-st} 0\, dt \right]$$

$$= \frac{1}{1 - e^{-2s}} \left[-\frac{e^{-s}}{s} + \frac{1 - e^{-s}}{s^2} \right] \tag{10}$$

$$= \frac{1 - (s + 1)e^{-s}}{s^2 (1 - e^{-2s})}.$$ ●

The result in (10) of Example 5 can be obtained without actually integrating by making use of the second translation theorem. If we define

$$g(t) = \begin{cases} t, & 0 \le t < 1 \\ 0, & t \ge 1, \end{cases}$$

then $f(t) = g(t)$ on the interval $[0, T]$, where $T = 2$. But we can express g in terms of unit step functions as

$$g(t) = t - t\mathcal{U}(t-1) = t - (t-1)\mathcal{U}(t-1) - \mathcal{U}(t-1).$$

Thus
$$\mathcal{L}\{f(t)\} = \frac{1}{1-e^{-2s}}\mathcal{L}\{g(t)\}$$

$$= \frac{1}{1-e^{-2s}}\mathcal{L}\{t - (t-1)\mathcal{U}(t-1) - \mathcal{U}(t-1)\}$$

$$= \frac{1}{1-e^{-2s}}\left[\frac{1}{s^2} - \frac{1}{s^2}e^{-s} - \frac{1}{s}e^{-s}\right].$$

Inspection of the expression inside the brackets reveals that it is identical to (10).

EXERCISES 5.4

Answers to odd-numbered problems begin on page AN-24.

1. Use the result $(d/dt)e^t = e^t$ and (1) of this section to evaluate $\mathcal{L}\{e^t\}$.
2. Use the result $(d/dt)\cos^2 t = -\sin 2t$ and (1) of this section to evaluate $\mathcal{L}\{\cos^2 t\}$.

In Problems 3 and 4 suppose a function $y(t)$ has the properties that $y(0) = 1$ and $y'(0) = -1$. Find the Laplace transform of the given expression.

3. $y'' + 3y'$
4. $y'' - 4y' + 5y$

In Problems 5 and 6 suppose a function $y(t)$ has the properties that $y(0) = 2$ and $y'(0) = 3$. Solve for the Laplace transform $\mathcal{L}\{y(t)\} = Y(s)$.

5. $y'' - 2y' + y = 0$
6. $y'' + y = 1$

In Problems 7–20 evaluate the given Laplace transform without evaluating the integral.

7. $\mathcal{L}\left\{\int_0^t e^\tau\, d\tau\right\}$
8. $\mathcal{L}\left\{\int_0^t \cos\tau\, d\tau\right\}$
9. $\mathcal{L}\left\{\int_0^t e^{-\tau}\cos\tau\, d\tau\right\}$
10. $\mathcal{L}\left\{\int_0^t \tau\sin\tau\, d\tau\right\}$
11. $\mathcal{L}\left\{\int_0^t \tau e^{t-\tau}\, d\tau\right\}$
12. $\mathcal{L}\left\{\int_0^t \sin\tau\cos(t-\tau)\, d\tau\right\}$
13. $\mathcal{L}\left\{t\int_0^t \sin\tau\, d\tau\right\}$
14. $\mathcal{L}\left\{t\int_0^t \tau e^{-\tau}\, d\tau\right\}$
15. $\mathcal{L}\{1 * t^3\}$
16. $\mathcal{L}\{1 * e^{-2t}\}$
17. $\mathcal{L}\{t^2 * t^4\}$
18. $\mathcal{L}\{t^2 * te^t\}$
19. $\mathcal{L}\{e^{-t} * e^t\cos t\}$
20. $\mathcal{L}\{e^{2t} * \sin t\}$

In Problems 21 and 22 suppose $\mathcal{L}^{-1}\{F(s)\} = f(t)$. Find the inverse Laplace transform of the given function.

21. $\dfrac{1}{s+5}F(s)$
22. $\dfrac{s}{s^2+4}F(s)$

In Problems 23–28 use (4) or (7) to find $f(t)$.

23. $\mathcal{L}^{-1}\left\{\dfrac{1}{s(s+1)}\right\}$
24. $\mathcal{L}^{-1}\left\{\dfrac{1}{s^3(s-1)}\right\}$
25. $\mathcal{L}^{-1}\left\{\dfrac{1}{(s+1)(s-2)}\right\}$
26. $\mathcal{L}^{-1}\left\{\dfrac{1}{(s+1)^2}\right\}$

27. $\mathscr{L}^{-1}\left\{\dfrac{s}{(s^2+4)^2}\right\}$

28. $\mathscr{L}^{-1}\left\{\dfrac{1}{(s^2+4s+5)^2}\right\}$

29. Prove the commutative property of the convolution integral $f * g = g * f$.

30. Prove the distributive property of the convolution integral $f * (g + h) = f * g + f * h$.

In Problems 31–38 use Theorem 5.10 to find the Laplace transform of the given periodic function.

31.

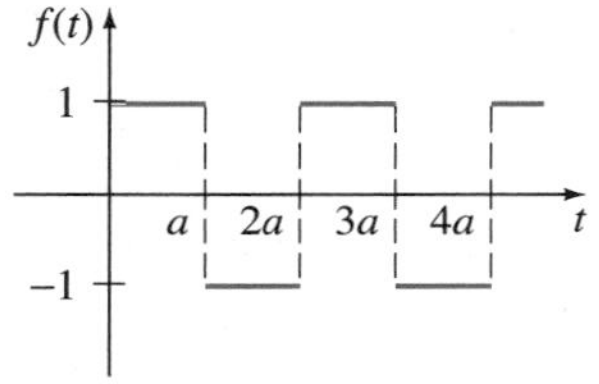

meander function

FIGURE 5.30

32.

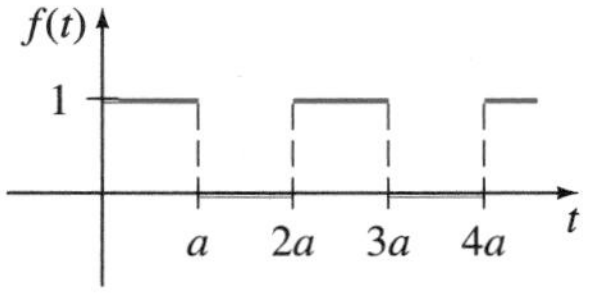

square wave

FIGURE 5.31

33.

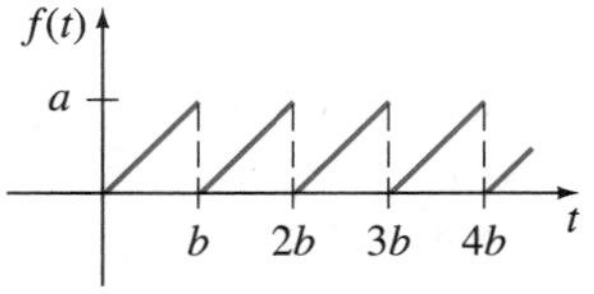

sawtooth function

FIGURE 5.32

34.

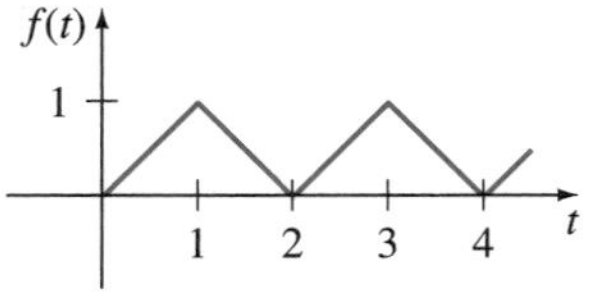

triangular wave

FIGURE 5.33

35.

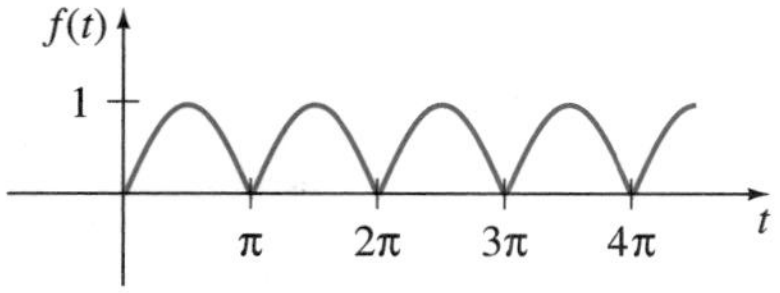

full-wave rectification of sin t

FIGURE 5.34

36.

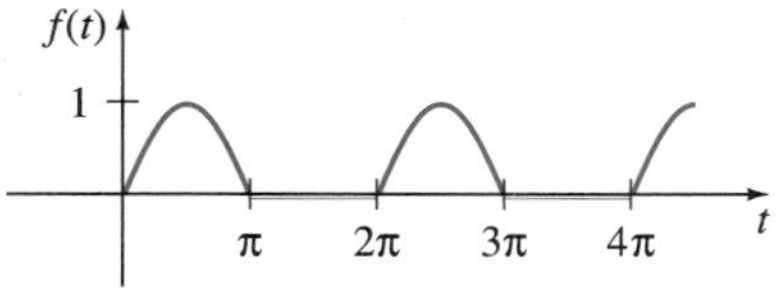

half-wave rectification of sin t

FIGURE 5.35

37. $f(t) = \sin t$
$f(t + 2\pi) = f(t)$

38. $f(t) = \cos t$
$f(t + 2\pi) = f(t)$

39. If f is piecewise continuous and of exponential order for $t \geq 0$, then show that

$$\mathscr{L}\left\{\int_a^t f(\tau)\,d\tau\right\} = \frac{F(s)}{s} - \frac{1}{s}\int_0^a f(\tau)\,d\tau,$$

where $F(s) = \mathscr{L}\{f(t)\}$ and a is any nonnegative real number.

40. *PROJECT PROBLEM*

(a) If f is piecewise continuous and of exponential order for $t \geq 0$ and $\lim_{t\to 0} f(t)/t$ exists, then show that

$$\mathscr{L}\left\{\frac{f(t)}{t}\right\} = \int_s^\infty F(u)\,du,$$

where $F(s) = \mathscr{L}\{f(t)\}$.

(b) Find $\mathscr{L}\left\{\dfrac{\sin t}{t}\right\}$.

(c) Show that the result in part (b) is identical to $\tan^{-1}\dfrac{1}{s}$.

(d) Show that $\displaystyle\int_0^\infty \frac{\sin t}{t}\,dt = \frac{\pi}{2}$.

(e) The **sine integral** function is defined as

$$\operatorname{Si}(t) = \int_0^t \frac{\sin u}{u}\,du.$$

Find $\mathscr{L}\{\operatorname{Si}(t)\}$.

5.5 APPLICATIONS

Since $\mathscr{L}\{y^{(n)}(t)\}$, $n > 1$, depends on $y(t)$ and its $n - 1$ derivatives evaluated at $t = 0$, the Laplace transform is ideally suited to initial-value problems for linear differential equations with constant coefficients. This kind of differential equation can be reduced to an *algebraic equation* in the transformed function $Y(s)$. To see this, consider the initial-value problem

$$a_n \frac{d^n y}{dt^n} + a_{n-1}\frac{d^{n-1}y}{dt^{n-1}} + \cdots + a_1 \frac{dy}{dt} + a_0 y = g(t)$$

$$y(0) = y_0, \quad y'(0) = y_1, \ldots, y^{(n-1)}(0) = y_{n-1},$$

where a_i, $i = 0, 1, \ldots, n$, and $y_0, y_1, \ldots, y_{n-1}$ are constants. By the linearity property of the Laplace transform, we can write

$$a_n \mathscr{L}\left\{\frac{d^n y}{dt^n}\right\} + a_{n-1}\mathscr{L}\left\{\frac{d^{n-1}y}{dt^{n-1}}\right\} + \cdots + a_0\mathscr{L}\{y\} = \mathscr{L}\{g(t)\}. \qquad \textbf{(1)}$$

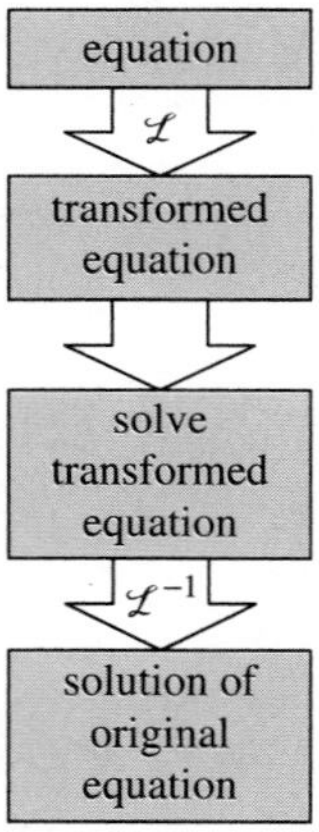

FIGURE 5.36

From Theorem 5.8, (1) becomes

$$a_n[s^nY(s) - s^{n-1}y(0) - \cdots - y^{(n-1)}(0)] +$$
$$a_{n-1}[s^{n-1}Y(s) - s^{n-2}y(0) - \cdots - y^{(n-2)}(0)] + \cdots + a_0Y(s) = G(s)$$

or

$$[a_ns^n + a_{n-1}s^{n-1} + \cdots + a_0]Y(s) = a_n[s^{n-1}y_0 + \cdots + y_{n-1}]$$
$$+ a_{n-1}[s^{n-2}y_0 + \cdots + y_{n-2}] + \cdots + G(s), \qquad \textbf{(2)}$$

where $Y(s) = \mathscr{L}\{y(t)\}$ and $G(s) = \mathscr{L}\{g(t)\}$. By solving (2) for $Y(s)$, we find $y(t)$ by determining the inverse transform

$$y(t) = \mathscr{L}^{-1}\{Y(s)\}.$$

The procedure is outlined in Figure 5.36. Note that this method incorporates the prescribed initial conditions directly into the solution. Hence there is no need for the separate operations of determining constants in the general solution of the differential equation.

EXAMPLE 1 DE Transformed into an Algebraic Equation

Solve
$$\frac{dy}{dt} - 3y = e^{2t}, \quad y(0) = 1.$$

Solution We first take the transform of each member of the given differential equation:

$$\mathscr{L}\left\{\frac{dy}{dt}\right\} - 3\mathscr{L}\{y\} = \mathscr{L}\{e^{2t}\}.$$

We then use $\mathcal{L}\{dy/dt\} = sY(s) - y(0) = sY(s) - 1$ and $\mathcal{L}\{e^{2t}\} = 1/(s-2)$.

Solving
$$sY(s) - 1 - 3Y(s) = \frac{1}{s-2}$$

for $Y(s)$ and carrying out partial fraction decomposition give

$$Y(s) = \frac{s-1}{(s-2)(s-3)} = \frac{-1}{s-2} + \frac{2}{s-3},$$

and so
$$y(t) = -\mathcal{L}^{-1}\left\{\frac{1}{s-2}\right\} + 2\mathcal{L}^{-1}\left\{\frac{1}{s-3}\right\}.$$

From part (c) of Theorem 5.3 it follows that

$$y(t) = -e^{2t} + 2e^{3t}.$$

EXAMPLE 2 An Initial-Value Problem

Solve
$$y'' - 6y' + 9y = t^2e^{3t}, \quad y(0) = 2, \quad y'(0) = 6.$$

Solution
$$\mathcal{L}\{y''\} - 6\mathcal{L}\{y'\} + 9\mathcal{L}\{y\} = \mathcal{L}\{t^2e^{3t}\}$$

$$\underbrace{s^2Y(s) - sy(0) - y'(0)}_{\mathcal{L}\{y''\}} - 6\underbrace{[sY(s) - y(0)]}_{\mathcal{L}\{y'\}} + 9\underbrace{Y(s)}_{\mathcal{L}\{y\}} = \underbrace{\frac{2}{(s-3)^3}}_{\mathcal{L}\{t^2e^{3t}\}}.$$

Using the initial conditions and simplifying give

$$(s^2 - 6s + 9)Y(s) = 2s - 6 + \frac{2}{(s-3)^3}$$

$$(s-3)^2Y(s) = 2(s-3) + \frac{2}{(s-3)^3}$$

$$Y(s) = \frac{2}{s-3} + \frac{2}{(s-3)^5}$$

Thus
$$y(t) = 2\mathcal{L}^{-1}\left\{\frac{1}{s-3}\right\} + \frac{2}{4!}\mathcal{L}^{-1}\left\{\frac{4!}{(s-3)^5}\right\}.$$

Recall from the first translation theorem that

$$\mathcal{L}^{-1}\left\{\frac{4!}{s^5}\bigg|_{s \to s-3}\right\} = t^4e^{3t}.$$

Hence we have
$$y(t) = 2e^{3t} + \frac{1}{12}t^4e^{3t}.$$

EXAMPLE 3 Using the First Translation Theorem

Solve $$y'' + 4y' + 6y = 1 + e^{-t}, \quad y(0) = 0, \quad y'(0) = 0.$$

Solution $$\mathscr{L}\{y''\} + 4\mathscr{L}\{y'\} + 6\mathscr{L}\{y\} = \mathscr{L}\{1\} + \mathscr{L}\{e^{-t}\}$$

$$s^2Y(s) - sy(0) - y'(0) + 4[sY(s) - y(0)] + 6Y(s) = \frac{1}{s} + \frac{1}{s+1}$$

$$(s^2 + 4s + 6)Y(s) = \frac{2s+1}{s(s+1)}$$

$$Y(s) = \frac{2s+1}{s(s+1)(s^2+4s+6)}.$$

The partial fraction decomposition for $Y(s)$ is

$$Y(s) = \frac{1/6}{s} + \frac{1/3}{s+1} + \frac{-s/2 - 5/3}{s^2+4s+6}.$$

In preparation for taking the inverse transform, we fix up $Y(s)$ in the following manner:

$$\begin{aligned} Y(s) &= \frac{1/6}{s} + \frac{1/3}{s+1} + \frac{(-1/2)(s+2) - 2/3}{(s+2)^2+2} \\ &= \frac{1/6}{s} + \frac{1/3}{s+1} - \frac{1}{2}\frac{s+2}{(s+2)^2+2} - \frac{2}{3}\frac{1}{(s+2)^2+2}. \end{aligned}$$

Finally, from parts (a) and (c) of Theorem 5.3 and the first translation theorem, we obtain

$$\begin{aligned} y(t) &= \frac{1}{6}\mathscr{L}^{-1}\left\{\frac{1}{s}\right\} + \frac{1}{3}\mathscr{L}^{-1}\left\{\frac{1}{s+1}\right\} - \frac{1}{2}\mathscr{L}^{-1}\left\{\frac{s+2}{(s+2)^2+2}\right\} \\ &\quad - \frac{2}{3\sqrt{2}}\mathscr{L}^{-1}\left\{\frac{\sqrt{2}}{(s+2)^2+2}\right\} \\ &= \frac{1}{6} + \frac{1}{3}e^{-t} - \frac{1}{2}e^{-2t}\cos\sqrt{2}t - \frac{\sqrt{2}}{3}e^{-2t}\sin\sqrt{2}t. \end{aligned}$$

EXAMPLE 4 Using Theorems 5.3/5.7

Solve $$x'' + 16x = \cos 4t, \quad x(0) = 0, \quad x'(0) = 1.$$

Solution Recall that this initial-value problem could describe the forced, undamped, and resonant motion of a mass on a spring. The mass starts with an initial velocity of 1 foot per second in the downward direction from the equilibrium position.

Transforming the equation gives

$$(s^2 + 16)X(s) = 1 + \frac{s}{s^2 + 16}$$

$$X(s) = \frac{1}{s^2 + 16} + \frac{s}{(s^2 + 16)^2}.$$

With the aid of part (d) of Theorem 5.3 and Theorem 5.7, we find

$$x(t) = \frac{1}{4}\mathscr{L}^{-1}\left\{\frac{4}{s^2 + 16}\right\} + \frac{1}{8}\mathscr{L}^{-1}\left\{\frac{8s}{(s^2 + 16)^2}\right\}$$

$$= \frac{1}{4}\sin 4t + \frac{1}{8}t\sin 4t.$$

EXAMPLE 5 Using a Unit Step Function

Solve $x'' + 16x = f(t), \quad x(0) = 0, \quad x'(0) = 1,$

where
$$f(t) = \begin{cases} \cos 4t, & 0 \le t < \pi \\ 0, & t \ge \pi. \end{cases}$$

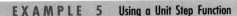

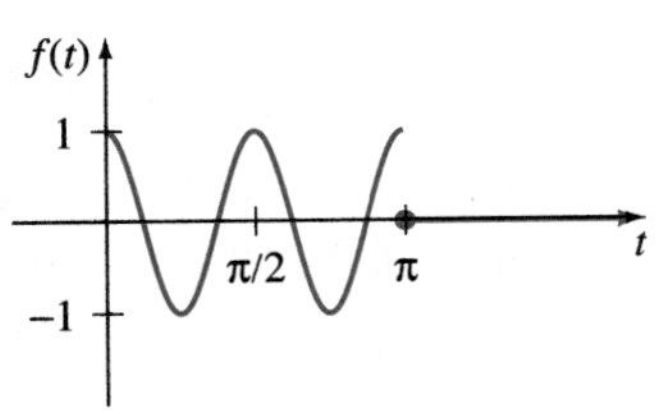

FIGURE 5.37

Solution The function $f(t)$ can be interpreted as an external force that is acting on a mechanical system for only a short period of time and then is removed. See Figure 5.37. Although this problem could be solved by conventional means, the procedure is not at all convenient when $f(t)$ is defined in a piecewise manner. With the aid of (2) and (3) of Section 5.3 and the periodicity of the cosine, we can rewrite f in terms of the unit step function as

$$f(t) = \cos 4t - \cos 4t\,\mathscr{U}(t - \pi) = \cos 4t - \cos 4(t - \pi)\mathscr{U}(t - \pi).$$

The second translation theorem then yields

$$\mathscr{L}\{x''\} + 16\mathscr{L}\{x\} = \mathscr{L}\{f(t)\}$$

$$s^2X(s) - sx(0) - x'(0) + 16X(s) = \frac{s}{s^2 + 16} - \frac{s}{s^2 + 16}e^{-\pi s}$$

$$(s^2 + 16)X(s) = 1 + \frac{s}{s^2 + 16} - \frac{s}{s^2 + 16}e^{-\pi s}$$

$$X(s) = \frac{1}{s^2 + 16} + \frac{s}{(s^2 + 16)^2} - \frac{s}{(s^2 + 16)^2}e^{-\pi s}.$$

From part (b) of Example 12 in Section 5.3 (with $k = 4$) along with (8) of that section, we find

$$x(t) = \frac{1}{4}\mathscr{L}^{-1}\left\{\frac{4}{s^2 + 16}\right\} + \frac{1}{8}\mathscr{L}^{-1}\left\{\frac{8s}{(s^2 + 16)^2}\right\} - \frac{1}{8}\mathscr{L}^{-1}\left\{\frac{8s}{(s^2 + 16)^2}e^{-\pi s}\right\}$$

$$= \frac{1}{4}\sin 4t + \frac{1}{8}t\sin 4t - \frac{1}{8}(t - \pi)\sin 4(t - \pi)\mathscr{U}(t - \pi).$$

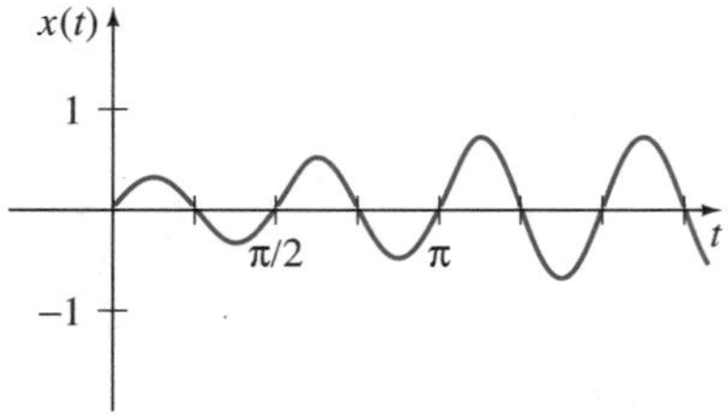

FIGURE 5.38

The foregoing solution is the same as

$$x(t) = \begin{cases} \dfrac{1}{4}\sin 4t + \dfrac{1}{8}t\sin 4t, & 0 \leq t < \pi \\ \dfrac{2+\pi}{8}\sin 4t, & t \geq \pi. \end{cases}$$

Observe from the graph of $x(t)$ in Figure 5.38 that the amplitudes of vibration become steady as soon as the external force is turned off. ●

Volterra Integral Equation

The convolution theorem is useful in solving other types of equations in which an unknown function appears under an integral sign. In the next example we solve a **Volterra integral equation**

$$f(t) = g(t) + \int_0^t f(\tau)h(t-\tau)\,d\tau$$

for $f(t)$. The functions $g(t)$ and $h(t)$ are known.

EXAMPLE 6 An Integral Equation

Solve $\quad f(t) = 3t^2 - e^{-t} - \int_0^t f(\tau)e^{t-\tau}\,d\tau \quad$ for $f(t)$.

Solution It follows from Theorem 5.9 that

$$\mathscr{L}\{f(t)\} = 3\mathscr{L}\{t^2\} - \mathscr{L}\{e^{-t}\} - \mathscr{L}\{f(t)\}\mathscr{L}\{e^t\}$$

$$F(s) = 3 \cdot \frac{2}{s^3} - \frac{1}{s+1} - F(s)\cdot\frac{1}{s-1}.$$

Solving the last equation for $F(s)$ gives

$$F(s) = \frac{6(s-1)}{s^4} - \frac{s-1}{s(s+1)}$$

$$= \frac{6}{s^3} - \frac{6}{s^4} + \frac{1}{s} - \frac{2}{s+1}. \qquad \leftarrow \text{termwise division and partial fractions}$$

The inverse transform is

$$f(t) = 3\mathscr{L}^{-1}\left\{\frac{2!}{s^3}\right\} - \mathscr{L}^{-1}\left\{\frac{3!}{s^4}\right\} + \mathscr{L}^{-1}\left\{\frac{1}{s}\right\} - 2\mathscr{L}^{-1}\left\{\frac{1}{s+1}\right\}$$

$$= 3t^2 - t^3 + 1 - 2e^{-t}.$$ ●

Series Circuits

In a single loop or series circuit, Kirchoff's second law states that the sum of the voltage drops across an inductor, resistor, and capacitor is equal to the

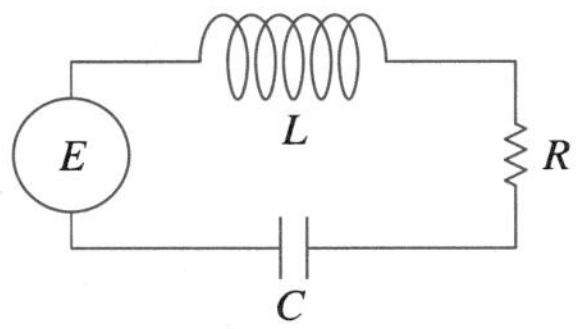

FIGURE 5.39

impressed voltage $E(t)$. Now it is known that the voltage drops across an inductor, resistor, and capacitor are, respectively,

$$L\frac{di}{dt}, \quad R\,i(t), \quad \text{and} \quad \frac{1}{C}\int_0^t i(\tau)\,d\tau,$$

where $i(t)$ is the current and L, R, and C are constants. It follows that the current in a circuit, such as that shown in Figure 5.39, is governed by the **integrodifferential equation**

$$L\frac{di}{dt} + Ri + \frac{1}{C}\int_0^t i(\tau)\,d\tau = E(t). \tag{3}$$

EXAMPLE 7 An Integrodifferential Equation

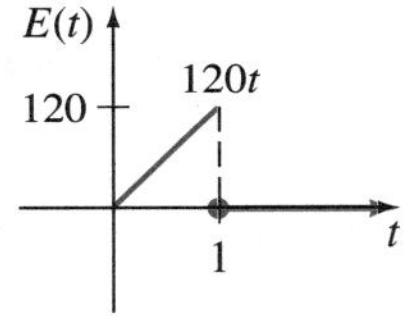

FIGURE 5.40

Determine the current $i(t)$ in a single loop L-R-C circuit when $L = 0.1$ henry, $R = 20$ ohms, $C = 10^{-3}$ farad, $i(0) = 0$, and the impressed voltage $E(t)$ is as given in Figure 5.40.

Solution Since the voltage is off for $t \geq 1$, we can write

$$E(t) = 120t - 120t\,\mathcal{U}(t-1). \tag{4}$$

But in order to use the second translation theorem we rewrite (4) as

$$E(t) = 120t - 120(t-1)\mathcal{U}(t-1) - 120\,\mathcal{U}(t-1).$$

Equation (3) then becomes

$$0.1\frac{di}{dt} + 20i + 10^3\int_0^t i(\tau)\,d\tau = 120t - 120(t-1)\mathcal{U}(t-1) - 120\,\mathcal{U}(t-1). \tag{5}$$

Now recall from (6) of Section 5.4, $\mathcal{L}\{\int_0^t i(\tau)\,d\tau\} = I(s)/s$, where $I(s) = \mathcal{L}\{i(t)\}$. Thus the transform of equation (5) is

$$0.1sI(s) + 20I(s) + 10^3\frac{I(s)}{s} = 120\left[\frac{1}{s^2} - \frac{1}{s^2}e^{-s} - \frac{1}{s}e^{-s}\right],$$

or, after multiplying by $10s$,

$$(s+100)^2 I(s) = 1200\left[\frac{1}{s} - \frac{1}{s}e^{-s} - e^{-s}\right]$$

$$I(s) = 1200\left[\frac{1}{s(s+100)^2} - \frac{1}{s(s+100)^2}e^{-s} - \frac{1}{(s+100)^2}e^{-s}\right].$$

By partial fractions we can write

$$I(s) = 1200\left[\frac{1/10{,}000}{s} - \frac{1/10{,}000}{s+100} - \frac{1/100}{(s+100)^2} - \frac{1/10{,}000}{s}e^{-s}\right.$$

$$\left. + \frac{1/10{,}000}{s+100}e^{-s} + \frac{1/100}{(s+100)^2}e^{-s} - \frac{1}{(s+100)^2}e^{-s}\right].$$

Employing the inverse form of the second translation theorem, we obtain

$$i(t) = \frac{3}{25}[1 - \mathcal{U}(t-1)] - \frac{3}{25}[e^{-100t} - e^{-100(t-1)}\mathcal{U}(t-1)]$$
$$-12te^{-100t} - 1188(t-1)e^{-100(t-1)}\mathcal{U}(t-1).$$

EXAMPLE 8 A Periodic Impressed Voltage

The differential equation for the current $i(t)$ in a single loop L-R series circuit is

$$L\frac{di}{dt} + Ri = E(t). \tag{6}$$

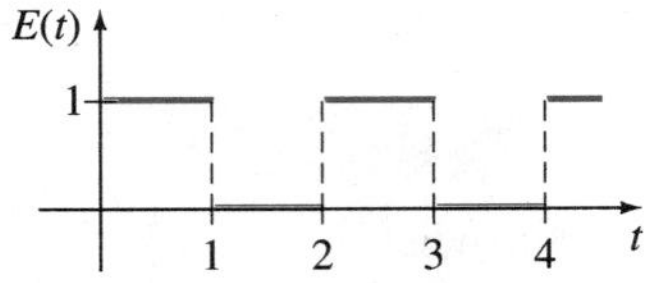

FIGURE 5.41

Determine the current $i(t)$ when $i(0) = 0$ and $E(t)$ is the square wave function shown in Figure 5.41.

Solution The Laplace transform of the equation is

$$LsI(s) + RI(s) = \mathscr{L}\{E(t)\}. \tag{7}$$

Since $E(t)$ is periodic with period $T = 2$, we use (7) of Section 5.4:

$$\mathscr{L}\{E(t)\} = \frac{1}{1-e^{-2s}}\left(\int_0^1 1\cdot e^{-st}\,dt + \int_1^2 0\cdot e^{-st}\,dt\right)$$

$$= \frac{1}{1-e^{-2s}}\,\frac{1-e^{-s}}{s} \qquad \leftarrow 1 - e^{-2s} = (1+e^{-s})(1-e^{-s})$$

$$= \frac{1}{s(1+e^{-s})}.$$

Hence from (7) we find

$$I(s) = \frac{1/L}{s(s+R/L)(1+e^{-s})}. \tag{8}$$

To find the inverse Laplace transform of this function we first make use of a geometric series. Recall, for $|x| < 1$,

$$\frac{1}{1+x} = 1 - x + x^2 - x^3 + \cdots.$$

With the identification $x = e^{-s}$ we then have for $s > 0$

$$\frac{1}{1+e^{-s}} = 1 - e^{-s} + e^{-2s} - e^{-3s} + \cdots.$$

By writing $$\frac{1}{s(s+R/L)} = \frac{L/R}{s} - \frac{L/R}{s+R/L}$$

(8) becomes

$$I(s) = \frac{1}{R}\left(\frac{1}{s} - \frac{1}{s + R/L}\right)(1 - e^{-s} + e^{-2s} - e^{-3s} + \cdots)$$

$$= \frac{1}{R}\left(\frac{1}{s} - \frac{e^{-s}}{s} + \frac{e^{-2s}}{s} - \frac{e^{-3s}}{s} + \cdots\right) - \frac{1}{R}\left(\frac{1}{s + R/L} - \frac{e^{-s}}{s + R/L} + \frac{e^{-2s}}{s + R/L} - \frac{e^{-3s}}{s + R/L} + \cdots\right).$$

By applying the inverse form of the second translation theorem to each term of both series, we obtain

$$i(t) = \frac{1}{R}(1 - \mathcal{U}(t-1) + \mathcal{U}(t-2) - \mathcal{U}(t-3) + \cdots)$$

$$- \frac{1}{R}(e^{-Rt/L} - e^{-R(t-1)/L}\mathcal{U}(t-1) + e^{-R(t-2)/L}\mathcal{U}(t-2) - e^{-R(t-3)/L}\mathcal{U}(t-3) + \cdots)$$

or equivalently

$$i(t) = \frac{1}{R}(1 - e^{-Rt/L}) + \frac{1}{R}\sum_{n=1}^{\infty}(-1)^n(1 - e^{-R(t-n)/L})\mathcal{U}(t-n).$$ ●

To interpret the solution in Example 8 let us assume for the sake of illustration that $R = 1$, $L = 1$, and $0 \le t < 4$. In this case

$$i(t) = 1 - e^{-t} - (1 - e^{t-1})\mathcal{U}(t-1) + (1 - e^{-(t-2)})\mathcal{U}(t-2) - (1 - e^{-(t-3)})\mathcal{U}(t-3);$$

in other words,

$$i(t) = \begin{cases} 1 - e^{-t}, & 0 \le t < 1 \\ -e^{-t} + e^{-(t-1)}, & 1 \le t < 2 \\ 1 - e^{-t} + e^{-(t-1)} - e^{-(t-2)}, & 2 \le t < 3 \\ -e^{-t} + e^{-(t-1)} - e^{-(t-2)} + e^{-(t-3)}, & 3 \le t < 4. \end{cases}$$

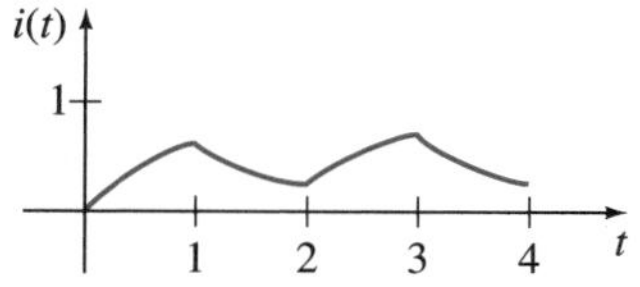

FIGURE 5.42

The graph of $i(t)$ on the interval $0 \le t < 4$ is given in Figure 5.42.

Beams

In Section 3.8 we saw that the static deflection $y(x)$ of a uniform beam of length L carrying a load $w(x)$ per unit length is found from the fourth-order differential equation

$$EI\frac{d^4y}{dx^4} = w(x), \tag{9}$$

where E is Young's modulus of elasticity and I is a moment of inertia of a cross section of the beam. The Laplace transform method is especially useful in solving (9) when $w(x)$ is piecewise defined. To apply the Laplace transform, we tacitly assume that $w(x)$ and $y(x)$ are defined on $(0, \infty)$ rather than on $(0, L)$. Note too, the next example is a boundary-value problem rather than an initial-value problem.

EXAMPLE 9 A Boundary-Value Problem

A beam of length L is embedded at both ends as shown in Figure 5.43. Find the deflection of the beam when the load is given by

$$w(x) = \begin{cases} w_0\left(1 - \dfrac{2}{L}x\right), & 0 < x < L/2 \\ 0, & L/2 < x < L. \end{cases}$$

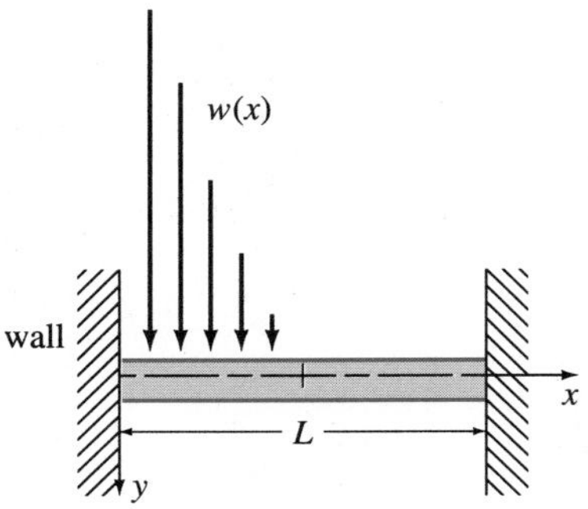

FIGURE 5.43

Solution Recall that since the beam is embedded at both ends, the boundary conditions are $y(0) = 0$, $y'(0) = 0$, $y(L) = 0$, $y'(L) = 0$. Also, you should verify

$$\begin{aligned} w(x) &= w_0\left(1 - \frac{2}{L}x\right) - w_0\left(1 - \frac{2}{L}x\right)\mathcal{U}\left(x - \frac{L}{2}\right) \\ &= \frac{2w_0}{L}\left[\frac{L}{2} - x + \left(x - \frac{L}{2}\right)\mathcal{U}\left(x - \frac{L}{2}\right)\right]. \end{aligned}$$

Transforming (9) with respect to the variable x gives

$$EI(s^4Y(s) - s^3y(0) - s^2y'(0) - sy''(0) - y'''(0)) = \frac{2w_0}{EIL}\left[\frac{L/2}{s} - \frac{1}{s^2} + \frac{1}{s^2}e^{-Ls/2}\right]$$

or

$$s^4Y(s) - sy''(0) - y'''(0) = \frac{2w_0}{EIL}\left[\frac{L/2}{s} - \frac{1}{s^2} + \frac{1}{s^2}e^{-Ls/2}\right].$$

If we let $c_1 = y''(0)$ and $c_2 = y'''(0)$, then

$$Y(s) = \frac{c_1}{s^3} + \frac{c_2}{s^4} + \frac{2w_0}{EIL}\left[\frac{L/2}{s^5} - \frac{1}{s^6} + \frac{1}{s^6}e^{-Ls/2}\right],$$

and consequently

$$\begin{aligned} y(x) &= \frac{c_1}{2!}\mathcal{L}^{-1}\left\{\frac{2!}{s^3}\right\} + \frac{c_2}{3!}\mathcal{L}^{-1}\left\{\frac{3!}{s^4}\right\} + \frac{2w_0}{EIL}\left[\frac{L/2}{4!}\mathcal{L}^{-1}\left\{\frac{4!}{s^5}\right\} - \frac{1}{5!}\mathcal{L}^{-1}\left\{\frac{5!}{s^6}\right\} + \frac{1}{5!}\mathcal{L}^{-1}\left\{\frac{5!}{s^6}e^{-Ls/2}\right\}\right] \\ &= \frac{c_1}{2}x^2 + \frac{c_2}{6}x^3 + \frac{w_0}{60EIL}\left[\frac{5L}{2}x^4 - x^5 + \left(x - \frac{L}{2}\right)^5\mathcal{U}\left(x - \frac{L}{2}\right)\right]. \end{aligned}$$

Applying the conditions $y(L) = 0$ and $y'(L) = 0$ to the last result yields a system of equations for c_1 and c_2:

$$\begin{aligned} c_1\frac{L^2}{2} + c_2\frac{L^3}{6} + \frac{49w_0L^4}{1920EI} &= 0 \\ c_1L + c_2\frac{L^2}{2} + \frac{85w_0L^3}{960EI} &= 0. \end{aligned}$$

Solving, we find $c_1 = 23w_0L^2/960EI$ and $c_2 = -9w_0L/40EI$. Thus the deflection is given by

$$y(x) = \frac{23w_0L^2}{1920EI}x^2 - \frac{9w_0L}{240EI}x^3 + \frac{w_0}{60EIL}\left[\frac{5L}{2}x^4 - x^5 + \left(x - \frac{L}{2}\right)^5 \mathcal{U}\left(x - \frac{L}{2}\right)\right]. \quad \bullet$$

EXERCISES 5.5

Answers to odd-numbered problems begin on page AN-25.

A table of the transforms of some basic functions is given in Appendix II.

In Problems 1–26 use the Laplace transform to solve the given differential equation subject to the indicated initial conditions. Where appropriate, write f in terms of unit step functions.

1. $\dfrac{dy}{dt} - y = 1, \quad y(0) = 0$
2. $\dfrac{dy}{dt} + 2y = t, \quad y(0) = -1$
3. $y' + 4y = e^{-4t}, \quad y(0) = 2$
4. $y' - y = \sin t, \quad y(0) = 0$
5. $y'' + 5y' + 4y = 0, \quad y(0) = 1, y'(0) = 0$
6. $y'' - 6y' + 13y = 0, \quad y(0) = 0, y'(0) = -3$
7. $y'' - 6y' + 9y = t, \quad y(0) = 0, y'(0) = 1$
8. $y'' - 4y' + 4y = t^3, \quad y(0) = 1, y'(0) = 0$
9. $y'' - 4y' + 4y = t^3e^{2t}, \quad y(0) = 0, y'(0) = 0$
10. $y'' - 2y' + 5y = 1 + t, \quad y(0) = 0, y'(0) = 4$
11. $y'' + y = \sin t, \quad y(0) = 1, y'(0) = -1$
12. $y'' + 16y = 1, \quad y(0) = 1, y'(0) = 2$
13. $y'' - y' = e^t \cos t, \quad y(0) = 0, y'(0) = 0$
14. $y'' - 2y' = e^t \sinh t, \quad y(0) = 0, y'(0) = 0$
15. $2y''' + 3y'' - 3y' - 2y = e^{-t},$
    $y(0) = 0, y'(0) = 0, y''(0) = 1$
16. $y''' + 2y'' - y' - 2y = \sin 3t,$
    $y(0) = 0, y'(0) = 0, y''(0) = 1$
17. $y^{(4)} - y = 0, \quad y(0) = 1, y'(0) = 0,$
    $y''(0) = -1, y'''(0) = 0$
18. $y^{(4)} - y = t, \quad y(0) = 0, y'(0) = 0, y''(0) = 0, y'''(0) = 0$
19. $y' + y = f(t), \quad y(0) = 0$
    where $f(t) = \begin{cases} 0, & 0 \le t < 1 \\ 5, & t \ge 1 \end{cases}$
20. $y' + y = f(t), \quad y(0) = 0$
    where $f(t) = \begin{cases} 1, & 0 \le t < 1 \\ -1, & t \ge 1 \end{cases}$
21. $y' + 2y = f(t), \quad y(0) = 0$
    where $f(t) = \begin{cases} t, & 0 \le t < 1 \\ 0, & t \ge 1 \end{cases}$
22. $y'' + 4y = f(t), \quad y(0) = 0, y'(0) = -1$
    where $f(t) = \begin{cases} 1, & 0 \le t < 1 \\ 0, & t \ge 1 \end{cases}$
23. $y'' + 4y = \sin t\,\mathcal{U}(t - 2\pi), \quad y(0) = 1, y'(0) = 0$
24. $y'' - 5y' + 6y = \mathcal{U}(t - 1), \quad y(0) = 0, y'(0) = 1$
25. $y'' + y = f(t), \quad y(0) = 0, y'(0) = 1$
    where $f(t) = \begin{cases} 0, & 0 \le t < \pi \\ 1, & \pi \le t < 2\pi \\ 0, & t \ge 2\pi \end{cases}$
26. $y'' + 4y' + 3y = 1 - \mathcal{U}(t - 2) - \mathcal{U}(t - 4) + \mathcal{U}(t - 6),$
    $y(0) = 0, y'(0) = 0$

In Problems 27 and 28 use the Laplace transform to solve the given differential equation subject to the indicated boundary conditions.

27. $y'' + 2y' + y = 0, \quad y'(0) = 2, y(1) = 2$
28. $y'' - 9y' + 20y = 1, \quad y(0) = 0, y'(1) = 0$

In Problems 29–38 use the Laplace transform to solve the given integral equation or integrodifferential equation.

29. $f(t) + \int_0^t (t - \tau)f(\tau)\,d\tau = t$
30. $f(t) = 2t - 4\int_0^t \sin \tau f(t - \tau)\,d\tau$
31. $f(t) = te^t + \int_0^t \tau f(t - \tau)\,d\tau$
32. $f(t) + 2\int_0^t f(\tau) \cos (t - \tau)\,d\tau = 4e^{-t} + \sin t$
33. $f(t) + \int_0^t f(\tau)\,d\tau = 1$

34. $f(t) = \cos t + \int_0^t e^{-\tau} f(t - \tau)\, d\tau$

35. $f(t) = 1 + t - \frac{8}{3}\int_0^t (\tau - t)^3 f(\tau)\, d\tau$

36. $t - 2f(t) = \int_0^t (e^{\tau} - e^{-\tau}) f(t - \tau)\, d\tau$

37. $y'(t) = 1 - \sin t - \int_0^t y(\tau)\, d\tau, \quad y(0) = 0$

38. $\frac{dy}{dt} + 6y(t) + 9\int_0^t y(\tau)\, d\tau = 1, \quad y(0) = 0$

39. Use equation (3) to determine the current $i(t)$ in a single-loop L-R-C circuit when $L = 0.005$ henry, $R = 1$ ohm, $C = 0.02$ farad, $E(t) = 100[1 - \mathcal{U}(t - 1)]$ volts, and $i(0) = 0$.

40. Solve Problem 39 when $E(t) = 100[t - (t - 1)\mathcal{U}(t - 1)]$.

41. Recall that the differential equation for the charge $q(t)$ on the capacitor in an R-C series circuit is

$$R\frac{dq}{dt} + \frac{1}{C}q = E(t),$$

where $E(t)$ is the impressed voltage. See Section 2.5. Use the Laplace transform to determine the charge $q(t)$ when $q(0) = 0$ and $E(t) = E_0 e^{-kt}$, $k > 0$. Consider two cases: $k \neq 1/RC$ and $k = 1/RC$.

42. Use the Laplace transform to determine the charge on the capacitor in an R-C series circuit if $q(0) = q_0$, $R = 10$ ohms, $C = 0.1$ farad, and $E(t)$ is as given in Figure 5.44.

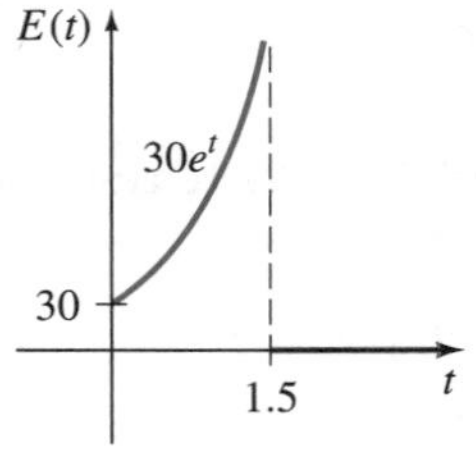

FIGURE 5.44

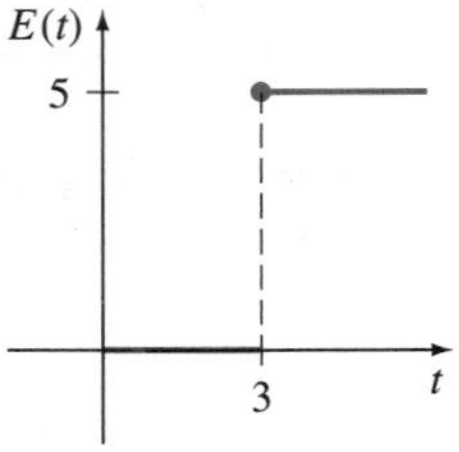

FIGURE 5.45

43. Use the Laplace transform to determine the charge on the capacitor in an R-C series circuit if $q(0) = 0$, $R = 2.5$ ohms, $C = 0.08$ farad, and $E(t)$ is as given in Figure 5.45.

44. (a) Use the Laplace transform to determine the charge $q(t)$ on the capacitor in an R-C series circuit when $q(0) = 0$, $R = 50$ ohms, $C = 0.01$ farad, and $E(t)$ is as given in Figure 5.46.

(b) Assume $E_0 = 100$ volts. Use a computer graphing program to graph $q(t)$ on the interval $0 \leq t \leq 6$. Use the graph to estimate q_{max}, the maximum value of the charge.

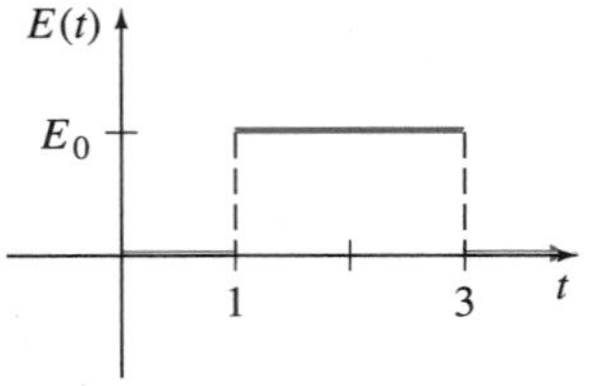

FIGURE 5.46

45. (a) Use the Laplace transform to determine the current $i(t)$ in a single loop L-R series circuit when $i(0) = 0$, $L = 1$ henry, $R = 10$ ohms, and $E(t)$ is as given in Figure 5.47.

(b) Use a computer graphing program to graph $i(t)$ on the interval $0 \leq t \leq 6$. Use the graph to estimate i_{max} and i_{min} the maximum and minimum values of the current.

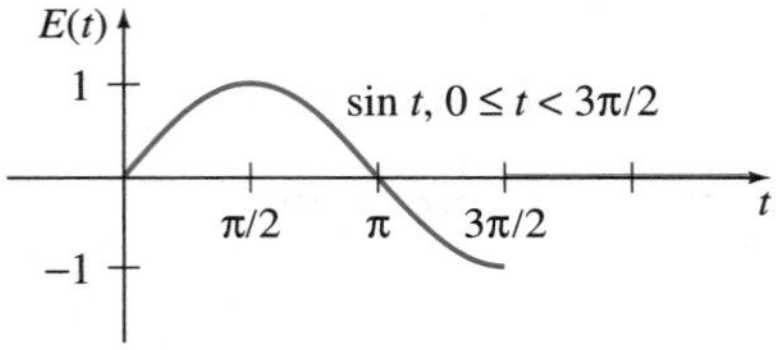

FIGURE 5.47

46. Solve equation (6) subject to $i(0) = 0$, where $E(t)$ is the meander function given in Figure 5.48. [*Hint:* See Problem 31 in Exercises 5.4.]

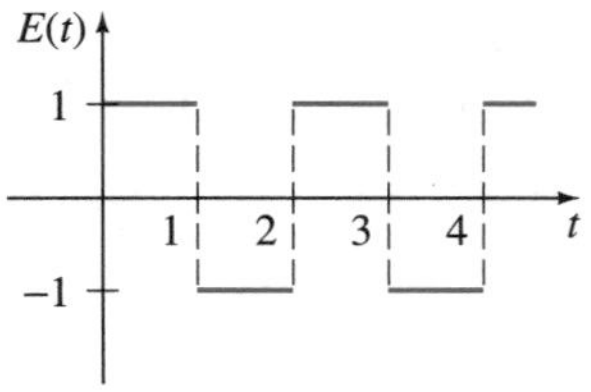

FIGURE 5.48

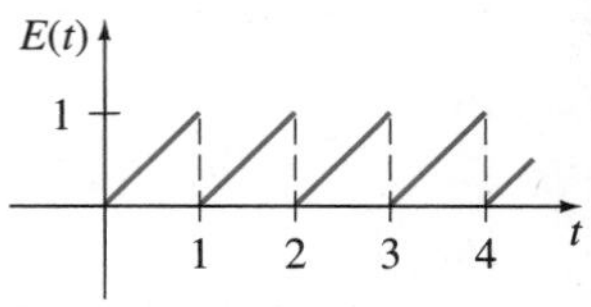

FIGURE 5.49

47. Solve equation (6) subject to $i(0) = 0$, where $E(t)$ is the sawtooth function given in Figure 5.49. Specify the solution for $0 \leq t < 2$. [*Hint:* See Problem 33 in Exercises 5.4.]

48. Recall that the differential equation for the instantaneous charge $q(t)$ on the capacitor in an L-R-C series circuit is

given by

$$L\frac{d^2q}{dt^2} + R\frac{dq}{dt} + \frac{1}{C}q = E(t). \quad \textbf{(10)}$$

See Section 3.7. Use the Laplace transform to determine $q(t)$ when $L = 1$ henry, $R = 20$ ohms, $C = 0.005$ farad, $E(t) = 150$ volts, $t > 0$, and $q(0) = 0$, $i(0) = 0$. What is the current $i(t)$? What is the charge $q(t)$ if the same constant voltage is turned off for $t \geq 2$?

49. Determine the charge $q(t)$ and current $i(t)$ for a series circuit in which $L = 1$ henry, $R = 20$ ohms, $C = 0.01$ farad, $E(t) = 120 \sin 10t$ volts, $q(0) = 0$, and $i(0) = 0$. What is the steady-state current?

50. Consider the battery of constant voltage, E_0 that charges the capacitor shown in Figure 5.50. If we divide by L and define $\lambda = R/2L$ and $\omega^2 = 1/LC$, then (10) becomes

$$\frac{d^2q}{dt^2} + 2\lambda\frac{dq}{dt} + \omega^2 q = \frac{E_0}{L}.$$

Use the Laplace transform to show that the solution of this equation, subject to $q(0) = 0$ and $i(0) = 0$, is

$$q(t) = \begin{cases} E_0C[1 - e^{-\lambda t}(\cosh\sqrt{\lambda^2 - \omega^2}\,t \\ \qquad + \dfrac{\lambda}{\sqrt{\lambda^2 - \omega^2}}\sinh\sqrt{\lambda^2 - \omega^2}\,t)], & \lambda > \omega \\ E_0C[1 - e^{-\lambda t}(1 + \lambda t)], & \lambda = \omega \\ E_0C[1 - e^{-\lambda t}(\cos\sqrt{\omega^2 - \lambda^2}\,t \\ \qquad + \dfrac{\lambda}{\sqrt{\omega^2 - \lambda^2}}\sin\sqrt{\omega^2 - \lambda^2}\,t)], & \lambda < \omega. \end{cases}$$

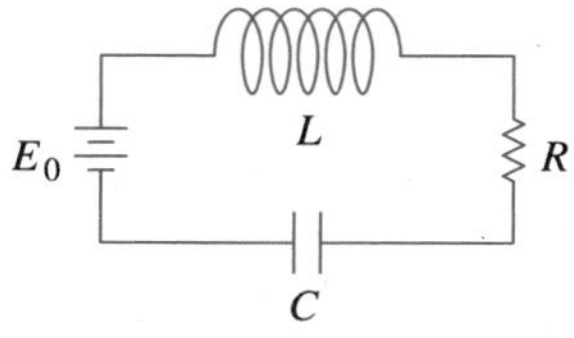

FIGURE 5.50

51. Use the Laplace transform to determine the charge $q(t)$ on the capacitor in an L-C series circuit when $q(0) = 0$, $i(0) = 0$, and $E(t) = E_0e^{-kt}$, $k > 0$.

52. Suppose a 32-lb weight stretches a spring 2 ft. If the weight is released from rest at the equilibrium position, determine the equation of motion if an impressed force $f(t) = \sin t$ acts on the system for $0 \leq t < 2\pi$ and is then removed. Ignore any damping forces. [*Hint:* Write the impressed force in terms of the unit step function.]

53. A 4-lb weight stretches a spring 2 ft. The weight is released from rest 18 in. above the equilibrium position, and the resulting motion takes place in a medium offering a damping force numerically equal to 7/8 times the instantaneous velocity. Use the Laplace transform to determine the equation of motion.

54. A 16-lb weight is attached to a spring whose constant is $k = 4.5$ lb/ft. Beginning at $t = 0$, a force equal to $f(t) = 4 \sin 3t + 2 \cos 3t$ acts on the system. Assuming that no damping forces are present, use the Laplace transform to find the equation of motion if the weight is released from rest from the equilibrium position.

55. A cantilever beam is embedded at its left end and free at its right end. Find the deflection $y(x)$ when the load is given by

$$w(x) = \begin{cases} w_0, & 0 < x < L/2 \\ 0, & L/2 < x < L. \end{cases}$$

56. Solve Problem 55 when the load is given by

$$w(x) = \begin{cases} 0, & 0 < x < L/3 \\ w_0, & L/3 < x < 2L/3 \\ 0, & 2L/3 < x < L. \end{cases}$$

57. Find the deflection $y(x)$ of a cantilever beam when the load is as given in Example 9.

58. A beam is embedded at its left end and simply supported at its right end. Find the deflection $y(x)$ when the load is as given in Problem 55.

59. *PROJECT PROBLEM* Consider a spring/mass system in which the mass m slides over a dry horizontal surface whose coefficient of kinetic friction is μ. The frictional force of constant magnitude $f_k = \mu mg$ acts opposite to the direction of motion. In the case when the mass is released from rest from a point $x(0) = a > 0$, then, in the absence of other external forces, differential equations describing the motion of the mass are

$$x'' + \omega^2 x = F, \qquad 0 < x < T/2,$$
$$x'' + \omega^2 x = -F, \quad T/2 < x < 3T/2,$$
$$x'' + \omega^2 x = F, \quad 3T/2 < x < 2T,$$

and so on, where $\omega^2 = k/m$, $F = f_k/m = \mu g$, and $T = 2\pi/\omega$. (See Problem 15, Exercises 3.11.)

(a) Explain why, in general, the initial displacement must satisfy $\omega^2|a| > F$.
(b) Explain why the interval $-F/\omega^2 \le x \le F/\omega^2$ is appropriately called the ''dead zone'' of the system.
(c) Use the Laplace transform and the concept of the meander function to solve for the displacement $x(t)$ for $t \ge 0$.
(d) Show that each successive oscillation is $2F/\omega^2$ shorter than the preceding one.
(e) Predict the long-term behavior of the system.

60. *PROJECT PROBLEM* A **difference equation** is the discrete analog of a differential equation. A solution of a difference equation is a sequence. The **z-transform** defined by $F(z) = \sum_{n=0}^{\infty} f_n z^n$ is the discrete analog of the Laplace transform $F(s) = \int_0^\infty e^{-st} f(t)\,dt$ and can be used to solve a difference equation subject to initial conditions.

(a) Write a short report on the important operational properties of the z-transform. Include in your discussion the z-transforms of f_{n+1} and f_{n+2} (these are the counterparts of the Laplace transforms of the first and second derivatives) and at least two examples of applications of difference equations.
(b) Find the z-transform of the sequence $f_n = a^n$, $n = 0, 1, 2, \ldots$, where a is a constant.
(c) Write out the first six terms of the sequence f_n defined by the second-order difference equation $f_{n+2} - \frac{1}{9}f_n = 0$ subject to $f_0 = 1$ and $f_1 = 0$.
(d) Use the z-transform to obtain a closed form solution of the initial-value problem in part (c).

61. *PROJECT PROBLEM* The Laplace transform is not well-suited to solving linear differential equations with variable coefficients; nevertheless, it can be used in some instances. Use the Laplace transform and appropriate operational properties to find a solution of $ty'' + 2ty' + 2y = t$, $y(0) = 0$.

62. *PROJECT PROBLEM* Consider a frictionless wire bent in the form of a smooth curve C. As shown in Figure 5.51, a bead with mass m starts from rest at the point $P(x,y)$ and slides down the wire under the influence of its own weight. We want to find the shape of the wire so that the time of descent to its lowest end (the origin in the figure) is the same constant T, regardless of the position of the starting point P. Let σ denote arc length along C.

(a) Show how the integral equation

$$T = \frac{1}{\sqrt{2g}} \int_0^y \frac{\sigma'(v)}{\sqrt{y - v}}\,dv$$

arises in this problem.
(b) Show how the Laplace transform is useful in the eventual solution of the problem. The curve C is called a **tautochrone** (from the Greek *tautos*, meaning *same*, and *chronos*, meaning *time*.)

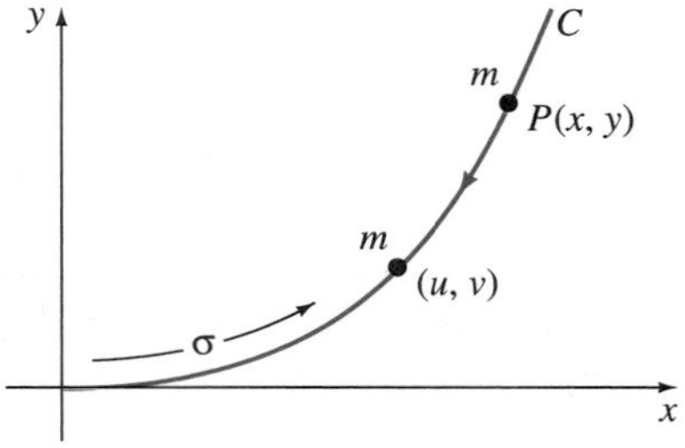

FIGURE 5.51

63. *PROJECT PROBLEM* In this problem you are asked to show how the convolution integral can be used to find a solution of a certain kind of initial-value problem.

(a) Use the Laplace transform to show that a solution of

$$ay'' + by' + cy = g(t), \qquad y(0) = 0, \quad y'(0) = 0$$

is given by $y_2(t) = \dfrac{1}{a}\, g * y_1$ where y_1 is a solution of

$$ay'' + by' + cy = 0, \qquad y(0) = 0, \quad y'(0) = 1.$$

(b) The interesting thing about the solution y_2 in part (a) is that it remains valid even though the Laplace transform of $g(t)$ *may not exist.* Use part (a) to find a solution of the initial-value problem

$$y'' + y = \sec t, \qquad y(0) = 0, \quad y'(0) = 0.$$

Note that $\sec t$ does not possess a Laplace transform.

64. *WRITING PROJECT* Write a report on the life, times, discoveries, and hard-luck career of Oliver Heaviside. Stress his contributions to the theory and applications of the Laplace transform. You might start with *Heaviside and the Operational Calculus,* J.L.B. Cooper, Mathematics Gazette, vol. 36, 1952, pp. 5–19.

5.6 DIRAC DELTA FUNCTION

Unit Impulse

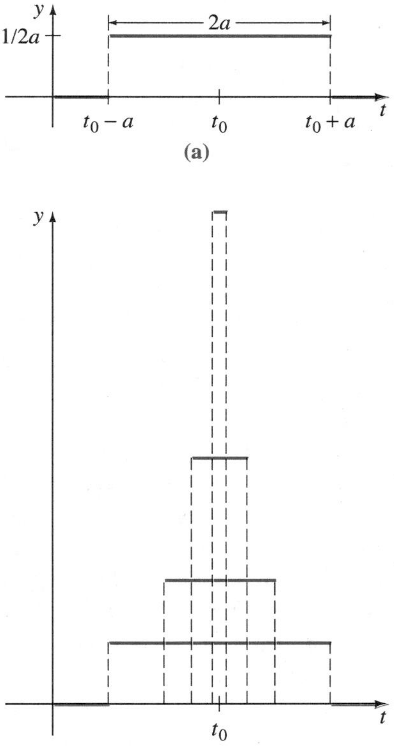

(a)

(b) behavior of δ_a as $a \to 0$

FIGURE 5.52

Mechanical systems are often acted upon by an external force (or emf in an electrical circuit) of large magnitude that acts only for a very short period of time. For example, a vibrating airplane wing could be struck by lightning, a mass on a spring could be given a sharp blow by a ball peen hammer, a ball (baseball, golf ball, tennis ball) could be sent soaring when struck violently by some kind of club (baseball bat, golf club, tennis racket). The function

$$\delta_a(t - t_0) = \begin{cases} 0, & 0 \le t < t_0 - a \\ \dfrac{1}{2a}, & t_0 - a \le t < t_0 + a \\ 0, & t \ge t_0 + a, \end{cases} \quad \textbf{(1)}$$

$a > 0$, $t_0 > 0$, shown in Figure 5.52(a), could serve as a mathematical model for such a force. For a small value of a, $\delta_a(t - t_0)$ is essentially a constant function of large magnitude that is ''on'' for just a very short period of time around t_0. The behavior of $\delta_a(t - t_0)$ as $a \to 0$ is illustrated in Figure 5.52(b). The function $\delta_a(t - t_0)$ is called a **unit impulse** since it possesses the integration property $\int_0^\infty \delta_a(t - t_0)\,dt = 1$.

Dirac Delta Function

In practice it is convenient to work with another type of unit impulse, a ''function'' that approximates $\delta_a(t - t_0)$ and is defined by the limit

$$\delta(t - t_0) = \lim_{a \to 0} \delta_a(t - t_0). \quad \textbf{(2)}$$

The latter expression, which is not a function at all, can be characterized by the two properties

$$\textbf{(i)} \quad \delta(t - t_0) = \begin{cases} \infty, & t = t_0 \\ 0, & t \ne t_0 \end{cases}, \quad \text{and} \quad \textbf{(ii)} \quad \int_0^\infty \delta(t - t_0)\,dt = 1.$$

The expression $\delta(t - t_0)$ is called the **Dirac delta function.**

It is possible to obtain the Laplace transform of the Dirac delta function by the formal assumption that $\mathscr{L}\{\delta(t - t_0)\} = \lim_{a \to 0} \mathscr{L}\{\delta_a(t - t_0)\}$.

THEOREM 5.11 Transform of the Dirac Delta Function

For $t_0 > 0$,
$$\mathscr{L}\{\delta(t - t_0)\} = e^{-st_0}. \quad \textbf{(3)}$$

Proof To begin, we can write $\delta_a(t - t_0)$ in terms of the unit step function by virtue of (4) and (5) of Section 5.3:

$$\delta_a(t - t_0) = \frac{1}{2a}[\mathscr{U}(t - (t_0 - a)) - \mathscr{U}(t - (t_0 + a))].$$

By linearity and (7) of Section 5.3 the Laplace transform of this last expression is

$$\mathscr{L}\{\delta_a(t-t_0)\} = \frac{1}{2a}\left[\frac{e^{-s(t_0-a)}}{s} - \frac{e^{-s(t_0+a)}}{s}\right] = e^{-st_0}\left(\frac{e^{sa}-e^{-sa}}{2sa}\right). \tag{4}$$

Since (4) has the indeterminate form 0/0 as $a \to 0$, we apply L'Hôpital's rule:

$$\mathscr{L}\{\delta(t-t_0)\} = \lim_{a\to 0}\mathscr{L}\{\delta_a(t-t_0)\} = e^{-st_0}\lim_{a\to 0}\left(\frac{e^{sa}-e^{-sa}}{2sa}\right) = e^{-st_0}. \qquad \bigcirc$$

Now when $t_0 = 0$, it seems plausible to conclude from (3) that

$$\mathscr{L}\{\delta(t)\} = 1.$$

The last result emphasizes the fact that $\delta(t)$ is not the usual type of function that we have been considering since we expect from Theorem 5.4 that $\mathscr{L}\{f(t)\} \to 0$ as $s \to \infty$.

EXAMPLE 1 Two Initial-Value Problems

Solve $y'' + y = 4\delta(t - 2\pi)$ subject to **(a)** $y(0) = 1$, $y'(0) = 0$, and **(b)** $y(0) = 0$, $y'(0) = 0$. The two initial-value problems could serve as models for describing the motion of a mass on a spring moving in a medium in which damping is negligible. At $t = 2\pi$ seconds the mass is given a sharp blow. In (a) the mass is released from rest 1 unit below the equilibrium position. In (b) the mass is at rest in the equilibrium position.

Solution

(a) From (3) the Laplace transform of the differential equation is

$$s^2Y(s) - s + Y(s) = 4e^{-2\pi s} \quad \text{or} \quad Y(s) = \frac{s}{s^2+1} + \frac{4e^{-2\pi s}}{s^2+1}.$$

Using the inverse form of the second translation theorem, we find

$$y(t) = \cos t + 4\sin(t - 2\pi)\mathscr{U}(t - 2\pi).$$

Since $\sin(t - 2\pi) = \sin t$, the foregoing solution can be written as

$$y(t) = \begin{cases} \cos t, & 0 \le t < 2\pi \\ \cos t + 4\sin t, & t \ge 2\pi. \end{cases} \tag{5}$$

In Figure 5.53 we see from the graph of (5) that the mass is exhibiting simple harmonic motion until it is struck at $t = 2\pi$. The influence of the unit impulse is to increase the amplitude of vibration to $\sqrt{17}$ for $t > 2\pi$.

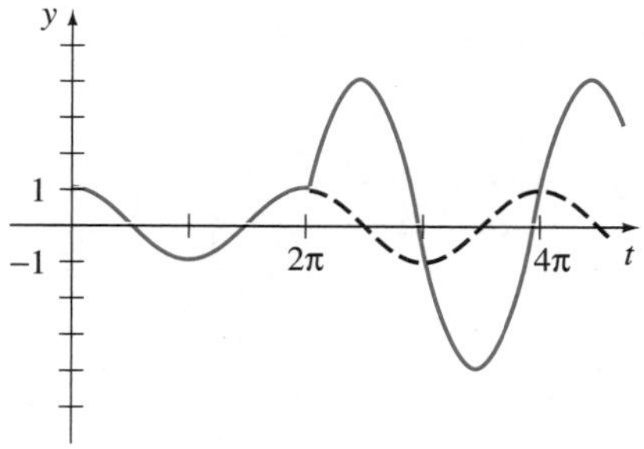

FIGURE 5.53

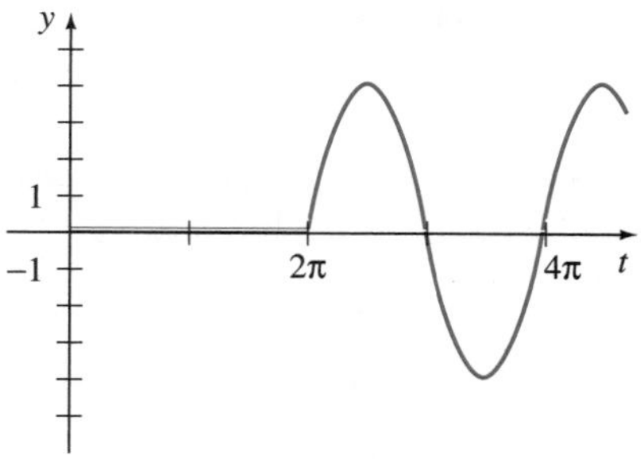

FIGURE 5.54

(b) In this case the transform of the equation is simply

$$Y(s) = \frac{4e^{-2\pi s}}{s^2 + 1}$$

and so
$$y(t) = 4\sin(t - 2\pi)\,\mathcal{U}(t - 2\pi)$$

$$= \begin{cases} 0 & 0 \le t < 2\pi \\ 4\sin t, & t \ge 2\pi. \end{cases} \qquad \textbf{(6)}$$

The graph of (6) in Figure 5.54 shows, as we would expect from the initial conditions, that the mass exhibits no motion until it is struck at $t = 2\pi$. ●

Remark If $\delta(t - t_0)$ were a function in the usual sense, then property (i) on page 293 would imply $\int_0^\infty \delta(t - t_0)\, dt = 0$ rather than $\int_0^\infty \delta(t - t_0)\, dt = 1$. Since the Dirac delta function did not ''behave'' like an ordinary function, even though its users produced correct results, it was met initially with great scorn by mathematicians. However, in the 1940s Dirac's controversial function was put on a rigorous footing by the French mathematician Laurent Schwartz in his book *La Théorie de distribution* and this, in turn, led to an entirely new branch of mathematics, known as the **theory of distributions** or **generalized functions**. See Problem 17 in Exercises 5.6. In this theory, (2) is not an accepted definition of $\delta(t - t_0)$, nor does one speak of a function whose values are either ∞ or 0. Although we shall not pursue this topic any further, suffice it to say that the Dirac delta function is best characterized by its effect on other functions. If f is a continuous function, then

$$\int_0^\infty f(t)\,\delta(t - t_0)\, dt = f(t_0) \qquad \textbf{(7)}$$

can be taken as the *definition* of $\delta(t - t_0)$. This result is known as the **sifting property** since $\delta(t - t_0)$ has the effect of sifting the value $f(t_0)$ out of the set of values of f on $[0, \infty)$. Note that property (ii) (with $f(t) = 1$) and (3) (with $f(t) = e^{-st}$) are consistent with (7).

EXERCISES 5.6

Answers to odd-numbered problems begin on page AN-26.

In Problems 1–12 use the Laplace transform to solve the given differential equation subject to the indicated initial conditions.

1. $y' - 3y = \delta(t-2), \quad y(0) = 0$
2. $y' + y = \delta(t-1), \quad y(0) = 2$
3. $y'' + y = \delta(t-2\pi), \quad y(0) = 0, y'(0) = 1$
4. $y'' + 16y = \delta(t-2\pi), \quad y(0) = 0, y'(0) = 0$
5. $y'' + y = \delta\left(t - \frac{\pi}{2}\right) + \delta\left(t - \frac{3\pi}{2}\right), \quad y(0) = 0, y'(0) = 0$
6. $y'' + y = \delta(t-2\pi) + \delta(t-4\pi), \quad y(0) = 1, y'(0) = 0$
7. $y'' + 2y' = \delta(t-1), \quad y(0) = 0, y'(0) = 1$
8. $y'' - 2y' = 1 + \delta(t-2), \quad y(0) = 0, y'(0) = 1$
9. $y'' + 4y' + 5y = \delta(t-2\pi), \quad y(0) = 0, y'(0) = 0$
10. $y'' + 2y' + y = \delta(t-1), \quad y(0) = 0, y'(0) = 0$
11. $y'' + 4y' + 13y = \delta(t-\pi) + \delta(t-3\pi)$,
$y(0) = 1, y'(0) = 0$
12. $y'' - 7y' + 6y = e^t + \delta(t-2) + \delta(t-4)$;
$y(0) = 0, y'(0) = 0$
13. A uniform beam of length L carries a concentrated load w_0 at $x = L/2$. The beam is embedded at its left end and is free at its right end. Use the Laplace transform to determine the deflection $y(x)$ from
$$EI\frac{d^4y}{dx^4} = w_0\delta\left(x - \frac{L}{2}\right),$$
where $y(0) = 0$, $y'(0) = 0$, $y''(L) = 0$, and $y'''(L) = 0$.
14. Solve the differential equation in Problem 13 subject to $y(0) = 0$, $y'(0) = 0$, $y(L) = 0$, $y'(L) = 0$. In this case the beam is embedded at both ends. See Figure 5.55.

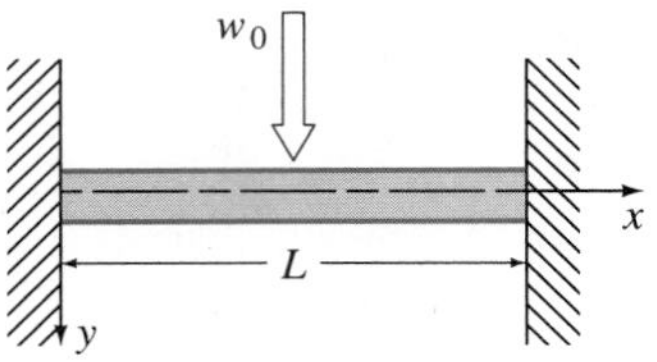

FIGURE 5.55

15. To emphasize the unusual nature of the Dirac delta function, show that the "solution" of the initial-value problem $y'' + \omega^2 y = \delta(t)$, $y(0) = 0$, $y'(0) = 0$ does not satisfy the initial condition $y'(0) = 0$.
16. *PROJECT PROBLEM*
   (a) Show formally that the Laplace transform of the nth derivative of $\delta(t - t_0)$ is given by $\mathscr{L}\{\delta^{(n)}(t - t_0)\} = s^n e^{-t_0 s}$, $t_0 \geq 0$.
   (b) Solve the initial-value problem $y'' + 9y = \delta''(t)$, $y(0) = 0$, $y'(0) = 0$.
   (c) Look up the mean value theorem for definite integrals in your calculus text and then give a formal proof of the sifting property (7).
   (d) Solve the initial-value problem $y'' + 2y' + 2y = \cos t\,\delta\left(t - \frac{3\pi}{2}\right)$, $y(0) = 1$, $y'(0) = -1$. Use a computer graphing program to obtain a graph of the solution.
17. *WRITING PROJECT* Write a report on the notion of **generalized functions**. A classic book in the field is *Fourier Analysis and Generalized Functions*, by M. J. Lighthill, Cambridge University Press, 1958.

5.7 SYSTEMS OF DIFFERENTIAL EQUATIONS

When initial conditions are specified, the Laplace transform reduces a system of linear differential equations with constant coefficients to a set of simultaneous algebraic equations in the transformed functions.

EXAMPLE 1 System of DEs Transformed into an Algebraic System

Solve

$$\begin{aligned} 2x' + y' - y &= t \\ x' + y' \quad &= t^2 \end{aligned} \tag{1}$$

subject to $x(0) = 1$, $y(0) = 0$.

Solution If $X(s) = \mathscr{L}\{x(t)\}$ and $Y(s) = \mathscr{L}\{y(t)\}$, then after transforming each equation, we obtain

$$2[sX(s) - x(0)] + sY(s) - y(0) - Y(s) = \frac{1}{s^2}$$

$$sX(s) - x(0) \ + sY(s) - y(0) \qquad = \frac{2}{s^3}$$

or

$$\begin{aligned} 2sX(s) + (s-1)Y(s) &= 2 + \frac{1}{s^2} \\ sX(s) + \qquad sY(s) &= 1 + \frac{2}{s^3}. \end{aligned} \tag{2}$$

Multiplying the second equation of (2) by 2 and subtracting yield

$$(-s-1)Y(s) = \frac{1}{s^2} - \frac{4}{s^3} \quad \text{or} \quad Y(s) = \frac{4-s}{s^3(s+1)}. \tag{3}$$

Now by partial fractions,

$$Y(s) = \frac{5}{s} - \frac{5}{s^2} + \frac{4}{s^3} - \frac{5}{s+1},$$

and so

$$\begin{aligned} y(t) &= 5\mathscr{L}^{-1}\left\{\frac{1}{s}\right\} - 5\mathscr{L}^{-1}\left\{\frac{1}{s^2}\right\} + 2\mathscr{L}^{-1}\left\{\frac{2!}{s^3}\right\} - 5\mathscr{L}^{-1}\left\{\frac{1}{s+1}\right\} \\ &= 5 - 5t + 2t^2 - 5e^{-t}. \end{aligned}$$

By the second equation of (2),

$$X(s) = -Y(s) + \frac{1}{s} + \frac{2}{s^4},$$

from which it follows that

$$\begin{aligned} x(t) &= -\mathscr{L}^{-1}\{Y(s)\} + \mathscr{L}^{-1}\left\{\frac{1}{s}\right\} + \frac{2}{3!}\mathscr{L}^{-1}\left\{\frac{3!}{s^4}\right\} \\ &= -4 + 5t - 2t^2 + \frac{1}{3}t^3 + 5e^{-t}. \end{aligned}$$

Hence we conclude that the solution of the given system (1) is

$$x(t) = -4 + 5t - 2t^2 + \frac{1}{3}t^3 + 5e^{-t}$$
$$y(t) = 5 - 5t + 2t^2 - 5e^{-t}. \tag{4}$$

Applications

Let us turn now to some elementary applications involving systems of differential equations. The solutions of the problems that we shall consider can be obtained either by the method of Section 3.9 or through the use of the Laplace transform.

Coupled Springs

Two masses m_1 and m_2 are connected to two springs A and B of negligible mass having spring constants k_1 and k_2, respectively. In turn, the two springs are attached as shown in Figure 5.56. Let $x_1(t)$ and $x_2(t)$ denote the vertical displacements of the masses from their equilibrium positions. When the system is in motion, spring B is subject to both an elongation and a compression; hence its net elongation is $x_2 - x_1$. Therefore it follows from Hooke's law that springs A and B exert forces $-k_1x_1$ and $k_2(x_2 - x_1)$, respectively, on m_1. If no external force is impressed on the system and if no damping force is present, then the net force on m_1 is $-k_1x_1 + k_2(x_2 - x_1)$. By Newton's second law we can write

$$m_1\frac{d^2x_1}{dt^2} = -k_1x_1 + k_2(x_2 - x_1).$$

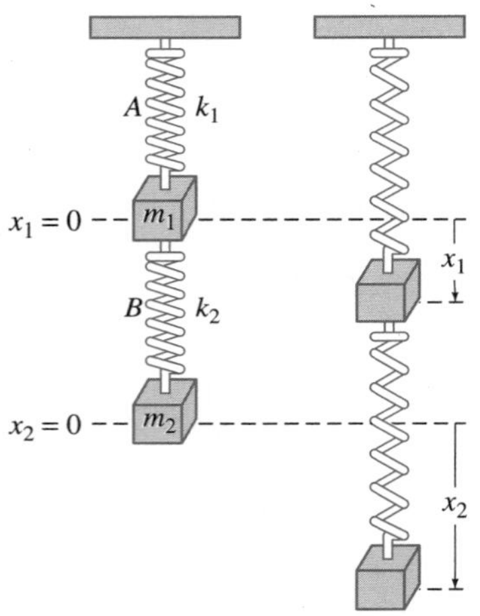

FIGURE 5.56

Similarly, the net force exerted on mass m_2 is due solely to the net elongation of B; that is $-k_2(x_2 - x_1)$. Hence we have

$$m_2\frac{d^2x_2}{dt^2} = -k_2(x_2 - x_1).$$

In other words, the motion of the coupled system is represented by the system of simultaneous second-order differential equations

$$m_1x_1'' = -k_1x_1 + k_2(x_2 - x_1)$$
$$m_2x_2'' = -k_2(x_2 - x_1). \tag{5}$$

In the next example we solve (5) under the assumption that $k_1 = 6$, $k_2 = 4$, $m_1 = 1$, $m_2 = 1$ and that the masses start from their equilibrium positions with opposite unit velocities.

EXAMPLE 2 Coupled Springs

Solve

$$x_1'' + 10x_1 \quad - 4x_2 = 0$$
$$-\ 4x_1 + x_2'' + 4x_2 = 0 \tag{6}$$

subject to $x_1(0) = 0$, $x_1'(0) = 1$, $x_2(0) = 0$, $x_2'(0) = -1$.

Solution The Laplace transform of each equation is

$$s^2X_1(s) - sx_1(0) - x_1'(0) + 10X_1(s) - 4X_2(s) = 0$$

$$-4X_1(s) + s^2X_2(s) - sx_2(0) - x_2'(0) + 4X_2(s) = 0,$$

where $X_1(s) = \mathscr{L}\{x_1(t)\}$ and $X_2(s) = \mathscr{L}\{x_2(t)\}$. The preceding system is the same as

$$\begin{aligned} (s^2 + 10)X_1(s) - \qquad 4X_2(s) &= 1 \\ -4X_1(s) + (s^2 + 4)X_2(s) &= -1. \end{aligned} \tag{7}$$

Solving (7) for $X_1(s)$ and using partial fractions on the result yield

$$X_1(s) = \frac{s^2}{(s^2 + 2)(s^2 + 12)} = -\frac{1/5}{s^2 + 2} + \frac{6/5}{s^2 + 12},$$

and therefore

$$\begin{aligned} x_1(t) &= -\frac{1}{5\sqrt{2}}\mathscr{L}^{-1}\left\{\frac{\sqrt{2}}{s^2 + 2}\right\} + \frac{6}{5\sqrt{12}}\mathscr{L}^{-1}\left\{\frac{\sqrt{12}}{s^2 + 12}\right\} \\ &= -\frac{\sqrt{2}}{10}\sin\sqrt{2}t + \frac{\sqrt{3}}{5}\sin 2\sqrt{3}t. \end{aligned}$$

Substituting the expression for $X_1(s)$ into the first equation of (7) gives us

$$X_2(s) = -\frac{s^2 + 6}{(s^2 + 2)(s^2 + 12)} = -\frac{2/5}{s^2 + 2} - \frac{3/5}{s^2 + 12}$$

and

$$\begin{aligned} x_2(t) &= -\frac{2}{5\sqrt{2}}\mathscr{L}^{-1}\left\{\frac{\sqrt{2}}{s^2 + 2}\right\} - \frac{3}{5\sqrt{12}}\mathscr{L}^{-1}\left\{\frac{\sqrt{12}}{s^2 + 12}\right\} \\ &= -\frac{\sqrt{2}}{5}\sin\sqrt{2}t - \frac{\sqrt{3}}{10}\sin 2\sqrt{3}t. \end{aligned}$$

Finally the solution to the given system (6) is

$$\begin{aligned} x_1(t) &= -\frac{\sqrt{2}}{10}\sin\sqrt{2}t + \frac{\sqrt{3}}{5}\sin 2\sqrt{3}t \\ x_2(t) &= -\frac{\sqrt{2}}{5}\sin\sqrt{2}t - \frac{\sqrt{3}}{10}\sin 2\sqrt{3}t. \end{aligned} \tag{8}$$

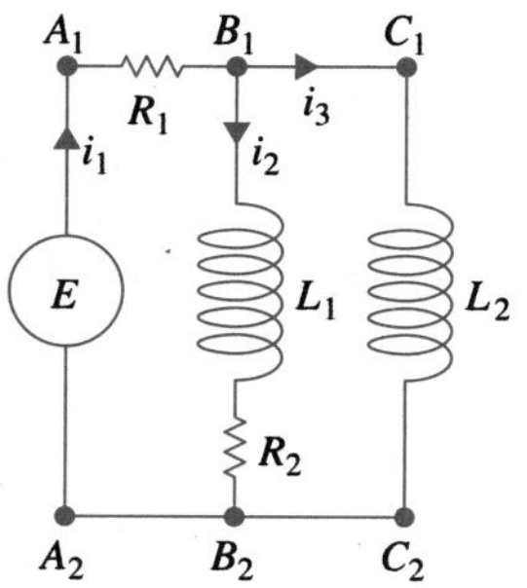

FIGURE 5.57

Networks

An electrical network having more than one loop also gives rise to simultaneous differential equations. As shown in Figure 5.57 the current $i_1(t)$ splits in the directions shown at point B_1 called a *branch point* of the network. By Kirchhoff's first law we can write

$$i_1(t) = i_2(t) + i_3(t). \quad \textbf{(9)}$$

In addition we can also apply **Kirchhoff's second law** to each loop. For loop $A_1B_1B_2A_2A_1$, summing the voltage drops across each part of the loop gives

$$E(t) = i_1R_1 + L_1\frac{di_2}{dt} + i_2R_2. \quad \textbf{(10)}$$

Similarly, for loop $A_1B_1C_1C_2B_2A_2A_1$ we find

$$E(t) = i_1R_1 + L_2\frac{di_3}{dt}. \quad \textbf{(11)}$$

Using (9) to eliminate i_1 in (10) and (11) yields two first-order equations for the currents $i_2(t)$ and $i_3(t)$:

$$\begin{aligned} L_1\frac{di_2}{dt} + (R_1 + R_2)i_2 + R_1i_3 &= E(t) \\ L_2\frac{di_3}{dt} + \qquad R_1i_2 + R_1i_3 &= E(t). \end{aligned} \quad \textbf{(12)}$$

Given the natural initial conditions $i_2(0) = 0$, $i_3(0) = 0$, the system (12) is amenable to solution by the Laplace transform.

We leave it as an exercise (see Problem 18) to show that the system of differential equations describing the currents $i_1(t)$ and $i_2(t)$ in the network containing a resistor, an inductor, and a capacitor shown in Figure 5.58 is

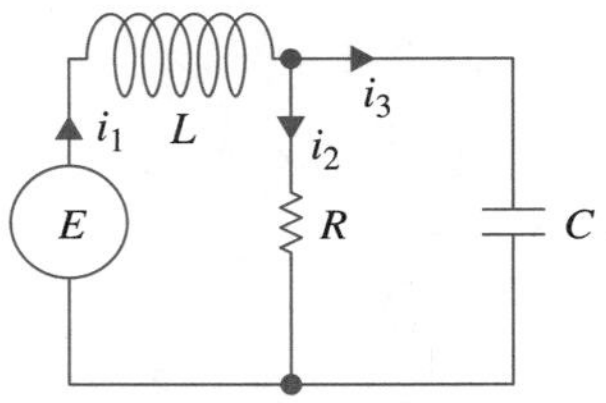

FIGURE 5.58

$$\begin{aligned} L\frac{di_1}{dt} + Ri_2 \qquad &= E(t) \\ RC\frac{di_2}{dt} + i_2 - i_1 &= 0. \end{aligned} \quad \textbf{(13)}$$

EXAMPLE 3 An Electrical Network

Solve the system (13) under the conditions $E = 60$ volts, $L = 1$ henry, $R = 50$ ohms, $C = 10^{-4}$ farad, and i_1 and i_2 are initially zero.

Solution We must solve

$$\begin{aligned} \frac{di_1}{dt} + 50i_2 \qquad &= 60 \\ 50(10^{-4})\frac{di_2}{dt} + i_2 - i_1 &= 0 \end{aligned}$$

subject to $i_1(0) = 0$, $i_2(0) = 0$.

Applying the Laplace transform to each equation of the system and simplifying give

$$sI_1(s) + \qquad 50I_2(s) = \frac{60}{s}$$

$$-200I_1(s) + (s + 200)I_2(s) = 0,$$

where $I_1(s) = \mathcal{L}\{i_1(t)\}$ and $I_2(s) = \mathcal{L}\{i_2(t)\}$. Solving the system for I_1 and I_2 and decomposing the results into partial fractions give

$$I_1(s) = \frac{60s + 12{,}000}{s(s+100)^2} = \frac{6/5}{s} - \frac{6/5}{s+100} - \frac{60}{(s+100)^2}$$

$$I_2(s) = \frac{12{,}000}{s(s+100)^2} = \frac{6/5}{s} - \frac{6/5}{s+100} - \frac{120}{(s+100)^2}.$$

Taking the inverse Laplace transform, we find the currents to be

$$i_1(t) = \frac{6}{5} - \frac{6}{5}e^{-100t} - 60te^{-100t}$$

$$i_2(t) = \frac{6}{5} - \frac{6}{5}e^{-100t} - 120te^{-100t}.$$

Note that both $i_1(t)$ and $i_2(t)$ in Example 3 tend toward the value $E/R = 6/5$ as $t \to \infty$. Furthermore since the current through the capacitor is $i_3(t) = i_1(t) - i_2(t) = 60te^{-100t}$, we observe that $i_3(t) \to 0$ as $t \to \infty$.

EXERCISES 5.7

Answers to odd-numbered problems begin on page AN-26.

In Problems 1–12 use the Laplace transform to solve the given system of differential equations.

1. $\dfrac{dx}{dt} = -x + y$
   $\dfrac{dy}{dt} = 2x$
   $x(0) = 0,\ y(0) = 1$

2. $\dfrac{dx}{dt} = 2y + e^t$
   $\dfrac{dy}{dt} = 8x - t$
   $x(0) = 1,\ y(0) = 1$

3. $\dfrac{dx}{dt} = x - 2y$
   $\dfrac{dy}{dt} = 5x - y$
   $x(0) = -1,\ y(0) = 2$

4. $\dfrac{dx}{dt} + 3x + \dfrac{dy}{dt} = 1$
   $\dfrac{dx}{dt} - x + \dfrac{dy}{dt} - y = e^t$
   $x(0) = 0,\ y(0) = 0$

5. $2\dfrac{dx}{dt} + \dfrac{dy}{dt} - 2x = 1$
   $\dfrac{dx}{dt} + \dfrac{dy}{dt} - 3x - 3y = 2$
   $x(0) = 0,\ y(0) = 0$

6. $\dfrac{dx}{dt} + x - \dfrac{dy}{dt} + y = 0$
   $\dfrac{dx}{dt} + \dfrac{dy}{dt} + 2y = 0$
   $x(0) = 0,\ y(0) = 1$

7. $\dfrac{d^2x}{dt^2} + x - y = 0$
   $\dfrac{d^2y}{dt^2} + y - x = 0$
   $x(0) = 0,\ x'(0) = -2$
   $y(0) = 0,\ y'(0) = 1$

8. $\dfrac{d^2x}{dt^2} + \dfrac{dx}{dt} + \dfrac{dy}{dt} = 0$
   $\dfrac{d^2y}{dt^2} + \dfrac{dy}{dt} - 4\dfrac{dx}{dt} = 0$
   $x(0) = 1,\ x'(0) = 0,$
   $y(0) = -1,\ y'(0) = 5$

9. $\dfrac{d^2x}{dt^2} + \dfrac{d^2y}{dt^2} = t^2$

$\dfrac{d^2x}{dt^2} - \dfrac{d^2y}{dt^2} = 4t$

$x(0) = 8,\ x'(0) = 0,$

$y(0) = 0,\ y'(0) = 0$

10. $\dfrac{dx}{dt} - 4x + \dfrac{d^3y}{dt^3} = 6\sin t$

$\dfrac{dx}{dt} + 2x - 2\dfrac{d^3y}{dt^3} = 0$

$x(0) = 0,\ y(0) = 0,$

$y'(0) = 0,\ y''(0) = 0$

11. $\dfrac{d^2x}{dt^2} + 3\dfrac{dy}{dt} + 3y = 0$

$\dfrac{d^2x}{dt^2} \qquad + 3y = te^{-t}$

$x(0) = 0,\ x'(0) = 2,\ y(0) = 0$

12. $\dfrac{dx}{dt} = 4x - 2y + 2\mathcal{U}(t-1)$

$\dfrac{dy}{dt} = 3x - \ y + \ \mathcal{U}(t-1)$

$x(0) = 0,\ y(0) = 1/2$

13. Solve system (5) when $k_1 = 3$, $k_2 = 2$, $m_1 = 1$, $m_2 = 1$ and $x_1(0) = 0$, $x_1'(0) = 1$, $x_2(0) = 1$, $x_2'(0) = 0$.

14. Derive the system of differential equations describing the straight-line vertical motion of the coupled springs shown in Figure 5.59. Use the Laplace transform to solve the system when $k_1 = 1$, $k_2 = 1$, $k_3 = 1$, $m_1 = 1$, $m_2 = 1$, and $x_1(0) = 0$, $x_1'(0) = -1$, $x_2(0) = 0$, $x_2'(0) = 1$.

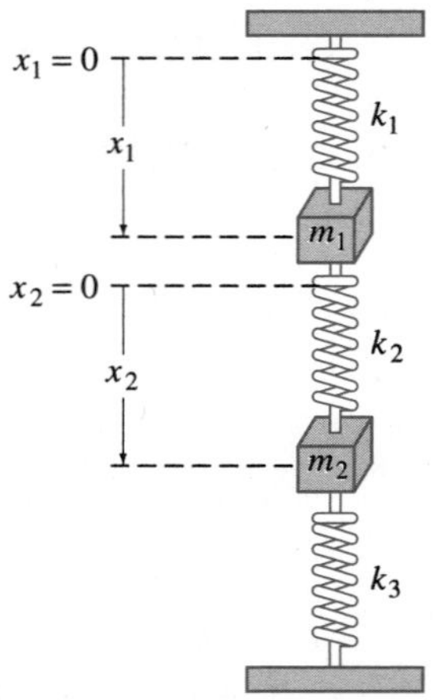

FIGURE 5.59

15. (a) Show that the system of differential equations for the currents $i_2(t)$ and $i_3(t)$ in the electrical network shown in Figure 5.60 is

$$L_1\frac{di_2}{dt} + Ri_2 + Ri_3 = E(t)$$

$$L_2\frac{di_3}{dt} + Ri_2 + Ri_3 = E(t).$$

(b) Solve the system in part (a) if $R = 5$ ohms, $L_1 = 0.01$ henry, $L_2 = 0.0125$ henry, $E = 100$ volts, $i_2(0) = 0$, and $i_3(0) = 0$.

(c) Determine the current $i_1(t)$.

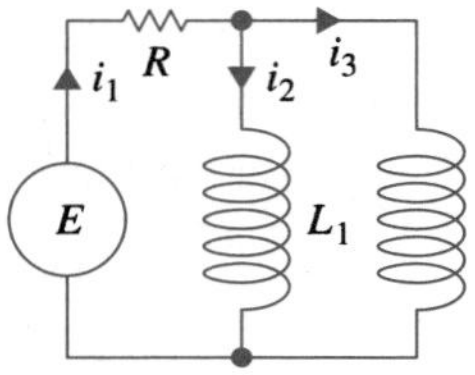

FIGURE 5.60

16. (a) Show that the system of differential equations for the currents $i_2(t)$ and $i_3(t)$ in the electrical network shown in Figure 5.61 is

$$L\frac{di_2}{dt} + L\frac{di_3}{dt} + R_1 i_2 = E(t)$$

$$-R_1\frac{di_2}{dt} + R_2\frac{di_3}{dt} + \frac{1}{C}i_3 = 0.$$

(b) Solve the system in part (a) if $R_1 = 10$ ohms, $R_2 = 5$ ohms, $L = 1$ henry, $C = 0.2$ farad,

$$E(t) = \begin{cases} 120, & 0 \le t < 2 \\ 0, & t \ge 2, \end{cases}$$

$i_2(0) = 0$, and $i_3(0) = 0$.

(c) Determine the current $i_1(t)$.

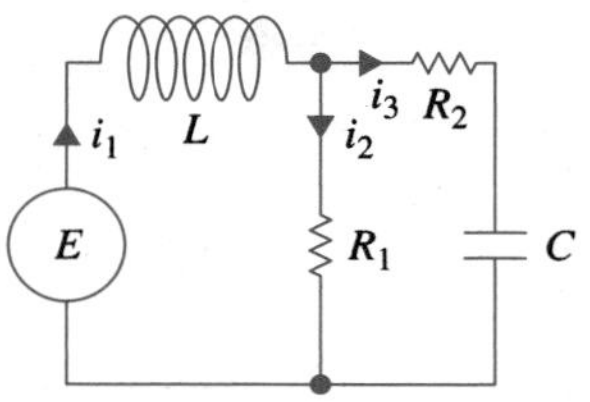

FIGURE 5.61

17. Solve the system given in (12) when $R_1 = 6$ ohms, $R_2 = 5$ ohms, $L_1 = 1$ henry, $L_2 = 1$ henry, $E(t) = 50 \sin t$ volts.

18. Derive the system of equations (13).

19. Solve (13) when $E = 60$ volts, $L = 1/2$ henry, $R = 50$ ohms, $C = 10^{-4}$ farad, $i_1(0) = 0$, $i_2(0) = 0$.

20. Solve (13) when $E = 60$ volts, $L = 2$ henrys, $R = 50$ ohms, $C = 10^{-4}$ farad, $i_1(0) = 0$, $i_2(0) = 0$.

21. (a) Show that the system of differential equations for the charge on the capacitor $q(t)$ and the current $i_3(t)$ in the electrical network shown in Figure 5.62 is

$$R_1 \frac{dq}{dt} + \frac{1}{C} q + R_1 i_3 = E(t)$$

$$L \frac{di_3}{dt} + R_2 i_3 - \frac{1}{C} q = 0.$$

(b) Find the charge on the capacitor when $L = 1$ henry, $R_1 = 1$ ohm, $R_2 = 1$ ohm, $C = 1$ farad,

$$E(t) = \begin{cases} 0, & 0 < t < 1 \\ 50e^{-t}, & t \geq 1, \end{cases}$$

$i_3(0) = 0$, and $q(0) = 0$.

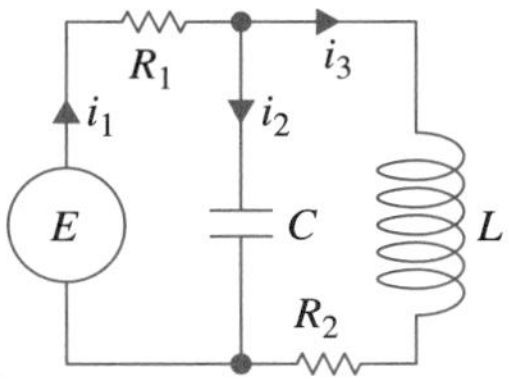

FIGURE 5.62

22. A double pendulum oscillates in a vertical plane under the influence of gravity. See Figure 5.63. For small displacements $\theta_1(t)$ and $\theta_2(t)$, it can be shown that the differential equations of motion are

$$(m_1 + m_2)l_1^2\theta_1'' + m_2 l_1 l_2 \theta_2'' + (m_1 + m_2) l_1 g \theta_1 = 0$$

$$m_2 l_2^2 \theta_2'' + m_2 l_1 l_2 \theta_1'' + m_2 l_2 g \theta_2 = 0.$$

Use the Laplace transform to solve the system when $m_1 = 3$, $m_2 = 1$, $l_1 = l_2 = 16$, $\theta_1(0) = 1$, $\theta_2(0) = -1$, $\theta_1'(0) = 0$, and $\theta_2'(0) = 0$.

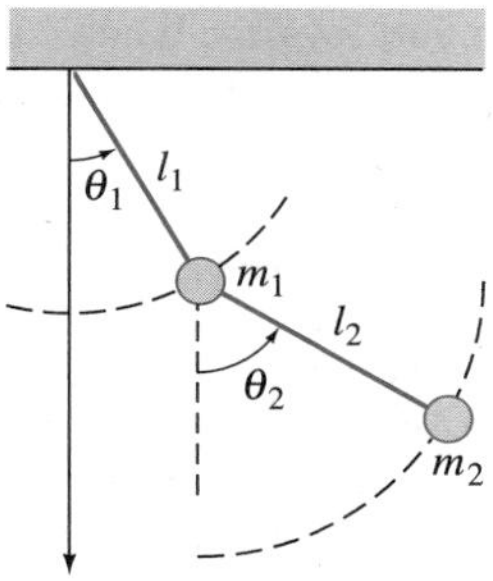

FIGURE 5.63

23. PROJECT PROBLEM Use the Laplace transform to solve the system of differential equations

$$\frac{dx}{dt} + x + \frac{d^2y}{dt^2} = e^t$$

$$\frac{dy}{dt} + x = -e^t$$

$x(0) = -1$, $y(0) = 1$, $y'(0) = 0$.

Examine your solution in careful detail. State, and try to explain, any peculiarities.

24. PROJECT PROBLEM When a radioactive substance decays it transmutes into another radioactive substance; this substance in turn transmutes yet into a third substance, and the process continues until a stable element is reached. For example, in the uranium series, U-238 → Th-234 → · · · · → Pb-206, where Pb-206 is a stable isotope of lead. The half-lives of the elements in a radioactive series can range from millions of years to a small fraction of a second.

Suppose that a radioactive series is described schematically by $A \xrightarrow{-\lambda_1} B \xrightarrow{-\lambda_2} C$, where $k_1 = -\lambda_1 < 0$ and $k_2 = -\lambda_2 < 0$ are the decay constants for substances A and B, respectively, and C is a stable element.

(a) Construct a mathematical model in the form of a system of linear differential equations for the amounts $x(t)$, $y(t)$, and $z(t)$ of the substances A, B, and C, respectively, remaining at any time based on the linear growth and decay models discussed in Section 2.5.

(b) Use the Laplace transform to solve your system in part (a). Assume that $x(0) = x_0$, $y(0) = 0$, and $z(0) = 0$.

(c) Suppose time is measured in days, the decay constants are $k_1 = -0.138629$ and $k_2 = -0.004951$, and that $x_0 = 20$. Use a computer graphing program to obtain the graphs of $x(t)$, $y(t)$, and $z(t)$ on the same set of coordinate axes. Use the graphs to approximate the half-lives of substances A and B. Use the graphs to approximate the times when **(i)** the amounts $x(t)$ and $y(t)$ are the same, **(ii)** the amounts $x(t)$ and $z(t)$ are the same, and **(iii)** the amounts $y(t)$ and $z(t)$ are the same. Why does the time determined for **(iii)** make intuitive sense?

25. *PROJECT PROBLEM* In this problem we apply the Laplace transform to the matrix form of a system of linear first-order differential equations with constant coefficients. Consider the initial value problem:

$$\mathbf{X}' = \mathbf{AX} + \mathbf{F}(t), \quad \mathbf{X}(0) = \mathbf{X}_0,$$

where $\mathbf{X}$, $\mathbf{F}$, and $\mathbf{X}_0$ are column matrices, and $\mathbf{A}$ is an $n \times n$ matrix of constants. See Section 4.1. Let $\mathcal{L}\{\mathbf{X}\} = \hat{\mathbf{X}}(s)$ and $\mathcal{L}\{\mathbf{F}(t)\} = \hat{\mathbf{F}}(s)$.

(a) Show that $\hat{\mathbf{X}}(s) = (s\mathbf{I} - \mathbf{A})^{-1}[\mathbf{X}_0 + \hat{\mathbf{F}}(s)]$.

(b) Use the result in part (a) to solve the initial-value problem

$$\mathbf{X}' = \begin{pmatrix} 1 & -2 & 2 \\ -1 & 1 & -2 \\ 2 & -2 & 1 \end{pmatrix}\mathbf{X}, \quad \mathbf{X}(0) = \begin{pmatrix} 1 \\ -1 \\ 0 \end{pmatrix}.$$

Use a CAS to find the matrices $(s\mathbf{I} - \mathbf{A})^{-1}$ and $(s\mathbf{I} - \mathbf{A})^{-1}[\mathbf{X}_0 + \hat{\mathbf{F}}(s)]$ as well as to find the inverse Laplace transform of the matrix $\hat{\mathbf{X}}(s)$.

(c) Repeat part (b) to solve

$$\mathbf{X}' = \begin{pmatrix} 1 & -2 & 2 \\ -1 & 1 & -2 \\ 2 & -2 & 1 \end{pmatrix}\mathbf{X} + \begin{pmatrix} e^{-t} \\ e^{2t} \\ e^{3t} \end{pmatrix}, \quad \mathbf{X}(0) = \begin{pmatrix} 1 \\ -1 \\ 0 \end{pmatrix}.$$

CHAPTER 6

SERIES SOLUTIONS

6.1 REVIEW OF POWER SERIES; POWER SERIES SOLUTIONS

Review of Power Series

Section 3.6 notwithstanding, most linear differential equations with variable coefficients cannot be solved in terms of elementary functions. A standard technique for solving higher-order linear differential equations with variable coefficients is to try to find a solution in the form of an infinite series. Often the solution can be found in the form of a power series. Because of this, it is appropriate to list some of the more important facts about power series. For an in-depth review of the infinite series concept you should consult a calculus text.

- **Definition of a Power Series** A **power series** in $x - a$ is an infinite series of the form $\sum_{n=0}^{\infty} c_n(x - a)^n$. A series such as this is also said to be a **power series centered at *a***. For example, $\sum_{n=1}^{\infty} \frac{(-1)^{n+1}}{n^2} x^n$ is a power series in x; the series is centered at zero.
- **Convergence** For a specified value of x a power series is a series of constants. If the series equals a finite real constant for the given x, then the series is said to **converge** at x. If the series does not converge at x, it is said to **diverge** at x.
- **Interval of Convergence** Every power series has an **interval of convergence**. The interval of convergence is the set of all numbers for which the series converges.
- **Radius of Convergence** Every interval of convergence has a **radius of convergence** R. For a power series $\sum_{n=0}^{\infty} c_n(x - a)^n$ we have just three possibilities:

  (i) The series converges only at its center a. In this case $R = 0$.

  (ii) The series converges for all x satisfying $|x - a| < R$, where $R > 0$. The series diverges for $|x - a| > R$.

  (iii) The series converges for all x. In this case we write $R = \infty$.
- **Convergence at an Endpoint** Recall, the absolute-value inequality $|x - a| < R$ is equivalent to $-R < x - a < R$, or $a - R < x < a + R$. If a power series converges for $|x - a| < R$, where $R > 0$, it may or may not converge at the endpoints of the interval $a - R < x < a + R$. Figure 6.1 shows four possible intervals of convergence.
- **Absolute Convergence** Within its interval of convergence a power series converges absolutely. In other words, for x in the interval of convergence the series of absolute values $\sum_{n=0}^{\infty} |c_n||(x - a)^n|$ converges.
- **Finding the Interval of Convergence** Convergence of a power series can often be determined by the **ratio test**:

$$\lim_{n \to \infty} \left| \frac{c_{n+1}}{c_n} \right| |x - a| = L.$$

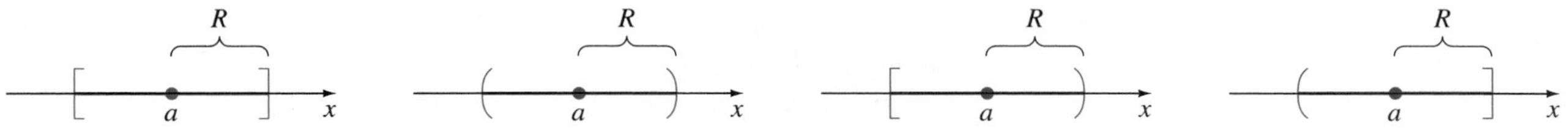

(a) $[a - R, a + R]$ Series converges at both endpoints.

(b) $(a - R, a + R)$ Series diverges at both endpoints.

(c) $[a - R, a + R)$ Series converges at $a - R$, diverges at $a + R$.

(d) $(a - R, a + R]$ Series diverges at $a - R$, converges at $a + R$.

FIGURE 6.1

The series will converge absolutely for those values of x for which $L < 1$. From this test we see that the radius of convergence is given by

$$R = \lim_{n \to \infty} \left| \frac{c_n}{c_{n+1}} \right| \tag{1}$$

provided the limit exists.

- **A Power Series Represents a Function** A power series represents a function

$$f(x) = \sum_{n=0}^{\infty} c_n(x - a)^n = c_0 + c_1(x - a) + c_2(x - a)^2 + c_3(x - a)^3 + \cdots$$

whose domain is the interval of convergence of the series. If the series has a radius of convergence $R > 0$, then f is continuous, differentiable, and integrable on the interval $(a - R, a + R)$. Moreover, $f'(x)$ and $\int f(x)\, dx$ can be found from term-by-term differentiation and integration:

$$f'(x) = c_1 + 2c_2(x - a) + 3c_3(x - a)^2 + \cdots = \sum_{n=1}^{\infty} nc_n(x - a)^{n-1}$$

$$\int f(x)\, dx = C + c_0(x - a) + c_1\frac{(x - a)^2}{2} + c_2\frac{(x - a)^3}{3} + \cdots = C + \sum_{n=0}^{\infty} c_n \frac{(x - a)^{n+1}}{n + 1}.$$

Although the radius of convergence for both these series is R, the interval of convergence may differ from the original series in that convergence at an endpoint may be either lost by differentiation or gained through integration.

- **Series That Are Identically Zero** If $\Sigma_{n=0}^{\infty} c_n(x - a)^n = 0$, $R > 0$, for all real numbers x in the interval of convergence, then $c_n = 0$ for all n.

- **Analytic at a Point** In calculus it is seen that functions such as e^x, $\cos x$, and $\ln(x - 1)$ can be represented by power series by expansions in either Maclaurin or Taylor series. We say that a function f is **analytic at point** $\boldsymbol{a}$ if it can be represented by a power series in $x - a$ with a positive radius of convergence. The notion of analyticity at a point will be important in Sections 6.2 and 6.3.

- **Arithmetic of Power Series** Power series can be combined through the operations of addition, multiplication, and division. The procedures for power series are similar to the way in which two polynomials are added, multiplied, and divided; that is, we add coefficients of like powers of x, use the distributive law and collect like terms, and perform long division. For example, if the power series $f(x) = \sum_{n=0}^{\infty} c_n x^n$ and $g(x) = \sum_{n=0}^{\infty} b_n x^n$ both converge for $|x| < R$, then

$$f(x) + g(x) = (c_0 + b_0) + (c_1 + b_1)x + (c_2 + b_2)x^2 + \cdots$$

$$f(x)g(x) = c_0 b_0 + (c_0 b_1 + c_1 b_0)x + (c_0 b_2 + c_1 b_1 + c_2 b_0)x^2 + \cdots.$$

EXAMPLE 1 Interval of Convergence

Find the interval of convergence of the power series $\displaystyle\sum_{n=1}^{\infty} \frac{(x-3)^n}{2^n n}$.

Solution The power series is centered at 3. From (1), the radius of convergence is

$$R = \lim_{n\to\infty} \frac{2^{n+1}(n+1)}{2^n n} = 2.$$

The series converges absolutely for $|x - 3| < 2$, or $1 < x < 5$. At the left endpoint $x = 1$ we find that the series of constants $\sum_{n=1}^{\infty}((-1)^n/n)$ is convergent by the alternating series test. At the right endpoint $x = 5$ we find that the series is the divergent harmonic series $\sum_{n=1}^{\infty}(1/n)$. Thus the interval of convergence is $[1,5)$. ●

EXAMPLE 2 Multiplication of Two Power Series

Find the first four terms of a power series in x for $e^x \cos x$.

Solution From calculus the Maclaurin series for e^x and $\cos x$ are, respectively,

$$e^x = 1 + x + \frac{x^2}{2} + \frac{x^3}{6} + \frac{x^4}{24} + \cdots \quad \text{and} \quad \cos x = 1 - \frac{x^2}{2} + \frac{x^4}{24} + \cdots.$$

Multiplying out and collecting like terms yield

$$e^x \cos x = \left(1 + x + \frac{x^2}{2} + \frac{x^3}{6} + \frac{x^4}{24} + \cdots\right)\left(1 - \frac{x^2}{2} + \frac{x^4}{24} - \cdots\right)$$

$$= 1 + (1)x + \left(-\frac{1}{2} + \frac{1}{2}\right)x^2 + \left(-\frac{1}{2} + \frac{1}{6}\right)x^3 + \left(\frac{1}{24} - \frac{1}{4} + \frac{1}{24}\right)x^4 + \cdots$$

$$= 1 + x - \frac{x^3}{3} - \frac{x^4}{6} + \cdots.$$ ●

In Example 2 the interval of convergence for the Maclaurin series for both e^x and $\cos x$ is $(-\infty, \infty)$. Consequently the interval of convergence for the power series for $e^x \cos x$ is also $(-\infty, \infty)$.

EXAMPLE 3 Division by a Power Series

Find the first four terms of a power series in x for $\sec x$.

Solution One way of proceeding is to use the Maclaurin series for $\cos x$ given in Example 2 and then to use long division. Since $\sec x = 1/\cos x$, we have

$$\begin{array}{r|l}
 & 1 + \frac{x^2}{2} + \frac{5x^4}{24} + \frac{61x^6}{720} + \cdots \\
\cos x = 1 - \frac{x^2}{2} + \frac{x^4}{24} - \frac{x^6}{720} + \cdots & 1 \\
 & 1 - \frac{x^2}{2} + \frac{x^4}{24} - \frac{x^6}{720} + \cdots \\
 & \frac{x^2}{2} - \frac{x^4}{24} + \frac{x^6}{720} - \cdots \\
 & \frac{x^2}{2} - \frac{x^4}{4} + \frac{x^6}{48} - \cdots \\
 & \frac{5x^4}{24} - \frac{7x^6}{360} + \cdots \\
 & \frac{5x^4}{24} - \frac{5x^6}{48} + \cdots \\
 & \frac{61x^6}{720} - \cdots
\end{array}$$

Thus
$$\sec x = 1 + \frac{x^2}{2} + \frac{5x^4}{24} + \frac{61x^6}{720} + \cdots. \tag{2}$$

The interval of convergence of this series is $(-\pi/2, \pi/2)$. (Why?) ●

The procedures illustrated in Examples 2 and 3 are obviously tedious to do by hand. Problems of this sort can be done with minimal fuss using a computer algebra system such as *Mathematica* or *Maple*. In *Mathematica* the division in Example 3 is avoided by using the command **Series[Sec[x], {x, 0, 8}]**. See Problems 11–14 in Exercises 6.1 and *Computer Lab Experiment* 1 for this section.

For the remainder of this section, as well as this chapter, it is important that you become adept at simplifying the sum of two or more power series, each series expressed in summation (sigma) notation, to an expression with a single Σ. This often requires a shift of the summation indices.

EXAMPLE 4 Adding Two Power Series

Write $\Sigma_{n=1}^{\infty} 2nc_n x^{n-1} + \Sigma_{n=0}^{\infty} 6c_n x^{n+1}$ as one series.

Solution In order to add the series, we require that both summation indices start with the same number and that the powers of x in each series be "in phase"; that is, if one series starts with a multiple of, say, x to the first power, then we want the other series to start with the same power. By writing

series starts with x for $n = 2$ ↓ series starts with x for $n = 0$ ↓

$$\sum_{n=1}^{\infty} 2nc_n x^{n-1} + \sum_{n=0}^{\infty} 6c_n x^{n+1} = 2 \cdot 1 \cdot c_1 x^0 + \sum_{n=2}^{\infty} 2nc_n x^{n-1} + \sum_{n=0}^{\infty} 6c_n x^{n+1}, \quad (3)$$

we have both series on the right side start with x^1. To get the same summation index we are inspired by the exponents of x; we let $k = n - 1$ in the first series and at the same time let $k = n + 1$ in the second series. Thus the right side of (3) becomes

$$2c_1 + \sum_{k=1}^{\infty} 2(k+1)c_{k+1}x^k + \sum_{k=1}^{\infty} 6c_{k-1}x^k. \quad (4)$$

Recall that the summation index is a "dummy" variable. The fact that $k = n - 1$ in one case and $k = n + 1$ in the other should cause no confusion if you keep in mind that it is the *value* of the summation index that is important. In both cases k takes on the same successive values 1, 2, 3, . . . for $n = 2, 3, 4, \ldots$ (for $k = n - 1$) and $n = 0, 1, 2, \ldots$ (for $k = n + 1$).

We are now in a position to add the series in (4) term by term:

$$\sum_{n=1}^{\infty} 2nc_n x^{n-1} + \sum_{n=0}^{\infty} 6c_n x^{n+1} = 2c_1 + \sum_{k=1}^{\infty} [2(k+1)c_{k+1} + 6c_{k-1}]x^k. \quad (5)$$

If you are not convinced, then write out a few terms on both sides of (5). ●

Powers Series Solution of a Differential Equation

We saw in Section 1.1 that the function $y = e^{x^2}$ is an explicit solution of the linear first-order differential equation

$$\frac{dy}{dx} - 2xy = 0. \quad (6)$$

By replacing x by x^2 in the Maclaurin series for e^x, we can write the solution of (6) as $y = \sum_{n=0}^{\infty} \frac{1}{n!} x^{2n}$. This last series converges for all real values of x. In other words, when we know the solution in advance, we can find an infinite series solution of the differential equation.

We now propose to obtain a **power series solution** of equation (6) directly; the method of attack is similar to the technique of undetermined coefficients.

EXAMPLE 5 Using a Power Series to Solve a DE

Find a solution of $dy/dx - 2xy = 0$ in the form of a power series in x.

Solution If we assume that a solution of the given equation exists in the form

$$y = \sum_{n=0}^{\infty} c_n x^n, \quad (7)$$

we pose the question: Can we determine coefficients c_n for which the power series converges to a function satisfying (6)? Formal* term-by-term differentiation of (7) gives

$$\frac{dy}{dx} = \sum_{n=0}^{\infty} nc_n x^{n-1} = \sum_{n=1}^{\infty} nc_n x^{n-1}.$$

Note that since the first term in the first series (corresponding to $n = 0$) is zero, we begin the summation with $n = 1$. Using the last result and assumption (7), we find

$$\frac{dy}{dx} - 2xy = \sum_{n=1}^{\infty} nc_n x^{n-1} - \sum_{n=0}^{\infty} 2c_n x^{n+1}. \tag{8}$$

We would like to add the two series in (8). To this end we write

$$\frac{dy}{dx} - 2xy = 1 \cdot c_1 x^0 + \sum_{n=2}^{\infty} nc_n x^{n-1} - \sum_{n=0}^{\infty} 2c_n x^{n+1} \tag{9}$$

and then proceed as in Example 4 by letting $k = n - 1$ in the first series and $k = n + 1$ in the second. The right side of (9) becomes

$$c_1 + \sum_{k=1}^{\infty} (k + 1)c_{k+1}x^k - \sum_{k=1}^{\infty} 2c_{k-1}x^k.$$

After we add the series termwise, it follows that

$$\frac{dy}{dx} - 2xy = c_1 + \sum_{k=1}^{\infty} [(k + 1)c_{k+1} - 2c_{k-1}]x^k = 0. \tag{10}$$

Hence in order to have (10) identically zero it is necessary that the coefficients of like powers of x be zero, that is,

$$c_1 = 0 \quad \text{and} \quad (k + 1)c_{k+1} - 2c_{k-1} = 0, \qquad k = 1, 2, 3, \ldots. \tag{11}$$

Equation (11) provides a **recurrence relation** that determines the c_k. Since $k + 1 \neq 0$ for all the indicated values of k, we can write (11) as

$$c_{k+1} = \frac{2c_{k-1}}{k + 1}. \tag{12}$$

Iteration of this formula then gives

$$k = 1, \qquad c_2 = \frac{2}{2}c_0 = c_0$$

$$k = 2, \qquad c_3 = \frac{2}{3}c_1 = 0$$

$$k = 3, \qquad c_4 = \frac{2}{4}c_2 = \frac{1}{2}c_0 = \frac{1}{2!}c_0$$

$$k = 4, \qquad c_5 = \frac{2}{5}c_3 = 0$$

* At this point we do not know the interval of convergence.

$$k = 5, \qquad c_6 = \frac{2}{6}c_4 = \frac{1}{3 \cdot 2!}c_0 = \frac{1}{3!}c_0$$

$$k = 6, \qquad c_7 = \frac{2}{7}c_5 = 0$$

$$k = 7, \qquad c_8 = \frac{2}{8}c_6 = \frac{1}{4 \cdot 3!}c_0 = \frac{1}{4!}c_0$$

and so on. Thus from the original assumption (7), we find

$$\begin{aligned} y = \sum_{n=0}^{\infty} c_n x^n &= c_0 + c_1 x + c_2 x^2 + c_3 x^3 + c_4 x^4 + c_5 x^5 + \cdots \\ &= c_0 + 0 + c_0 x^2 + 0 + \frac{1}{2!}c_0 x^4 + 0 + \frac{1}{3!}c_0 x^6 + 0 + \cdots \\ &= c_0\left[1 + x^2 + \frac{1}{2!}x^4 + \frac{1}{3!}x^6 + \cdots\right] = c_0 \sum_{n=0}^{\infty} \frac{x^{2n}}{n!}. \end{aligned} \tag{13}$$

Since the iteration of (12) leaves c_0 completely undetermined, we have in fact found the general solution of (6). ●

The differential equation in Example 5, like the differential equation in the following example, can be easily solved by prior methods. The point of these two examples is to prepare you for the techniques considered in Sections 6.2 and 6.3.

EXAMPLE 6 Using a Power Series to Solve a DE

Find solutions of $4y'' + y = 0$ in the form of power series in x.

Solution If $y = \sum_{n=0}^{\infty} c_n x^n$, we have seen that $y' = \sum_{n=1}^{\infty} nc_n x^{n-1}$ and so

$$y'' = \sum_{n=1}^{\infty} n(n-1)c_n x^{n-2} = \sum_{n=2}^{\infty} n(n-1)c_n x^{n-2}.$$

Substituting the expressions for y'' and y back into the differential equation gives

$$4y'' + y = \underbrace{\sum_{n=2}^{\infty} 4n(n-1)c_n x^{n-2} + \sum_{n=0}^{\infty} c_n x^n}_{\text{both series start with } x^0}.$$

If we substitute $k = n - 2$ in the first series and $k = n$ in the second, we get (after using, in turn, $n = k + 2$ and $n = k$)

$$\begin{aligned} 4y'' + y &= \sum_{k=0}^{\infty} 4(k+2)(k+1)c_{k+2}x^k + \sum_{k=0}^{\infty} c_k x^k \\ &= \sum_{k=0}^{\infty} [4(k+2)(k+1)c_{k+2} + c_k]x^k = 0. \end{aligned}$$

From this last identity it is apparent that for $k = 0, 1, 2, \ldots$

$$4(k+2)(k+1)c_{k+2} + c_k = 0 \quad \text{or} \quad c_{k+2} = \frac{-c_k}{4(k+2)(k+1)}.$$

By iterating the last formula we get

$$c_2 = \frac{-c_0}{4 \cdot 2 \cdot 1} = -\frac{c_0}{2^2 \cdot 2!}$$

$$c_3 = \frac{-c_1}{4 \cdot 3 \cdot 2} = -\frac{c_1}{2^2 \cdot 3!}$$

$$c_4 = \frac{-c_2}{4 \cdot 4 \cdot 3} = \frac{c_0}{2^4 \cdot 4!}$$

$$c_5 = \frac{-c_3}{4 \cdot 5 \cdot 4} = \frac{c_1}{2^4 \cdot 5!}$$

$$c_6 = \frac{-c_4}{4 \cdot 6 \cdot 5} = -\frac{c_0}{2^6 \cdot 6!}$$

$$c_7 = \frac{-c_5}{4 \cdot 7 \cdot 6} = -\frac{c_1}{2^6 \cdot 7!}$$

and so forth. This iteration leaves both c_0 and c_1 arbitrary. From the original assumption we have

$$y = c_0 + c_1 x + c_2 x^2 + c_3 x^3 + c_4 x^4 + c_5 x^5 + c_6 x^6 + c_7 x^7 + \cdots$$

$$= c_0 + c_1 x - \frac{c_0}{2^2 \cdot 2!}x^2 - \frac{c_1}{2^2 \cdot 3!}x^3 + \frac{c_0}{2^4 \cdot 4!}x^4 + \frac{c_1}{2^4 \cdot 5!}x^5 - \frac{c_0}{2^6 \cdot 6!}x^6 - \frac{c_1}{2^6 \cdot 7!}x^7 + \cdots$$

or

$$y = c_0\left[1 - \frac{1}{2^2 \cdot 2!}x^2 + \frac{1}{2^4 \cdot 4!}x^4 - \frac{1}{2^6 \cdot 6!}x^6 + \cdots\right]$$

$$+ c_1\left[x - \frac{1}{2^2 \cdot 3!}x^3 + \frac{1}{2^4 \cdot 5!}x^5 - \frac{1}{2^6 \cdot 7!}x^7 + \cdots\right]$$

is a general solution. When the series are written in summation notation,

$$y_1(x) = c_0 \sum_{k=0}^{\infty} \frac{(-1)^k}{(2k)!}\left(\frac{x}{2}\right)^{2k} \quad \text{and} \quad y_2(x) = 2c_1 \sum_{k=0}^{\infty} \frac{(-1)^k}{(2k+1)!}\left(\frac{x}{2}\right)^{2k+1},$$

the ratio test can be applied to show that both series converge for all x. You might also recognize the Maclaurin series as $y_1(x) = c_0 \cos(x/2)$ and $y_2(x) = 2c_1 \sin(x/2)$. ●

Use of Taylor Series

In some instances the solution of an initial-value problem, in which the conditions are specified at a point x_0, can be found in the form of a Taylor series centered at x_0.

EXAMPLE 7 Taylor Series Solution of an IVP

In Sections 1.2 and 2.1 we saw that a solution of the initial-value problem

$$y' = x^2 + y^2, \quad y(0) = 1 \tag{14}$$

exists. If we further assume that the solution $y(x)$ of the problem is analytic at 0, then $y(x)$ possesses a Taylor series expansion centered at 0:

$$y(x) = y(0) + \frac{y'(0)}{1!}x + \frac{y''(0)}{2!}x^2 + \frac{y'''(0)}{3!}x^3 + \cdots. \tag{15}$$

The value of the first term in the series (15) is known since that is the specified initial condition. Moreover, by using the initial condition and the differential equation itself we get the value of the first derivative: $y'(0) = 0^2 + 1^2 = 1$. We can then find expressions for the higher derivatives y'', y''', . . . , by calculating the successive derivatives of the differential equation:

$$y''(x) = \frac{d}{dx}(x^2 + y^2) = 2x + 2y(x)y'(x), \tag{16}$$

$$y'''(x) = \frac{d}{dx}(2x + 2y(x)y'(x)) = 2 + 2y(x)y''(x) + 2(y'(x))^2, \tag{17}$$

and so on. Now using $y(0) = 1$ and $y'(0) = 1$, we find $y''(0) = 2$ from (16). Using this last bit of information, we find $y'''(0) = 8$ from (17). Hence from (15) the first four terms of a series solution of the initial-value problem (14) are

$$y(x) = 1 + x + x^2 + \tfrac{4}{3}x^3 + \cdots.$$

EXERCISES 6.1

Answers to odd-numbered problems begin on page AN-26.

In Problems 1–10 find the interval of convergence of the given power series.

1. $\sum_{n=1}^{\infty} \frac{(-1)^n}{n} x^n$
2. $\sum_{n=1}^{\infty} \frac{x^n}{n^2}$
3. $\sum_{k=1}^{\infty} \frac{2^k}{k} x^k$
4. $\sum_{k=0}^{\infty} \frac{5^k}{k!} x^k$
5. $\sum_{n=1}^{\infty} \frac{(x-3)^n}{n^3}$
6. $\sum_{n=1}^{\infty} \frac{(x+7)^n}{\sqrt{n}}$
7. $\sum_{k=1}^{\infty} \frac{(-1)^k}{10^k} (x-5)^k$
8. $\sum_{k=1}^{\infty} \frac{k}{(k+2)^2} (x-4)^k$
9. $\sum_{k=0}^{\infty} k!2^k x^k$
10. $\sum_{k=2}^{\infty} \frac{k-1}{k^{2k}} x^k$

In Problems 11–14 find the first four terms of a power series in x for the given function. Calculate the series by hand or use a CAS as instructed. See *Computer Lab Experiment* 1 for this section for additional problems.

11. $e^x \sin x$
12. $e^{-x} \cos x$
13. $\sin x \cos x$
14. $e^x \ln(1 - x)$

In Problems 15–24 solve each differential equation in the manner of the previous chapters and then compare the results with the solutions obtained by assuming a power series solution $y = \sum_{n=0}^{\infty} c_n x^n$.

15. $y' + y = 0$
16. $y' = 2y$
17. $y' - x^2 y = 0$
18. $y' + x^3 y = 0$
19. $(1 - x)y' - y = 0$
20. $(1 + x)y' - 2y = 0$
21. $y'' + y = 0$
22. $y'' - y = 0$
23. $y'' = y'$
24. $2y'' + y' = 0$

In Problems 25 and 26 use the Taylor series method illustrated in Example 7 to find the first four nonzero terms of a series solution of the given initial-value problem. Also see *Computer Lab Experiment* 2 for this section.

25. $y' = (x - y)^2, \quad y(0) = \frac{1}{2}$

26. $y' = xy + \sqrt{y}, \quad y(0) = 1$

27. In Example 1 of Section 3.11 we considered the initial-value problem

$$y'' + y + y^3 = 0, \qquad y(0) = 2, \quad y'(0) = -3.$$

The differential equation is a model of a nonlinear hard spring. Use the Taylor series method to find the first four nonzero terms of a series solution of the problem.

28. In Example 2 of Section 3.11 we considered the initial-value problem

$$y'' + \sin y = 0, \qquad y(0) = \tfrac{1}{2}, \quad y'(0) = 2.$$

The differential equation is a model of a nonlinear pendulum. Use the Taylor series method to find the first four nonzero terms of a series solution of the problem.

6.2 SOLUTIONS ABOUT ORDINARY POINTS

Suppose the linear second-order differential equation

$$a_2(x)y'' + a_1(x)y' + a_0(x)y = 0 \tag{1}$$

is put into the standard form

$$y'' + P(x)y' + Q(x)y = 0 \tag{2}$$

by dividing by the leading coefficient $a_2(x)$. We make the following definition.

DEFINITION 6.1 Ordinary and Singular Points

A point x_0 is said to be an **ordinary point** of the differential equation (1) if both $P(x)$ and $Q(x)$ are analytic at x_0. A point that is not an ordinary point is said to be a **singular point** of the equation.

EXAMPLE 1 Ordinary Points

Every finite value of x is an ordinary point of $y'' + (e^x)y' + (\sin x)y = 0$. In particular we see that $x = 0$ is an ordinary point since both e^x and $\sin x$ are analytic at this point; that is, both functions can be represented by power series centered at 0. Recall from calculus that

$$e^x = 1 + \frac{x}{1!} + \frac{x^2}{2!} + \cdots \quad \text{and} \quad \sin x = x - \frac{x^3}{3!} + \frac{x^5}{5!} - \cdots$$

converge for all finite values of x. ●

EXAMPLE 2 Ordinary/Singular Points

(a) The differential equation $xy'' + (\sin x)y = 0$ has an ordinary point at $x = 0$ since $Q(x) = (\sin x)/x$ possesses the power series expansion

$$Q(x) = 1 - \frac{x^2}{3!} + \frac{x^4}{5!} - \frac{x^6}{7!} + \cdots$$

that converges for all finite values of x.

(b) The differential equation $y'' + (\ln x)y = 0$ has a singular point at $x = 0$ because $Q(x) = \ln x$ possesses no power series in x. ●

Polynomial Coefficients

Primarily we shall be concerned with the case when (1) has *polynomial* coefficients. As a consequence of Definition 6.1, we note that when $a_2(x)$, $a_1(x)$, and $a_0(x)$ are polynomials with *no common factors*, a point $x = x_0$ is

(i) an ordinary point if $a_2(x_0) \neq 0$, or (ii) a singular point if $a_2(x_0) = 0$.

EXAMPLE 3 Singular/Ordinary Points

(a) The singular points of the equation $(x^2 - 1)y'' + 2xy' + 6y = 0$ are the solutions of $x^2 - 1 = 0$ or $x = \pm 1$. All other finite values of x are ordinary points.

(b) Singular points need not be real numbers. The equation $(x^2 + 1)y'' + xy' - y = 0$ has singular points at the solutions of $x^2 + 1 = 0$, namely, $x = \pm i$. All other finite values of x, real or complex, are ordinary points.

(c) The Cauchy-Euler equation $ax^2y'' + bxy' + cy = 0$, where a, b, and c are constants, has a singular point at $x = 0$. All other finite values of x, real or complex, are ordinary points. ●

For our purpose ordinary points and singular points will always be finite points. It is possible for a differential equation to have, say, a singular point at infinity. (See Problem 31 in Exercises 6.3.)

We state the following theorem about the existence of power series solutions without proof.

THEOREM 6.1 Existence of Power Series Solutions

If $x = x_0$ is an ordinary point of the differential equation (2), we can always find two linearly independent solutions in the form of power series centered at x_0:

$$y = \sum_{n=0}^{\infty} c_n(x - x_0)^n. \tag{3}$$

A series solution converges at least for $|x - x_0| < R$, where R is the distance from x_0 to the closest singular point (real or complex).

A solution of a differential equation of the form given in (3) is said to be a solution *about* the ordinary point x_0. The distance R given in Theorem 6.1 is the minimum value for the radius of convergence. A differential equation could have a finite singular point and yet a solution could be valid for all x; for example, the differential equation may possess a polynomial solution.

To solve a linear second-order equation such as (2) we find two sets of coefficients c_n so that we have two distinct power series $y_1(x)$ and $y_2(x)$, both expanded about the same ordinary point x_0. The procedure used to solve a second-order equation is the same as that used in Example 6 of Section 6.1; that is, we assume a solution $y = \sum_{n=0}^{\infty} c_n(x - x_0)^n$ and then determine the c_n. The general solution of the differential equation is $y = C_1y_1(x) + C_2y_2(x)$; in fact, it can be shown that $C_1 = c_0$ and $C_2 = c_1$, where c_0 and c_1 are arbitrary.

Note For the sake of simplicity, we assume an ordinary point is always located at $x = 0$, since, if not, the substitution $t = x - x_0$ translates the value $x = x_0$ to $t = 0$.

EXAMPLE 4 Power Series Solution About an Ordinary Point

Solve $\quad y'' + xy = 0.$

Solution We see that $x = 0$ is an ordinary point of the equation. Since there are no finite singular points, Theorem 6.1 guarantees two power series solutions, centered at 0, convergent for $|x| < \infty$. Substituting

$$y = \sum_{n=0}^{\infty} c_nx^n \quad \text{and} \quad y'' = \sum_{n=2}^{\infty} n(n-1)c_nx^{n-2}$$

into the differential equation gives

$$\begin{aligned} y'' + xy &= \sum_{n=2}^{\infty} n(n-1)c_nx^{n-2} + \sum_{n=0}^{\infty} c_nx^{n+1} \\ &= 2 \cdot 1c_2x^0 + \underbrace{\sum_{n=3}^{\infty} n(n-1)c_nx^{n-2} + \sum_{n=0}^{\infty} c_nx^{n+1}}_{\text{both series start with } x}. \end{aligned}$$

Letting $k = n - 2$ in the first series and $k = n + 1$ in the second, we have

$$\begin{aligned} y'' + xy &= 2c_2 + \sum_{k=1}^{\infty} (k+2)(k+1)c_{k+2}x^k + \sum_{k=1}^{\infty} c_{k-1}x^k \\ &= 2c_2 + \sum_{k=1}^{\infty} [(k+2)(k+1)c_{k+2} + c_{k-1}]x^k = 0. \end{aligned}$$

We must then have $2c_2 = 0$, which implies $c_2 = 0$, and

$$(k+2)(k+1)c_{k+2} + c_{k-1} = 0.$$

The last expression is the same as

$$c_{k+2} = -\frac{c_{k-1}}{(k+2)(k+1)}, \qquad k = 1, 2, 3, \ldots .$$

Iteration gives

$$c_3 = -\frac{c_0}{3 \cdot 2}$$

$$c_4 = -\frac{c_1}{4 \cdot 3}$$

$$c_5 = -\frac{c_2}{5 \cdot 4} = 0$$

$$c_6 = -\frac{c_3}{6 \cdot 5} = \frac{1}{6 \cdot 5 \cdot 3 \cdot 2} c_0$$

$$c_7 = -\frac{c_4}{7 \cdot 6} = \frac{1}{7 \cdot 6 \cdot 4 \cdot 3} c_1$$

$$c_8 = -\frac{c_5}{8 \cdot 7} = 0$$

$$c_9 = -\frac{c_6}{9 \cdot 8} = -\frac{1}{9 \cdot 8 \cdot 6 \cdot 5 \cdot 3 \cdot 2} c_0$$

$$c_{10} = -\frac{c_7}{10 \cdot 9} = -\frac{1}{10 \cdot 9 \cdot 7 \cdot 6 \cdot 4 \cdot 3} c_1$$

$$c_{11} = -\frac{c_8}{11 \cdot 10} = 0$$

and so on. It should be apparent that both c_0 and c_1 are arbitrary. Now

$$\begin{aligned} y &= c_0 + c_1 x + c_2 x^2 + c_3 x^3 + c_4 x^4 + c_5 x^5 + c_6 x^6 + c_7 x^7 + c_8 x^8 \\ &\quad + c_9 x^9 + c_{10} x^{10} + c_{11} x^{11} + \cdots \\ &= c_0 + c_1 x + 0 - \frac{1}{3 \cdot 2} c_0 x^3 - \frac{1}{4 \cdot 3} c_1 x^4 + 0 + \frac{1}{6 \cdot 5 \cdot 3 \cdot 2} c_0 x^6 \\ &\quad + \frac{1}{7 \cdot 6 \cdot 4 \cdot 3} c_1 x^7 + 0 - \frac{1}{9 \cdot 8 \cdot 6 \cdot 5 \cdot 3 \cdot 2} c_0 x^9 \\ &\quad - \frac{1}{10 \cdot 9 \cdot 7 \cdot 6 \cdot 4 \cdot 3} c_1 x^{10} + 0 + \cdots \\ &= c_0 \left[1 - \frac{1}{3 \cdot 2} x^3 + \frac{1}{6 \cdot 5 \cdot 3 \cdot 2} x^6 - \frac{1}{9 \cdot 8 \cdot 6 \cdot 5 \cdot 3 \cdot 2} x^9 + \cdots \right] \\ &\quad + c_1 \left[x - \frac{1}{4 \cdot 3} x^4 + \frac{1}{7 \cdot 6 \cdot 4 \cdot 3} x^7 - \frac{1}{10 \cdot 9 \cdot 7 \cdot 6 \cdot 4 \cdot 3} x^{10} + \cdots \right]. \end{aligned}$$

Although the pattern of the coefficients in Example 4 should be clear, it is sometimes useful to write the solutions in terms of summation notation. By using the properties of the factorial, we can write

$$y_1(x) = c_0 \left[1 + \sum_{k=1}^{\infty} \frac{(-1)^k [1 \cdot 4 \cdot 7 \cdots (3k-2)]}{(3k)!} x^{3k} \right]$$

and $$y_2(x) = c_1\left[x + \sum_{k=1}^{\infty} \frac{(-1)^k[2 \cdot 5 \cdot 8 \cdots (3k-1)]}{(3k+1)!} x^{3k+1}\right].$$

In this form the ratio test can be used to show that each series converges for $|x| < \infty$.

The differential equation in Example 4 is called **Airy's equation** and is encountered in the study of diffraction of light, diffraction of radio waves around the surface of the Earth, aerodynamics, and the deflection of a uniform thin vertical column that bends under its own weight. Other common forms of Airy's equation are $y'' - xy = 0$ and $y'' + \alpha^2 xy = 0$. See Exercises 6.4 for an application of the last equation.

EXAMPLE 5 Power Series Solution About an Ordinary Point

Solve $(x^2 + 1)y'' + xy' - y = 0$.

Solution Since the singular points are $x = \pm i$, a power series solution will converge at least for $|x| < 1$.* The assumption $y = \sum_{n=0}^{\infty} c_n x^n$ leads to

$$(x^2+1)\sum_{n=2}^{\infty} n(n-1)c_n x^{n-2} + x\sum_{n=1}^{\infty} nc_n x^{n-1} - \sum_{n=0}^{\infty} c_n x^n$$

$$= \sum_{n=2}^{\infty} n(n-1)c_n x^n + \sum_{n=2}^{\infty} n(n-1)c_n x^{n-2} + \sum_{n=1}^{\infty} nc_n x^n - \sum_{n=0}^{\infty} c_n x^n$$

$$= 2c_2x^0 - c_0x^0 + 6c_3x + c_1x - c_1x + \underbrace{\sum_{n=2}^{\infty} n(n-1)c_n x^n}_{k=n}$$

$$+ \underbrace{\sum_{n=4}^{\infty} n(n-1)c_n x^{n-2}}_{k=n-2} + \underbrace{\sum_{n=2}^{\infty} nc_n x^n}_{k=n} - \underbrace{\sum_{n=2}^{\infty} c_n x^n}_{k=n}$$

$$= 2c_2 - c_0 + 6c_3x$$

$$+ \sum_{k=2}^{\infty} [k(k-1)c_k + (k+2)(k+1)c_{k+2} + kc_k - c_k]x^k$$

$$= 2c_2 - c_0 + 6c_3x$$

$$+ \sum_{k=2}^{\infty} [(k+1)(k-1)c_k + (k+2)(k+1)c_{k+2}]x^k = 0.$$

---

* The **modulus** or magnitude of the complex number $x = i$ is $|x| = 1$. If $x = a + bi$ is a singular point, then $|x| = \sqrt{a^2 + b^2}$.

Thus
$$2c_2 - c_0 = 0, \quad c_3 = 0$$
$$(k + 1)(k - 1)c_k + (k + 2)(k + 1)c_{k+2} = 0$$
or
$$c_2 = \tfrac{1}{2}c_0, \quad c_3 = 0$$
$$c_{k+2} = \frac{1-k}{k+2}c_k, \quad k = 2, 3, 4, \ldots .$$

Iteration of the last formula gives

$$c_4 = -\frac{1}{4}c_2 = -\frac{1}{2 \cdot 4}c_0 = -\frac{1}{2^2 2!}c_0$$

$$c_5 = -\frac{2}{5}c_3 = 0$$

$$c_6 = -\frac{3}{6}c_4 = \frac{3}{2 \cdot 4 \cdot 6}c_0 = \frac{1 \cdot 3}{2^3 3!}c_0$$

$$c_7 = -\frac{4}{7}c_5 = 0$$

$$c_8 = -\frac{5}{8}c_6 = -\frac{3 \cdot 5}{2 \cdot 4 \cdot 6 \cdot 8}c_0 = -\frac{1 \cdot 3 \cdot 5}{2^4 4!}c_0$$

$$c_9 = -\frac{6}{9}c_7 = 0$$

$$c_{10} = -\frac{7}{10}c_8 = \frac{3 \cdot 5 \cdot 7}{2 \cdot 4 \cdot 6 \cdot 8 \cdot 10}c_0 = \frac{1 \cdot 3 \cdot 5 \cdot 7}{2^5 5!}c_0$$

and so on. Therefore

$$y = c_0 + c_1x + c_2x^2 + c_3x^3 + c_4x^4 + c_5x^5 + c_6x^6 + c_7x^7 + c_8x^8 + \cdots$$

$$= c_1x + c_0\left[1 + \frac{1}{2}x^2 - \frac{1}{2^2 2!}x^4 + \frac{1 \cdot 3}{2^3 3!}x^6 - \frac{1 \cdot 3 \cdot 5}{2^4 4!}x^8 + \frac{1 \cdot 3 \cdot 5 \cdot 7}{2^5 5!}x^{10} - \cdots\right].$$

The solutions are the polynomial $y_2(x) = c_1x$ and the series

$$y_1(x) = c_0\left[1 + \frac{1}{2}x^2 + \sum_{n=2}^{\infty}(-1)^{n-1}\frac{1 \cdot 3 \cdot 5 \cdots (2n-3)}{2^n n!}x^{2n}\right], \quad |x| < 1.$$ ●

EXAMPLE 6 Three-Term Recurrence Relation

If we seek a solution $y = \Sigma_{n=0}^{\infty} c_n x^n$ for the equation

$$y'' - (1 + x)y = 0,$$

we obtain $c_2 = c_0/2$ and the three-term recurrence relation

$$c_{k+2} = \frac{c_k + c_{k-1}}{(k+1)(k+2)}, \qquad k = 1, 2, 3, \ldots.$$

To simplify the iteration we can first choose $c_0 \neq 0$, $c_1 = 0$; this yields one solution. The other solution follows from next choosing $c_0 = 0$, $c_1 \neq 0$. With the first assumption we find

$$c_2 = \frac{1}{2}c_0$$

$$c_3 = \frac{c_1 + c_0}{2 \cdot 3} = \frac{c_0}{2 \cdot 3} = \frac{1}{6}c_0$$

$$c_4 = \frac{c_2 + c_1}{3 \cdot 4} = \frac{c_0}{2 \cdot 3 \cdot 4} = \frac{1}{24}c_0$$

$$c_5 = \frac{c_3 + c_2}{4 \cdot 5} = \frac{c_0}{4 \cdot 5}\left[\frac{1}{2 \cdot 3} + \frac{1}{2}\right] = \frac{1}{30}c_0$$

and so on. Thus one solution is

$$y_1(x) = c_0\left[1 + \frac{1}{2}x^2 + \frac{1}{6}x^3 + \frac{1}{24}x^4 + \frac{1}{30}x^5 + \cdots\right].$$

Similarly if we choose $c_0 = 0$, then

$$c_2 = 0$$

$$c_3 = \frac{c_1 + c_0}{2 \cdot 3} = \frac{c_1}{2 \cdot 3} = \frac{1}{6}c_1$$

$$c_4 = \frac{c_2 + c_1}{3 \cdot 4} = \frac{c_1}{3 \cdot 4} = \frac{1}{12}c_1$$

$$c_5 = \frac{c_3 + c_2}{4 \cdot 5} = \frac{c_1}{2 \cdot 3 \cdot 4 \cdot 5} = \frac{1}{120}c_1$$

and so on. Hence another solution is

$$y_2(x) = c_1\left[x + \frac{1}{6}x^3 + \frac{1}{12}x^4 + \frac{1}{120}x^5 + \cdots\right].$$

Each series converges for all finite values of x. ●

Nonpolynomial Coefficients

The next example illustrates how to find a power series solution about an ordinary point of a differential equation when its coefficients are not polynomials. In this example we see an application of multiplication of two power series that was discussed in Section 6.1.

EXAMPLE 7 DE with Nonpolynomial Coefficients

Solve $y'' + (\cos x)y = 0.$

Solution Since $\cos x = 1 - \frac{x^2}{2!} + \frac{x^4}{4!} - \frac{x^6}{6!} + \cdots$, it is seen that $x = 0$ is an ordinary point. Thus the assumption $y = \Sigma_{n=0}^{\infty} c_n x^n$ leads to

$$y'' + (\cos x)y = \sum_{n=2}^{\infty} n(n-1)c_n x^{n-2} + \left(1 - \frac{x^2}{2!} + \frac{x^4}{4!} - \cdots\right)\sum_{n=0}^{\infty} c_n x^n$$

$$= (2c_2 + 6c_3 x + 12c_4 x^2 + 20c_5 x^3 + \cdots)$$

$$+ \left(1 - \frac{x^2}{2} + \frac{x^4}{24} - \cdots\right)(c_0 + c_1 x + c_2 x^2 + c_3 x^3 + \cdots)$$

$$= 2c_2 + c_0 + (6c_3 + c_1)x + \left(12c_4 + c_2 - \frac{1}{2}c_0\right)x^2 + \left(20c_5 + c_3 - \frac{1}{2}c_1\right)x^3 + \cdots.$$

Since the last line is to be identically zero, we must have

$$2c_2 + c_0 = 0$$

$$6c_3 + c_1 = 0$$

$$12c_4 + c_2 - \frac{1}{2}c_0 = 0$$

$$20c_5 + c_3 - \frac{1}{2}c_1 = 0$$

and so on. Since c_0 and c_1 are arbitrary, we find

$$y_1(x) = c_0\left[1 - \frac{1}{2}x^2 + \frac{1}{12}x^4 - \cdots\right] \quad \text{and} \quad y_2(x) = c_1\left[x - \frac{1}{6}x^3 + \frac{1}{30}x^5 - \cdots\right].$$

Since the differential equation has no singular points, both series converge for all finite values of x. ●

EXERCISES 6.2

Answers to odd-numbered problems begin on page AN-26.

In Problems 1–14 for each differential equation find two linearly independent power series solutions about the ordinary point $x = 0$.

1. $y'' - xy = 0$
2. $y'' + x^2y = 0$
3. $y'' - 2xy' + y = 0$
4. $y'' - xy' + 2y = 0$
5. $y'' + x^2y' + xy = 0$
6. $y'' + 2xy' + 2y = 0$
7. $(x - 1)y'' + y' = 0$
8. $(x + 2)y'' + xy' - y = 0$
9. $(x^2 - 1)y'' + 4xy' + 2y = 0$
10. $(x^2 + 1)y'' - 6y = 0$
11. $(x^2 + 2)y'' + 3xy' - y = 0$
12. $(x^2 - 1)y'' + xy' - y = 0$
13. $y'' - (x + 1)y' - y = 0$
14. $y'' - xy' - (x + 2)y = 0$

In Problems 15–18 use the power series method to solve the given differential equation subject to the indicated initial conditions.

15. $(x - 1)y'' - xy' + y = 0, \quad y(0) = -2, y'(0) = 6$

16. $(x + 1)y'' - (2 - x)y' + y = 0, \quad y(0) = 2, y'(0) = -1$

17. $y'' - 2xy' + 8y = 0, \quad y(0) = 3, y'(0) = 0$

18. $(x^2 + 1)y'' + 2xy' = 0, \quad y(0) = 0, y'(0) = 1$

In Problems 19–22 use the procedure illustrated in Example 7 to find two power series solutions of the given differential equation about the ordinary point $x = 0$.

19. $y'' + (\sin x)y = 0$

20. $xy'' + (\sin x)y = 0$ [*Hint*: See Example 2.]

21. $y'' + e^{-x}y = 0$

22. $y'' + e^{x}y' - y = 0$

In Problems 23 and 24 use the power series method to solve the nonhomogeneous equation.

23. $y'' - xy = 1$

24. $y'' - 4xy' - 4y = e^x$

25. *PROJECT PROBLEM* Note that $x = 0$ is an ordinary point of the differential equation $y'' + xy' + y = 0$.

(a) Write a short explanation of why you think it is not a good idea to find two power series solutions about $x = 0$ for

$$y'' + xy' + y = 0, \qquad y(1) = -5, \quad y'(1) = 2.$$

(b) Find a way, more convenient than that mentioned in part (a), to solve the given initial-value problem.

26. *PROJECT PROBLEM* Historically many interesting linear second-order differential equations with variable coefficients arose from someone's study of a physical problem. The study of the solutions of these equations, and other functions related to these solutions, eventually gave rise to a separate branch of mathematics known as **special functions**. Moreover, some of these differential equations possess polynomial solutions obtained, of course, by trying to find power series solutions centered at the ordinary point $x = 0$.

(a) Let $n \geq 0$ be an integer. Find polynomial solutions of the following equations corresponding to $n = 0, 1, 2, 3$. For convenience choose $c_0 = 1, c_1 = 1$ when iterating your recurrence relation.

$y'' - 2xy' + 2ny = 0$ **Hermite's equation**

$(1 - x^2)y'' - 2xy' + n(n + 1)y = 0$ **Legendre's equation**

$(1 - x^2)y'' - xy' + n^2y = 0$ **Chebyshev's equation**

(b) For the rest of this problem let us focus on the polynomials obtained from Hermite's equation. Denote your solutions by $h_0(x)$, $h_1(x)$, $h_2(x)$, and $h_3(x)$. The polynomial solutions $h_n(x)$ are not what is commonly accepted as *the* **Hermite polynomials** $H_0(x)$, $H_1(x)$, $H_2(x)$, and $H_3(x)$. In general the H_n are constant multiples of the h_n, that is, $H_n(x) = C_nh_n(x)$, where, as indicated, C_n depends on the value of $n \geq 0$. There are several ways of defining the H_n; one is by means of the **generating function** $e^{2xt - t^2}$. It can be shown that

$$e^{2xt - t^2} = \sum_{n=0}^{\infty} \frac{H_n(x)}{n!} t^n. \tag{4}$$

Obtain the first four terms of a power series expansion of $e^{2xt - t^2}$ centered at $t = 0$ (Maclaurin series) to obtain $H_0(x)$, $H_1(x)$, $H_2(x)$, and $H_3(x)$. Use this result to find C_0, C_1, C_2, and C_3. Can you discern a rule for choosing C_n?

(c) The Hermite polynomials $H_n(x)$ can also be obtained from a differential relation known as a **Rodrigues formula**:

$$H_n(x) = (-1)^n e^{x^2} \frac{d^n}{dx^n} e^{-x^2}.$$

Use this formula to obtain $H_0(x)$, $H_1(x)$, $H_2(x)$, and $H_3(x)$.

(d) Derive the Rodrigues formula in part (c) from (4). Start by observing that $e^{2xt - t^2} = e^{x^2}e^{-(x - t)^2}$ and use the general form of a Maclaurin series.

6.3 SOLUTIONS ABOUT SINGULAR POINTS

Introduction

In the preceding section we saw that there is no basic problem in finding two linearly independent power series solutions of

$$a_2(x)y'' + a_1(x)y' + a_0(x)y = 0 \tag{1}$$

about an ordinary point $x = x_0$. However, when $x = x_0$ is a singular point, it is not always possible to find a solution in the form of $y = \sum_{n=0}^{\infty} c_n (x - x_0)^n$; it turns out that we *may* be able to find a series solution of the form $y = \sum_{n=0}^{\infty} c_n (x - x_0)^{n+r}$, where r is a constant to be determined. If r is found to be a number that is not a nonnegative integer, then the last series is not a power series.

Regular and Irregular Singular Points

Singular points are further classified as either regular or irregular. To define these concepts we again put (1) into the standard form

$$y'' + P(x)y' + Q(x)y = 0. \tag{2}$$

DEFINITION 6.2 Regular and Irregular Singular Points

A singular point $x = x_0$ of equation (1) is said to be a **regular singular point** if both $(x - x_0)P(x)$ and $(x - x_0)^2Q(x)$ are analytic at x_0. A singular point that is not regular is said to be an **irregular singular point** of the equation.

Polynomial Coefficients

In the case in which the coefficients in (1) are polynomials with no common factors, Definition 6.2 is equivalent to the following.

> Let $a_2(x_0) = 0$. Form $P(x)$ and $Q(x)$ by reducing $a_1(x)/a_2(x)$ and $a_0(x)/a_2(x)$ to lowest terms, respectively. If the factor $(x - x_0)$ appears *at most* to the first power in the denominator of $P(x)$ and *at most* to the second power in the denominator of $Q(x)$, then $x = x_0$ is a regular singular point.

EXAMPLE 1 Classification of Singular Points

It should be clear that $x = -2$ and $x = 2$ are singular points of the equation

$$(x^2 - 4)^2y'' + (x - 2)y' + y = 0.$$

Dividing the equation by $(x^2 - 4)^2 = (x - 2)^2(x + 2)^2$, we find that

$$P(x) = \frac{1}{(x-2)(x+2)^2} \quad \text{and} \quad Q(x) = \frac{1}{(x-2)^2(x+2)^2}.$$

We now test $P(x)$ and $Q(x)$ at each singular point.

In order that $x = -2$ be a regular singular point, the factor $x + 2$ can appear at most to the first power in the denominator of $P(x)$, and can appear at most to the second power in the denominator of $Q(x)$. Inspection of $P(x)$ and

$Q(x)$ shows that the first condition does not obtain, and so we conclude that $x = -2$ is an irregular singular point.

In order that $x = 2$ be a regular singular point, the factor $x - 2$ can appear at most to the first power in the denominator of $P(x)$ and can appear at most to the second power in the denominator of $Q(x)$. Further inspection of $P(x)$ and $Q(x)$ shows that both these conditions are satisfied, so $x = 2$ is a regular singular point. ●

EXAMPLE 2 Classification of Singular Points

Both $x = 0$ and $x = -1$ are singular points of the differential equation

$$x^2(x+1)^2y'' + (x^2-1)y' + 2y = 0.$$

Inspection of $\quad P(x) = \dfrac{x-1}{x^2(x+1)} \quad$ and $\quad Q(x) = \dfrac{2}{x^2(x+1)^2}$

shows that $x = 0$ is an irregular singular point since $(x - 0)$ appears to the second power in the denominator of $P(x)$. Note, however, that $x = -1$ is a regular singular point. ●

EXAMPLE 3 Classification of Singular Points

(a) $x = 1$ and $x = -1$ are regular singular points of

$$(1-x^2)y'' - 2xy' + 30y = 0.$$

(b) $x = 0$ is an irregular singular point of $x^3y'' - 2xy' + 5y = 0$ since

$$Q(x) = \frac{5}{x^3}.$$

(c) $x = 0$ is a regular singular point of $xy'' - 2xy' + 5y = 0$ since

$$P(x) = -2 \quad \text{and} \quad Q(x) = \frac{5}{x}.$$ ●

In part (c) of Example 3 notice that $(x - 0)$ and $(x - 0)^2$ do not even appear in the denominators of $P(x)$ and $Q(x)$, respectively. Remember, these factors can appear at most in this fashion. For a singular point $x = x_0$, any nonnegative power of $(x - x_0)$ less than one (namely, zero) and nonnegative power less than two (namely, zero and one) in the denominators of $P(x)$ and $Q(x)$, respectively, imply x_0 is a regular singular point.

Also, recall that singular points can be complex numbers. It should be apparent that both $x = 3i$ and $x = -3i$ are regular singular points of the equation $(x^2 + 9)y'' - 3xy' + (1 - x)y = 0$ since

$$P(x) = \frac{-3x}{(x-3i)(x+3i)} \quad \text{and} \quad Q(x) = \frac{1-x}{(x-3i)(x+3i)}.$$

EXAMPLE 4 Cauchy-Euler Equation

From our discussion of the Cauchy-Euler equation in Section 3.6, we can show that $y_1 = x^2$ and $y_2 = x^2 \ln x$ are solutions of the equation $x^2y'' - 3xy' + 4y = 0$ on the interval $(0, \infty)$. If the procedure of Theorem 6.1 is attempted at the regular singular point $x = 0$ (that is, an assumed solution of the form $y = \sum_{n=0}^{\infty} c_n x^n$), we would succeed in obtaining only the solution $y_1 = x^2$. The fact that we would not obtain the second solution is not really surprising since $\ln x$ does not possess a Taylor series expansion about $x = 0$. It follows that $y_2 = x^2 \ln x$ does not have a power series in x.

EXAMPLE 5 DE with No Power Series Solution

The differential equation $6x^2y'' + 5xy' + (x^2 - 1)y = 0$ has a regular singular point at $x = 0$ but does not possess any solution of the form $y = \sum_{n=0}^{\infty} c_n x^n$. By the procedure that we shall now consider it can be shown, however, that there exist two series solutions of the form

$$y = \sum_{n=0}^{\infty} c_n x^{n+1/2} \quad \text{and} \quad y = \sum_{n=0}^{\infty} c_n x^{n-1/3}.$$

Method of Frobenius

To solve a differential equation such as (1) about a regular singular point we employ the following theorem due to Frobenius.

THEOREM 6.2 Frobenius' Theorem

If $x = x_0$ is a regular singular point of equation (1), then there exists at least one series solution of the form

$$y = (x - x_0)^r \sum_{n=0}^{\infty} c_n (x - x_0)^n = \sum_{n=0}^{\infty} c_n (x - x_0)^{n+r}, \tag{3}$$

where the number r is a constant that must be determined. The series will converge at least on some interval $0 < x - x_0 < R$.

Note, the words *at least* in the second line of Theorem 6.2. This means that, in contrast to Theorem 6.1, we are *not* guaranteed two solutions of the indicated form. The **method of Frobenius** consists of identifying a regular singular point x_0, substituting $y = \sum_{n=0}^{\infty} c_n (x - x_0)^{n+r}$ into the differential equation, and determining the unknown exponent r and the coefficients c_n.

As in the preceding section, for the sake of simplicity we shall always assume $x_0 = 0$.

EXAMPLE 6 Series Solution About a Regular Singular Point

Since $x = 0$ is a regular singular point of the differential equation

$$3xy'' + y' - y = 0, \tag{4}$$

we try a solution of the form $y = \sum_{n=0}^{\infty} c_n x^{n+r}$. Now

$$y' = \sum_{n=0}^{\infty} (n + r)c_n x^{n+r-1} \quad \text{and} \quad y'' = \sum_{n=0}^{\infty} (n + r)(n + r - 1)c_n x^{n+r-2},$$

so that

$$\begin{aligned} 3xy'' + y' - y &= 3\sum_{n=0}^{\infty} (n + r)(n + r - 1)c_n x^{n+r-1} \\ &\quad + \sum_{n=0}^{\infty} (n + r)c_n x^{n+r-1} - \sum_{n=0}^{\infty} c_n x^{n+r} \\ &= \sum_{n=0}^{\infty} (n + r)(3n + 3r - 2)c_n x^{n+r-1} - \sum_{n=0}^{\infty} c_n x^{n+r} \\ &= x^r \left[r(3r - 2)c_0 x^{-1} + \underbrace{\sum_{n=1}^{\infty} (n + r)(3n + 3r - 2)c_n x^{n-1}}_{k = n - 1} - \underbrace{\sum_{n=0}^{\infty} c_n x^n}_{k = n} \right] \\ &= x^r \left[r(3r - 2)c_0 x^{-1} + \sum_{k=0}^{\infty} [(k + r + 1)(3k + 3r + 1)c_{k+1} - c_k] x^k \right] = 0, \end{aligned}$$

which implies $\qquad r(3r - 2)c_0 = 0$

$$(k + r + 1)(3k + 3r + 1)c_{k+1} - c_k = 0, \qquad k = 0, 1, 2, \ldots. \tag{5}$$

Since nothing is gained by taking $c_0 = 0$, we must then have

$$r(3r - 2) = 0 \tag{6}$$

and

$$c_{k+1} = \frac{c_k}{(k + r + 1)(3k + 3r + 1)}, \qquad k = 0, 1, 2, \ldots. \tag{7}$$

The two values of r that satisfy (6), $r_1 = 2/3$ and $r_2 = 0$, when substituted in (7), give two different recurrence relations:

$$r_1 = \tfrac{2}{3}, \qquad c_{k+1} = \frac{c_k}{(3k + 5)(k + 1)}, \quad k = 0, 1, 2, \ldots, \tag{8}$$

and

$$r_2 = 0, \qquad c_{k+1} = \frac{c_k}{(k + 1)(3k + 1)}, \quad k = 0, 1, 2, \ldots. \tag{9}$$

Iteration of (8) gives
$$c_1 = \frac{c_0}{5 \cdot 1}$$
$$c_2 = \frac{c_1}{8 \cdot 2} = \frac{c_0}{2!5 \cdot 8}$$
$$c_3 = \frac{c_2}{11 \cdot 3} = \frac{c_0}{3!5 \cdot 8 \cdot 11}$$
$$c_4 = \frac{c_3}{14 \cdot 4} = \frac{c_0}{4!5 \cdot 8 \cdot 11 \cdot 14}$$
$$\vdots$$
$$c_n = \frac{c_0}{n!5 \cdot 8 \cdot 11 \cdots (3n+2)}, \qquad n = 1, 2, 3, \ldots,$$

whereas iteration of (9) yields
$$c_1 = \frac{c_0}{1 \cdot 1}$$
$$c_2 = \frac{c_1}{2 \cdot 4} = \frac{c_0}{2!1 \cdot 4}$$
$$c_3 = \frac{c_2}{3 \cdot 7} = \frac{c_0}{3!1 \cdot 4 \cdot 7}$$
$$c_4 = \frac{c_3}{4 \cdot 10} = \frac{c_0}{4!1 \cdot 4 \cdot 7 \cdot 10}$$
$$\vdots$$
$$c_n = \frac{c_0}{n!1 \cdot 4 \cdot 7 \cdots (3n-2)}, \qquad n = 1, 2, 3, \ldots.$$

Thus we obtain two series solutions
$$y_1 = c_0 x^{2/3}\left[1 + \sum_{n=1}^{\infty} \frac{1}{n!5 \cdot 8 \cdot 11 \cdots (3n+2)} x^n\right] \tag{10}$$

and
$$y_2 = c_0 x^0\left[1 + \sum_{n=1}^{\infty} \frac{1}{n!1 \cdot 4 \cdot 7 \cdots (3n-2)} x^n\right]. \tag{11}$$

By the ratio test it can be demonstrated that both (10) and (11) converge for all finite values of x. Also it should be clear from the form of (10) and (11) that neither series is a constant multiple of the other and, therefore, $y_1(x)$ and $y_2(x)$ are linearly independent solutions on the x-axis. Hence by the superposition principle

$$y = C_1y_1(x) + C_2y_2(x) = C_1\left[x^{2/3} + \sum_{n=1}^{\infty} \frac{1}{n!5 \cdot 8 \cdot 11 \cdots (3n+2)} x^{n+2/3}\right]$$

$$+ C_2\left[1 + \sum_{n=1}^{\infty} \frac{1}{n!1 \cdot 4 \cdot 7 \cdots (3n-2)} x^n\right], \qquad |x| < \infty.$$

is another solution of (4). On any interval not containing the origin, this combination represents the general solution of the differential equation. ●

Although Example 6 illustrates the general procedure for using the method of Frobenius, we hasten to point out that we may not always be able to find two solutions so readily or for that matter find two solutions that are infinite series consisting entirely of powers of x.

Indicial Equation

Equation (6) is called the **indicial equation** of the problem, and the values $r_1 = 2/3$ and $r_2 = 0$ are called the **indicial roots**, or **exponents**, of the singularity. In general, if $x = 0$ is a regular singular point of (1), then the functions $xP(x)$ and $x^2Q(x)$ obtained from (2) are analytic at zero; that is, the expansions

$$xP(x) = p_0 + p_1x + p_2x^2 + \cdots \quad \text{and} \quad x^2Q(x) = q_0 + q_1x + q_2x^2 + \cdots \tag{12}$$

are valid on intervals that have a positive radius of convergence. After substituting $y = \sum_{n=0}^{\infty} c_nx^{n+r}$ in (1) or (2) and simplifying, the indicial equation is a quadratic equation in r that results from equating the *total coefficient of the lowest power of x to zero.* It is readily shown that the general indicial equation is

$$r(r-1) + p_0r + q_0 = 0. \tag{13}$$

We then solve the latter equation for the two values of the exponents and substitute these values into a recurrence relation such as (7). Theorem 6.2 guarantees that at least one solution of the assumed series form can be found.

Cases of Indicial Roots

When using the method of Frobenius, we usually distinguish three cases corresponding to the nature of the indicial roots. For the sake of discussion let us suppose that r_1 and r_2 are the *real* solutions of the indicial equation and that, when appropriate, r_1 *denotes the largest root.*

CASE I **Roots Not Differing by an Integer** If r_1 and r_2 are distinct and do not differ by an integer, then there exist two linearly independent solutions of equation

(1) of the form

$$y_1 = \sum_{n=0}^{\infty} c_n x^{n+r_1}, \qquad c_0 \neq 0, \tag{14a}$$

$$y_2 = \sum_{n=0}^{\infty} b_n x^{n+r_2}, \qquad b_0 \neq 0. \tag{14b}$$

EXAMPLE 7 Case I: Two Solutions of Form (3)

Solve
$$2xy'' + (1 + x)y' + y = 0. \tag{15}$$

Solution If $y = \sum_{n=0}^{\infty} c_n x^{n+r}$, then

$$2xy'' + (1 + x)y' + y = 2\sum_{n=0}^{\infty} (n + r)(n + r - 1)c_n x^{n+r-1} + \sum_{n=0}^{\infty} (n + r)c_n x^{n+r-1}$$

$$+ \sum_{n=0}^{\infty} (n + r)c_n x^{n+r} + \sum_{n=0}^{\infty} c_n x^{n+r}$$

$$= \sum_{n=0}^{\infty} (n + r)(2n + 2r - 1)c_n x^{n+r-1} + \sum_{n=0}^{\infty} (n + r + 1)c_n x^{n+r}$$

$$= x^r \Bigg[r(2r - 1)c_0 x^{-1} + \underbrace{\sum_{n=1}^{\infty} (n + r)(2n + 2r - 1)c_n x^{n-1}}_{k = n - 1} + \underbrace{\sum_{n=0}^{\infty} (n + r + 1)c_n x^n}_{k = n} \Bigg]$$

$$= x^r \Bigg[r(2r - 1)c_0 x^{-1} + \sum_{k=0}^{\infty} [(k + r + 1)(2k + 2r + 1)c_{k+1} + (k + r + 1)c_k]x^k \Bigg] = 0,$$

which implies
$$r(2r - 1) = 0 \tag{16}$$

$$(k + r + 1)(2k + 2r + 1)c_{k+1} + (k + r + 1)c_k = 0, \qquad k = 0, 1, 2, \ldots. \tag{17}$$

From (16) we see that $r_1 = \frac{1}{2}$ and $r_2 = 0$. Because the difference of the indicial roots $r_1 - r_2$ is not an integer we are guaranteed, as indicated in (14a) and (14b), two linearly independent solutions of the form $y_1 = \sum_{n=0}^{\infty} c_n x^{n+1/2}$ and $y_2 = \sum_{n=0}^{\infty} c_n x^n$.

For $r_1 = \frac{1}{2}$, we can divide by $k + \frac{3}{2}$ in (17) to obtain

$$c_{k+1} = \frac{-c_k}{2(k + 1)}$$

$$c_1 = \frac{-c_0}{2 \cdot 1}$$

$$c_2 = \frac{-c_1}{2 \cdot 2} = \frac{c_0}{2^2 \cdot 2!}$$

$$c_3 = \frac{-c_2}{2 \cdot 3} = \frac{-c_0}{2^3 \cdot 3!}$$

$$\vdots$$

$$c_n = \frac{(-1)^n c_0}{2^n n!}, \qquad n = 1, 2, 3, \ldots .$$

Thus $$y_1 = c_0 x^{1/2}\left[1 + \sum_{n=1}^{\infty} \frac{(-1)^n}{2^n n!} x^n\right] = c_0 \sum_{n=0}^{\infty} \frac{(-1)^n}{2^n n!} x^{n+1/2}, \tag{18}$$

which converges for $x \geq 0$. As given, the series is not meaningful for $x < 0$ because of the presence of $x^{1/2}$.

Now for $r_2 = 0$, (17) becomes

$$c_{k+1} = \frac{-c_k}{2k+1}$$

$$c_1 = \frac{-c_0}{1}$$

$$c_2 = \frac{-c_1}{3} = \frac{c_0}{1 \cdot 3}$$

$$c_3 = \frac{-c_2}{5} = \frac{-c_0}{1 \cdot 3 \cdot 5}$$

$$c_4 = \frac{-c_3}{7} = \frac{c_0}{1 \cdot 3 \cdot 5 \cdot 7}$$

$$\vdots$$

$$c_n = \frac{(-1)^n c_0}{1 \cdot 3 \cdot 5 \cdot 7 \cdots (2n-1)}, \qquad n = 1, 2, 3, \ldots .$$

We conclude that a second solution of (15) is

$$y_2 = c_0\left[1 + \sum_{n=1}^{\infty} \frac{(-1)^n}{1 \cdot 3 \cdot 5 \cdot 7 \cdots (2n-1)} x^n\right], \qquad |x| < \infty. \tag{19}$$

On the interval $(0, \infty)$, the general solution is $y = C_1 y_1(x) + C_2 y_2(x)$. ●

When the roots of the indicial equation differ by a positive integer, we may or may not be able to find two solutions of (1) having form (3). If not, then one solution corresponding to the smaller root contains a logarithmic term. When the indicial roots are equal, a second solution *always* contains a logarithm. This latter situation is analogous to the solutions of the Cauchy-Euler differential equation when the roots of the auxiliary equation are equal. We have the next two cases.

CASE II **Roots Differing by a Positive Integer** If $r_1 - r_2 = N$, where N is a positive integer, then there exist two linearly independent solutions of equation (1) of the form

$$y_1 = \sum_{n=0}^{\infty} c_n x^{n+r_1}, \qquad c_0 \neq 0 \tag{20a}$$

$$y_2 = Cy_1(x)\ln x + \sum_{n=0}^{\infty} b_n x^{n+r_2}, \qquad b_0 \neq 0, \tag{20b}$$

where C is a constant that could be zero.

CASE III **Equal Indicial Roots** If $r_1 = r_2$, there always exist two linearly independent solutions of equation (1) of the form

$$y_1 = \sum_{n=0}^{\infty} c_n x^{n+r_1}, \qquad c_0 \neq 0 \tag{21a}$$

$$y_2 = y_1(x)\ln x + \sum_{n=1}^{\infty} b_n x^{n+r_1}. \tag{21b}$$

EXAMPLE 8 Case II: Two Solutions of Form (3)

Solve
$$xy'' + (x-6)y' - 3y = 0. \tag{22}$$

Solution The assumption $y = \sum_{n=0}^{\infty} c_n x^{n+r}$ leads to

$$xy'' + (x-6)y' - 3y$$

$$= \sum_{n=0}^{\infty}(n+r)(n+r-1)c_n x^{n+r-1} - 6\sum_{n=0}^{\infty}(n+r)c_n x^{n+r-1} + \sum_{n=0}^{\infty}(n+r)c_n x^{n+r} - 3\sum_{n=0}^{\infty} c_n x^{n+r}$$

$$= x^r\Big[r(r-7)c_0x^{-1} + \underbrace{\sum_{n=1}^{\infty}(n+r)(n+r-7)c_n x^{n-1}}_{k=n-1} + \underbrace{\sum_{n=0}^{\infty}(n+r-3)c_n x^n}_{k=n}\Big]$$

$$= x^r\Big[r(r-7)c_0x^{-1} + \sum_{k=0}^{\infty}[(k+r+1)(k+r-6)c_{k+1} + (k+r-3)c_k]x^k\Big] = 0.$$

Thus $r(r-7) = 0$ so that $r_1 = 7$, $r_2 = 0$, $r_1 - r_2 = 7$, and

$$(k+r+1)(k+r-6)c_{k+1} + (k+r-3)c_k = 0, \qquad k = 0, 1, 2, \ldots. \tag{23}$$

For the smaller root $r_2 = 0$, (23) becomes

$$(k+1)(k-6)c_{k+1} + (k-3)c_k = 0. \tag{24}$$

Since $k - 6 = 0$ when $k = 6$, we do not divide by this term until $k > 6$. We find

$$\begin{aligned}
1 \cdot (-6)c_1 + (-3)c_0 &= 0 \\
2 \cdot (-5)c_2 + (-2)c_1 &= 0 \\
3 \cdot (-4)c_3 + (-1)c_2 &= 0 \\
\left.\begin{aligned}
4 \cdot (-3)c_4 + 0 \cdot c_3 &= 0 \\
5 \cdot (-2)c_5 + 1 \cdot c_4 &= 0 \\
6 \cdot (-1)c_6 + 2 \cdot c_5 &= 0 \\
7 \cdot 0c_7 + 3 \cdot c_6 &= 0
\end{aligned}\right\} &\leftarrow \text{implies } c_4 = c_5 = c_6 = 0 \text{ but } c_0 \text{ and } c_7 \text{ can be chosen arbitrarily}
\end{aligned}$$

Hence

$$\begin{aligned}
c_1 &= -\frac{1}{2}c_0 \\
c_2 &= -\frac{1}{5}c_1 = \frac{1}{10}c_0 \\
c_3 &= -\frac{1}{12}c_2 = -\frac{1}{120}c_0.
\end{aligned} \tag{25}$$

Now for $k \geq 7$,

$$c_{k+1} = \frac{-(k-3)}{(k+1)(k-6)}c_k.$$

Iterating this last formula gives

$$\begin{aligned}
c_8 &= \frac{-4}{8 \cdot 1}c_7 \\
c_9 &= \frac{-5}{9 \cdot 2}c_8 = \frac{4 \cdot 5}{2!8 \cdot 9}c_7 \\
c_{10} &= \frac{-6}{10 \cdot 3}c_9 = \frac{-4 \cdot 5 \cdot 6}{3!8 \cdot 9 \cdot 10}c_7 \\
&\vdots \\
c_n &= \frac{(-1)^{n+1}4 \cdot 5 \cdot 6 \cdots (n-4)}{(n-7)!8 \cdot 9 \cdot 10 \cdots n}c_7, \qquad n = 8, 9, 10, \ldots.
\end{aligned} \tag{26}$$

If we choose $c_7 = 0$ and $c_0 \neq 0$, we obtain the polynomial solution

$$y_1 = c_0\left[1 - \frac{1}{2}x + \frac{1}{10}x^2 - \frac{1}{120}x^3\right], \tag{27}$$

but when $c_7 \neq 0$ and $c_0 = 0$, it follows that a second, though infinite series, solution is

$$y_2 = c_7\left[x^7 + \sum_{n=8}^{\infty} \frac{(-1)^{n+1}4 \cdot 5 \cdot 6 \cdots (n-4)}{(n-7)!8 \cdot 9 \cdot 10 \cdots n} x^n\right]$$

$$= c_7\left[x^7 + \sum_{k=1}^{\infty} \frac{(-1)^k 4 \cdot 5 \cdot 6 \cdots (k+3)}{k!8 \cdot 9 \cdot 10 \cdots (k+7)} x^{k+7}\right], \qquad |x| < \infty. \tag{28}$$

Finally, the general solution of (22) on the interval $(0, \infty)$ is

$$y = C_1 y_1(x) + C_2 y_2(x)$$

$$= C_1\left[1 - \frac{1}{2}x + \frac{1}{10}x^2 - \frac{1}{120}x^3\right] + C_2\left[x^7 + \sum_{k=1}^{\infty} \frac{(-1)^k 4 \cdot 5 \cdot 6 \cdots (k+3)}{k!8 \cdot 9 \cdot 10 \cdots (k+7)} x^{k+7}\right]. \quad ●$$

It is interesting to observe that in Example 8 we did not use the larger root $r_1 = 7$. Had we done so, we would have obtained a series solution of the form* $y = \sum_{n=0}^{\infty} c_n x^{n+7}$, where the c_n are defined by (23) with $r_1 = 7$:

$$c_{k+1} = \frac{-(k+4)}{(k+8)(k+1)} c_k, \qquad k = 0, 1, 2, \ldots .$$

Iteration of this latter recurrence relation then would yield only *one* solution, namely, the solution given by (28) (with c_0 playing the part of c_7).

When the roots of the indicial equation differ by a positive integer, the second solution *may* contain a logarithm. In practice this is something we do not know in advance but is determined after we have found the indicial roots and have carefully examined the recurrence relation that defines the coefficients c_n. As the foregoing example shows, we just may be lucky enough to find two solutions that involve only powers of x. On the other hand, if we fail to find a second series-type solution, we can always use the fact that

$$y_2 = y_1(x) \int \frac{e^{-\int P(x)\,dx}}{y_1^2(x)}\,dx \tag{29}$$

is also a solution of the equation $y'' + P(x)y' + Q(x)y = 0$, whenever y_1 is a known solution (see Section 3.2).

EXAMPLE 9 Case II: One Solution of Form (3)

Find the general solution of $xy'' + 3y' - y = 0$.

Solution You should verify that the indicial roots are $r_1 = 0$, $r_2 = -2$, $r_1 - r_2 = 2$, and that the method of Frobenius yields only one solution:

$$y_1 = \sum_{n=0}^{\infty} \frac{2}{n!(n+2)!} x^n = 1 + \frac{1}{3}x + \frac{1}{24}x^2 + \frac{1}{360}x^3 + \cdots . \tag{30}$$

---

* Observe that both (28) and this series start with the power x^7. In Case II it is always a good idea to work with the smaller root first.

From (29) we obtain a second solution:

$$y_2 = y_1(x)\int \frac{e^{-\int (3/x)\,dx}}{y_1^2(x)}\,dx = y_1(x)\int \frac{dx}{x^3\left[1 + \frac{1}{3}x + \frac{1}{24}x^2 + \frac{1}{360}x^3 + \cdots\right]^2}$$

$$= y_1(x)\int \frac{dx}{x^3\left[1 + \frac{2}{3}x + \frac{7}{36}x^2 + \frac{1}{30}x^3 + \cdots\right]} \qquad \leftarrow \text{squaring}$$

$$= y_1(x)\int \frac{1}{x^3}\left[1 - \frac{2}{3}x + \frac{1}{4}x^2 - \frac{19}{270}x^3 + \cdots\right]dx \qquad \leftarrow \text{long division}$$

$$= y_1(x)\int \left[\frac{1}{x^3} - \frac{2}{3x^2} + \frac{1}{4x} - \frac{19}{270} + \cdots\right]dx$$

$$= y_1(x)\left[-\frac{1}{2x^2} + \frac{2}{3x} + \frac{1}{4}\ln x - \frac{19}{270}x + \cdots\right]$$

or

$$y_2 = \frac{1}{4}y_1(x)\ln x + y_1(x)\left[-\frac{1}{2x^2} + \frac{2}{3x} - \frac{19}{270}x + \cdots\right]. \tag{31}$$

Hence on the interval $(0, \infty)$, the general solution is

$$y = C_1y_1(x) + C_2\left[\frac{1}{4}y_1(x)\ln x + y_1(x)\left(-\frac{1}{2x^2} + \frac{2}{3x} - \frac{19}{270}x + \cdots\right)\right], \tag{32}$$

where $y_1(x)$ is defined by (30). ●

EXAMPLE 10 Case III: Finding the Second Solution

Find the general solution of $xy'' + y' - 4y = 0$.

Solution The assumption $y = \sum_{n=0}^{\infty} c_nx^{n+r}$ leads to

$$xy'' + y' - 4y = \sum_{n=0}^{\infty}(n + r)(n + r - 1)c_nx^{n+r-1} + \sum_{n=0}^{\infty}(n + r)c_nx^{n+r-1} - 4\sum_{n=0}^{\infty}c_nx^{n+r}$$

$$= \sum_{n=0}^{\infty}(n + r)^2c_nx^{n+r-1} - 4\sum_{n=0}^{\infty}c_nx^{n+r}$$

$$= x^r\left[r^2c_0x^{-1} + \underbrace{\sum_{n=1}^{\infty}(n + r)^2c_nx^{n-1}}_{k = n-1} - 4\underbrace{\sum_{n=0}^{\infty}c_nx^n}_{k = n}\right]$$

$$= x^r\left[r^2c_0x^{-1} + \sum_{k=0}^{\infty}[(k + r + 1)^2c_{k+1} - 4c_k]x^k\right] = 0.$$

Therefore $r^2 = 0$, and so the indicial roots are equal: $r_1 = r_2 = 0$. Moreover, we have

$$(k + r + 1)^2 c_{k+1} - 4c_k = 0, \qquad k = 0, 1, 2, \ldots. \tag{33}$$

Clearly the root $r_1 = 0$ will only yield one solution corresponding to the coefficients defined by the iteration of

$$c_{k+1} = \frac{4c_k}{(k+1)^2}, \qquad k = 0, 1, 2, \ldots.$$

The result is
$$y_1 = c_0 \sum_{n=0}^{\infty} \frac{4^n}{(n!)^2} x^n, \qquad |x| < \infty. \tag{34}$$

To obtain the second linearly independent solution we set $c_0 = 1$ in (34) and then use (29):

$$y_2 = y_1(x) \int \frac{e^{-\int (1/x)\,dx}}{y_1^2(x)}\,dx = y_1(x) \int \frac{dx}{x\left[1 + 4x + 4x^2 + \frac{16}{9}x^3 + \cdots\right]^2}$$

$$= y_1(x) \int \frac{dx}{x\left[1 + 8x + 24x^2 + \frac{16}{9}x^3 + \cdots\right]}$$

$$= y_1(x) \int \frac{1}{x}\left[1 - 8x + 40x^2 - \frac{1472}{9}x^3 + \cdots\right] dx$$

$$= y_1(x) \int \left[\frac{1}{x} - 8 + 40x - \frac{1472}{9}x^2 + \cdots\right] dx$$

$$= y_1(x) \left[\ln x - 8x + 20x^2 - \frac{1472}{27}x^3 + \cdots\right].$$

Thus on the interval $(0, \infty)$ the general solution is

$$y = C_1 y_1(x) + C_2 \left[y_1(x) \ln x + y_1(x)\left(-8x + 20x^2 - \frac{1472}{27}x^3 + \cdots\right)\right],$$

where $y_1(x)$ is defined by (34). ●

Use of Computers

The operations performed in Examples 9 and 10—squaring a series, long division by a series, and integration of the quotient—can, of course, be carried out by hand. But life can be simplified since all these operations, including the indicated multiplication in (31), can be done with relative ease with the help of a CAS such as *Mathematica*, *Maple*, or *Derive*. See *Computer Lab Experiment* 1 for this section.

Remarks (i) We deliberately have not addressed several additional complications of solving a differential equation such as (1) about a singular point x_0. The indicial roots r_1 and r_2 could be complex numbers. In this case the inequality $r_1 > r_2$ is meaningless and must be replaced with $\text{Re}(r_1) > \text{Re}(r_2)$ (if $r = \alpha + i\beta$, then $\text{Re}(r) = \alpha$). In particular, when the indicial equation has real coefficients, the complex roots will be a conjugate pair $r_1 = \alpha + i\beta$, $r_2 = \alpha - i\beta$, and $r_1 - r_2 = 2i\beta \neq$ integer. Thus for $x_0 = 0$ there always exist two solutions, $y_1 = \sum_{n=0}^{\infty} c_n x^{n+r_1}$ and $y_2 = \sum_{n=0}^{\infty} c_n x^{n+r_2}$. Both solutions give complex values of y for each real choice of x. We can overcome this difficulty by using the superposition principle and forming appropriate linear combinations of $y_1(x)$ and $y_2(x)$ to yield real solutions (see Case III in Section 3.6).

(ii) If $x_0 = 0$ is an irregular singular point, we may not be able to find *any* solution of the form $y = \sum_{n=0}^{\infty} c_n x^{n+r}$.

(iii) In the advanced study of differential equations it is sometimes important to examine the nature of a singular point at ∞. See Problems 31 and 32.

EXERCISES 6.3

Answers to odd-numbered problems begin on page AN-26.

In Problems 1–10 determine the singular points of each differential equation. Classify each singular point as regular or irregular.

1. $x^3y'' + 4x^2y' + 3y = 0$
2. $xy'' - (x + 3)^{-2}y = 0$
3. $(x^2 - 9)^2y'' + (x + 3)y' + 2y = 0$
4. $y'' - \dfrac{1}{x}y' + \dfrac{1}{(x-1)^3}y = 0$
5. $(x^3 + 4x)y'' - 2xy' + 6y = 0$
6. $x^2(x - 5)^2y'' + 4xy' + (x^2 - 25)y = 0$
7. $(x^2 + x - 6)y'' + (x + 3)y' + (x - 2)y = 0$
8. $x(x^2 + 1)^2y'' + y = 0$
9. $x^3(x^2 - 25)(x - 2)^2y'' + 3x(x - 2)y' + 7(x + 5)y = 0$
10. $(x^3 - 2x^2 - 3x)^2y'' + x(x - 3)^2y' - (x + 1)y = 0$

In Problems 11–22 show that the indicial roots do not differ by an integer. Use the method of Frobenius to obtain two linearly independent series solutions about the regular singular point $x_0 = 0$. Form the general solution on $(0, \infty)$.

11. $2xy'' - y' + 2y = 0$
12. $2xy'' + 5y' + xy = 0$
13. $4xy'' + \frac{1}{2}y' + y = 0$
14. $2x^2y'' - xy' + (x^2 + 1)y = 0$
15. $3xy'' + (2 - x)y' - y = 0$
16. $x^2y'' - (x - \frac{2}{9})y = 0$
17. $2xy'' - (3 + 2x)y' + y = 0$
18. $x^2y'' + xy' + (x^2 - \frac{4}{9})y = 0$
19. $9x^2y'' + 9x^2y' + 2y = 0$
20. $2x^2y'' + 3xy' + (2x - 1)y = 0$
21. $2x^2y'' - x(x - 1)y' - y = 0$
22. $x(x - 2)y'' + y' - 2y = 0$

In Problems 23–30 show that the indicial roots differ by an integer. Use the method of Frobenius to obtain two linearly independent series solutions about the regular singular point $x_0 = 0$. Form the general solution on $(0, \infty)$.

23. $xy'' + 2y' - xy = 0$
24. $x^2y'' + xy' + (x^2 - \frac{1}{4})y = 0$
25. $x(x - 1)y'' + 3y' - 2y = 0$
26. $y'' + \dfrac{3}{x}y' - 2y = 0$
27. $xy'' + (1 - x)y' - y = 0$
28. $xy'' + y = 0$
29. $xy'' + y' + y = 0$
30. $xy'' - xy' + y = 0$
31. ***PROJECT PROBLEM*** A differential equation is said to have a singular point at ∞ if,

after substituting $z = 1/x$, the resulting equation has a singular point at $z = 0$.

(a) Show that the equation $x^4y'' + y = 0$ has a regular singular point at ∞.

(b) If ∞ is a regular singular point, then there exists at least one solution of the equation of the form

$$y = x^r \sum_{n=0}^{\infty} \frac{c_n}{x^n} = x^r\left(c_0 + \frac{c_1}{x} + \frac{c_2}{x^2} + \cdots\right), \quad c_0 \neq 0$$

that is valid for some region defined by $|x| > R$. Use the assumption

$$y = x^r \sum_{n=0}^{\infty} \frac{c_n}{x^n}$$

to find two series solutions about ∞ for $x^4y'' + y = 0$.

(c) Explain how the series solutions obtianed in part (b) can be useful.

32. *PROJECT PROBLEM* If ∞ is an irregular singular point of a differential equation (see Problem 31), then the assumption

$$y = e^{mx}x^r \sum_{n=0}^{\infty} \frac{c_n}{x^n}$$

may give some useful results.

(a) Show that the equation $xy'' + y' + xy = 0$ has an irregular singular point at ∞.

(b) Use the assumption

$$y = e^{mx}x^r \sum_{n=0}^{\infty} \frac{c_n}{x^n}$$

to obtain two formal series solutions about ∞ of the differential equation in part (a).

(c) Convince yourself that both series obtained in part (b) are divergent.

(d) Write a short report on **asymptotic series** (also called **semiconvergent series**). Explain why the "solutions" defined by the divergent series in part (b) are still of some importance.

33. *PROJECT PROBLEM* The regular singular points of the **hypergeometric equation**

$$x(1 - x)y'' + [\gamma - (\alpha + \beta + 1)x]y' - \alpha\beta y = 0,$$

α, β, γ real constants, are 0, 1, and ∞. This equation has been extensively studied and is important because any linear second-order differential equation with three distinct regular singular points can be transformed into the hypergeometric equation.

(a) Show that the indicial roots are 0 and $1 - \gamma$.

(b) Show that a solution corresponding to the indicial root 0 is given by

$$y_1(x) = 1 + \frac{\alpha\beta}{1!\gamma}x + \frac{\alpha(\alpha+1)\beta(\beta+1)}{2!\gamma(\gamma+1)}x^2$$

$$+ \frac{\alpha(\alpha+1)(\alpha+2)\beta(\beta+1)(\beta+2)}{3!\gamma(\gamma+1)(\gamma+2)}x^3 + \cdots,$$

provided γ is not zero or a negative integer. This solution is known as the **hypergeometric series** and is denoted by $F(a, b, c, x)$. The series converges for at least $|x| < 1$.

(c) For γ not a positive integer, express the solution $y_2(x)$ corresponding to the indicial root $1 - \gamma$ in terms of F. For γ not an integer, express the general solution of the differential equation in terms of F.

(d) Identify the functions defined by $F(1, b, b, x)$ and $F(-k, b, b, -x)$.

(e) Use the substitution $\frac{1}{2}(1 - x) = t$ to show that a particular solution of **Legendre's equation**

$$(1 - x^2)y'' - 2xy' + n(n + 1)y = 0$$

is given by $y = F(n + 1, -n, 1, \frac{1}{2}(1 - x))$. Find particular solutions of Legendre's equation for $n = 1$ and for $n = 2$.

6.4 BESSEL'S EQUATION

The differential equation

$$x^2y'' + xy' + (x^2 - \nu^2)y = 0 \quad \textbf{(1)}$$

is called **Bessel's equation** and occurs frequently in advanced studies in applied mathematics, physics, and engineering. In solving (1) we seek a series solution about the regular singular point $x = 0$; in addition we assume that $\nu \geq 0$.

Solution of Bessel's Equation

If we assume that $y = \sum_{n=0}^{\infty} c_n x^{n+r}$, then

$$x^2y'' + xy' + (x^2 - \nu^2)y = \sum_{n=0}^{\infty} c_n(n+r)(n+r-1)x^{n+r} + \sum_{n=0}^{\infty} c_n(n+r)x^{n+r} + \sum_{n=0}^{\infty} c_n x^{n+r+2}$$

$$- \nu^2 \sum_{n=0}^{\infty} c_n x^{n+r}$$

$$= c_0(r^2 - r + r - \nu^2)x^r + x^r \sum_{n=1}^{\infty} c_n[(n+r)(n+r-1) + (n+r) - \nu^2]x^n$$

$$+ x^r \sum_{n=0}^{\infty} c_n x^{n+2}$$

$$= c_0(r^2 - \nu^2)x^r + x^r \sum_{n=1}^{\infty} c_n[(n+r)^2 - \nu^2]x^n + x^r \sum_{n=0}^{\infty} c_n x^{n+2}. \tag{2}$$

From (2) we see that the indicial equation is $r^2 - \nu^2 = 0$, so that the indicial roots are $r_1 = \nu$ and $r_2 = -\nu$. When $r_1 = \nu$, (2) becomes

$$x^\nu \sum_{n=1}^{\infty} c_n n(n+2\nu)x^n + x^\nu \sum_{n=0}^{\infty} c_n x^{n+2}$$

$$= x^\nu \Bigg[(1+2\nu)c_1 x + \underbrace{\sum_{n=2}^{\infty} c_n n(n+2\nu)x^n}_{k\,=\,n-2} + \underbrace{\sum_{n=0}^{\infty} c_n x^{n+2}}_{k\,=\,n}\Bigg]$$

$$= x^\nu \left[(1+2\nu)c_1 x + \sum_{k=0}^{\infty} [(k+2)(k+2+2\nu)c_{k+2} + c_k]x^{k+2}\right] = 0.$$

Therefore, by the usual argument we can write $(1 + 2\nu)c_1 = 0$ and

$$c_{k+2} = \frac{-c_k}{(k+2)(k+2+2\nu)}, \qquad k = 0, 1, 2, \ldots. \tag{3}$$

The choice $c_1 = 0$ in (3) implies that $c_3 = c_5 = c_7 = \cdots = 0$, so for $k = 0, 2, 4, \ldots$, we find, after letting $k + 2 = 2n$, $n = 1, 2, 3, \ldots$, that

$$c_{2n} = -\frac{c_{2n-2}}{2^2 n(n+\nu)}. \tag{4}$$

Thus $\quad c_2 = -\dfrac{c_0}{2^2 \cdot 1 \cdot (1+\nu)}$

$$c_4 = -\frac{c_2}{2^2 \cdot 2(2+\nu)} = \frac{c_0}{2^4 \cdot 1 \cdot 2(1+\nu)(2+\nu)}$$

$$c_6 = -\frac{c_4}{2^2 \cdot 3(3+\nu)} = -\frac{c_0}{2^6 \cdot 1 \cdot 2 \cdot 3(1+\nu)(2+\nu)(3+\nu)}$$

$$\vdots$$

$$c_{2n} = \frac{(-1)^n c_0}{2^{2n} n!(1+\nu)(2+\nu)\cdots(n+\nu)}, \qquad n = 1, 2, 3, \ldots. \tag{5}$$

It is standard practice to choose c_0 to be a specific value, namely,

$$c_0 = \frac{1}{2^\nu \Gamma(1+\nu)},$$

where $\Gamma(1+\nu)$ is the Gamma function. See Problem 42 in Exercises 5.1. Since this latter function possesses the convenient property $\Gamma(1+\alpha) = \alpha\Gamma(\alpha)$, we can reduce the indicated product in the denominator of (5) to one term. For example,

$$\Gamma(1+\nu+1) = (1+\nu)\Gamma(1+\nu)$$

$$\Gamma(1+\nu+2) = (2+\nu)\Gamma(2+\nu) = (2+\nu)(1+\nu)\Gamma(1+\nu).$$

Hence we can write (5) as

$$c_{2n} = \frac{(-1)^n}{2^{2n+\nu} n!(1+\nu)(2+\nu)\cdots(n+\nu)\Gamma(1+\nu)} = \frac{(-1)^n}{2^{2n+\nu} n!\Gamma(1+\nu+n)}$$

for $n = 0, 1, 2, \ldots$.

Bessel Functions of the First Kind

The series solution $y = \sum_{n=0}^{\infty} c_{2n} x^{2n+\nu}$ is usually denoted by $J_\nu(x)$:

$$J_\nu(x) = \sum_{n=0}^{\infty} \frac{(-1)^n}{n!\Gamma(1+\nu+n)} \left(\frac{x}{2}\right)^{2n+\nu}. \tag{6}$$

If $\nu \geq 0$, the series converges at least on the interval $[0, \infty)$. Also, for the second exponent $r_2 = -\nu$, we obtain, in exactly the same manner,

$$J_{-\nu}(x) = \sum_{n=0}^{\infty} \frac{(-1)^n}{n!\Gamma(1-\nu+n)} \left(\frac{x}{2}\right)^{2n-\nu}. \tag{7}$$

The functions $J_\nu(x)$ and $J_{-\nu}(x)$ are called **Bessel functions of the first kind** of order ν and $-\nu$, respectively. Depending on the value of ν, (7) may contain negative powers of x and hence converges on $(0, \infty)$.*

Now some care must be taken in writing the general solution of (1). When $\nu = 0$, it is apparent that (6) and (7) are the same. If $\nu > 0$ and $r_1 - r_2 = \nu - (-\nu) = 2\nu$ is not a positive integer, it follows from Case I of Section 6.3 that $J_\nu(x)$ and $J_{-\nu}(x)$ are linearly independent solutions of (1) on $(0, \infty)$, and so the

---

* When we replace x by $|x|$, the series given in (7) and (8) converge for $0 < |x| < \infty$.

general solution of the interval would be $y = c_1J_\nu(x) + c_2J_{-\nu}(x)$. But we also know from Case II of Section 6.3 that when $r_1 - r_2 = 2\nu$ is a positive integer, a second series solution of (1) *may* exist. In this second case we distinguish two possibilities. When $\nu = m =$ positive integer, $J_{-m}(x)$ defined by (7) and $J_m(x)$ are not linearly independent solutions. It can be shown that J_{-m} is a constant multiple of J_m (see Property (i) on page 342). In addition, $r_1 - r_2 = 2\nu$ can be a positive integer when ν is half an odd positive integer. It can be shown in this latter event that $J_\nu(x)$ and $J_{-\nu}(x)$ are linearly independent. In other words, the general solution of (1) on $(0, \infty)$ is

$$y = c_1J_\nu(x) + c_2J_{-\nu}(x), \qquad \nu \neq \text{integer}. \tag{8}$$

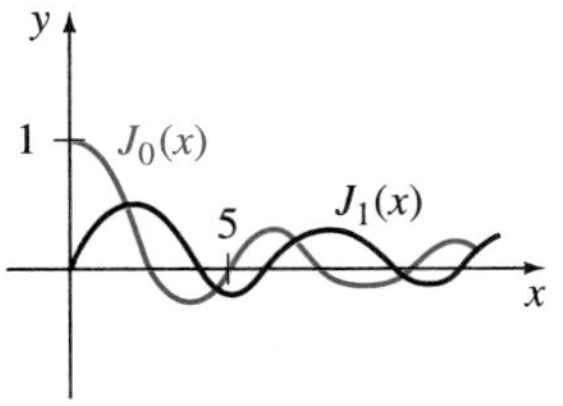

FIGURE 6.2

The graphs of $y = J_0(x)$ and $y = J_1(x)$ are given in Figure 6.2.

EXAMPLE 1 General Solution: ν Not an Integer

By identifying $\nu^2 = \frac{1}{4}$ and $\nu = \frac{1}{2}$, we see from (8) that the general solution of the equation $x^2y'' + xy' + (x^2 - \frac{1}{4})y = 0$ on $(0, \infty)$ is $y = c_1J_{1/2}(x) + c_2J_{-1/2}(x)$.

Bessel Functions of the Second Kind

If $\nu \neq$ integer, the function defined by the linear combination

$$Y_\nu(x) = \frac{\cos \nu\pi J_\nu(x) - J_{-\nu}(x)}{\sin \nu\pi} \tag{9}$$

and the function $J_\nu(x)$ are linearly independent solutions of (1). Thus another form of the general solution of (1) is $y = c_1J_\nu(x) + c_2Y_\nu(x)$, provided $\nu \neq$ integer. As $\nu \to m$, m an integer, (9) has the indeterminate form 0/0. However, it can be shown by L'Hôpital's rule that $\lim_{\nu\to m} Y_\nu(x)$ exists. Moreover, the function

$$Y_m(x) = \lim_{\nu\to m} Y_\nu(x)$$

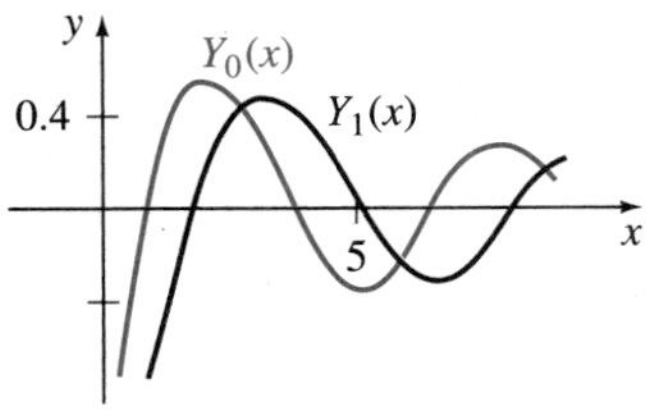

and $J_m(x)$ are linearly independent solutions of $x^2y'' + xy' + (x^2 - m^2)y = 0$. Hence for *any* value of ν the general solution of (1) on $(0, \infty)$ can be written as

$$y = c_1J_\nu(x) + c_2Y_\nu(x). \tag{10}$$

FIGURE 6.3

$Y_\nu(x)$ is called the **Bessel function of the second kind** of order ν. Figure 6.3 shows the graphs of $Y_0(x)$ and $Y_1(x)$.

EXAMPLE 2 General Solution: ν an Integer

By identifying $\nu^2 = 9$ and $\nu = 3$, we see from (10) that the general solution of the equation $x^2y'' + xy' + (x^2 - 9)y = 0$ on $(0, \infty)$ is $y = c_1J_3(x) + c_2Y_3(x)$.

Parametric Bessel Equation

By replacing x by λx in (1) and using the chain rule, we obtain an alternative form of Bessel's equation known as the **parametric Bessel equation**:

$$x^2y'' + xy' + (\lambda^2x^2 - \nu^2)y = 0. \tag{11}$$

The general solution of (11) is

$$y = c_1J_\nu(\lambda x) + c_2Y_\nu(\lambda x). \tag{12}$$

Properties

We list below a few of the more useful properties of Bessel functions of order m, $m = 0, 1, 2, \ldots$:

$$\text{(i) } J_{-m}(x) = (-1)^mJ_m(x) \qquad \text{(ii) } J_m(-x) = (-1)^mJ_m(x)$$

$$\text{(iii) } J_m(0) = \begin{cases} 0, & m > 0 \\ 1, & m = 0 \end{cases} \qquad \text{(iv) } \lim_{x \to 0^+} Y_m(x) = -\infty$$

Note that Property (ii) indicates that $J_m(x)$ is an even function if m is an even integer and an odd function if m is an odd integer. The graphs of $Y_0(x)$ and $Y_1(x)$ in Figure 6.3 illustrate Property (iv): $Y_m(x)$ is unbounded at the origin. This last fact is not obvious from (9). It can be shown either from (9) or by the methods of Section 6.3 that for $x > 0$,

$$Y_0(x) = \frac{2}{\pi}J_0(x)\left[\gamma + \ln\frac{x}{2}\right] - \frac{2}{\pi}\sum_{k=1}^{\infty}\frac{(-1)^k}{(k!)^2}\left(1 + \frac{1}{2} + \cdots + \frac{1}{k}\right)\left(\frac{x}{2}\right)^{2k},$$

where $\gamma = 0.57721566\ldots$ is **Euler's constant** (see page 272).

Numerical Values of $J_0(x)$, $J_1(x)$, $Y_0(x)$, and $Y_1(x)$

Some functional values of $J_0(x)$, $J_1(x)$, $Y_0(x)$, and $Y_1(x)$ for selected values of x are given in Table 6.1. The first five nonnegative zeros of $J_0(x)$, $J_1(x)$, $Y_0(x)$, and $Y_1(x)$ are given in Table 6.2.

TABLE 6.1 Numerical Values of J_0, J_1, Y_0, and Y_1

x	$J_0(x)$	$J_1(x)$	$Y_0(x)$	$Y_1(x)$
0	1.0000	0.0000	—	—
1	0.7652	0.4401	0.0883	−0.7812
2	0.2239	0.5767	0.5104	−0.1070
3	−0.2601	0.3391	0.3769	0.3247
4	−0.3971	−0.0660	−0.0169	0.3979
5	−0.1776	−0.3276	−0.3085	0.1479
6	0.1506	−0.2767	−0.2882	−0.1750
7	0.3001	−0.0047	−0.0259	−0.3027
8	0.1717	0.2346	0.2235	−0.1581
9	−0.0903	0.2453	0.2499	0.1043
10	−0.2459	0.0435	0.0557	0.2490
11	−0.1712	−0.1768	−0.1688	0.1637
12	0.0477	−0.2234	−0.2252	−0.0571
13	0.2069	−0.0703	−0.0782	−0.2101
14	0.1711	0.1334	0.1272	−0.1666
15	−0.0142	0.2051	0.2055	0.0211

TABLE 6.2 Zeros of J_0, J_1, Y_0, and Y_1

$J_0(x)$	$J_1(x)$	$Y_0(x)$	$Y_1(x)$
2.4048	0.0000	0.8936	2.1971
5.5201	3.8317	3.9577	5.4297
8.6537	7.0156	7.0861	8.5960
11.7915	10.1735	10.2223	11.7492
14.9309	13.3237	13.3611	14.8974

Differential Recurrence Relation

Recurrence formulas that relate Bessel functions of different orders are important in theory and in applications. In the next example we derive a **differential recurrence relation**.

EXAMPLE 3 Derivation Using the Series Definition

Derive the formula $xJ'_\nu(x) = \nu J_\nu(x) - xJ_{\nu+1}(x)$.

Solution It follows from (6) that

$$xJ'_\nu(x) = \sum_{n=0}^{\infty} \frac{(-1)^n(2n+\nu)}{n!\Gamma(1+\nu+n)}\left(\frac{x}{2}\right)^{2n+\nu}$$

$$= \nu \sum_{n=0}^{\infty} \frac{(-1)^n}{n!\Gamma(1+\nu+n)}\left(\frac{x}{2}\right)^{2n+\nu} + 2\sum_{n=0}^{\infty} \frac{(-1)^n n}{n!\Gamma(1+\nu+n)}\left(\frac{x}{2}\right)^{2n+\nu}$$

$$= \nu J_\nu(x) + x \underbrace{\sum_{n=1}^{\infty} \frac{(-1)^n}{(n-1)!\Gamma(1+\nu+n)}\left(\frac{x}{2}\right)^{2n+\nu-1}}_{k = n-1}$$

$$= \nu J_\nu(x) - x\sum_{k=0}^{\infty} \frac{(-1)^k}{k!\Gamma(2+\nu+k)}\left(\frac{x}{2}\right)^{2k+\nu+1} = \nu J_\nu(x) - xJ_{\nu+1}(x).$$

The result in Example 3 can be written in an alternative form. Dividing $xJ'_\nu(x) - \nu J_\nu(x) = -xJ_{\nu+1}(x)$ by x gives

$$J'_\nu(x) - \frac{\nu}{x}J_\nu(x) = -J_{\nu+1}(x).$$

This last expression is recognized as a linear first-order differential equation in $J_\nu(x)$. Multiplying both sides of the equality by the integrating factor $x^{-\nu}$ then yields

$$\frac{d}{dx}[x^{-\nu}J_\nu(x)] = -x^{-\nu}J_{\nu+1}(x). \tag{13}$$

It can be shown in a similar manner that

$$\frac{d}{dx}[x^{\nu}J_\nu(x)] = x^{\nu}J_{\nu-1}(x). \tag{14}$$

See Problem 20. The differential recurrence relations (13) and (14) are also valid for the Bessel function of the second kind $Y_\nu(x)$. Observe that when $\nu = 0$ it follows from (13) that

$$J'_0(x) = -J_1(x) \quad \text{and} \quad Y'_0(x) = -Y_1(x). \tag{15}$$

An application of these results is given in Problem 24.

EXERCISES 6.4

Answers to odd-numbered problems begin on page AN-28.

In Problems 1–8 find the general solution of the given differential equation on $(0, \infty)$.

1. $x^2y'' + xy' + (x^2 - \frac{1}{9})y = 0$
2. $x^2y'' + xy' + (x^2 - 1)y = 0$
3. $4x^2y'' + 4xy' + (4x^2 - 25)y = 0$
4. $16x^2y'' + 16xy' + (16x^2 - 1)y = 0$
5. $xy'' + y' + xy = 0$
6. $\dfrac{d}{dx}[xy'] + \left(x - \dfrac{4}{x}\right)y = 0$
7. $x^2y'' + xy' + (9x^2 - 4)y = 0$
8. $x^2y'' + xy' + (36x^2 - \frac{1}{4})y = 0$
9. Use the change of variables $y = x^{-1/2}w(x)$ to find the general solution of the equation
$$x^2y'' + 2xy' + \lambda^2x^2y = 0, \qquad x > 0.$$
10. Verify that the differential equation
$$xy'' + (1 - 2n)y' + xy = 0, \qquad x > 0,$$
possesses the particular solution $y = x^nJ_n(x)$.
11. Verify that the differential equation
$$xy'' + (1 + 2n)y' + xy = 0, \qquad x > 0,$$
possesses the particular solution $y = x^{-n}J_n(x)$.
12. Verify that the differential equation
$$x^2y'' + (\lambda^2x^2 - \nu^2 + \tfrac{1}{4})y = 0, \qquad x > 0,$$
possesses the particular solution $y = \sqrt{x}J_\nu(\lambda x)$, where $\lambda > 0$.

In Problems 13–18 use the results of Problems 10, 11, and 12 to find a particular solution of the given differential equation on $(0, \infty)$.

13. $y'' + y = 0$
14. $xy'' - y' + xy = 0$
15. $xy'' + 3y' + xy = 0$
16. $4x^2y'' + (16x^2 + 1)y = 0$
17. $x^2y'' + (x^2 - 2)y = 0$
18. $xy'' - 5y' + xy = 0$

In Problems 19–22 derive the given recurrence relation.

19. $xJ'_\nu(x) = -\nu J_\nu(x) + xJ_{\nu-1}(x)$
[*Hint*: $2n + \nu = 2(n + \nu) - \nu$.]
20. $\dfrac{d}{dx}[x^\nu J_\nu(x)] = x^\nu J_{\nu-1}(x)$
21. $2\nu J_\nu(x) = xJ_{\nu+1}(x) + xJ_{\nu-1}(x)$
22. $2J'_\nu(x) = J_{\nu-1}(x) - J_{\nu+1}(x)$
23. (a) Show that $y = x^{1/2}\,w(\frac{2}{3}\,\alpha x^{3/2})$ is a solution of **Airy's differential equation** $y'' + \alpha^2xy = 0$, $x > 0$ whenever w is a solution of Bessel's equation $t^2w'' + tw' + (t^2 - \frac{1}{9})w = 0$, $t > 0$. [*Hint*: After differentiating, substituting, and simplifying, let $t = \frac{2}{3}\,\alpha x^{3/2}$.]
(b) Use the result of part (a) to express the general solution of Airy's equation for $x > 0$ in terms of Bessel functions.
24. *PROJECT PROBLEM*
(a) In Section 3.7 we saw that one mathematical model for the free undamped motion of a mass on an aging spring is given by $mx'' + ke^{-\alpha t}x = 0$, $\alpha > 0$. Use the change of variables
$$s = \frac{2}{\alpha}\sqrt{\frac{k}{m}}\,e^{-\alpha t/2}$$
to show that the differential equation of the aging spring becomes
$$s^2\frac{d^2x}{ds^2} + s\frac{dx}{ds} + s^2x = 0.$$
(b) Use the result in part (a) to show that the general solution of the original differential equation is
$$x(t) = c_1J_0\left(\frac{2}{\alpha}\sqrt{\frac{k}{m}}\,e^{-\alpha t/2}\right) + c_2Y_0\left(\frac{2}{\alpha}\sqrt{\frac{k}{m}}\,e^{-\alpha t/2}\right).$$
(c) Use the solution in part (b) to solve the initial-value problem
$$4x'' + e^{-0.1t}x = 0, \qquad x(0) = 1, \quad x'(0) = -\tfrac{1}{2}.$$
[*Hint*: Use Table 6.1 and (15).]
25. *PROJECT PROBLEM* Suppose that a uniform thin column is positioned vertically by embedding one end in the ground. Let L denote the length of the column above the ground. If the length of the column is greater than a certain critical height it will deflect or bend away from the vertical under the influence of its own weight. The angular deflection $\theta(x)$ of the column from the vertical at a point $P(x)$ can be shown to satisfy the following boundary-value problem:
$$EI\frac{d^2\theta}{dx^2} + \delta g(L - x)\theta = 0, \qquad \theta(0) = 0, \quad \theta'(L) = 0,$$

where E is Young's modulus, I is the cross-sectional moment of inertia, δ is the constant linear density, and x is distance along the column measured from its base. See Figure 6.4. Find the critical length of a solid steel rod of radius $r = 0.05$ in., $\delta g = 0.28\ A$ lb/in., $E = 2.6 \times 10^7$ lb/in.^{2}, $A = \pi r^2$, and $I = \frac{1}{4}\pi r^4$. [*Hint*: See Problem 23.]

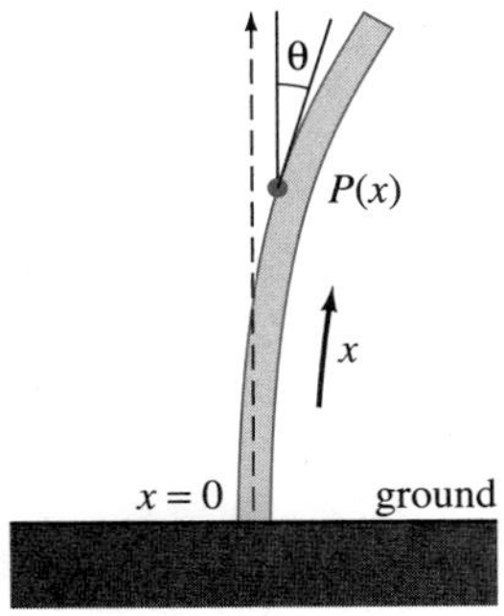

FIGURE 6.4

26. **PROJECT PROBLEM**

(a) Use the fact that $\Gamma(\frac{1}{2}) = \sqrt{\pi}$ and (7) to show that $J_{1/2}(x) = \sqrt{\dfrac{2}{\pi x}}\sin x$.

(b) Show that $J_{-1/2}(x) = \sqrt{\dfrac{2}{\pi x}}\cos x$.

(c) From (a) and (b) and the result in Problem 21 it follows that when ν = half an odd integer, $J_\nu(x)$ can be expressed in terms of sin x, cos x, and powers of x. Such Bessel functions are called **spherical Bessel functions**. Express $J_{3/2}(x)$ in terms of sin x, cos x, and powers of x.

27. **PROJECT PROBLEM** In Example 2 of Section 1.2 we indicated that the differential equation $dy/dx = x^2 + y^2$ cannot be solved in terms of elementary functions. This first-order equation can, however, be solved in terms of Bessel functions.

(a) Show that the substitution $y = -\dfrac{1}{u}\dfrac{du}{dx}$ leads to the equation $u'' + x^2u = 0$.

(b) Show that $u = x^{1/2}\, w(\frac{1}{2}x^2)$ is a solution of $u'' + x^2u = 0$ whenever w is a solution of Bessel's equation $t^2w'' + tw' + (t^2 - \frac{1}{16})\,w = 0$.

(c) Use $J'_\nu(x) = \dfrac{\nu}{x}J_\nu(x) - J_{\nu+1}(x)$ (Example 3)

$$J'_\nu(x) = -\frac{\nu}{x}J_\nu(x) + J_{\nu-1}(x) \quad \text{(Problem 19)}$$

as an aid to show that a one-parameter family of solutions of $dy/dx = x^2 + y^2$ is given by

$$y = x\frac{J_{3/4}(\frac{1}{2}x^2) - cJ_{-3/4}(\frac{1}{2}x^2)}{cJ_{1/4}(\frac{1}{2}x^2) + J_{-1/4}(\frac{1}{2}x^2)}.$$

(d) We saw in Figure 1.8 and in Example 5 of Section 2.1 that the initial-value problem $dy/dx = x^2 + y^2$, $y(0) = 1$ possesses a solution. Use the family of solutions in part (c) along with (6) and (7) to show that the initial condition $y(0) = 1$ yields $c = -\Gamma(\frac{1}{4})/2\Gamma(\frac{3}{4})$.

28. **PROJECT PROBLEM** The Bessel function $y = J_0(t)$ defined by (6) is a particular solution of equation (1) when $\nu = 0$. Verify this fact in an alternative manner by using the Laplace transform to solve

$$ty'' + y' + ty = 0, \quad y(0) = 1$$

29. **PROJECT PROBLEM** Bessel functions $J_n(t)$, n an integer, are often defined by means of a **generating function**. Show that

$$e^{(x/2)(t - 1/t)} = \sum_{n=-\infty}^{\infty} J_n(x)t^n.$$

30. **WRITING PROJECT** Write a short report on why $J_0(x)$ and $J_1(x)$ could be called "almost periodic functions." Discuss the connection between the nonnegative zeros of $J_0(x)$ and $J_1(x)$ and the concept of "almost periodic."

Appendices

I INTRODUCTION TO MATRICES

A.1 Basic Definitions and Theory

DEFINITION A.1 Matrix

A **matrix A** is any rectangular array of numbers or functions:

$$\mathbf{A} = \begin{pmatrix} a_{11} & a_{12} & \cdots & a_{1n} \\ a_{21} & a_{22} & \cdots & a_{2n} \\ \vdots & \vdots & & \vdots \\ a_{m1} & a_{m2} & \cdots & a_{mn} \end{pmatrix}. \tag{1}$$

If a matrix has m rows and n columns, we say that its **size** is m by n (written $m \times n$). An $n \times n$ matrix is called a **square** matrix of order n.

The element, or entry, in the ith row and jth column of an $m \times n$ matrix $\mathbf{A}$ is written a_{ij}. An $m \times n$ matrix $\mathbf{A}$ is then abbreviated as $\mathbf{A} = (a_{ij})_{m \times n}$, or simply $\mathbf{A} = (a_{ij})$. A 1×1 matrix is simply one constant or function.

DEFINITION A.2 Equality of Matrices

Two $m \times n$ matrices $\mathbf{A}$ and $\mathbf{B}$ are **equal** if $a_{ij} = b_{ij}$ for each i and j.

DEFINITION A.3 Column Matrix

A **column matrix X** is any matrix having n rows and one column:

$$\mathbf{X} = \begin{pmatrix} b_{11} \\ b_{21} \\ \vdots \\ b_{n1} \end{pmatrix} = (b_{i1})_{n \times 1}.$$

A column matrix is also called a **column vector** or simply a **vector**.

DEFINITION A.4 Multiples of Matrices

A **multiple** of a matrix **A** is defined to be

$$k\mathbf{A} = \begin{pmatrix} ka_{11} & ka_{12} & \cdots & ka_{1n} \\ ka_{21} & ka_{22} & \cdots & ka_{2n} \\ \vdots & & & \vdots \\ ka_{m1} & ka_{m2} & \cdots & ka_{mn} \end{pmatrix} = (ka_{ij})_{m \times n},$$

where k is a constant or function.

EXAMPLE 1 Multiples of Matrices

$$\textbf{(a)}\ 5\begin{pmatrix} 2 & -3 \\ 4 & -1 \\ \frac{1}{5} & 6 \end{pmatrix} = \begin{pmatrix} 10 & -15 \\ 20 & -5 \\ 1 & 30 \end{pmatrix} \qquad \textbf{(b)}\ e^t\begin{pmatrix} 1 \\ -2 \\ 4 \end{pmatrix} = \begin{pmatrix} e^t \\ -2e^t \\ 4e^t \end{pmatrix}$$

We note in passing that for any matrix **A**, the product $k\mathbf{A}$ is the same as $\mathbf{A}k$. For example,

$$e^{-3t}\begin{pmatrix} 2 \\ 5 \end{pmatrix} = \begin{pmatrix} 2e^{-3t} \\ 5e^{-3t} \end{pmatrix} = \begin{pmatrix} 2 \\ 5 \end{pmatrix}e^{-3t}.$$

DEFINITION A.5 Addition of Matrices

The **sum** of two $m \times n$ matrices **A** and **B** is defined to be the matrix

$$\mathbf{A} + \mathbf{B} = (a_{ij} + b_{ij})_{m \times n}.$$

In other words, when adding two matrices of the same size, we add the corresponding elements.

EXAMPLE 2 Matrix Addition

The sum of $\mathbf{A} = \begin{pmatrix} 2 & -1 & 3 \\ 0 & 4 & 6 \\ -6 & 10 & -5 \end{pmatrix}$ and $\mathbf{B} = \begin{pmatrix} 4 & 7 & -8 \\ 9 & 3 & 5 \\ 1 & -1 & 2 \end{pmatrix}$ is

$$\mathbf{A} + \mathbf{B} = \begin{pmatrix} 2+4 & -1+7 & 3+(-8) \\ 0+9 & 4+3 & 6+5 \\ -6+1 & 10+(-1) & -5+2 \end{pmatrix} = \begin{pmatrix} 6 & 6 & -5 \\ 9 & 7 & 11 \\ -5 & 9 & -3 \end{pmatrix}.$$

EXAMPLE 3 A Matrix Written as a Sum of Column Matrices

The single matrix $\begin{pmatrix} 3t^2 - 2e^t \\ t^2 + 7t \\ 5t \end{pmatrix}$ can be written as the sum of three column vectors

$$\begin{pmatrix} 3t^2 - 2e^t \\ t^2 + 7t \\ 5t \end{pmatrix} = \begin{pmatrix} 3t^2 \\ t^2 \\ 0 \end{pmatrix} + \begin{pmatrix} 0 \\ 7t \\ 5t \end{pmatrix} + \begin{pmatrix} -2e^t \\ 0 \\ 0 \end{pmatrix} = \begin{pmatrix} 3 \\ 1 \\ 0 \end{pmatrix} t^2 + \begin{pmatrix} 0 \\ 7 \\ 5 \end{pmatrix} t + \begin{pmatrix} -2 \\ 0 \\ 0 \end{pmatrix} e^t.$$

The **difference** of two $m \times n$ matrices is defined in the usual manner: $\mathbf{A} - \mathbf{B} = \mathbf{A} + (-\mathbf{B})$, where $-\mathbf{B} = (-1)\mathbf{B}$.

DEFINITION A.6 Multiplication of Matrices

Let **A** be a matrix having m rows and n columns and **B** be a matrix having n rows and p columns. We define the **product AB** to be the $m \times p$ matrix

$$\mathbf{AB} = \begin{pmatrix} a_{11} & a_{12} & \cdots & a_{1n} \\ a_{21} & a_{22} & \cdots & a_{2n} \\ \vdots & & & \vdots \\ a_{m1} & a_{m2} & \cdots & a_{mn} \end{pmatrix} \begin{pmatrix} b_{11} & b_{12} & \cdots & b_{1p} \\ b_{21} & b_{22} & \cdots & b_{2p} \\ \vdots & & & \vdots \\ b_{n1} & b_{n2} & \cdots & b_{np} \end{pmatrix}$$

$$= \begin{pmatrix} a_{11}b_{11} + a_{12}b_{21} + \cdots + a_{1n}b_{n1} & \cdots & a_{11}b_{1p} + a_{12}b_{2p} + \cdots + a_{1n}b_{np} \\ a_{21}b_{11} + a_{22}b_{21} + \cdots + a_{2n}b_{n1} & \cdots & a_{21}b_{1p} + a_{22}b_{2p} + \cdots + a_{2n}b_{np} \\ \vdots & & \vdots \\ a_{m1}b_{11} + a_{m2}b_{21} + \cdots + a_{mn}b_{n1} & \cdots & a_{m1}b_{1p} + a_{m2}b_{2p} + \cdots + a_{mn}b_{np} \end{pmatrix}$$

$$= \left(\sum_{k=1}^{n} a_{ik}b_{kj} \right)_{m \times p}.$$

Note carefully in Definition A.6 that the product $\mathbf{AB} = \mathbf{C}$ is defined only when the number of columns in the matrix **A** is the same as the number of rows in **B**. The size of the product can be determined from

$$\mathbf{A}_{m \times n}\mathbf{B}_{n \times p} = \mathbf{C}_{m \times p}.$$

Also, you might recognize that the entries in, say, the ith row of the final matrix **AB** are formed by using the component definition of the inner or dot product of the ith row of **A** with each of the columns of **B**.

EXAMPLE 4 Multiplication of Matrices

(a) For $\mathbf{A} = \begin{pmatrix} 4 & 7 \\ 3 & 5 \end{pmatrix}$ and $\mathbf{B} = \begin{pmatrix} 9 & -2 \\ 6 & 8 \end{pmatrix}$,

$$\mathbf{AB} = \begin{pmatrix} 4 \cdot 9 + 7 \cdot 6 & 4 \cdot (-2) + 7 \cdot 8 \\ 3 \cdot 9 + 5 \cdot 6 & 3 \cdot (-2) + 5 \cdot 8 \end{pmatrix} = \begin{pmatrix} 78 & 48 \\ 57 & 34 \end{pmatrix}.$$

(b) For $\mathbf{A} = \begin{pmatrix} 5 & 8 \\ 1 & 0 \\ 2 & 7 \end{pmatrix}$ and $\mathbf{B} = \begin{pmatrix} -4 & -3 \\ 2 & 0 \end{pmatrix}$,

$$\mathbf{AB} = \begin{pmatrix} 5 \cdot (-4) + 8 \cdot 2 & 5 \cdot (-3) + 8 \cdot 0 \\ 1 \cdot (-4) + 0 \cdot 2 & 1 \cdot (-3) + 0 \cdot 0 \\ 2 \cdot (-4) + 7 \cdot 2 & 2 \cdot (-3) + 7 \cdot 0 \end{pmatrix} = \begin{pmatrix} -4 & -15 \\ -4 & -3 \\ 6 & -6 \end{pmatrix}.$$

In general, *matrix multiplication is not commutative*; that is, $\mathbf{AB} \neq \mathbf{BA}$. Observe in part (a) of Example 4 that $\mathbf{BA} = \begin{pmatrix} 30 & 53 \\ 48 & 82 \end{pmatrix}$, whereas in part (b) the product **BA** is not defined since Definition A.6 requires that the first matrix, (in this case **B**) have the same number of columns as the second matrix has rows.

We are particularly interested in the product of a square matrix and a column vector.

EXAMPLE 5 Multiplication of Matrices

(a) $$\begin{pmatrix} 2 & -1 & 3 \\ 0 & 4 & 5 \\ 1 & -7 & 9 \end{pmatrix} \begin{pmatrix} -3 \\ 6 \\ 4 \end{pmatrix} = \begin{pmatrix} 2 \cdot (-3) + (-1) \cdot 6 + 3 \cdot 4 \\ 0 \cdot (-3) + 4 \cdot 6 + 5 \cdot 4 \\ 1 \cdot (-3) + (-7) \cdot 6 + 9 \cdot 4 \end{pmatrix} = \begin{pmatrix} 0 \\ 44 \\ -9 \end{pmatrix}$$

(b) $$\begin{pmatrix} -4 & 2 \\ 3 & 8 \end{pmatrix} \begin{pmatrix} x \\ y \end{pmatrix} = \begin{pmatrix} -4x + 2y \\ 3x + 8y \end{pmatrix}$$

Multiplicative Identity

For a given positive integer n, the $n \times n$ matrix

$$\mathbf{I} = \begin{pmatrix} 1 & 0 & 0 & \cdots & 0 \\ 0 & 1 & 0 & \cdots & 0 \\ \vdots & \vdots & & & \vdots \\ 0 & 0 & 0 & \cdots & 1 \end{pmatrix}$$

is called the **multiplicative identity matrix**. It follows from Definition A.6 that for any $n \times n$ matrix $\mathbf{A}$,

$$\mathbf{AI} = \mathbf{IA} = \mathbf{A}.$$

Also, it is readily verified that if $\mathbf{X}$ is an $n \times 1$ column matrix, then $\mathbf{IX} = \mathbf{X}$.

Zero Matrix

A matrix consisting of all zero entries is called a **zero matrix** and is denoted by $\mathbf{0}$. For example,

$$\mathbf{0} = \begin{pmatrix} 0 \\ 0 \end{pmatrix}, \quad \mathbf{0} = \begin{pmatrix} 0 & 0 \\ 0 & 0 \end{pmatrix}, \quad \mathbf{0} = \begin{pmatrix} 0 & 0 \\ 0 & 0 \\ 0 & 0 \end{pmatrix},$$

and so on. If $\mathbf{A}$ and $\mathbf{0}$ are $m \times n$ matrices, then

$$\mathbf{A} + \mathbf{0} = \mathbf{0} + \mathbf{A} = \mathbf{A}.$$

Associative Law

Although we shall not prove it, matrix multiplication is **associative**. If $\mathbf{A}$ is an $m \times p$ matrix, $\mathbf{B}$ a $p \times r$ matrix, and $\mathbf{C}$ an $r \times n$ matrix, then

$$\mathbf{A}(\mathbf{BC}) = (\mathbf{AB})\mathbf{C}$$

is an $m \times n$ matrix.

Distributive Law

If all products are defined, multiplication is **distributive** over addition,

$$\mathbf{A}(\mathbf{B} + \mathbf{C}) = \mathbf{AB} + \mathbf{AC} \quad \text{and} \quad (\mathbf{B} + \mathbf{C})\mathbf{A} = \mathbf{BA} + \mathbf{CA}.$$

Determinant of a Matrix

Associated with every *square* matrix $\mathbf{A}$ of constants, there is a number called the **determinant of the matrix**, which is denoted by det $\mathbf{A}$.

EXAMPLE 6 Determinant of a Square Matrix

For $\mathbf{A} = \begin{pmatrix} 3 & 6 & 2 \\ 2 & 5 & 1 \\ -1 & 2 & 4 \end{pmatrix}$ we expand det $\mathbf{A}$ by cofactors of the first row:

$$\det \mathbf{A} = \begin{vmatrix} 3 & 6 & 2 \\ 2 & 5 & 1 \\ -1 & 2 & 4 \end{vmatrix} = 3\begin{vmatrix} 5 & 1 \\ 2 & 4 \end{vmatrix} - 6\begin{vmatrix} 2 & 1 \\ -1 & 4 \end{vmatrix} + 2\begin{vmatrix} 2 & 5 \\ -1 & 2 \end{vmatrix}$$

$$= 3(20 - 2) - 6(8 + 1) + 2(4 + 5) = 18.$$

It can be proved that a determinant det **A** can be expanded by cofactors using any row or column. If det **A** has a row (or a column) containing many zero entries, then wisdom dictates that we expand the determinant by that row (or column).

DEFINITION A.7 Transpose of a Matrix

The **transpose** of the $m \times n$ matrix (1) is the $n \times m$ matrix $\mathbf{A}^T$ given by

$$\mathbf{A}^T = \begin{pmatrix} a_{11} & a_{21} & \cdots & a_{m1} \\ a_{12} & a_{22} & \cdots & a_{m2} \\ \vdots & \vdots & & \vdots \\ a_{1n} & a_{2n} & \cdots & a_{mn} \end{pmatrix}.$$

In other words, the rows of a matrix **A** become the columns of its transpose $\mathbf{A}^T$.

EXAMPLE 7 Transpose of a Matrix

(a) The transpose of $\mathbf{A} = \begin{pmatrix} 3 & 6 & 2 \\ 2 & 5 & 1 \\ -1 & 2 & 4 \end{pmatrix}$ is $\mathbf{A}^T = \begin{pmatrix} 3 & 2 & -1 \\ 6 & 5 & 2 \\ 2 & 1 & 4 \end{pmatrix}$.

(b) If $\mathbf{X} = \begin{pmatrix} 5 \\ 0 \\ 3 \end{pmatrix}$, then $\mathbf{X}^T = (5 \quad 0 \quad 3)$.

DEFINITION A.8 Multiplicative Inverse of a Matrix

Let **A** be an $n \times n$ matrix. If there exists an $n \times n$ matrix **B** such that

$$\mathbf{AB} = \mathbf{BA} = \mathbf{I},$$

where **I** is the multiplicative identity, then **B** is said to be the **multiplicative inverse of A** and is denoted by $\mathbf{B} = \mathbf{A}^{-1}$.

DEFINITION A.9 Nonsingular/Singular Matrices

Let **A** be an $n \times n$ matrix. If det $\mathbf{A} \neq 0$, then **A** is said to be **nonsingular**. If det $\mathbf{A} = 0$, then **A** is said to be **singular**.

The following gives a necessary and sufficient condition for a square matrix to have a multiplicative inverse.

THEOREM A.1 Nonsingularity Implies A has an Inverse

An $n \times n$ matrix $\mathbf{A}$ has a multiplicative inverse $\mathbf{A}^{-1}$ if and only if $\mathbf{A}$ is nonsingular.

The following theorem gives one way of finding the multiplicative inverse for a nonsingular matrix.

THEOREM A.2 A Formula for the Inverse of a Matrix

Let $\mathbf{A}$ be an $n \times n$ nonsingular matrix and let $C_{ij} = (-1)^{i+j}M_{ij}$, where M_{ij} is the determinant of the $(n-1) \times (n-1)$ matrix obtained by deleting the ith row and jth column from $\mathbf{A}$. Then

$$\mathbf{A}^{-1} = \frac{1}{\det \mathbf{A}}(C_{ij})^T. \tag{2}$$

Each C_{ij} in Theorem A.2 is simply the **cofactor** (signed minor) of the corresponding entry a_{ij} in $\mathbf{A}$. Note that the transpose is utilized in formula (2).

For future reference we observe in the case of a 2×2 nonsingular matrix

$$\mathbf{A} = \begin{pmatrix} a_{11} & a_{12} \\ a_{21} & a_{22} \end{pmatrix}$$

that $C_{11} = a_{22}$, $C_{12} = -a_{21}$, $C_{21} = -a_{12}$, and $C_{22} = a_{11}$. Thus

$$\mathbf{A}^{-1} = \frac{1}{\det \mathbf{A}}\begin{pmatrix} a_{22} & -a_{21} \\ -a_{12} & a_{11} \end{pmatrix}^T = \frac{1}{\det \mathbf{A}}\begin{pmatrix} a_{22} & -a_{12} \\ -a_{21} & a_{11} \end{pmatrix}. \tag{3}$$

For a 3×3 nonsingular matrix

$$\mathbf{A} = \begin{pmatrix} a_{11} & a_{12} & a_{13} \\ a_{21} & a_{22} & a_{23} \\ a_{31} & a_{32} & a_{33} \end{pmatrix},$$

$$C_{11} = \begin{vmatrix} a_{22} & a_{23} \\ a_{32} & a_{33} \end{vmatrix}, \qquad C_{12} = -\begin{vmatrix} a_{21} & a_{23} \\ a_{31} & a_{33} \end{vmatrix}, \qquad C_{13} = \begin{vmatrix} a_{21} & a_{22} \\ a_{31} & a_{32} \end{vmatrix},$$

and so on. Carrying out the transposition gives

$$\mathbf{A}^{-1} = \frac{1}{\det \mathbf{A}}\begin{pmatrix} C_{11} & C_{21} & C_{31} \\ C_{12} & C_{22} & C_{32} \\ C_{13} & C_{23} & C_{33} \end{pmatrix}. \tag{4}$$

EXAMPLE 8 Inverse of a 2 × 2 Matrix

Find the multiplicative inverse for $\mathbf{A} = \begin{pmatrix} 1 & 4 \\ 2 & 10 \end{pmatrix}$.

Solution Since $\det \mathbf{A} = 10 - 8 = 2 \neq 0$, $\mathbf{A}$ is nonsingular. It follows from Theorem A.1 that $\mathbf{A}^{-1}$ exists. From (3) we find

$$\mathbf{A}^{-1} = \frac{1}{2}\begin{pmatrix} 10 & -4 \\ -2 & 1 \end{pmatrix} = \begin{pmatrix} 5 & -2 \\ -1 & \frac{1}{2} \end{pmatrix}.$$

Not every square matrix has a multiplicative inverse. The matrix $\mathbf{A} = \begin{pmatrix} 2 & 2 \\ 3 & 3 \end{pmatrix}$ is singular since $\det \mathbf{A} = 0$. Hence $\mathbf{A}^{-1}$ does not exist.

EXAMPLE 9 Inverse of a 3 × 3 Matrix

Find the multiplicative inverse for $\mathbf{A} = \begin{pmatrix} 2 & 2 & 0 \\ -2 & 1 & 1 \\ 3 & 0 & 1 \end{pmatrix}$.

Solution Since $\det \mathbf{A} = 12 \neq 0$, the given matrix is nonsingular. The cofactors corresponding to the entries in each row of $\det \mathbf{A}$ are

$$C_{11} = \begin{vmatrix} 1 & 1 \\ 0 & 1 \end{vmatrix} = 1 \qquad C_{12} = -\begin{vmatrix} -2 & 1 \\ 3 & 1 \end{vmatrix} = 5 \qquad C_{13} = \begin{vmatrix} -2 & 1 \\ 3 & 0 \end{vmatrix} = -3$$

$$C_{21} = -\begin{vmatrix} 2 & 0 \\ 0 & 1 \end{vmatrix} = -2 \qquad C_{22} = \begin{vmatrix} 2 & 0 \\ 3 & 1 \end{vmatrix} = 2 \qquad C_{23} = -\begin{vmatrix} 2 & 2 \\ 3 & 0 \end{vmatrix} = 6$$

$$C_{31} = \begin{vmatrix} 2 & 0 \\ 1 & 1 \end{vmatrix} = 2 \qquad C_{32} = -\begin{vmatrix} 2 & 0 \\ -2 & 1 \end{vmatrix} = -2 \qquad C_{33} = \begin{vmatrix} 2 & 2 \\ -2 & 1 \end{vmatrix} = 6$$

It follows from (4) that

$$\mathbf{A}^{-1} = \frac{1}{12}\begin{pmatrix} 1 & -2 & 2 \\ 5 & 2 & -2 \\ -3 & 6 & 6 \end{pmatrix} = \begin{pmatrix} \frac{1}{12} & -\frac{1}{6} & \frac{1}{6} \\ \frac{5}{12} & \frac{1}{6} & -\frac{1}{6} \\ -\frac{1}{4} & \frac{1}{2} & \frac{1}{2} \end{pmatrix}.$$

You are urged to verify that $\mathbf{A}^{-1}\mathbf{A} = \mathbf{A}\mathbf{A}^{-1} = \mathbf{I}$.

Formula (2) presents obvious difficulties for nonsingular matrices larger than 3 × 3. For example, to apply (2) to a 4 × 4 matrix we would have to calculate *sixteen* 3 × 3 determinants.* In the case of a large matrix, there are

---

* Strictly speaking, a determinant is a number, but it is sometimes convenient to refer to a determinant as if it were an array.

more efficient ways of finding $\mathbf{A}^{-1}$. The curious reader is referred to any text in linear algebra.

Since our goal is to apply the concept of a matrix to systems of linear differential equations in normal form, we need the following definitions.

DEFINITION A.10 Derivative of a Matrix of Functions

If $\mathbf{A}(t) = (a_{ij}(t))_{m \times n}$ is a matrix whose entries are functions differentiable on a common interval, then

$$\frac{d\mathbf{A}}{dt} = \left(\frac{d}{dt} a_{ij}\right)_{m \times n}.$$

DEFINITION A.11 Integral of a Matrix of Functions

If $\mathbf{A}(t) = (a_{ij}(t))_{m \times n}$ is a matrix whose entries are functions continuous on a common interval containing t and t_0, then

$$\int_{t_0}^{t} \mathbf{A}(s)\, ds = \left(\int_{t_0}^{t} a_{ij}(s)\, ds\right)_{m \times n}.$$

To differentiate (integrate) a matrix of functions we simply differentiate (integrate) each entry. The derivative of a matrix is also denoted by $\mathbf{A}'(t)$.

EXAMPLE 10 Derivative/Integral of a Matrix

If $$\mathbf{X}(t) = \begin{pmatrix} \sin 2t \\ e^{3t} \\ 8t - 1 \end{pmatrix}, \quad \text{then} \quad \mathbf{X}'(t) = \begin{pmatrix} \dfrac{d}{dt}\sin 2t \\ \dfrac{d}{dt} e^{3t} \\ \dfrac{d}{dt}(8t - 1) \end{pmatrix} = \begin{pmatrix} 2\cos 2t \\ 3e^{3t} \\ 8 \end{pmatrix}$$

and $$\int_0^t \mathbf{X}(s)\, ds = \begin{pmatrix} \int_0^t \sin 2s\, ds \\ \int_0^t e^{3s}\, ds \\ \int_0^t (8s - 1)\, ds \end{pmatrix} = \begin{pmatrix} -\frac{1}{2}\cos 2t + \frac{1}{2} \\ \frac{1}{3}e^{3t} - \frac{1}{3} \\ 4t^2 - t \end{pmatrix}.$$

A.2 Gaussian and Gauss-Jordan Elimination

Matrices are an invaluable aid in solving algebraic systems of n linear equations in n unknowns

$$\begin{aligned} a_{11}x_1 + a_{12}x_2 + \cdots + a_{1n}x_n &= b_1 \\ a_{21}x_1 + a_{22}x_2 + \cdots + a_{2n}x_n &= b_2 \\ \vdots \qquad\qquad\qquad &\quad \vdots \\ a_{n1}x_1 + a_{n2}x_2 + \cdots + a_{nn}x_n &= b_n. \end{aligned} \tag{5}$$

If **A** denotes the matrix of coefficients in (5), we know that Cramer's rule could be used to solve the system whenever det $\mathbf{A} \neq 0$. However, that rule requires a herculean effort if **A** is larger than 3×3. The procedure that we shall now consider has the distinct advantage of being not only an efficient way of handling large systems but also a means of solving consistent systems (5) in which det $\mathbf{A} = 0$ and a means of solving m linear equations in n unknowns.

DEFINITION A.12 Augmented Matrix

The **augmented matrix** of the system (5) is the $n \times (n + 1)$ matrix

$$\left(\begin{array}{cccc|c} a_{11} & a_{12} & \cdots & a_{1n} & b_1 \\ a_{21} & a_{22} & \cdots & a_{2n} & b_2 \\ \vdots & \vdots & & & \vdots \\ a_{n1} & a_{n2} & \cdots & a_{nn} & b_n \end{array}\right).$$

If **B** is the column matrix of the b_i, $i = 1, 2, \ldots, n$, the augmented matrix of (5) is denoted by $(\mathbf{A} \mid \mathbf{B})$.

Elementary Row Operations

Recall from algebra that we can transform an algebraic system of equations into an equivalent system (that is, one having the same solution) by multiplying an equation by a nonzero constant, interchanging the positions of any two equations in a system, and adding a nonzero constant multiple of an equation to another equation. These operations on equations in a system are, in turn, equivalent to **elementary row operations** on an augmented matrix:

(i) Multiply a row by a nonzero constant.

(ii) Interchange any two rows.

(iii) Add a nonzero constant multiple of one row to any other row.

Elimination Methods

To solve a system such as (5) using an augmented matrix we use either **Gaussian elimination** or the **Gauss-Jordan elimination method**. In the former method we carry out a succession of elementary row operations until we arrive at an augmented matrix in **row-echelon form**:

(i) The first nonzero entry in a nonzero row is 1.

(ii) In consecutive nonzero rows, the first entry 1 in the lower row appears to the right of the first 1 in the higher row.

(iii) Rows consisting of all 0's are at the bottom of the matrix.

In the Gauss-Jordan method the row operations are continued until we obtain an augmented matrix that is in **reduced row-echelon form**. A reduced row-echelon matrix has the same three properties listed above in addition to

(iv) A column containing a first entry 1 has 0's everywhere else.

EXAMPLE 11 Row-Echelon/Reduced Row-Echelon Form

(a) The augmented matrices

$$\left(\begin{array}{ccc|c} 1 & 5 & 0 & 2 \\ 0 & 1 & 0 & -1 \\ 0 & 0 & 0 & 0 \end{array}\right) \quad \text{and} \quad \left(\begin{array}{ccccc|c} 0 & 0 & 1 & -6 & 2 & 2 \\ 0 & 0 & 0 & 0 & 1 & 4 \end{array}\right)$$

are in row-echelon form. You should verify that the three criteria are satisfied.

(b) The augmented matrices

$$\left(\begin{array}{ccc|c} 1 & 0 & 0 & 7 \\ 0 & 1 & 0 & -1 \\ 0 & 0 & 0 & 0 \end{array}\right) \quad \text{and} \quad \left(\begin{array}{ccccc|c} 0 & 0 & 1 & -6 & 0 & -6 \\ 0 & 0 & 0 & 0 & 1 & 4 \end{array}\right)$$

are in reduced row-echelon form. Note that the remaining entries in the columns containing a leading entry 1 are all 0's. ●

Note that in Gaussian elimination, we stop once we have obtained *an* augmented matrix in row-echelon form. In other words, by using different sequences of row operations we may arrive at different row-echelon forms. This method then requires the use of back-substitution. In Gauss-Jordan elimination we stop when we have obtained *the* augmented matrix in reduced row-echelon form. Any sequence of row operations will lead to the same augmented matrix in reduced row-echelon form. This method does not require back-substitution; the solution of the system will be apparent by inspection of the final matrix. In terms of the equations of the original system, our goal in both methods is simply to make the coefficient of x_1 in the first equation* equal to 1 and then use multiples of that equation to eliminate x_1 from other equations. The process is repeated on the other variables.

---

* We can always interchange equations so that the first equation contains the variable x_1.

To keep track of the row operations on an augmented matrix, we utilize the following notation:

Symbol	Meaning
R_{ij}	Interchange rows i and j
cR_i	Multiply the ith row by the nonzero constant c
$cR_i + R_j$	Multiply the ith row by c and add to the jth row

EXAMPLE 12 Solution by Elimination

Solve
$$\begin{aligned} 2x_1 + 6x_2 + x_3 &= 7 \\ x_1 + 2x_2 - x_3 &= -1 \\ 5x_1 + 7x_2 - 4x_3 &= 9 \end{aligned}$$

using **(a)** Gaussian elimination and **(b)** Gauss-Jordan elimination.

Solution

(a) Using row operations on the augmented matrix of the system, we obtain

$$\left(\begin{array}{ccc|c} 2 & 6 & 1 & 7 \\ 1 & 2 & -1 & -1 \\ 5 & 7 & -4 & 9 \end{array}\right) \xrightarrow{R_{12}} \left(\begin{array}{ccc|c} 1 & 2 & -1 & -1 \\ 2 & 6 & 1 & 7 \\ 5 & 7 & -4 & 9 \end{array}\right) \xrightarrow[-5R_1 + R_3]{-2R_1 + R_2} \left(\begin{array}{ccc|c} 1 & 2 & -1 & -1 \\ 0 & 2 & 3 & 9 \\ 0 & -3 & 1 & 14 \end{array}\right)$$

$$\xrightarrow{\frac{1}{2}R_2} \left(\begin{array}{ccc|c} 1 & 2 & -1 & -1 \\ 0 & 1 & \frac{3}{2} & \frac{9}{2} \\ 0 & -3 & 1 & 14 \end{array}\right) \xrightarrow{3R_2 + R_3} \left(\begin{array}{ccc|c} 1 & 2 & -1 & -1 \\ 0 & 1 & \frac{3}{2} & \frac{9}{2} \\ 0 & 0 & \frac{11}{2} & \frac{55}{2} \end{array}\right) \xrightarrow{\frac{2}{11}R_3} \left(\begin{array}{ccc|c} 1 & 2 & -1 & -1 \\ 0 & 1 & \frac{3}{2} & \frac{9}{2} \\ 0 & 0 & 1 & 5 \end{array}\right).$$

The last matrix is in row-echelon form and represents the system

$$\begin{aligned} x_1 + 2x_2 - x_3 &= -1 \\ x_2 + \frac{3}{2}x_3 &= \frac{9}{2} \\ x_3 &= 5. \end{aligned}$$

Substituting $x_3 = 5$ into the second equation then gives $x_2 = -3$. Substituting both these values back into the first equation finally yields $x_1 = 10$.

(b) We start with the last matrix above. Since the first entries in the second and third rows are 1's, we must, in turn, make the remaining entries in the second and third columns 0's:

$$\left(\begin{array}{ccc|c} 1 & 2 & -1 & -1 \\ 0 & 1 & \frac{3}{2} & \frac{9}{2} \\ 0 & 0 & 1 & 5 \end{array}\right) \xrightarrow{-2R_2 + R_1} \left(\begin{array}{ccc|c} 1 & 0 & -4 & -10 \\ 0 & 1 & \frac{3}{2} & \frac{9}{2} \\ 0 & 0 & 1 & 5 \end{array}\right) \xrightarrow[-\frac{3}{2}R_3 + R_2]{4R_3 + R_1} \left(\begin{array}{ccc|c} 1 & 0 & 0 & 10 \\ 0 & 1 & 0 & -3 \\ 0 & 0 & 1 & 5 \end{array}\right).$$

The last matrix is now in reduced row-echelon form. Because of what the matrix means in terms of equations, it is evident that the solution of the system is $x_1 = 10$, $x_2 = -3$, $x_3 = 5$. ●

EXAMPLE 13 Gauss-Jordan Elimination

Solve

$$\begin{aligned} x + 3y - 2z &= -7 \\ 4x + y + 3z &= 5 \\ 2x - 5y + 7z &= 19. \end{aligned}$$

Solution We solve the system using Gauss-Jordan elimination:

$$\left(\begin{array}{ccc|c} 1 & 3 & -2 & -7 \\ 4 & 1 & 3 & 5 \\ 2 & -5 & 7 & 19 \end{array}\right) \xrightarrow[-2R_1 + R_3]{-4R_1 + R_2} \left(\begin{array}{ccc|c} 1 & 3 & -2 & -7 \\ 0 & -11 & 11 & 33 \\ 0 & -11 & 11 & 33 \end{array}\right)$$

$$\xrightarrow[-\frac{1}{11}R_3]{-\frac{1}{11}R_2} \left(\begin{array}{ccc|c} 1 & 3 & -2 & -7 \\ 0 & 1 & -1 & -3 \\ 0 & 1 & -1 & -3 \end{array}\right) \xrightarrow[-R_2 + R_3]{-3R_2 + R_1} \left(\begin{array}{ccc|c} 1 & 0 & 1 & 2 \\ 0 & 1 & -1 & -3 \\ 0 & 0 & 0 & 0 \end{array}\right).$$

In this case the last matrix in reduced row-echelon form implies that the original system of three equations in three unknowns is really equivalent to two equations in three unknowns. Since only z is common to both equations (the nonzero rows), we can assign its values arbitrarily. If we let $z = t$, where t represents any real number, then we see that the system has infinitely many solutions: $x = 2 - t$, $y = -3 + t$, $z = t$. Geometrically, these equations are the parametric equations for the line of intersection of the planes $x + 0y + z = 2$ and $0x + y - z = -3$. ●

A.3 The Eigenvalue Problem

Gauss-Jordan elimination can be used to find the **eigenvectors** of a square matrix.

DEFINITION A.13 Eigenvalues and Eigenvectors

Let **A** be an $n \times n$ matrix. A number λ is said to be an **eigenvalue** of **A** if there exists a *nonzero* solution vector **K** of the linear system

$$\mathbf{AK} = \lambda\mathbf{K}. \tag{6}$$

The solution vector **K** is said to be an **eigenvector** corresponding to the eigenvalue λ.

The word *eigenvalue* is a combination of German and English terms adapted from the German word *eigenwert*, which, translated literally, is "proper value." Eigenvalues and eigenvectors are also called **characteristic values** and **characteristic vectors**, respectively.

EXAMPLE 14 Eigenvector of a Matrix

Verify that $\mathbf{K} = \begin{pmatrix} 1 \\ -1 \\ 1 \end{pmatrix}$ is an eigenvector of the matrix $\mathbf{A} = \begin{pmatrix} 0 & -1 & -3 \\ 2 & 3 & 3 \\ -2 & 1 & 1 \end{pmatrix}$.

Solution By carrying out the multiplication **AK**, we see that

$$\mathbf{AK} = \begin{pmatrix} 0 & -1 & -3 \\ 2 & 3 & 3 \\ -2 & 1 & 1 \end{pmatrix}\begin{pmatrix} 1 \\ -1 \\ 1 \end{pmatrix} = \begin{pmatrix} -2 \\ 2 \\ -2 \end{pmatrix} = (-2)\begin{pmatrix} 1 \\ -1 \\ 1 \end{pmatrix} = \overset{\text{eigenvalue}}{\overset{\downarrow}{(-2)}}\mathbf{K}.$$

We see from the preceding line and Definition A.13 that $\lambda = -2$ is an eigenvalue of **A**. ●

Using properties of matrix algebra, we can write (6) in the alternative form

$$(\mathbf{A} - \lambda\mathbf{I})\mathbf{K} = \mathbf{0}, \tag{7}$$

where **I** is the multiplicative identity. If we let

$$\mathbf{K} = \begin{pmatrix} k_1 \\ k_2 \\ \vdots \\ k_n \end{pmatrix},$$

then (7) is the same as

$$\begin{aligned} (a_{11} - \lambda)k_1 + \quad a_{12}k_2 + \cdots + \quad a_{1n}k_n &= 0 \\ a_{21}k_1 + (a_{22} - \lambda)k_2 + \cdots + \quad a_{2n}k_n &= 0 \\ \vdots \qquad\qquad\qquad\qquad & \vdots \\ a_{n1}k_1 + \quad a_{n2}k_2 + \cdots + (a_{nn} - \lambda)k_n &= 0. \end{aligned} \tag{8}$$

Although an obvious solution of (8) is $k_1 = 0$, $k_2 = 0, \ldots, k_n = 0$, we are seeking only nontrivial solutions. Now it is known that a homogeneous system of n linear equations in n unknowns (that is, $b_i = 0$, $i = 1, 2, \ldots, n$, in (5)) has a nontrivial solution if and only if the determinant of the coefficient matrix is equal to zero. Thus to find a nonzero solution **K** for (7), we must have

$$\det(\mathbf{A} - \lambda\mathbf{I}) = 0. \tag{9}$$

Inspection of (8) shows that the expansion of $\det(\mathbf{A} - \lambda\mathbf{I})$ by cofactors results in an nth-degree polynomial in λ. The equation (9) is called the **characteristic equation** of **A**. Thus *the eigenvalues of* **A** *are the roots of the characteristic* *equation.* To find an eigenvector corresponding to an eigenvalue λ we simply solve the system of equations $(\mathbf{A} - \lambda\mathbf{I})\mathbf{K} = \mathbf{0}$ by applying Gauss-Jordan elimination to the augmented matrix $(\mathbf{A} - \lambda\mathbf{I}|\mathbf{0})$.

EXAMPLE 15 Eigenvalues/Eigenvectors

Find the eigenvalues and eigenvectors of $\mathbf{A} = \begin{pmatrix} 1 & 2 & 1 \\ 6 & -1 & 0 \\ -1 & -2 & -1 \end{pmatrix}$.

Solution To expand the determinant in the characteristic equation we use the cofactors of the second row:

$$\det(\mathbf{A} - \lambda\mathbf{I}) = \begin{vmatrix} 1-\lambda & 2 & 1 \\ 6 & -1-\lambda & 0 \\ -1 & -2 & -1-\lambda \end{vmatrix} = -\lambda^3 - \lambda^2 + 12\lambda = 0.$$

From $-\lambda^3 - \lambda^2 + 12\lambda = -\lambda(\lambda + 4)(\lambda - 3) = 0$ we see that the eigenvalues are $\lambda_1 = 0$, $\lambda_2 = -4$, and $\lambda_3 = 3$. To find the eignevectors we must now reduce $(\mathbf{A} - \lambda\mathbf{I}|\mathbf{0})$ three times corresponding to the three distinct eigenvalues.

For $\lambda_1 = 0$ we have

$$(\mathbf{A} - 0\mathbf{I}|\mathbf{0}) = \left(\begin{array}{ccc|c} 1 & 2 & 1 & 0 \\ 6 & -1 & 0 & 0 \\ -1 & -2 & -1 & 0 \end{array}\right) \xrightarrow[R_1 + R_3]{-6R_1 + R_2} \left(\begin{array}{ccc|c} 1 & 2 & 1 & 0 \\ 0 & -13 & -6 & 0 \\ 0 & 0 & 0 & 0 \end{array}\right)$$

$$\xrightarrow{-\frac{1}{13}R_2} \left(\begin{array}{ccc|c} 1 & 2 & 1 & 0 \\ 0 & 1 & \frac{6}{13} & 0 \\ 0 & 0 & 0 & 0 \end{array}\right) \xrightarrow{-2R_2 + R_1} \left(\begin{array}{ccc|c} 1 & 0 & \frac{1}{13} & 0 \\ 0 & 1 & \frac{6}{13} & 0 \\ 0 & 0 & 0 & 0 \end{array}\right).$$

Thus we see that $k_1 = (-1/13)k_3$ and $k_2 = (-6/13)k_3$. Choosing $k_3 = -13$, we get the eigenvector*

$$\mathbf{K}_1 = \begin{pmatrix} 1 \\ 6 \\ -13 \end{pmatrix}.$$

For $\lambda_2 = -4$,

$$(\mathbf{A} +, 4\mathbf{I}|\mathbf{0}) = \left(\begin{array}{ccc|c} 5 & 2 & 1 & 0 \\ 6 & 3 & 0 & 0 \\ -1 & -2 & 3 & 0 \end{array}\right) \xrightarrow[R_{31}]{-R_3} \left(\begin{array}{ccc|c} 1 & 2 & -3 & 0 \\ 6 & 3 & 0 & 0 \\ 5 & 2 & 1 & 0 \end{array}\right)$$

$$\xrightarrow[-5R_1 + R_3]{-6R_1 + R_2} \left(\begin{array}{ccc|c} 1 & 2 & -3 & 0 \\ 0 & -9 & 18 & 0 \\ 0 & -8 & 16 & 0 \end{array}\right) \xrightarrow[-\frac{1}{8}R_3]{-\frac{1}{9}R_2} \left(\begin{array}{ccc|c} 1 & 2 & -3 & 0 \\ 0 & 1 & -2 & 0 \\ 0 & 1 & -2 & 0 \end{array}\right) \xrightarrow[-R_2 + R_3]{-2R_2 + R_1} \left(\begin{array}{ccc|c} 1 & 0 & 1 & 0 \\ 0 & 1 & -2 & 0 \\ 0 & 0 & 0 & 0 \end{array}\right)$$

* Of course k_3 could be chosen as any nonzero number. In other words, a nonzero constant multiple of an eigenvector is also an eigenvector.

implies $k_1 = -k_3$ and $k_2 = 2k_3$. Choosing $k_3 = 1$ then yields the second eigenvector

$$\mathbf{K}_2 = \begin{pmatrix} -1 \\ 2 \\ 1 \end{pmatrix}.$$

Finally, for $\lambda_3 = 3$ Gauss-Jordan elimination gives

$$(\mathbf{A} - 3\mathbf{I}|\mathbf{0}) = \left(\begin{array}{ccc|c} -2 & 2 & 1 & 0 \\ 6 & -4 & 0 & 0 \\ -1 & -2 & -4 & 0 \end{array}\right) \xrightarrow{\text{row operations}} \left(\begin{array}{ccc|c} 1 & 0 & 1 & 0 \\ 0 & 1 & \frac{3}{2} & 0 \\ 0 & 0 & 0 & 0 \end{array}\right),$$

and so $k_1 = -k_3$ and $k_2 = (-3/2)k_3$. The choice of $k_3 = -2$ leads to the third eigenvector:

$$\mathbf{K}_3 = \begin{pmatrix} 2 \\ 3 \\ -2 \end{pmatrix}.$$

When an $n \times n$ matrix $\mathbf{A}$ possesses n distinct eigenvalues $\lambda_1, \lambda_2, \ldots, \lambda_n$, it can be proved that a set of n linearly independent* eigenvectors $\mathbf{K}_1, \mathbf{K}_2, \ldots, \mathbf{K}_n$ can be found. However, when the characteristic equation has repeated roots, it may not be possible to find n linearly independent eigenvectors for $\mathbf{A}$.

EXAMPLE 16 Eigenvalues/Eigenvectors

Find the eigenvalues and eigenvectors of $\mathbf{A} = \begin{pmatrix} 3 & 4 \\ -1 & 7 \end{pmatrix}$.

Solution From the characteristic equation

$$\det(\mathbf{A} - \lambda\mathbf{I}) = \begin{vmatrix} 3-\lambda & 4 \\ -1 & 7-\lambda \end{vmatrix} = (\lambda - 5)^2 = 0,$$

we see that $\lambda_1 = \lambda_2 = 5$ is an eigenvalue of multiplicity two. In the case of a 2×2 matrix there is no need to use Gauss-Jordan elimination. To find the eigenvector(s) corresponding to $\lambda_1 = 5$ we resort to the system $(\mathbf{A} - 5\mathbf{I}|\mathbf{0})$ in its equivalent form

$$-2k_1 + 4k_2 = 0$$
$$-k_1 + 2k_2 = 0.$$

It is apparent from this system that $k_1 = 2k_2$. Thus if we choose $k_2 = 1$, we find the single eigenvector

$$\mathbf{K}_1 = \begin{pmatrix} 2 \\ 1 \end{pmatrix}.$$

* Linear independence of column vectors is defined in exactly the same manner as for functions.

EXAMPLE 17 Eigenvalues/Eigenvectors

Find the eigenvalues and eigenvectors of $\mathbf{A} = \begin{pmatrix} 9 & 1 & 1 \\ 1 & 9 & 1 \\ 1 & 1 & 9 \end{pmatrix}$.

Solution The characteristic equation

$$\det(\mathbf{A} - \lambda\mathbf{I}) = \begin{vmatrix} 9-\lambda & 1 & 1 \\ 1 & 9-\lambda & 1 \\ 1 & 1 & 9-\lambda \end{vmatrix} = -(\lambda - 11)(\lambda - 8)^2 = 0$$

shows that $\lambda_1 = 11$ and that $\lambda_2 = \lambda_3 = 8$ is an eigenvalue of multiplicity two.

For $\lambda_1 = 11$ Gauss-Jordan elimination gives

$$(\mathbf{A} - 11\mathbf{I}|\mathbf{0}) = \left(\begin{array}{ccc|c} -2 & 1 & 1 & 0 \\ 1 & -2 & 1 & 0 \\ 1 & 1 & -2 & 0 \end{array}\right) \xrightarrow{\text{row operations}} \left(\begin{array}{ccc|c} 1 & 0 & -1 & 0 \\ 0 & 1 & -1 & 0 \\ 0 & 0 & 0 & 0 \end{array}\right).$$

Hence $k_1 = k_3$ and $k_2 = k_3$. If $k_3 = 1$, then

$$\mathbf{K}_1 = \begin{pmatrix} 1 \\ 1 \\ 1 \end{pmatrix}.$$

Now for $\lambda_2 = 8$ we have

$$(\mathbf{A} - 8\mathbf{I}|\mathbf{0}) = \left(\begin{array}{ccc|c} 1 & 1 & 1 & 0 \\ 1 & 1 & 1 & 0 \\ 1 & 1 & 1 & 0 \end{array}\right) \xrightarrow{\text{row operations}} \left(\begin{array}{ccc|c} 1 & 1 & 1 & 0 \\ 0 & 0 & 0 & 0 \\ 0 & 0 & 0 & 0 \end{array}\right).$$

In the equation $k_1 + k_2 + k_3 = 0$, we are free to select two of the variables arbitrarily. Choosing, on the one hand, $k_2 = 1$, $k_3 = 0$ and, on the other, $k_2 = 0$, $k_3 = 1$, we obtain two linearly independent eigenvectors

$$\mathbf{K}_2 = \begin{pmatrix} -1 \\ 1 \\ 0 \end{pmatrix} \quad \text{and} \quad \mathbf{K}_3 = \begin{pmatrix} -1 \\ 0 \\ 1 \end{pmatrix}.$$

APPENDIX I EXERCISES

Answers to odd-numbered problems begin on page AN-28.

A.1

1. If $\mathbf{A} = \begin{pmatrix} 4 & 5 \\ -6 & 9 \end{pmatrix}$ and $\mathbf{B} = \begin{pmatrix} -2 & 6 \\ 8 & -10 \end{pmatrix}$, find
(a) $\mathbf{A} + \mathbf{B}$, (b) $\mathbf{B} - \mathbf{A}$, (c) $2\mathbf{A} + 3\mathbf{B}$.

2. If $\mathbf{A} = \begin{pmatrix} -2 & 0 \\ 4 & 1 \\ 7 & 3 \end{pmatrix}$ and $\mathbf{B} = \begin{pmatrix} 3 & -1 \\ 0 & 2 \\ -4 & -2 \end{pmatrix}$, find
(a) $\mathbf{A} - \mathbf{B}$, (b) $\mathbf{B} - \mathbf{A}$, (c) $2(\mathbf{A} + \mathbf{B})$.

3. If $\mathbf{A} = \begin{pmatrix} 2 & -3 \\ -5 & 4 \end{pmatrix}$ and $\mathbf{B} = \begin{pmatrix} -1 & 6 \\ 3 & 2 \end{pmatrix}$, find **(a)** $\mathbf{AB}$, **(b)** $\mathbf{BA}$, **(c)** $\mathbf{A}^2 = \mathbf{AA}$, **(d)** $\mathbf{B}^2 = \mathbf{BB}$.

4. If $\mathbf{A} = \begin{pmatrix} 1 & 4 \\ 5 & 10 \\ 8 & 12 \end{pmatrix}$ and $\mathbf{B} = \begin{pmatrix} -4 & 6 & -3 \\ 1 & -3 & 2 \end{pmatrix}$, find **(a)** $\mathbf{AB}$, **(b)** $\mathbf{BA}$.

5. If $\mathbf{A} = \begin{pmatrix} 1 & -2 \\ -2 & 4 \end{pmatrix}$, $\mathbf{B} = \begin{pmatrix} 6 & 3 \\ 2 & 1 \end{pmatrix}$, and $\mathbf{C} = \begin{pmatrix} 0 & 2 \\ 3 & 4 \end{pmatrix}$, find **(a)** $\mathbf{BC}$, **(b)** $\mathbf{A(BC)}$, **(c)** $\mathbf{C(BA)}$, **(d)** $\mathbf{A(B + C)}$.

6. If $\mathbf{A} = (5 \quad -6 \quad 7)$, $\mathbf{B} = \begin{pmatrix} 3 \\ 4 \\ -1 \end{pmatrix}$, and $\mathbf{C} = \begin{pmatrix} 1 & 2 & 4 \\ 0 & 1 & -1 \\ 3 & 2 & 1 \end{pmatrix}$, find **(a)** $\mathbf{AB}$, **(b)** $\mathbf{BA}$, **(c)** $\mathbf{(BA)C}$, **(d)** $\mathbf{(AB)C}$.

7. If $\mathbf{A} = \begin{pmatrix} 4 \\ 8 \\ -10 \end{pmatrix}$ and $\mathbf{B} = (2 \quad 4 \quad 5)$, find **(a)** $\mathbf{A}^T\mathbf{A}$, **(b)** $\mathbf{B}^T\mathbf{B}$, **(c)** $\mathbf{A} + \mathbf{B}^T$.

8. If $\mathbf{A} = \begin{pmatrix} 1 & 2 \\ 2 & 4 \end{pmatrix}$ and $\mathbf{B} = \begin{pmatrix} -2 & 3 \\ 5 & 7 \end{pmatrix}$, find **(a)** $\mathbf{A} + \mathbf{B}^T$, **(b)** $2\mathbf{A}^T - \mathbf{B}^T$, **(c)** $\mathbf{A}^T(\mathbf{A} - \mathbf{B})$.

9. If $\mathbf{A} = \begin{pmatrix} 3 & 4 \\ 8 & 1 \end{pmatrix}$ and $\mathbf{B} = \begin{pmatrix} 5 & 10 \\ -2 & -5 \end{pmatrix}$, find **(a)** $(\mathbf{AB})^T$, **(b)** $\mathbf{B}^T\mathbf{A}^T$.

10. If $\mathbf{A} = \begin{pmatrix} 5 & 9 \\ -4 & 6 \end{pmatrix}$ and $\mathbf{B} = \begin{pmatrix} -3 & 11 \\ -7 & 2 \end{pmatrix}$, find **(a)** $\mathbf{A}^T + \mathbf{B}^T$, **(b)** $(\mathbf{A} + \mathbf{B})^T$.

In Problems 11–14 write the given sum as a single column matrix.

11. $4\begin{pmatrix} -1 \\ 2 \end{pmatrix} - 2\begin{pmatrix} 2 \\ 8 \end{pmatrix} + 3\begin{pmatrix} -2 \\ 3 \end{pmatrix}$

12. $3t\begin{pmatrix} 2 \\ t \\ -1 \end{pmatrix} + (t-1)\begin{pmatrix} -1 \\ -t \\ 3 \end{pmatrix} - 2\begin{pmatrix} 3t \\ 4 \\ -5t \end{pmatrix}$

13. $\begin{pmatrix} 2 & -3 \\ 1 & 4 \end{pmatrix}\begin{pmatrix} -2 \\ 5 \end{pmatrix} - \begin{pmatrix} -1 & 6 \\ -2 & 3 \end{pmatrix}\begin{pmatrix} -7 \\ 2 \end{pmatrix}$

14. $\begin{pmatrix} 1 & -3 & 4 \\ 2 & 5 & -1 \\ 0 & -4 & -2 \end{pmatrix}\begin{pmatrix} t \\ 2t-1 \\ -t \end{pmatrix} + \begin{pmatrix} -t \\ 1 \\ 4 \end{pmatrix} - \begin{pmatrix} 2 \\ 8 \\ -6 \end{pmatrix}$

In Problems 15–22 determine whether the given matrix is singular or nonsingular. If nonsingular, find $\mathbf{A}^{-1}$.

15. $\mathbf{A} = \begin{pmatrix} -3 & 6 \\ -2 & 4 \end{pmatrix}$

16. $\mathbf{A} = \begin{pmatrix} 2 & 5 \\ 1 & 4 \end{pmatrix}$

17. $\mathbf{A} = \begin{pmatrix} 4 & 8 \\ -3 & -5 \end{pmatrix}$

18. $\mathbf{A} = \begin{pmatrix} 7 & 10 \\ 2 & 2 \end{pmatrix}$

19. $\mathbf{A} = \begin{pmatrix} 2 & 1 & 0 \\ -1 & 2 & 1 \\ 1 & 2 & 1 \end{pmatrix}$

20. $\mathbf{A} = \begin{pmatrix} 3 & 2 & 1 \\ 4 & 1 & 0 \\ -2 & 5 & -1 \end{pmatrix}$

21. $\mathbf{A} = \begin{pmatrix} 2 & 1 & 1 \\ 1 & -2 & -3 \\ 3 & 2 & 4 \end{pmatrix}$

22. $\mathbf{A} = \begin{pmatrix} 4 & 1 & -1 \\ 6 & 2 & -3 \\ -2 & -1 & 2 \end{pmatrix}$

In Problems 23 and 24 show that the given matrix is nonsingular for every real value of t. Find $\mathbf{A}^{-1}(t)$.

23. $\mathbf{A}(t) = \begin{pmatrix} 2e^{-t} & e^{4t} \\ 4e^{-t} & 3e^{4t} \end{pmatrix}$

24. $\mathbf{A}(t) = \begin{pmatrix} 2e^t \sin t & -2e^t \cos t \\ e^t \cos t & e^t \sin t \end{pmatrix}$

In Problems 25–28 find $d\mathbf{X}/dt$.

25. $\mathbf{X} = \begin{pmatrix} 5e^{-t} \\ 2e^{-t} \\ -7e^{-t} \end{pmatrix}$

26. $\mathbf{X} = \begin{pmatrix} \frac{1}{2}\sin 2t - 4\cos 2t \\ -3\sin 2t + 5\cos 2t \end{pmatrix}$

27. $\mathbf{X} = 2\begin{pmatrix} 1 \\ -1 \end{pmatrix}e^{2t} + 4\begin{pmatrix} 2 \\ 1 \end{pmatrix}e^{-3t}$

28. $\mathbf{X} = \begin{pmatrix} 5te^{2t} \\ t\sin 3t \end{pmatrix}$

29. Let $\mathbf{A}(t) = \begin{pmatrix} e^{4t} & \cos \pi t \\ 2t & 3t^2 - 1 \end{pmatrix}$.

Find **(a)** $\dfrac{d\mathbf{A}}{dt}$, **(b)** $\int_0^2 \mathbf{A}(t)\,dt$, **(c)** $\int_0^t \mathbf{A}(s)\,ds$.

30. Let $\mathbf{A}(t) = \begin{pmatrix} \dfrac{1}{t^2+1} & 3t \\ t^2 & t \end{pmatrix}$ and $\mathbf{B}(t) = \begin{pmatrix} 6t & 2 \\ 1/t & 4t \end{pmatrix}$.

Find (a) $\dfrac{d\mathbf{A}}{dt}$, (b) $\dfrac{d\mathbf{B}}{dt}$, (c) $\int_0^1 \mathbf{A}(t)\,dt$, (d) $\int_1^2 \mathbf{B}(t)\,dt$,

(e) $\mathbf{A}(t)\mathbf{B}(t)$, (f) $\dfrac{d}{dt}\mathbf{A}(t)\mathbf{B}(t)$, (g) $\int_1^t \mathbf{A}(s)\mathbf{B}(s)\,ds$.

A.2

In Problems 31–38 solve the given system of equations by either Gaussian elimination or Gauss-Jordan elimination.

31. $\begin{aligned} x + y - 2z &= 14 \\ 2x - y + z &= 0 \\ 6x + 3y + 4z &= 1 \end{aligned}$

32. $\begin{aligned} 5x - 2y + 4z &= 10 \\ x + y + z &= 9 \\ 4x - 3y + 3z &= 1 \end{aligned}$

33. $\begin{aligned} y + z &= -5 \\ 5x + 4y - 16z &= -10 \\ x - y - 5z &= 7 \end{aligned}$

34. $\begin{aligned} 3x + y + z &= 4 \\ 4x + 2y - z &= 7 \\ x + y - 3z &= 6 \end{aligned}$

35. $\begin{aligned} 2x + y + z &= 4 \\ 10x - 2y + 2z &= -1 \\ 6x - 2y + 4z &= 8 \end{aligned}$

36. $\begin{aligned} x + 2z &= 8 \\ x + 2y - 2z &= 4 \\ 2x + 5y - 6z &= 6 \end{aligned}$

37. $\begin{aligned} x_1 + x_2 - x_3 - x_4 &= -1 \\ x_1 + x_2 + x_3 + x_4 &= 3 \\ x_1 - x_2 + x_3 - x_4 &= 3 \\ 4x_1 + x_2 - 2x_3 + x_4 &= 0 \end{aligned}$

38. $\begin{aligned} 2x_1 + x_2 + x_3 &= 0 \\ x_1 + 3x_2 + x_3 &= 0 \\ 7x_1 + x_2 + 3x_3 &= 0 \end{aligned}$

In Problems 39 and 40 use Gauss-Jordan elimination to demonstrate that the given system of equations has no solution.

39. $\begin{aligned} x + 2y + 4z &= 2 \\ 2x + 4y + 3z &= 1 \\ x + 2y - z &= 7 \end{aligned}$

40. $\begin{aligned} x_1 + x_2 - x_3 + 3x_4 &= 1 \\ x_2 - x_3 - 4x_4 &= 0 \\ x_1 + 2x_2 - 2x_3 - x_4 &= 6 \\ 4x_1 + 7x_2 - 7x_3 &= 9 \end{aligned}$

A.3

In Problems 41–48 find the eigenvalues and eigenvectors of the given matrix.

41. $\begin{pmatrix} -1 & 2 \\ -7 & 8 \end{pmatrix}$

42. $\begin{pmatrix} 2 & 1 \\ 2 & 1 \end{pmatrix}$

43. $\begin{pmatrix} -8 & -1 \\ 16 & 0 \end{pmatrix}$

44. $\begin{pmatrix} 1 & 1 \\ \frac{1}{4} & 1 \end{pmatrix}$

45. $\begin{pmatrix} 5 & -1 & 0 \\ 0 & -5 & 9 \\ 5 & -1 & 0 \end{pmatrix}$

46. $\begin{pmatrix} 3 & 0 & 0 \\ 0 & 2 & 0 \\ 4 & 0 & 1 \end{pmatrix}$

47. $\begin{pmatrix} 0 & 4 & 0 \\ -1 & -4 & 0 \\ 0 & 0 & -2 \end{pmatrix}$

48. $\begin{pmatrix} 1 & 6 & 0 \\ 0 & 2 & 1 \\ 0 & 1 & 2 \end{pmatrix}$

In Problems 49 and 50 show that the given matrix has complex eigenvalues. Find the eigenvectors of the matrix.

49. $\begin{pmatrix} -1 & 2 \\ -5 & 1 \end{pmatrix}$

50. $\begin{pmatrix} 2 & -1 & 0 \\ 5 & 2 & 4 \\ 0 & 1 & 2 \end{pmatrix}$

51. If $\mathbf{A}(t)$ is a 2×2 matrix of differentiable functions and $\mathbf{X}(t)$ is a 2×1 column matrix of differentiable functions, prove the product rule

$$\frac{d}{dt}[\mathbf{A}(t)\mathbf{X}(t)] = \mathbf{A}(t)\mathbf{X}'(t) + \mathbf{A}'(t)\mathbf{X}(t).$$

52. Derive formula (3). [*Hint*: Find a matrix $\mathbf{B} = \begin{pmatrix} b_{11} & b_{12} \\ b_{21} & b_{22} \end{pmatrix}$ for which $\mathbf{AB} = \mathbf{I}$. Solve for b_{11}, b_{12}, b_{21}, and b_{22}. Then show that $\mathbf{BA} = \mathbf{I}$.]

53. If $\mathbf{A}$ is nonsingular and $\mathbf{AB} = \mathbf{AC}$, show that $\mathbf{B} = \mathbf{C}$.

54. If $\mathbf{A}$ and $\mathbf{B}$ are nonsingular, show that $(\mathbf{AB})^{-1} = \mathbf{B}^{-1}\mathbf{A}^{-1}$.

55. Let $\mathbf{A}$ and $\mathbf{B}$ be $n \times n$ matrices. In general, is $(\mathbf{A} + \mathbf{B})^2 = \mathbf{A}^2 + 2\mathbf{AB} + \mathbf{B}^2$?

II LAPLACE TRANSFORMS

$f(t)$	$\mathscr{L}\{f(t)\} = F(s)$
1. 1	$\frac{1}{s}$
2. t	$\frac{1}{s^2}$
3. t^n	$\frac{n!}{s^{n+1}}$, n a positive integer
4. $t^{-1/2}$	$\sqrt{\frac{\pi}{s}}$
5. $t^{1/2}$	$\frac{\sqrt{\pi}}{2s^{3/2}}$
6. t^α	$\frac{\Gamma(\alpha+1)}{s^{\alpha+1}}$, $\alpha > -1$
7. $\sin kt$	$\frac{k}{s^2+k^2}$
8. $\cos kt$	$\frac{s}{s^2+k^2}$
9. $\sin^2 kt$	$\frac{2k^2}{s(s^2+4k^2)}$
10. $\cos^2 kt$	$\frac{s^2+2k^2}{s(s^2+4k^2)}$
11. e^{at}	$\frac{1}{s-a}$
12. $\sinh kt$	$\frac{k}{s^2-k^2}$
13. $\cosh kt$	$\frac{s}{s^2-k^2}$
14. $\sinh^2 kt$	$\frac{2k^2}{s(s^2-4k^2)}$
15. $\cosh^2 kt$	$\frac{s^2-2k^2}{s(s^2-4k^2)}$
16. te^{at}	$\frac{1}{(s-a)^2}$
17. $t^n e^{at}$	$\frac{n!}{(s-a)^{n+1}}$, n a positive integer
18. $e^{at}\sin kt$	$\frac{k}{(s-a)^2+k^2}$

$f(t)$	$\mathscr{L}\{f(t)\} = F(s)$
19. $e^{at} \cos kt$	$\dfrac{s-a}{(s-a)^2 - k^2}$
20. $e^{at} \sinh kt$	$\dfrac{k}{(s-a)^2 - k^2}$
21. $e^{at} \cosh kt$	$\dfrac{s-a}{(s-a)^2 - k^2}$
22. $t \sin kt$	$\dfrac{2ks}{(s^2+k^2)^2}$
23. $t \cos kt$	$\dfrac{s^2-k^2}{(s^2+k^2)^2}$
24. $\sin kt + kt \cos kt$	$\dfrac{2ks^2}{(s^2+k^2)^2}$
25. $\sin kt - kt \cos kt$	$\dfrac{2k^3}{(s^2+k^2)^2}$
26. $t \sinh kt$	$\dfrac{2ks}{(s^2-k^2)^2}$
27. $t \cosh kt$	$\dfrac{s^2+k^2}{(s^2-k^2)^2}$
28. $\dfrac{e^{at}-e^{bt}}{a-b}$	$\dfrac{1}{(s-a)(s-b)}$
29. $\dfrac{ae^{at}-be^{bt}}{a-b}$	$\dfrac{s}{(s-a)(s-b)}$
30. $1 - \cos kt$	$\dfrac{k^2}{s(s^2+k^2)}$
31. $kt - \sin kt$	$\dfrac{k^3}{s^2(s^2+k^2)}$
32. $\dfrac{a \sin bt - b \sin at}{ab(a^2-b^2)}$	$\dfrac{1}{(s^2+a^2)(s^2+b^2)}$
33. $\dfrac{\cos bt - \cos at}{a^2-b^2}$	$\dfrac{s}{(s^2+a^2)(s^2+b^2)}$
34. $\sin kt \sinh kt$	$\dfrac{2k^2 s}{s^4+4k^4}$
35. $\sin kt \cosh kt$	$\dfrac{k(s^2+2k^2)}{s^4+4k^4}$
36. $\cos kt \sinh kt$	$\dfrac{k(s^2-2k^2)}{s^4+4k^4}$
37. $\cos kt \cosh kt$	$\dfrac{s^3}{s^4+4k^4}$

$f(t)$	$\mathscr{L}\{f(t)\} = F(s)$
38. $J_0(kt)$	$\dfrac{1}{\sqrt{s^2 + k^2}}$
39. $\dfrac{e^{bt} - e^{at}}{t}$	$\ln \dfrac{s - a}{s - b}$
40. $\dfrac{2(1 - \cos kt)}{t}$	$\ln \dfrac{s^2 + k^2}{s^2}$
41. $\dfrac{2(1 - \cosh kt)}{t}$	$\ln \dfrac{s^2 - k^2}{s^2}$
42. $\dfrac{\sin at}{t}$	$\arctan \left(\dfrac{a}{s}\right)$
43. $\dfrac{\sin at \cos bt}{t}$	$\dfrac{1}{2} \arctan \dfrac{a + b}{s} + \dfrac{1}{2} \arctan \dfrac{a - b}{s}$
44. $\delta(t)$	1
45. $\delta(t - t_0)$	e^{-st_0}
46. $e^{at}f(t)$	$F(s - a)$
47. $f(t - a)\mathscr{U}(t - a)$	$e^{-as}F(s)$
48. $\mathscr{U}(t - a)$	$\dfrac{e^{-as}}{s}$
49. $f^{(n)}(t)$	$s^nF(s) - s^{(n-1)}f(0) - \cdots - f^{(n-1)}(0)$
50. $t^nf(t)$	$(-1)^n \dfrac{d^n}{ds^n} F(s)$
51. $\int_0^t f(\tau)g(t - \tau)\, d\tau$	$F(s)G(s)$

Answers to Odd-Numbered Problems

EXERCISES 1.1, Page 9

1. linear, second-order **3.** nonlinear, first-order

5. linear, fourth-order **7.** nonlinear, second-order

9. linear, third-order

11. $2y' + y = 2(-\frac{1}{2})e^{-x/2} + e^{-x/2} = 0$

13. $\dfrac{dy}{dx} - 2y - e^{3x} = (3e^{3x} + 20e^{2x})$
$-2(e^{3x} + 10e^{2x}) - e^{3x} = 0$

15. $y' - 25 - y^2 = 25 \sec^2 5x - 25(1 + \tan^2 5x)$
$= 25 \sec^2 5x - 25 \sec^2 5x = 0$

17. $y' + y - \sin x = \frac{1}{2} \cos x + \frac{1}{2} \sin x$
$-10e^{-x} + \frac{1}{2} \sin x - \frac{1}{2} \cos x + 10e^{-x} - \sin x = 0$

19. $\dfrac{d}{dt} \ln \dfrac{2 - X}{1 - X} = 1$

$$\left[\frac{-1}{2 - X} + \frac{1}{1 - X}\right] \frac{dX}{dt} = 1$$

simplifies to $\dfrac{dX}{dt} = (2 - X)(1 - X)$

21. The differential of $c_1 = xe^{y/x}/(x + y)^2$ is
$\{(x + y)^2[xe^{y/x}(x\,dy - y\,dx/x^2) + e^{y/x}\,dx]$
$-\ xe^{y/x}\, 2(x + y)(dx + dy)\}/(x + y)^4 = 0.$
Multiplying by $-x^2(x + y)^3 e^{-y/x}$ and simplifying yield
$(x^2 + y^2)\,dx + (x^2 - xy)\,dy = 0.$

23. $y'' - 6y' + 13y = 5e^{3x} \cos 2x - 12e^{3x} \sin 2x$
$+ 12e^{3x} \sin 2x - 18e^{3x} \cos 2x + 13e^{3x} \cos 2x = 0$

25. $y'' = \cosh x + \sinh x = y$

27. $x\dfrac{d^2y}{dx^2} + 2\dfrac{dy}{dx} = x(2c_2x^{-3}) + 2(-c_2x^{-2}) = 0$

29. $x^2y'' - 3xy' + 4y$
$= x^2(5 + 2 \ln x) - 3x(3x + 2x \ln x)$
$+ 4(x^2 + x^2 \ln x) = 9x^2 - 9x^2$
$+ 6x^2 \ln x - 6x^2 \ln x = 0$

31. $y''' - 3y'' + 3y' - y$
$= x^2e^x + 6xe^x + 6e^x - 3x^2e^x - 12xe^x$
$-6e^x + 3x^2e^x + 6xe^x - x^2e^x = 0$

33. For $x < 0$,
$xy' - 2y = x(-2x) - 2(-x^2) = 0;$
for $x \geq 0$,
$xy' - 2y = x(2x) - 2(x^2) = 0.$

35. $y = -1$

37. $m = 2$ and $m = 3$

39. $m = \dfrac{1 \pm \sqrt{5}}{2}$

41. $x' = -2e^{-2t} + 18e^{6t}$,
$x + 3y = -2e^{-2t} + 18e^{6t}$;
$y' = 2e^{-2t} + 30e^{6t}$,
$5x + 3y = 2e^{-2t} + 30e^{6t}$

EXERCISES 1.2, Page 18

1.

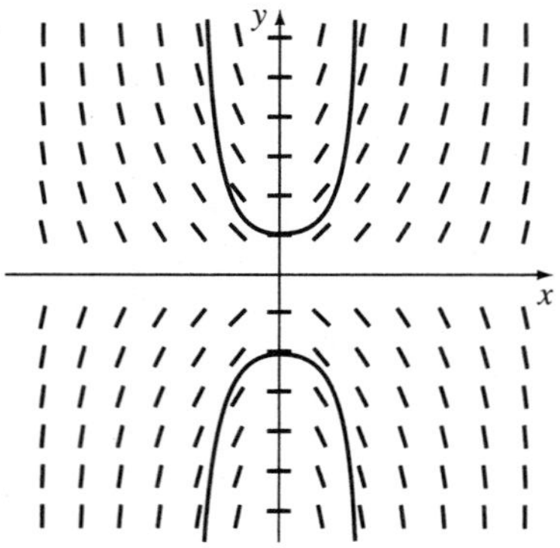

3.

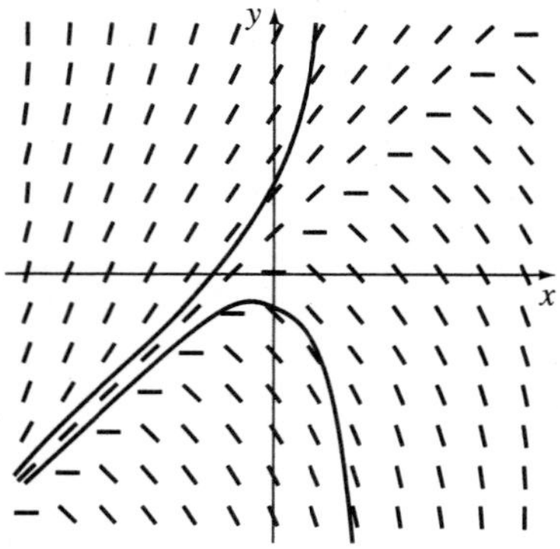

5.

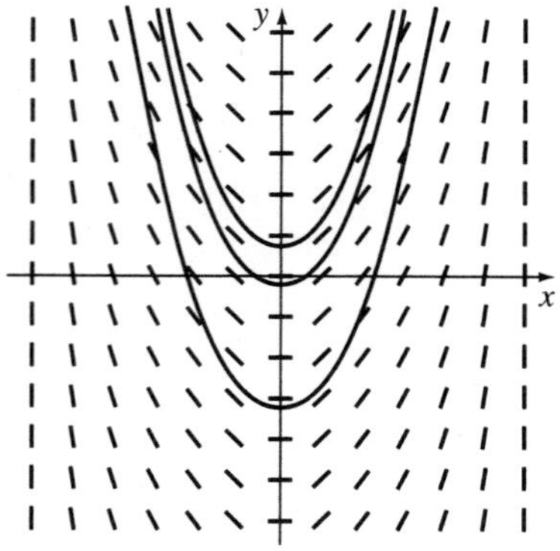

7.

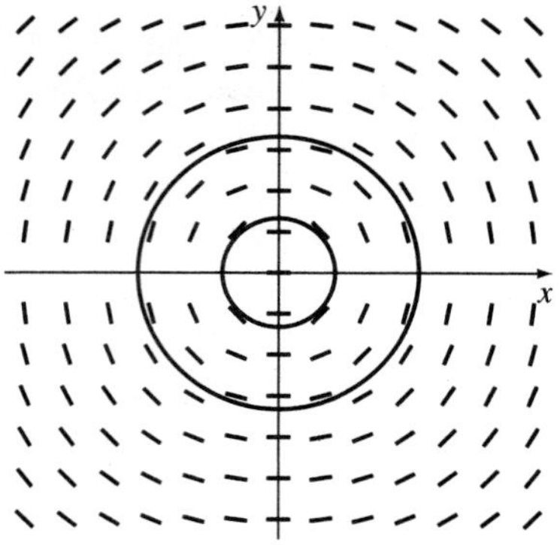

9.

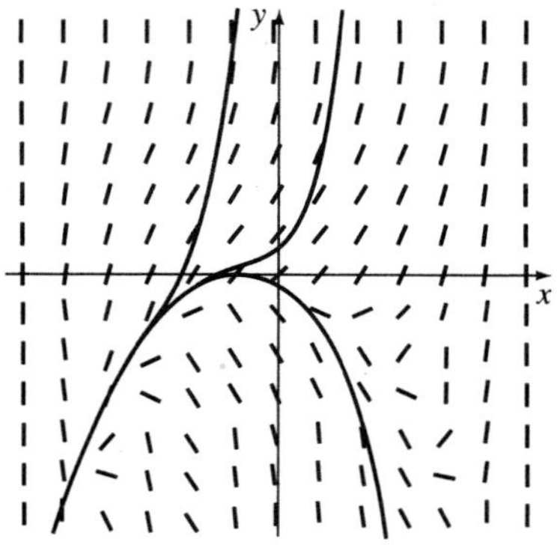

11.

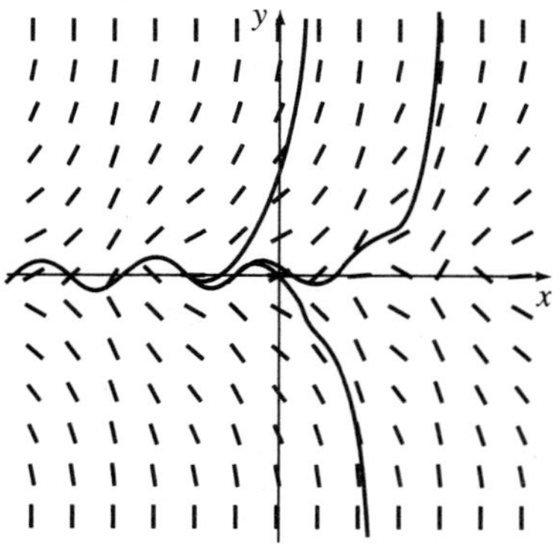

13. $x = 0, \quad x = 3$

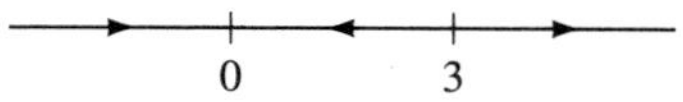

15. $x = 2$

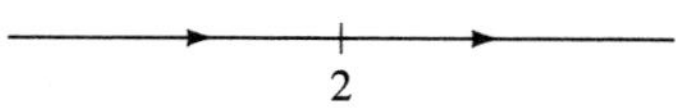

17. $x = 0, \quad x = -2, \quad x = 2$

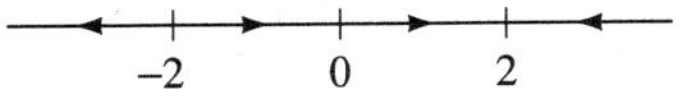

19. $x = -1, \quad x = 0$

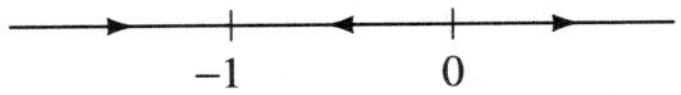

21.

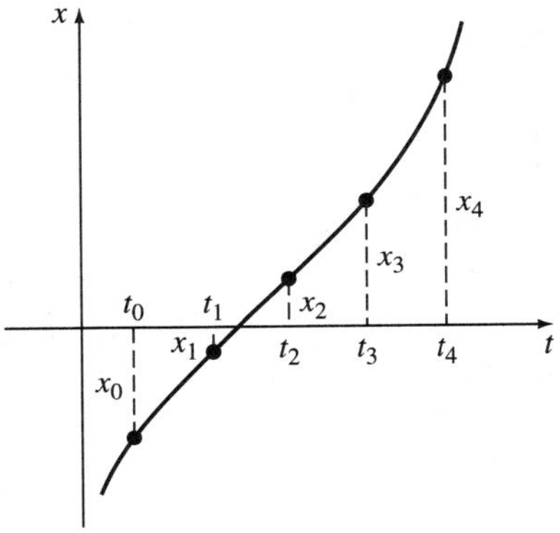

23.

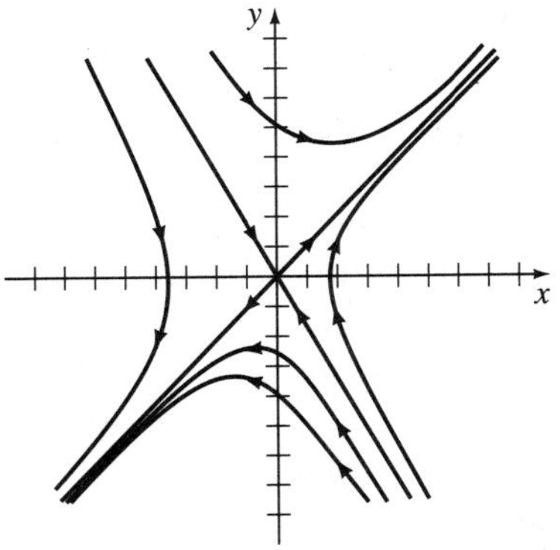

25. $x(t) = e^t$ and $x(t) = 10e^t$ correspond to the same trajectory, namely $y = x$, $x > 0$.

27.

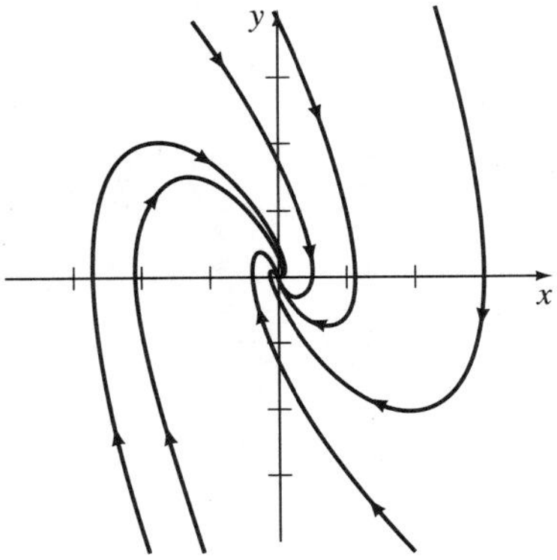

29.

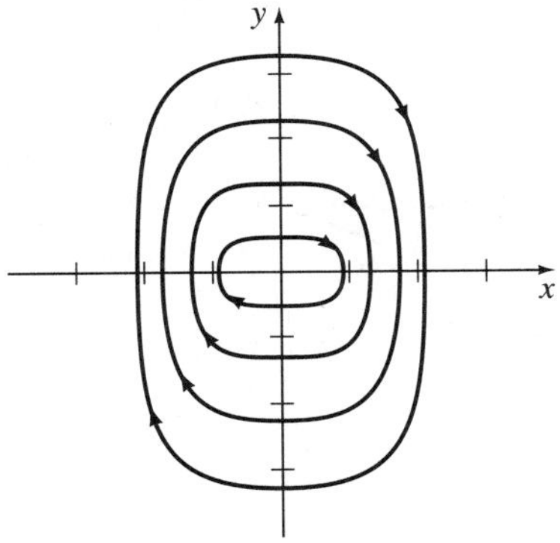

EXERCISES 2.1, Page 26

1. half-planes defined by either $y > 0$ or $y < 0$

3. half-planes defined by either $x > 0$ or $x < 0$

5. the regions defined by either $y > 2$, $y < -2$, or $-2 < y < 2$

7. any region not containing (0, 0)

9. the entire xy-plane **11.** $y = 0$, $y = x^3$

13. There is some interval around $x = 0$ on which the unique solution is $y = 0$.

15. $y = 0$, $y = x$. No, the given function is nondifferentiable at $x = 0$.

17. yes **19.** no

21. (a)

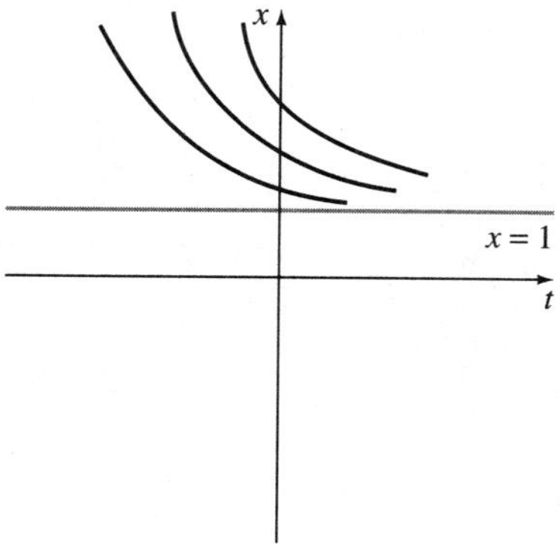

(b)

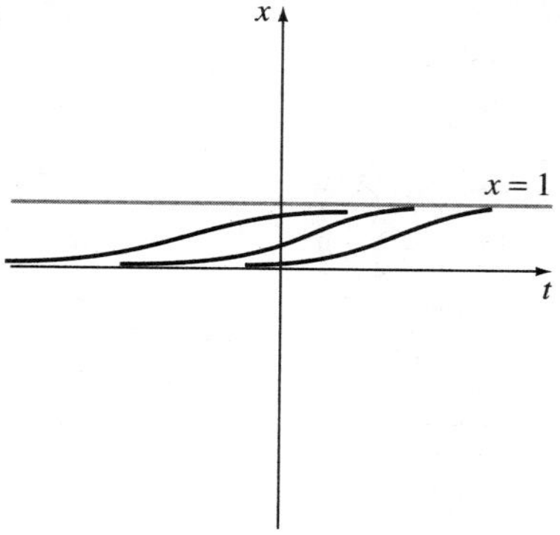

(c)

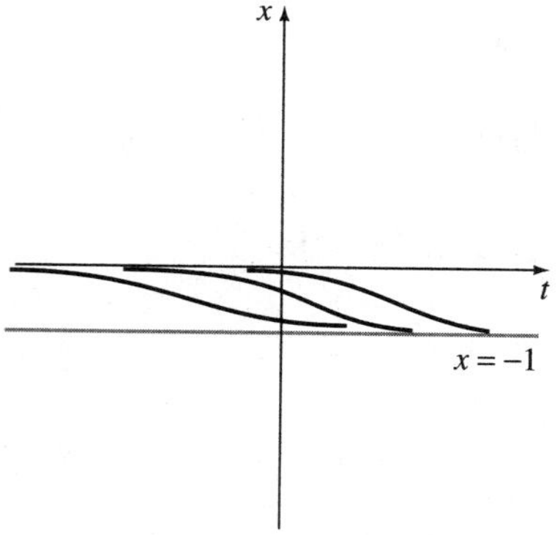

(d)

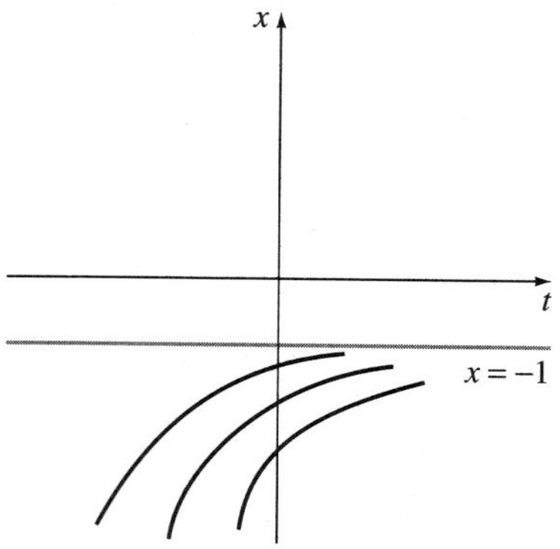

23. From the differential equation written as

$$\frac{dv}{dt} = \frac{k}{m}\left(\frac{mg}{k} - v\right)$$

we see that mg/k is a critical point. From the phase portrait of the differential equation we see that mg/k is an asymptotically stable critical point, so $\lim_{t\to\infty} v = mg/k$.

25. From the differential equation written as

$$\frac{dX}{dt} = B\left(\frac{A}{B} - X\right)$$

we see that A/B is a critical point. From the phase portrait of the differential equation we see that A/B is an asymptotically stable critical point. In addition, from the figure we see that $\lim_{t\to\infty} X = A/B$.

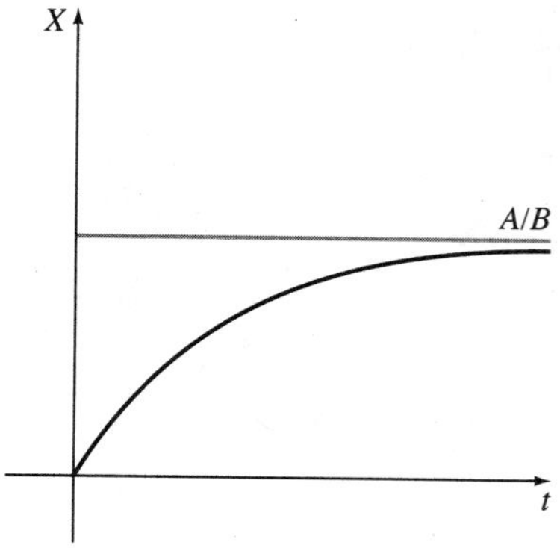

27. (a)

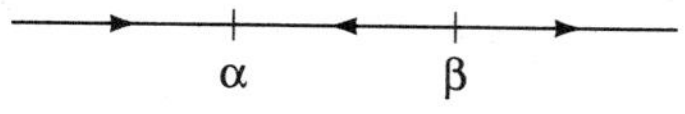

Let $X(0) = X_0$. If $X_0 < \alpha$, then $X(t) \to \alpha$ as $t \to \infty$. If $\alpha < X_0 < \beta$, then $X(t) \to \alpha$ as $t \to \infty$. If $X_0 > \beta$, then $X(t)$ increases in an unbounded manner. However, the behavior of $X(t)$ as $t \to \infty$ is not known.

(b) 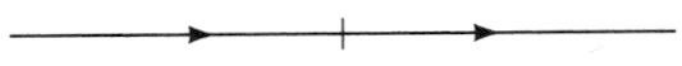

Let $X(0) = X_0$. If $X_0 < \alpha$, then $X(t) \to \alpha$ as $t \to \infty$. If $X_0 > \alpha$, then $X(t)$ increases in an unbounded manner. This could happen in finite time; that is, the phase portrait does not indicate that $X(t)$ becomes unbounded as $t \to \infty$.

(c) For $X(0) = \alpha/2$, $X(t) = \alpha - \dfrac{1}{t + 2/\alpha}$; for $X(0) = 2\alpha$, $X(t) = \alpha - \dfrac{1}{t - 1/\alpha}$.

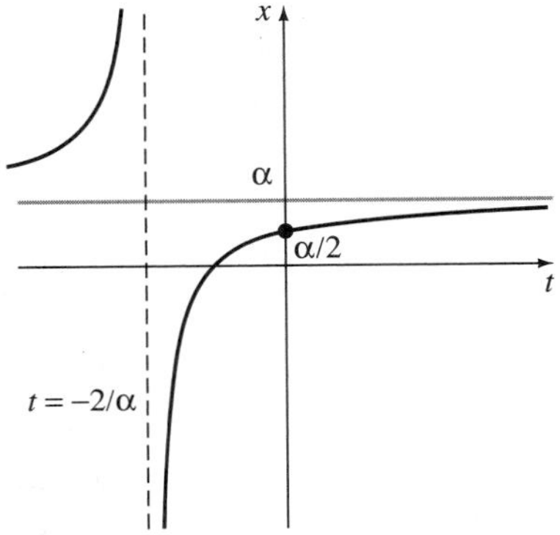

For $X_0 > \alpha$, $X(t)$ increases without bound up to $t = 1/\alpha$. For $t > 1/\alpha$, $X(t)$ increases but $X(t) \to \alpha$ as $t \to \infty$.

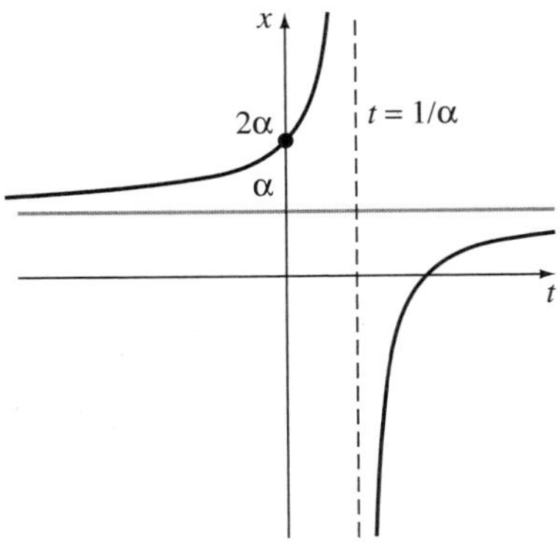

EXERCISES 2.2, Page 36

1. $y = -\frac{1}{5}\cos 5x + c$ **3.** $y = \frac{1}{3}e^{-3x} + c$

5. $y = x + 5\ln|x + 1| + c$ **7.** $y = cx^4$

9. $y^{-2} = 2x^{-1} + c$ **11.** $-3 + 3x\ln|x| = xy^3 + cx$

13. $-3e^{-2y} = 2e^{3x} + c$ **15.** $2 + y^2 = c(4 + x^2)$

17. $y^2 = x - \ln|x + 1| + c$

19. $\dfrac{x^3}{3}\ln x - \dfrac{1}{9}x^3 = \dfrac{y^2}{2} + 2y + \ln|y| + c$

21. $S = ce^{kr}$ **23.** $\dfrac{P}{1 - P} = ce^t$ or $P = \dfrac{ce^t}{1 + ce^t}$

25. $4\cos y = 2x + \sin 2x + c$

27. $-2\cos x + e^y + ye^{-y} + e^{-y} = c$

29. $(e^x + 1)^{-2} + 2(e^y + 1)^{-1} = c$

31. $(y + 1)^{-1} + \ln|y + 1| = \dfrac{1}{2}\ln\left|\dfrac{x + 1}{x - 1}\right| + c$

33. $y - 5\ln|y + 3| = x - 5\ln|x + 4| + c$ or $\left(\dfrac{y + 3}{x + 4}\right)^5 = c_1 e^{y - x}$

35. $x = \tan(4y - 3\pi/4)$ **37.** $xy = e^{-(1 + 1/x)}$

39. **(a)** $y = 3\dfrac{1 - e^{6x}}{1 + e^{6x}}$; **(b)** $y = 3$;

(c) $y = 3\dfrac{2 - e^{6x - 2}}{2 + e^{6x - 2}}$

41. $x\ln|x| + y = cx$

43. $(x - y)\ln|x - y| = y + c(x - y)$

45. $x + y\ln|x| = cy$

47. $\ln(x^2 + y^2) + 2\tan^{-1}(y/x) = c$

49. $4x = y(\ln|y| - c)^2$

51. $y = \ln|\cos(c_1 - x)| + c_2$

53. $y = c_1 + c_2 x^2$

55. $\frac{1}{3}y^3 - c_1 y = x + c_2$

57. $y = \pm\sin(x + c)$, but $y = 1$ and $y = -1$ are also solutions.

59. $y^2 - 9x^2 = c$

61. $2x^2 + 2y^2 - x^4 = c$

EXERCISES 2.3, Page 43

1. $x^2 - x + \frac{3}{2}y^2 + 7y = c$ **3.** $\frac{5}{2}x^2 + 4xy - 2y^4 = c$

5. $x^2y^2 - 3x + 4y = c$

7. not exact, but is homogeneous

9. $xy^3 + y^2 \cos x - \frac{1}{2}x^2 = c$ 11. not exact

13. $xy - 2xe^x + 2e^x - 2x^3 = c$

15. $x + y + xy - 3 \ln|xy| = c$

17. $x^3y^3 - \tan^{-1}3x = c$

19. $-\ln|\cos x| + \cos x \sin y = c$

21. $y - 2x^2y - y^2 - x^4 = c$

23. $x^4y - 5x^3 - xy + y^3 = c$

25. $\frac{1}{3}x^3 + x^2y + xy^2 - y = \frac{4}{3}$

27. $4xy + x^2 - 5x + 3y^2 - y = 8$

29. $y^2 \sin x - x^3y - x^2 + y \ln y - y = 0$

31. $k = 10$ 33. $k = 1$

35. $M(x, y) = ye^{xy} + y^2 - y/x^2 + h(x)$

EXERCISES 2.4, Page 51

1. $y = ce^{5x}, \quad -\infty < x < \infty$

3. $y = \frac{1}{3} + ce^{-4x}, \quad -\infty < x < \infty$

5. $y = \frac{1}{4}e^{3x} + ce^{-x}, \quad -\infty < x < \infty$

7. $y = \frac{1}{3} + ce^{-x^3}, \quad -\infty < x < \infty$

9. $y = x^{-1} \ln x + cx^{-1}, \quad 0 < x < \infty$

11. $x = -\frac{4}{5}y^2 + cy^{-1/2}, \quad 0 < y < \infty$

13. $y = -\cos x + \dfrac{\sin x}{x} + \dfrac{c}{x}, \quad 0 < x < \infty$

15. $y = \dfrac{c}{e^x + 1}, \quad -\infty < x < \infty$

17. $y = \sin x + c \cos x, \quad -\pi/2 < x < \pi/2$

19. $y = \frac{1}{7}x^3 - \frac{1}{5}x + cx^{-4}, \quad 0 < x < \infty$

21. $y = \dfrac{1}{2x^2}e^x + \dfrac{c}{x^2}e^{-x}, \quad 0 < x < \infty$

23. $y = \sec x + c \csc x, \quad 0 < x < \pi/2$

25. $x = \dfrac{1}{2}e^y - \dfrac{1}{2y}e^y + \dfrac{1}{4y^2}e^y + \dfrac{c}{y^2}e^{-y}, \quad 0 < y < \infty$

27. $y = e^{-3x} + \dfrac{c}{x}e^{-3x}, \quad 0 < x < \infty$

29. $x = 2y^6 + cy^4, \quad 0 < y < \infty$

31. $y = e^{-x} \ln(e^x + e^{-x}) + ce^{-x}, \quad -\infty < x < \infty$

33. $x = \dfrac{1}{y} + \dfrac{c}{y}e^{-y^2}, \quad 0 < y < \infty$

35. $(\sec \theta + \tan \theta)r = \theta - \cos \theta + c, \quad -\pi/2 < \theta < \pi/2$

37. $y = \frac{5}{3}(x + 2)^{-1} + c(x + 2)^{-4}, \quad -2 < x < \infty$

39. $y = 10 + ce^{-\sinh x}, \quad -\infty < x < \infty$

41. $y = 4 - 2e^{-5x}, \quad -\infty < x < \infty$

43. $i(t) = E/R + (i_0 - E/R)e^{-Rt/L}, \quad -\infty < t < \infty$

45. $y = \sin x \cos x - \cos x, \quad -\pi/2 < x < \pi/2$

47. $T(t) = 50 + 150e^{kt}, \quad -\infty < t < \infty$

49. $(x + 1)y = x \ln x - x + 21, \quad 0 < x < \infty$

51. $y = \begin{cases} \frac{1}{2}(1 - e^{-2x}), & 0 \le x \le 3 \\ \frac{1}{2}(e^6 - 1)e^{-2x}, & x > 3 \end{cases}$

53. $y = \begin{cases} \frac{1}{2} + \frac{3}{2}e^{-x^2}, & 0 \le x < 1 \\ (\frac{1}{2}e + \frac{3}{2})e^{-x^2}, & x \ge 1 \end{cases}$

55. $y^3 = 1 + cx^{-3}$ 57. $y^{-3} = x + \frac{1}{3} + ce^{3x}$

59. $y^{-3} = -\frac{9}{5}x^{-1} + \frac{49}{5}x^{-6}$

61. $x = x_0e^{-\lambda_1 t}$,

$$y = \frac{\lambda_1 x_0}{\lambda_2 - \lambda_1}e^{-\lambda_1 t} + \left(y_0 - \frac{\lambda_1 x_0}{\lambda_2 - \lambda_1}\right)e^{-\lambda_2 t}$$

EXERCISES 2.5, PAGE 66

1. 7.9 years; 10 years 3. 760 5. 11 hours

7. 136.5 hours

9. $I(15) = 0.00098I_0$, or $I(15)$ is approximately 0.1% of I_0.

11. 15,600 years

13. $T(1) = 36.67$ degrees; approximately 3.06 minutes

15. $i(t) = \frac{3}{5} - \frac{3}{5}e^{-500t}$; $i \to \frac{3}{5}$ as $t \to \infty$

17. $q(t) = \frac{1}{100} - \frac{1}{100}e^{-50t}$; $i(t) = \frac{1}{2}e^{-50t}$

19. $i(t) = \begin{cases} 60 - 60e^{-t/10}, & 0 \le t \le 20 \\ 60(e^2 - 1)e^{-t/10}, & t > 20 \end{cases}$

21. $A(t) = 200 - 170e^{-t/50}$

23. $A(t) = 1000 - 1000e^{-t/100}$ 25. 64.38 lb

33. 1834; 2000 35. 1,000,000; 52.9 months

37. For $\alpha \neq \beta$,

$$\frac{1}{\alpha - \beta}[\ln|\alpha - X| - \ln|\beta - X|] = kt + c$$

or

$$\frac{1}{\alpha - \beta}\ln\left|\frac{\alpha - X}{\beta - X}\right| = kt + c.$$

For $\alpha = \beta$,

$$(\alpha - X)^{-1} = kt + c$$

or

$$X = \alpha - \frac{1}{kt + c}.$$

39. **(a)** $v^2 = (2gR^2/y) + v_0^2 - 2gR$.
We note that as y increases, v decreases. In particular, if $v_0^2 - 2gR < 0$, then there must be some value of y for $v = 0$; the rocket stops and returns to earth under the influence of gravity. However, if $v_0^2 - 2gR \geq 0$, then $v > 0$ for all values of y. Hence we should have $v_0 \geq \sqrt{2gR}$.

(b) Using the values $R = 4000$ miles, $g = 32$ ft/s^2, 1 ft = 1/5280 mi, and 1 s = 1/3600 h, it follows that $v_0 \geq 25{,}067$ mi/h.

(c) 5291 mi/h

41. **(a)** $m\dfrac{dv}{dt} = mg - kv^2 - \rho V$, where ρ is the weight density of water, and V is the volume of water displaced.

(b) $v(t) = \sqrt{\dfrac{mg - \rho V}{k}} \tanh\left(\dfrac{\sqrt{kmg - k\rho V}}{m} t + c_1\right)$

(c) $\sqrt{\dfrac{mg - \rho V}{k}}$

EXERCISES 2.6, Page 80

1. **(a)** $y = 1 - x + \tan(x + \pi/4)$

(b) $h = 0.1$

x_n	y_n	True Value
0.00	2.0000	2.0000
0.10	2.1000	2.1230
0.20	2.2440	2.3085
0.30	2.4525	2.5958
0.40	2.7596	3.0650
0.50	3.2261	3.9082

$h = 0.05$

x_n	y_n	True Value
0.00	2.0000	2.0000
0.05	2.0500	2.0554
0.10	2.1105	2.1230
0.15	2.1838	2.2061
0.20	2.2727	2.3085
0.25	2.3812	2.4358
0.30	2.5142	2.5958
0.35	2.6788	2.7997
0.40	2.8845	3.0650
0.45	3.1455	3.4189
0.50	3.4823	3.9082

3. **(a)**

x_n	y_n
1.00	5.0000
1.10	3.8000
1.20	2.9800
1.30	2.4260
1.40	2.0582
1.50	1.8207

(b)

x_n	y_n
1.00	5.0000
1.05	4.4000
1.10	3.8950
1.15	3.4707
1.20	3.1151
1.25	2.8179
1.30	2.5702
1.35	2.3647
1.40	2.1950
1.45	2.0557
1.50	1.9424

5. **(a)**

x_n	y_n
0.00	0.0000
0.10	0.1000
0.20	0.2010
0.30	0.3050
0.40	0.4143
0.50	0.5315

(b)

x_n	y_n
0.00	0.0000
0.05	0.0500
0.10	0.1001
0.15	0.1506
0.20	0.2018
0.25	0.2538
0.30	0.3070
0.35	0.3617
0.40	0.4183
0.45	0.4770
0.50	0.5384

7. **(a)**

x_n	y_n
0.00	0.0000
0.10	0.1000
0.20	0.1905
0.30	0.2731
0.40	0.3492
0.50	0.4198

(b)

x_n	y_n
0.00	0.0000
0.05	0.0500
0.10	0.0976
0.15	0.1429
0.20	0.1863
0.25	0.2278
0.30	0.2676
0.35	0.3058
0.40	0.3427
0.45	0.3782
0.50	0.4124

9. (a)

x_n	y_n
0.00	0.5000
0.10	0.5250
0.20	0.5431
0.30	0.5548
0.40	0.5613
0.50	0.5639

(b)

x_n	y_n
0.00	0.5000
0.05	0.5125
0.10	0.5232
0.15	0.5322
0.20	0.5395
0.25	0.5452
0.30	0.5496
0.35	0.5527
0.40	0.5547
0.45	0.5559
0.50	0.5565

11. (a)

x_n	y_n
1.00	1.0000
1.10	1.0000
1.20	1.0191
1.30	1.0588
1.40	1.1231
1.50	1.2194

(b)

x_n	y_n
1.00	1.0000
1.05	1.0000
1.10	1.0049
1.15	1.0147
1.20	1.0298
1.25	1.0506
1.30	1.0775
1.35	1.1115
1.40	1.1538
1.45	1.2057
1.50	1.2696

13. (a) $h = 0.1$

x_n	y_n
1.00	5.0000
1.10	3.9900
1.20	3.2545
1.30	2.7236
1.40	2.3451
1.50	2.0801

$h = 0.05$

x_n	y_n
1.00	5.0000
1.05	4.4475
1.10	3.9763
1.15	3.5751
1.20	3.2342
1.25	2.9452
1.30	2.7009
1.35	2.4952
1.40	2.3226
1.45	2.1786
1.50	2.0592

(b) $h = 0.1$

x_n	y_n
0.00	0.0000
0.10	0.1005
0.20	0.2030
0.30	0.3098
0.40	0.4234
0.50	0.5470

$h = 0.05$

x_n	y_n
0.00	0.0000
0.05	0.0501
0.10	0.1004
0.15	0.1512
0.20	0.2028
0.25	0.2554
0.30	0.3095
0.35	0.3652
0.40	0.4230
0.45	0.4832
0.50	0.5465

(c) $h = 0.1$

x_n	y_n
0.00	0.0000
0.10	0.0952
0.20	0.1822
0.30	0.2622
0.40	0.3363
0.50	0.4053

$h = 0.05$

x_n	y_n
0.00	0.0000
0.05	0.0488
0.10	0.0953
0.15	0.1397
0.20	0.1823
0.25	0.2231
0.30	0.2623
0.35	0.3001
0.40	0.3364
0.45	0.3715
0.50	0.4054

(d) $h = 0.1$

x_n	y_n
0.00	0.5000
0.10	0.5215
0.20	0.5362
0.30	0.5449
0.40	0.5490
0.50	0.5503

$h = 0.05$

x_n	y_n
0.00	0.5000
0.05	0.5116
0.10	0.5214
0.15	0.5294
0.20	0.5359
0.25	0.5408
0.30	0.5444
0.35	0.5469
0.40	0.5484
0.45	0.5492
0.50	0.5495

(e) $h = 0.1$

x_n	y_n
1.00	1.0000
1.10	1.0095
1.20	1.0404
1.30	1.0967
1.40	1.1866
1.50	1.3260

$h = 0.05$

x_n	y_n
1.00	1.0000
1.05	1.0024
1.10	1.0100
1.15	1.0228
1.20	1.0414
1.25	1.0663
1.30	1.0984
1.35	1.1389
1.40	1.1895
1.45	1.2526
1.50	1.3315

15. (a) The appearance of the graph will depend on the ODE solver used. The following graph was obtained using *Mathematica* on the interval [1, 1.3556].

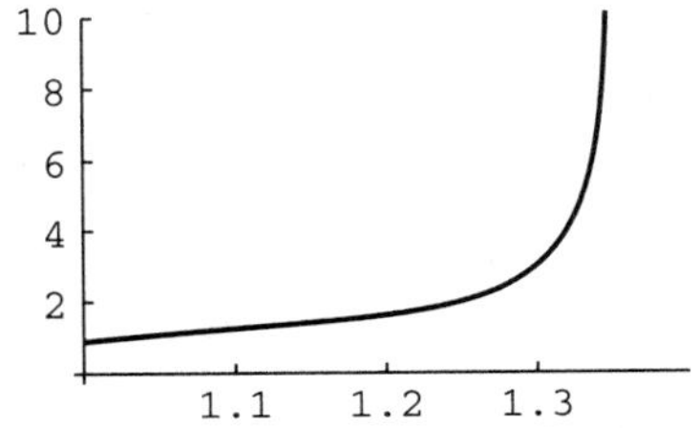

(b) x_n	Euler	Improved Euler
1.0	1.0000	1.0000
1.1	1.2000	1.2469
1.2	1.4938	1.6668
1.3	1.9711	2.6427
1.4	2.9060	8.7989

17. (a) $y_1 = 1.2$

(b) $y''(c)\dfrac{h^2}{2} = 4e^{2c}\dfrac{(0.1)^2}{2} = 0.02\,e^{2c} \le 0.02\,e^{0.2} = 0.0244$

(c) Actual value is $y(0.1) = 1.2214$.
Error is 0.0214.

(d) If $h = 0.05$, $y_2 = 1.21$.

(e) Error with $h = 0.1$ is 0.0214.
Error with $h = 0.05$ is 0.0114.

19. (a) $y_1 = 0.8$

(b) $y''(c)\dfrac{h^2}{2} = 5e^{-2c}\dfrac{(0.1)^2}{2} = 0.025\,e^{-2c} \le 0.025$ for $0 \le c \le 0.1$.

(c) Actual value is $y(0.1) = 0.8234$.
Error is 0.0234.

(d) If $h = 0.05$, $y_2 = 0.8125$.

(e) Error with $h = 0.1$ is 0.0234.
Error with $h = 0.05$ is 0.0109.

21. (a) Error is $19h^2e^{-3(c-1)}$.

(b) $y''(c)\dfrac{h^2}{2} \le 19(0.1)^2(1) = 0.19$

(c) If $h = 0.1$, $y_5 = 1.8207$.
If $h = 0.05$, $y_{10} = 1.9424$

(d) Error with $h = 0.1$ is 0.2325.
Error with $h = 0.05$ is 0.1109.

23. (a) Error is $\dfrac{1}{(c+1)^2}\dfrac{h^2}{2}$.

(b) $\left| y''(c)\dfrac{h^2}{2} \right| \le (1)\dfrac{(0.1)^2}{2} = 0.005$

(c) If $h = 0.1$, $y_5 = 0.4198$.
If $h = 0.05$, $y_{10} = 0.4124$

(d) Error with $h = 0.1$ is 0.0143.
Error with $h = 0.05$ is 0.0069.

EXERCISES 2.7, Page 86

1. x_n	y_n	True Value
0.00	2.0000	2.0000
0.10	2.1230	2.1230
0.20	2.3085	2.3085
0.30	2.5958	2.5958
0.40	3.0649	3.0650
0.50	3.9078	3.9082

3.

x_n	y_n
1.00	5.0000
1.10	3.9724
1.20	3.2284
1.30	2.6945
1.40	2.3163
1.50	2.0533

5.

x_n	y_n
0.00	0.0000
0.10	0.1003
0.20	0.2027
0.30	0.3093
0.40	0.4228
0.50	0.5463

7.

x_n	y_n
0.00	0.0000
0.10	0.0953
0.20	0.1823
0.30	0.2624
0.40	0.3365
0.50	0.4055

9.

x_n	y_n
0.00	0.5000
0.10	0.5213
0.20	0.5358
0.30	0.5443
0.40	0.5482
0.50	0.5493

11.

x_n	y_n
1.00	1.0000
1.10	1.0101
1.20	1.0417
1.30	1.0989
1.40	1.1905
1.50	1.3333

13. **(a)** $v(5) = 35.7678$

(b)

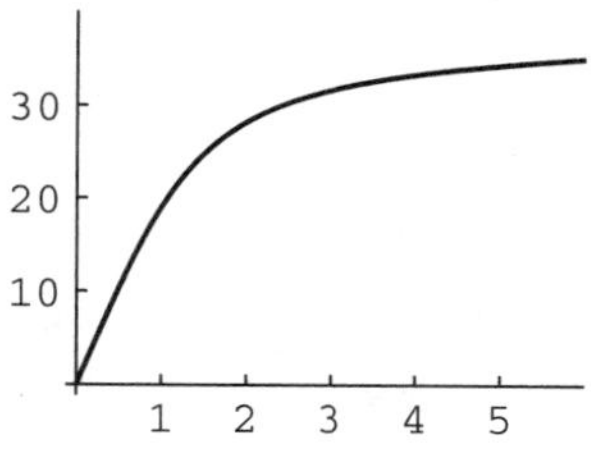

(c) $v(t) = \sqrt{\frac{mg}{k}} \tanh \sqrt{\frac{kg}{m}}\, t$; $v(5) = 35.7678$

15. **(a)** $h = 0.1$

x_n	y_n
1.00	1.0000
1.10	1.2511
1.20	1.6934
1.30	2.9425
1.40	903.0282

$h = 0.05$

x_n	y_n
1.00	1.0000
1.05	1.1112
1.10	1.2511
1.15	1.4348
1.20	1.6934
1.25	2.1047
1.30	2.9560
1.35	7.8981
1.40	1.1E + 15

(b) The appearance of the graph will depend on the ODE solver used. The following graph was obtained using *Mathematica* on the interval [1, 1.3556].

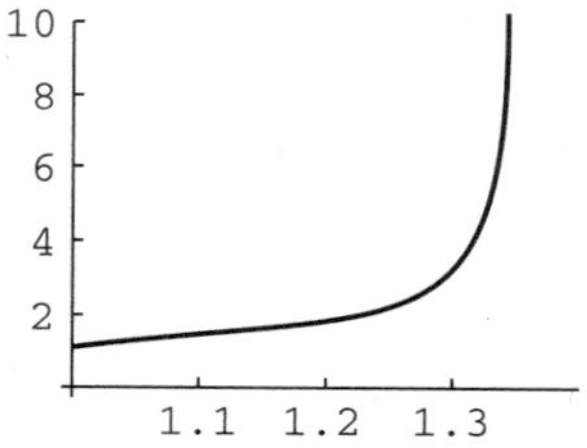

17. **(a)** $y_1 = 0.82341667$

(b) $y^{(5)}(c)\frac{h^5}{5!} = 40e^{-2c}\frac{h^5}{5!} \le 40e^{2(0)}\frac{(0.1)^5}{5!}$

$= 3.333 \times 10^{-6}$

(c) Actual value is $y(0.1) = 0.8234134413$. Error is $3.225 \times 10^{-6} \le 3.333 \times 10^{-6}$.

(d) If $h = 0.05$, $y_2 = 0.82341363$.

(e) Error with $h = 0.1$ is 3.225×10^{-6}. Error with $h = 0.05$ is 1.854×10^{-7}.

19. **(a)** $y^{(5)}(c)\frac{h^5}{5!} = \frac{24}{(c+1)^5}\frac{h^5}{5!}$

(b) $\frac{24}{(c+1)^5}\frac{h^3}{5!} \le 24\frac{(0.1)^3}{5!} = 2.0000 \times 10^{-6}$

(c) From calculation with $h = 0.1$, $y_5 = 0.40546517$. From calculation with $h = 0.05$, $y_{10} = 0.40546511$.

EXERCISES 2.8, Page 90

1. $y(x) = -x + e^x$; $y(0.2) = 1.0214$, $y(0.4) = 1.0918$, $y(0.6) = 1.2221$, $y(0.8) = 1.4255$

3.

x_n	y_n
0.00	1.0000
0.20	0.7328
0.40	0.6461
0.60	0.6585
0.80	0.7232

5.

x_n	y_n
0.00	0.0000
0.20	0.2027
0.40	0.4228
0.60	0.6841
0.80	1.0297
1.00	1.5569

x_n	y_n
0.00	0.0000
0.10	0.1003
0.20	0.2027
0.30	0.3093
0.40	0.4228
0.50	0.5463
0.60	0.6842
0.70	0.8423
0.80	1.0297
0.90	1.2603
1.00	1.5576

7.

x_n	y_n
0.00	0.0000
0.20	0.0026
0.40	0.0201
0.60	0.0630
0.80	0.1360
1.00	0.2385

x_n	y_n
0.00	0.0000
0.10	0.0003
0.20	0.0026
0.30	0.0087
0.40	0.0200
0.50	0.0379
0.60	0.0629
0.70	0.0956
0.80	0.1360
0.90	0.1837
1.00	0.2384

EXERCISES 3.1, Page 103

1. $y = \frac{1}{2}e^x - \frac{1}{2}e^{-x}$ **3.** $y = \frac{3}{5}e^{4x} + \frac{2}{5}e^{-x}$

5. $y = 3x - 4x \ln x$ **7.** $y = 0$, $y = x^2$

9. **(a)** $y = e^x \cos x - e^x \sin x$

(b) no solution

(c) $y = e^x \cos x + e^{-\pi/2}e^x \sin x$

(d) $y = c_2 e^x \sin x$, where c_2 is arbitrary

11. $(-\infty, 2)$

15. dependent **17.** dependent **19.** dependent

21. independent

23. The functions satisfy the differential equation and are linearly independent on the interval since $W(e^{-3x}, e^{4x}) = 7e^x \neq 0$; $y = c_1e^{-3x} + c_2e^{4x}$.

25. The functions satisfy the differential equation and are linearly independent on the interval since $W(e^x \cos 2x, e^x \sin 2x) = 2e^{2x} \neq 0$; $y = c_1e^x \cos 2x + c_2e^x \sin 2x$.

27. The functions satisfy the differential equation and are linearly independent on the interval since $W(x^3, x^4) = x^6 \neq 0$; $y = c_1x^3 + c_2x^4$.

29. The functions satisfy the differential equation and are linearly independent on the interval since $W(x, x^{-2}, x^{-2} \ln x) = 9x^{-6} \neq 0$; $y = c_1x + c_2x^{-2} + c_3x^{-2} \ln x$.

33. e^{2x} and e^{5x} form a fundamental set of solutions of the homogeneous equation; $6e^x$ is a particular solution of the nonhomogeneous equation.

35. e^{2x} and xe^{2x} form a fundamental set of solutions of the homogeneous equation; $x^2e^{2x} + x - 2$ is a particular solution of the nonhomogeneous equation.

37. $y_p = x^2 + 3x + 3e^{2x}$;
$y_p = -2x^2 - 6x - \frac{1}{3}e^{2x}$

EXERCISES 3.2, Page 108

1. $y_2 = e^{-5x}$ **3.** $y_2 = xe^{2x}$ **5.** $y_2 = \sin 4x$

7. $y_2 = \sinh x$ **9.** $y_2 = xe^{2x/3}$ **11.** $y_2 = x^4 \ln|x|$

13. $y_2 = 1$ **15.** $y_2 = x^2 + x + 2$

17. $y_2 = x \cos(\ln x)$ **19.** $y_2 = x$ **21.** $y_2 = x \ln x$

23. $y_2 = x^3$ **25.** $y_2 = e^{2x}$, $y_p = -\frac{1}{2}$

27. $y_2 = e^{2x}$, $y_p = \frac{5}{2}e^{3x}$

EXERCISES 3.3, Page 120

1. $y = c_1 + c_2e^{-x/4}$ **3.** $y = c_1e^{-6x} + c_2e^{6x}$

5. $y = c_1 \cos 3x + c_2 \sin 3x$ **7.** $y = c_1e^{3x} + c_2e^{-2x}$

9. $y = c_1e^{-4x} + c_2xe^{-4x}$

11. $y = c_1e^{(-3+\sqrt{29})x/2} + c_2e^{(-3-\sqrt{29})x/2}$

13. $y = c_1e^{2x/3} + c_2e^{-x/4}$

15. $y = e^{2x}(c_1 \cos x + c_2 \sin x)$

17. $y = e^{-x/3}\left(c_1 \cos \frac{\sqrt{2}}{3}x + c_2 \sin \frac{\sqrt{2}}{3}x\right)$

19. $y = c_1 + c_2e^{-x} + c_3e^{5x}$

21. $y = c_1e^{x} + e^{-x/2}\left(c_2 \cos \frac{\sqrt{3}}{2}x + c_3 \sin \frac{\sqrt{3}}{2}x\right)$

23. $y = c_1e^{-x} + c_2e^{3x} + c_3xe^{3x}$

25. $y = c_1e^{x} + e^{-x}(c_2 \cos x + c_3 \sin x)$

27. $y = c_1e^{-x} + c_2xe^{-x} + c_3x^2e^{-x}$

29. $y = c_1 + c_2x + e^{-x/2}\left(c_3 \cos \frac{\sqrt{3}}{2}x + c_4 \sin \frac{\sqrt{3}}{2}x\right)$

31. $y = c_1 \cos \frac{\sqrt{3}}{2}x + c_2 \sin \frac{\sqrt{3}}{2}x$
$+ c_3x \cos \frac{\sqrt{3}}{2}x + c_4x \sin \frac{\sqrt{3}}{2}x$

33. $y = c_1 + c_2e^{-2x} + c_3e^{2x} + c_4 \cos 2x + c_5 \sin 2x$

35. $y = c_1e^{x} + c_2xe^{x} + c_3e^{-x} + c_4xe^{-x} + c_5e^{-5x}$

37. $y = 2 \cos 4x - \frac{1}{2} \sin 4x$

39. $y = -\frac{3}{4}e^{-5x} + \frac{3}{4}e^{-x}$

41. $y = -e^{x/2} \cos(x/2) + e^{x/2} \sin(x/2)$ **43.** $y = 0$

45. $y = e^{2(x-1)} - e^{x-1}$

47. $y = \frac{5}{36} - \frac{5}{36}e^{-6x} + \frac{1}{6}xe^{-6x}$

49. $y = -\frac{1}{6}e^{2x} + \frac{1}{6}e^{-x} \cos \sqrt{3}\,x - \frac{\sqrt{3}}{6}e^{-x} \sin \sqrt{3}\,x$

51. $y = 2 - 2e^{x} + 2xe^{x} - \frac{1}{2}x^2e^{x}$ **53.** $y = e^{5x} - xe^{5x}$

55. $y = -2 \cos x$

57. $\frac{d^3y}{dx^3} + 6\frac{d^2y}{dx^2} - 15\frac{dy}{dx} - 100y = 0$

59. $y = c_1e^{x} + e^{4x}(c_2 \cos x + c_3 \sin x)$

61. $y = c_1e^{-0.270534\,x} + c_2e^{0.658675\,x} + c_3e^{5.61186\,x}$

63. $y = c_1e^{-1.74806\,x} + c_2e^{0.501219\,x} + c_3e^{0.62342x} \cos(0.588965\,x)$
$+ c_4e^{0.62342\,x} \sin(0.588965\,x)$

67. unstable saddle point

69. unstable node

71. unstable spiral point

73. asymptotically stable node

75. (a)

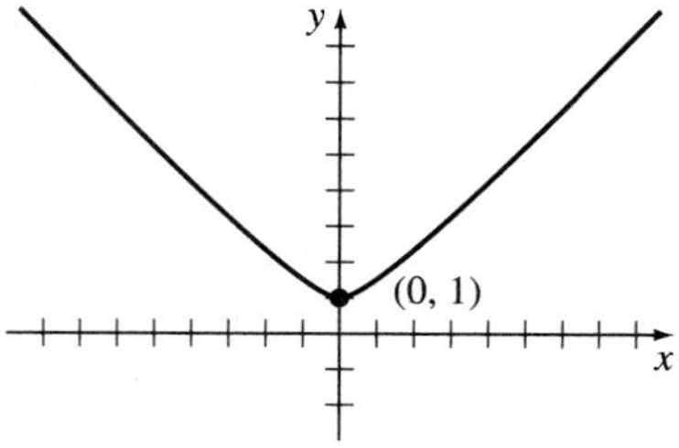

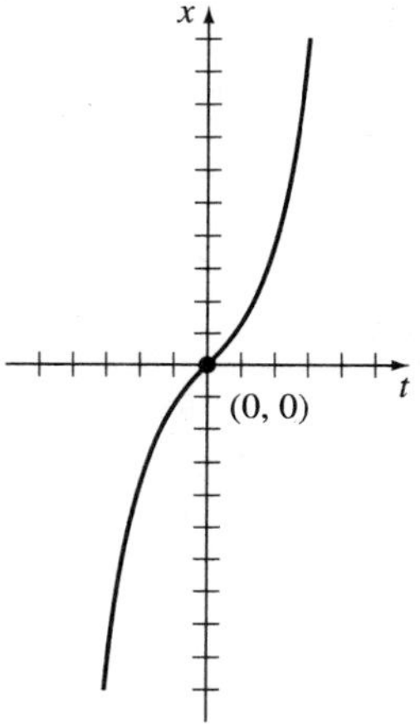

(b)

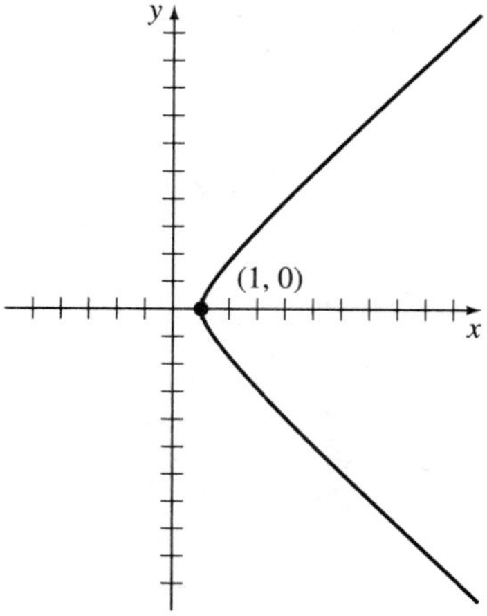

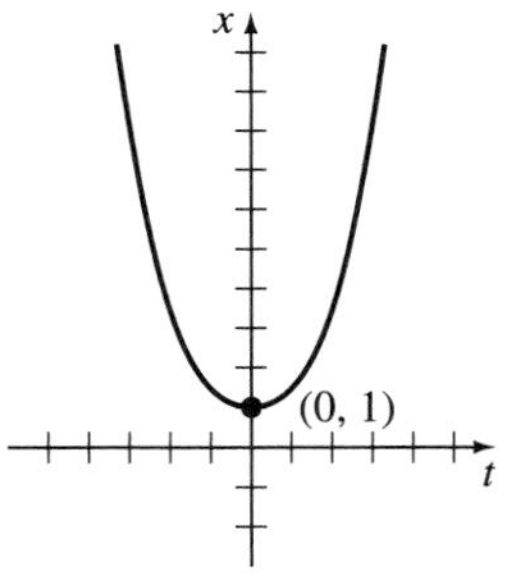

(c)

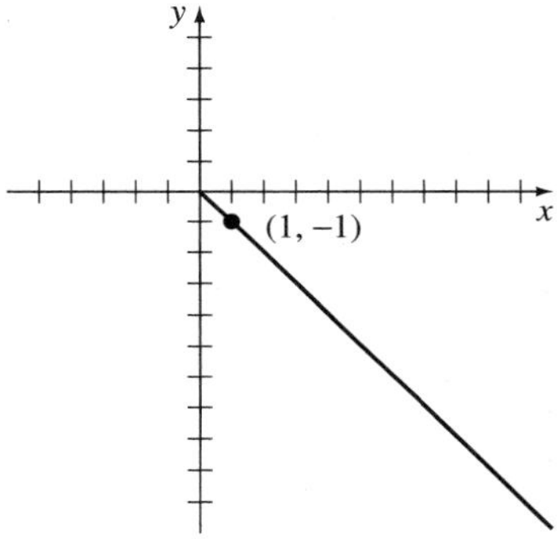

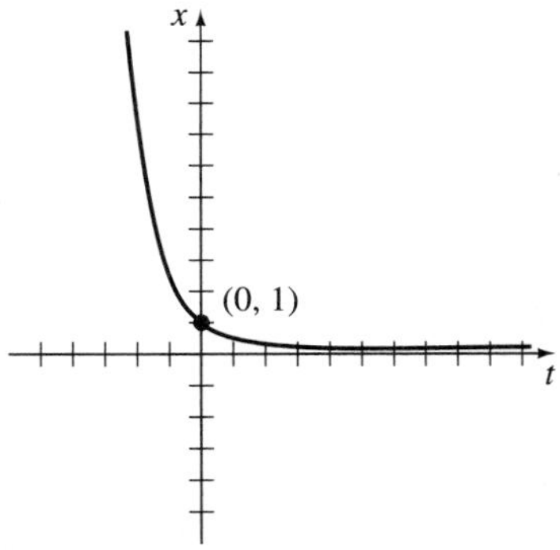

77. A family of horizontal lines in the phase plane.

79. The system that is equivalent to $x'' + x' = 0$ is

$$\frac{dx}{dt} = y$$

$$\frac{dy}{dt} = -y.$$

Inspection of this system shows that *all* points $(x, 0)$ are critical points of the differential equation. Thus there is no circular neighborhood of $(0, 0)$ that is free of critical points.

81. (b) The only other type of autonomous linear second-order differential equation is $ax'' + bx' + cx = d$. But because a change of variable can translate the single critical point of this equation to the origin, we can without any loss of generality only consider equations of the form $ax'' + bx' + cx = 0$, $a \neq 0$, $c \neq 0$.

EXERCISES 3.4, Page 132

1. $y = c_1e^{-x} + c_2e^{-2x} + 3$

3. $y = c_1e^{5x} + c_2xe^{5x} + \frac{6}{5}x + \frac{3}{5}$

5. $y = c_1e^{-2x} + c_2xe^{-2x} + x^2 - 4x + \frac{7}{2}$

7. $y = c_1 \cos \sqrt{3}\,x + c_2 \sin \sqrt{3}\,x$
$+ (-4x^2 + 4x - \frac{4}{3})e^{3x}$

9. $y = c_1 + c_2e^x + 3x$

11. $y = c_1e^{x/2} + c_2xe^{x/2} + 12 + \frac{1}{2}x^2e^{x/2}$

13. $y = c_1 \cos 2x + c_2 \sin 2x - \frac{3}{4}x \cos 2x$

15. $y = c_1 \cos x + c_2 \sin x - \frac{1}{2}x^2 \cos x + \frac{1}{2}x \sin x$

17. $y = c_1e^x \cos 2x + c_2e^x \sin 2x + \frac{1}{4}xe^x \sin 2x$

19. $y = c_1e^{-x} + c_2xe^{-x} - \frac{1}{2} \cos x + \frac{12}{25} \sin 2x - \frac{9}{25} \cos 2x$

21. $y = c_1 + c_2x + c_3e^{6x} - \frac{1}{4}x^2 - \frac{6}{37} \cos x + \frac{1}{37} \sin x$

23. $y = c_1e^x + c_2xe^x + c_3x^2e^x - x - 3 - \frac{2}{3}x^3e^x$

25. $y = c_1 \cos x + c_2 \sin x + c_3x \cos x$
$+ c_4x \sin x + x^2 - 2x - 3$

27. $y = \sqrt{2} \sin 2x - \frac{1}{2}$

29. $y = -200 + 200e^{-x/5} - 3x^2 + 30x$

31. $y = -10e^{-2x} \cos x + 9e^{-2x} \sin x + 7e^{-4x}$

33. $x = \dfrac{F_0}{2\omega^2} \sin \omega t - \dfrac{F_0}{2\omega} t \cos \omega t$

35. $y = 11 - 11e^x + 9xe^x + 2x - 12x^2e^x + \frac{1}{2}e^{5x}$

EXERCISES 3.5, Page 138

1. $y = c_1 \cos x + c_2 \sin x + x \sin x$
$+ \cos x \ln|\cos x|$; $(-\pi/2, \pi/2)$

3. $y = c_1 \cos x + c_2 \sin x + \frac{1}{2} \sin x - \frac{1}{2}x \cos x$
$= c_1 \cos x + c_3 \sin x - \frac{1}{2}x \cos x$; $(-\infty, \infty)$

5. $y = c_1 \cos x + c_2 \sin x + \frac{1}{2} - \frac{1}{6} \cos 2x$; $(-\infty, \infty)$

7. $y = c_1e^x + c_2e^{-x} + \frac{1}{4}xe^x - \frac{1}{4}xe^{-x}$
$= c_1e^x + c_2e^{-x} + \frac{1}{2}x \sinh x$; $(-\infty, \infty)$

9. $y = c_1e^{2x} + c_2e^{-2x} + \frac{1}{4}\left(e^{2x}\ln|x| - e^{-2x}\int_{x_0}^{x}\frac{e^{4t}}{t}dt\right)$, $x_0 > 0$; $(0, \infty)$

11. $y = c_1e^{-x} + c_2e^{-2x} + (e^{-x} + e^{-2x}) \times \ln(1 + e^x)$; $(-\infty, \infty)$

13. $y = c_1e^{-2x} + c_2e^{-x} - e^{-2x}\sin e^x$; $(-\infty, \infty)$

15. $y = c_1e^x + c_2xe^x - \frac{1}{2}e^x\ln(1 + x^2) + xe^x\tan^{-1}x$; $(-\infty, \infty)$

17. $y = c_1e^{-x} + c_2xe^{-x} + \frac{1}{2}x^2e^{-x}\ln x - \frac{3}{4}x^2e^{-x}$; $(0, \infty)$

19. $y = c_1e^x\cos 3x + c_2e^x\sin x - \frac{1}{27}e^x\cos 3x\ln|\sec 3x + \tan 3x|$; $(-\pi/6, \pi/6)$

21. $y = c_1 + c_2\cos x + c_3\sin x - \ln|\cos x| - \sin x\ln|\sec x + \tan x|$; $(-\pi/2, \pi/2)$

23. $y = c_1e^x + c_2e^{2x} + c_3e^{-x} + \frac{1}{8}e^{3x}$; $(-\infty, \infty)$

25. $y = \frac{1}{4}e^{-x/2} + \frac{3}{4}e^{x/2} + \frac{1}{8}x^2e^{x/2} - \frac{1}{4}xe^{x/2}$

27. $y = \frac{4}{9}e^{-4x} + \frac{25}{36}e^{2x} - \frac{1}{4}e^{-2x} + \frac{1}{9}e^{-x}$

29. $y = c_1x^{-1/2}\cos x + c_2x^{-1/2}\sin x + x^{-1/2}$

EXERCISES 3.6, Page 145

1. $y = c_1x^{-1} + c_2x^2$ **3.** $y = c_1 + c_2\ln x$

5. $y = c_1\cos(2\ln x) + c_2\sin(2\ln x)$

7. $y = c_1x^{(2-\sqrt{6})} + c_2x^{(2+\sqrt{6})}$

9. $y_1 = c_1\cos(\frac{1}{5}\ln x) + c_2\sin(\frac{1}{5}\ln x)$

11. $y = c_1x^{-2} + c_2x^{-2}\ln x$

13. $y = x[c_1\cos(\ln x) + c_2\sin(\ln x)]$

15. $y = x^{-1/2}\left[c_1\cos\left(\frac{\sqrt{3}}{6}\ln x\right) + c_2\sin\left(\frac{\sqrt{3}}{6}\ln x\right)\right]$

17. $y = c_1x^3 + c_2\cos(\sqrt{2}\ln x) + c_3\sin(\sqrt{2}\ln x)$

19. $y = c_1x^{-1} + c_2x^2 + c_3x^4$

21. $y = c_1 + c_2x + c_3x^2 + c_4x^{-3}$

23. $y = 2 - 2x^{-2}$

25. $y = \cos(\ln x) + 2\sin(\ln x)$

27. $y = 2(-x)^{1/2} - 5(-x)^{1/2}\ln(-x)$

29. $y = c_1 + c_2\ln x + \frac{x^2}{4}$

31. $y = c_1x^{-1/2} + c_2x^{-1} + \frac{1}{15}x^2 - \frac{1}{6}x$

33. $y = c_1x + c_2x\ln x + x(\ln x)^2$

35. $y = c_1x^{-1} + c_2x^{-8} + \frac{1}{30}x^2$

37. $y = x^2[c_1\cos(3\ln x) + c_2\sin(3\ln x)] + \frac{4}{13} + \frac{3}{10}x$

39. $y = c_1x^2 + c_2x^{-10} - \frac{1}{7}x^{-3}$

EXERCISES 3.7, Page 161

1. $\sqrt{2}\,\pi/8$ **3.** $x(t) = -\frac{1}{4}\cos 4\sqrt{6}\,t$

5. **(a)** $x(\pi/12) = -1/4$; $x(\pi/8) = -1/2$; $x(\pi/6) = -1/4$; $x(\pi/4) = 1/2$; $x(9\pi/32) = \sqrt{2}/4$

(b) 4 ft/s; downward

(c) $t = (2n + 1)\pi/16$, $n = 0, 1, 2, \ldots$

7. **(a)** the 20-kg mass

(b) the 20-kg mass; the 50-kg mass

(c) $t = n\pi$, $n = 0, 1, 2, \ldots$; at the equilibrium position; the 50-kg mass is moving upward whereas the 20-kg mass is moving upward when n is even and downward when n is odd.

9. $x(t) = \frac{1}{2}\cos 2t + \frac{3}{4}\sin 2t = \frac{\sqrt{13}}{4}\sin(2t + 0.5880)$

11. **(a)** $x(t) = -\frac{2}{3}\cos 10t + \frac{1}{2}\sin 10t = \frac{5}{6}\sin(10t - 0.927)$

(b) 5/6 ft; $\pi/5$

(c) 15 cycles

(d) 0.721 s

(e) $(2n + 1)\pi/20 + 0.0927$, $n = 0, 1, 2, \ldots$

(f) $x(3) = -0.597$ ft

(g) $x'(3) = -5.814$ ft/s

(h) $x''(3) = 59.702$ ft/s^2

(i) $\pm 8\frac{1}{3}$ ft/s

(j) $0.1451 + n\pi/5$; $0.3545 + n\pi/5$, $n = 0, 1, 2, \ldots$

(k) $0.3545 + n\pi/5$, $n = 0, 1, 2, \ldots$

13. 120 lb/ft; $x(t) = \frac{\sqrt{3}}{12}\sin 8\sqrt{3}\,t$

15. **(a)** above; **(b)** heading upward

17. **(a)** below; **(b)** heading upward

19. $\frac{1}{4}$s; $\frac{1}{2}$s, $x(\frac{1}{2}) = e^{-2}$; that is, the weight is approximately 0.14 ft below the equilibrium position.

21. **(a)** $x(t) = \frac{4}{3}e^{-2t} - \frac{1}{3}e^{-8t}$

(b) $x(t) = -\frac{2}{3}e^{-2t} + \frac{5}{3}e^{-8t}$

23. **(a)** $x(t) = e^{-2t}[-\cos 4t - \frac{1}{2}\sin 4t]$

(b) $x(t) = \dfrac{\sqrt{5}}{2}e^{-2t}\sin(4t + 4.249)$

(c) $t = 1.294$ s

25. **(a)** $\beta > 5/2$; **(b)** $\beta = 5/2$; **(c)** $0 < \beta < 5/2$

27. $$x(t) = e^{-t/2}\left(-\frac{4}{3}\cos\frac{\sqrt{47}}{2}t - \frac{64}{3\sqrt{47}}\sin\frac{\sqrt{47}}{2}t\right) + \frac{10}{3}(\cos 3t + \sin 3t)$$

29. $x(t) = \frac{1}{4}e^{-4t} + te^{-4t} - \frac{1}{4}\cos 4t$

31. $x(t) = -\frac{1}{2}\cos 4t + \frac{9}{4}\sin 4t + \frac{1}{2}e^{-2t}\cos 4t - 2e^{-2t}\sin 4t$

33. **(a)** $m\dfrac{d^2x}{dt^2} = -k(x-h) - \beta\dfrac{dx}{dt}$ or

$$\frac{d^2x}{dt^2} + 2\lambda\frac{dx}{dt} + \omega^2 x = \omega^2 h(t), \quad \text{where}$$

$2\lambda = \beta/m$ and $\omega^2 = k/m$

(b) $x(t) = e^{-2t}(-\frac{56}{13}\cos 2t - \frac{72}{13}\sin 2t) + \frac{56}{13}\cos t + \frac{32}{13}\sin t$

39. 4.568 coulombs; 0.0509 s

41. $q(t) = 10 - 10e^{-3t}(\cos 3t + \sin 3t)$;
$i(t) = 60e^{-3t}\sin 3t$; 10.432 coulombs

43. $q_p = \frac{100}{13}\sin t + \frac{150}{13}\cos t$
$i_p = \frac{100}{13}\cos t - \frac{150}{13}\sin t$

47. $q(t) = -\frac{1}{2}e^{-10t}(\cos 10t + \sin 10t) + \frac{3}{2}$; $\frac{3}{2}$ coulombs

49. Show that $dZ/dC = 0$ when $C = 1/L\gamma^2$. At this value, Z is a minimum and, correspondingly, the amplitude E_0/Z is a maximum.

51. $$q(t) = \left(q_0 - \frac{E_0C}{1-\gamma^2LC}\right)\cos\frac{t}{\sqrt{LC}} + \sqrt{LC}i_0\sin\frac{t}{\sqrt{LC}} + \frac{E_0C}{1-\gamma^2LC}\cos\gamma t;$$

$$i(t) = i_0\cos\frac{t}{\sqrt{LC}} - \frac{1}{\sqrt{LC}}\left(q_0 - \frac{E_0C}{1-\gamma^2LC}\right)\sin\frac{t}{\sqrt{LC}} - \frac{E_0C\gamma}{1-\gamma^2LC}\sin\gamma t$$

EXERCISES 3.8, Page 172

1. **(a)** $y(x) = \dfrac{w_0}{24EI}(6L^2x^2 - 4Lx^3 + x^4)$

(b)

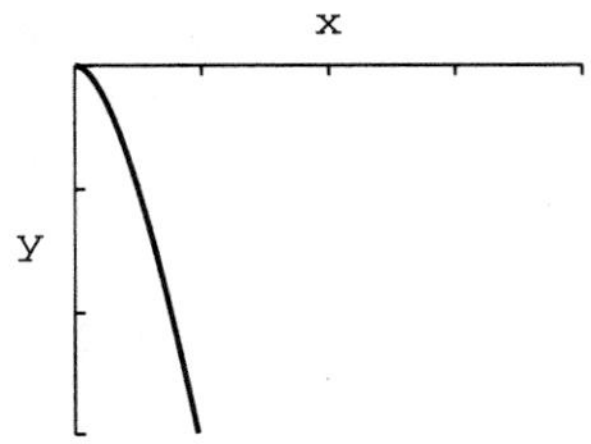

3. **(a)** $y(x) = \dfrac{w_0}{48EI}(3L^2x^2 - 5Lx^3 + 2x^4)$

(b)

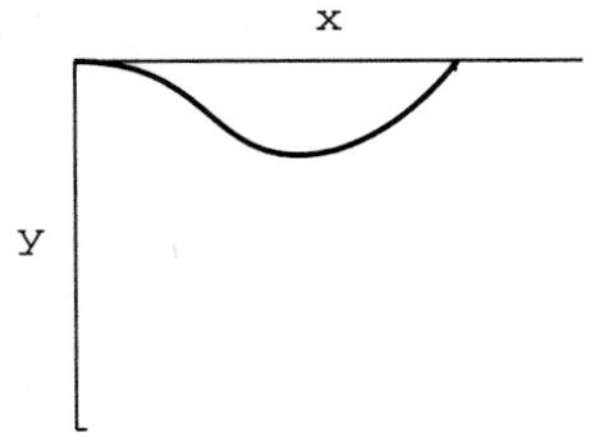

5. **(a)** $y_{\max} = \dfrac{w_0L^4}{8EI}$

(b) $\frac{1}{16}$th of the maximum deflection in part (a)

7. $$y(x) = -\frac{w_0EI}{P^2}\cosh\sqrt{\frac{P}{EI}}x + \left(\frac{w_0EI}{P^2}\sinh\sqrt{\frac{P}{EI}}L - \frac{w_0L\sqrt{EI}}{P\sqrt{P}}\right)\frac{\sinh\sqrt{\frac{P}{EI}}x}{\cosh\sqrt{\frac{P}{EI}}L} + \frac{w_0}{2P}x^2 + \frac{w_0EI}{P^2}$$

9. $\lambda = n^2, n = 1, 2, 3, \ldots$;
$y = \sin nx$

11. $\lambda = (2n-1)^2\pi^2/4L^2, n = 1, 2, 3, \ldots$;
$y = \cos((2n-1)\pi x/2L)$

13. $\lambda = n^2, n = 0, 1, 2, \ldots$;
$y = \cos nx$

15. $\lambda = n^2\pi^2/25, n = 1, 2, 3, \ldots$;
$y = e^{-x}\sin(n\pi x/5)$

17. $\lambda = n\pi/L, n = 1, 2, 3, \ldots$;
$y = \sin(n\pi x/L)$

19. $\lambda = n^2, n = 1, 2, 3, \ldots$;
$y = \sin(n \ln x)$

21. $\lambda = 0$; $y = 1$ and for
$\lambda = n^2\pi^2/4, n = 1, 2, 3, \ldots$;
$y = \cos((n\pi/2)\ln x)$

23. **(b)** From $\lambda_n = x_n^2$ we see no new eigenvalues result when x_n is negative. For $\lambda = 0$, the family of solutions for $y'' = 0$ is $y = c_1x + c_2$. The only member of this family that satisfies the given boundary conditions is $y = 0$.

(c) $\lambda_1 = 4.1159$, $\lambda_2 = 24.1393$, $\lambda_3 = 63.9652$, $\lambda_4 = 122.8883$

25. $\omega_n = n\pi\sqrt{T}/L\sqrt{\rho}$, $n = 1, 2, 3, \ldots$; $y = \sin(n\pi x/L)$

27. $u(r) = \left(\dfrac{u_0 - u_1}{b - a}\right)\dfrac{ab}{r} + \dfrac{u_1 b - u_0 a}{b - a}$

EXERCISES 3.9, Page 181

1. $x = c_1e^t + c_2te^t$
$y = (c_1 - c_2)e^t + c_2te^t$

3. $x = c_1 \cos t + c_2 \sin t + t + 1$
$y = c_1 \sin t - c_2 \cos t + t - 1$

5. $x = \frac{1}{2}c_1 \sin t + \frac{1}{2}c_2 \cos t - 2c_3 \sin\sqrt{6}t - 2c_4 \cos\sqrt{6}t$
$y = c_1 \sin t + c_2 \cos t + c_3 \sin\sqrt{6}t + c_4 \cos\sqrt{6}t$

7. $x = c_1e^{2t} + c_2e^{-2t} + c_3 \sin 2t + c_4 \cos 2t + \frac{1}{5}e^t$
$y = c_1e^{2t} + c_2e^{-2t} - c_3 \sin 2t - c_4 \cos 2t - \frac{1}{5}e^t$

9. $x = c_1 - c_2 \cos t + c_3 \sin t + \frac{17}{15}e^{3t}$
$y = c_1 + c_2 \sin t + c_3 \cos t - \frac{4}{15}e^{3t}$

11. $x = c_1e^t + c_2e^{-t/2} \cos\dfrac{\sqrt{3}}{2}t + c_3e^{-t/2} \sin\dfrac{\sqrt{3}}{2}t$

$$y = \left(-\frac{3}{2}c_2 - \frac{\sqrt{3}}{2}c_3\right)e^{-t/2} \cos\frac{\sqrt{3}}{2}t + \left(\frac{\sqrt{3}}{2}c_2 - \frac{3}{2}c_3\right)e^{-t/2} \sin\frac{\sqrt{3}}{2}t$$

13. $x = c_1e^{4t} + \frac{4}{3}e^t$
$y = -\frac{3}{4}c_1e^{4t} + c_2 + 5e^t$

15. $x = c_1 + c_2t + c_3e^t + c_4e^{-t} - \frac{1}{2}t^2$
$y = (c_1 - c_2 + 2) + (c_2 + 1)t + c_4e^{-t} - \frac{1}{2}t^2$

17. $x = c_1e^t + c_2e^{-t/2} \sin\dfrac{\sqrt{3}}{2}t + c_3e^{-t/2} \cos\dfrac{\sqrt{3}}{2}t$

$$y = c_1e^t + \left(-\frac{1}{2}c_2 - \frac{\sqrt{3}}{2}c_3\right)e^{-t/2} \sin\frac{\sqrt{3}}{2}t + \left(\frac{\sqrt{3}}{2}c_2 - \frac{1}{2}c_3\right)e^{-t/2} \cos\frac{\sqrt{3}}{2}t$$

$$z = c_1e^t + \left(-\frac{1}{2}c_2 + \frac{\sqrt{3}}{2}c_3\right)e^{-t/2} \sin\frac{\sqrt{3}}{2}t + \left(-\frac{\sqrt{3}}{2}c_2 - \frac{1}{2}c_3\right)e^{-t/2} \cos\frac{\sqrt{3}}{2}t$$

19. $x = -6c_1e^{-t} - 3c_2e^{-2t} + 2c_3e^{3t}$
$y = c_1e^{-t} + c_2e^{-2t} + c_3e^{3t}$
$z = 5c_1e^{-t} + c_2e^{-2t} + c_3e^{3t}$

21. $x = -c_1e^{-t} + c_2 + \frac{1}{3}t^3 - 2t^2 + 5t$
$y = c_1e^{-t} + 2t^2 - 5t + 5$

23. $x = e^{-3t+3} - te^{-3t+3}$
$y = -e^{-3t+3} + 2te^{-3t+3}$

EXERCISES 3.10, Page 187

1. $y(x) = -2e^{2x} + 5xe^{2x}$; $y(0.2) = -1.4918$, $y_2 = -1.6800$

3. $y_1 = -1.4928$, $y_2 = -1.4919$

5. $y_1 = 1.4640$, $y_2 = 1.4640$

7. $x_1 = 8.3055$, $y_1 = 3.4199$;
$x_2 = 8.3055$, $y_2 = 3.4199$

9. $x_1 = -3.9123$, $y_1 = 4.2857$;
$x_2 = -3.9123$, $y_2 = 4.2857$

11. $x_1 = 0.4179$, $y_1 = -2.1824$;
$x_2 = 0.4173$, $y_2 = -2.1821$

EXERCISES 3.11, Page 193

1. **(a)**

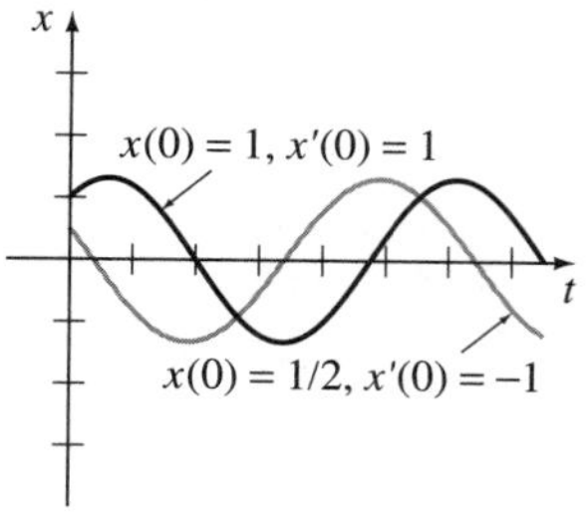

For the first IVP the period T is approximately 6; for the second IVP the period T is approximately 6.3.

(b) (0, 0)

(c)

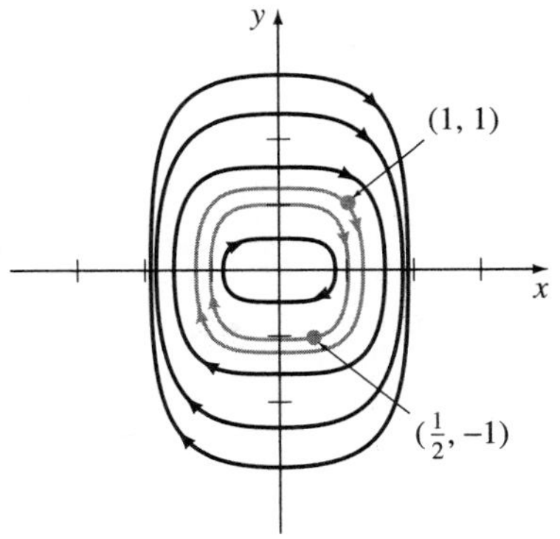

(0, 0) is a (stable) center

3. (a)

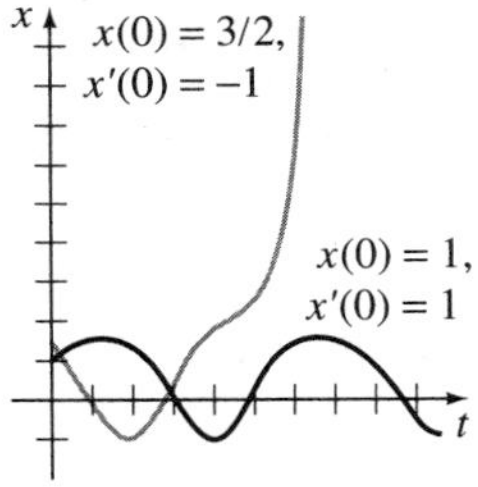

For the first IVP the period T is approximately 6; the solution of the second IVP appears not to be periodic.

(b) (0, 0), (2, 0)

(c)

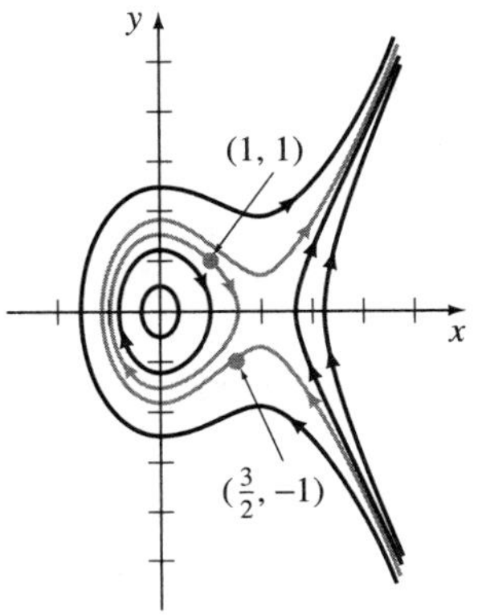

(0, 0) is a (stable) center;
(2, 0) is a (unstable) saddle point.

5. (a) $|x_1| \approx 1.2$

(b) $-0.8 \leq x \leq 1.1$

7. $\dfrac{d^2x}{dt^2} + x = 0$

9. (a) expect $x \to 0$ as $t \to \infty$

(b)

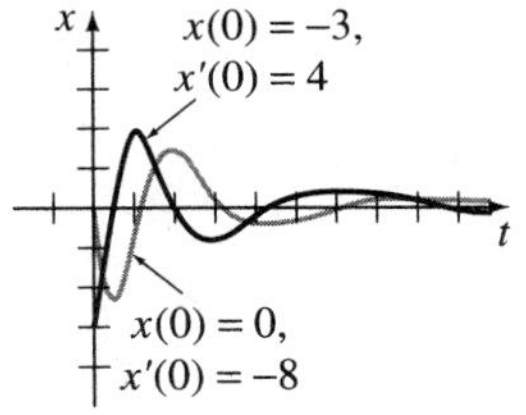

(c) (0, 0)

(d)

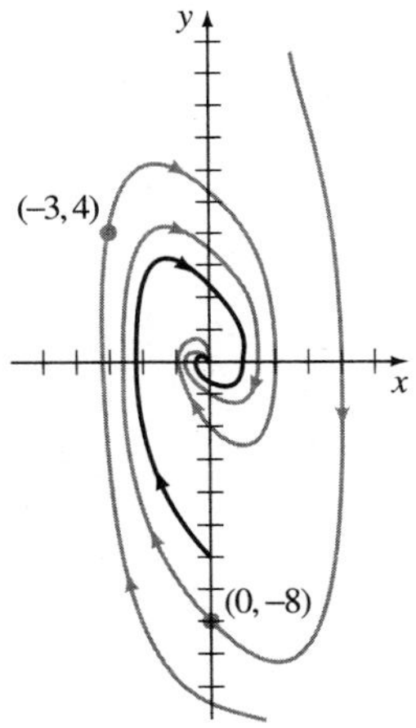

(0, 0) is an asymptotically stable spiral point.

11. The system is always underdamped.
(0, 0) is an asymptotically stable spiral point.

13. (c)

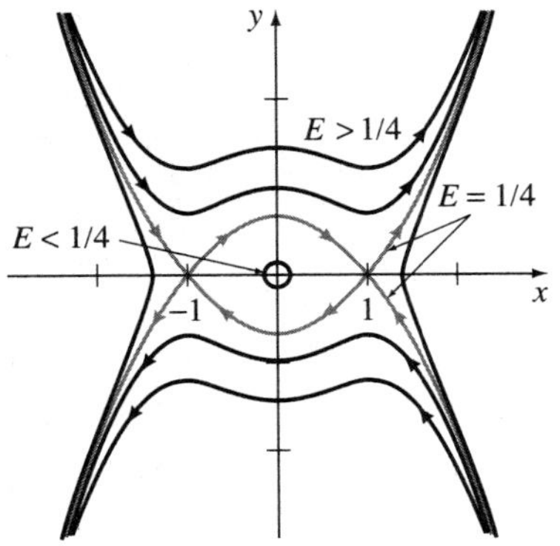

EXERCISES 4.1, Page 205

1. $\mathbf{X}' = \begin{pmatrix} 3 & -5 \\ 4 & 8 \end{pmatrix}\mathbf{X}$, where $\mathbf{X} = \begin{pmatrix} x \\ y \end{pmatrix}$

3. $\mathbf{X}' = \begin{pmatrix} -3 & 4 & -9 \\ 6 & -1 & 0 \\ 10 & 4 & 3 \end{pmatrix}\mathbf{X}$, where $\mathbf{X} = \begin{pmatrix} x \\ y \\ z \end{pmatrix}$

5. $\mathbf{X}' = \begin{pmatrix} 1 & -1 & 1 \\ 2 & 1 & -1 \\ 1 & 1 & 1 \end{pmatrix}\mathbf{X} + \begin{pmatrix} 0 \\ -3t^2 \\ t^2 \end{pmatrix} + \begin{pmatrix} t \\ 0 \\ -t \end{pmatrix}$

$+ \begin{pmatrix} -1 \\ 0 \\ 2 \end{pmatrix}$, where $\mathbf{X} = \begin{pmatrix} x \\ y \\ z \end{pmatrix}$

7. $\dfrac{dx}{dt} = 4x + 2y + e^t$

$\dfrac{dy}{dt} = -x + 3y - e^t$

9. $\dfrac{dx}{dt} = x - y + 2z + e^{-t} - 3t$

$\dfrac{dy}{dt} = 3x - 4y + z + 2e^{-t} + t$

$\dfrac{dz}{dt} = -2x + 5y + 6z + 2e^{-t} - t$

11. $\mathbf{X}' = \begin{pmatrix} -5 \\ -10 \end{pmatrix} e^{-5t}$

$\begin{pmatrix} 3 & -4 \\ 4 & -7 \end{pmatrix}\mathbf{X} = \begin{pmatrix} 3 - 8 \\ 4 - 14 \end{pmatrix} e^{-5t} = \begin{pmatrix} -5 \\ -10 \end{pmatrix} e^{-5t} = \mathbf{X}'$

13. $\mathbf{X}' = \begin{pmatrix} \frac{3}{2} \\ -3 \end{pmatrix} e^{-3t/2}$

$\begin{pmatrix} -1 & \frac{1}{4} \\ 1 & -1 \end{pmatrix}\mathbf{X} = \begin{pmatrix} 1 + \frac{1}{2} \\ -1 - 2 \end{pmatrix} e^{-3t/2} = \begin{pmatrix} \frac{3}{2} \\ -3 \end{pmatrix} e^{-3t/2} = \mathbf{X}'$

15. $\dfrac{d\mathbf{X}}{dt} = \begin{pmatrix} 0 \\ 0 \\ 0 \end{pmatrix}$

$\begin{pmatrix} 1 & 2 & 1 \\ 6 & -1 & 0 \\ -1 & -2 & -1 \end{pmatrix}\mathbf{X} = \begin{pmatrix} 1 + 12 - 13 \\ 6 - 6 \\ -1 - 12 + 13 \end{pmatrix} = \begin{pmatrix} 0 \\ 0 \\ 0 \end{pmatrix} = \dfrac{d\mathbf{X}}{dt}$

17. Yes; $W(\mathbf{X}_1, \mathbf{X}_2) = -2e^{-8t} \neq 0$ implies $\mathbf{X}_1$ and $\mathbf{X}_2$ are linearly independent on $(-\infty, \infty)$.

19. No; $W(\mathbf{X}_1, \mathbf{X}_2, \mathbf{X}_3) =$

$$\begin{vmatrix} 1 + t & 1 & 3 + 2t \\ -2 + 2t & -2 & -6 + 4t \\ 4 + 2t & 4 & 12 + 4t \end{vmatrix} = 0 \text{ for every } t.$$

The solution vectors are linearly dependent on $(-\infty, \infty)$. Note that $\mathbf{X}_3 = 2\mathbf{X}_1 + \mathbf{X}_2$.

21. $\dfrac{d\mathbf{X}_p}{dt} = \begin{pmatrix} 2 \\ -1 \end{pmatrix}$

$\begin{pmatrix} 1 & 4 \\ 3 & 2 \end{pmatrix}\mathbf{X}_p + \begin{pmatrix} 2 \\ -4 \end{pmatrix} t - \begin{pmatrix} 7 \\ 18 \end{pmatrix}$

$= \begin{pmatrix} (2 - 4)t + 9 + 2t - 7 \\ (6 - 2)t + 17 - 4t - 18 \end{pmatrix} = \begin{pmatrix} 2 \\ -1 \end{pmatrix} = \dfrac{d\mathbf{X}_p}{dt}$

23. $\mathbf{X}_p' = \begin{pmatrix} 2e^t + te^t \\ -te^t \end{pmatrix}$

$\begin{pmatrix} 2 & 1 \\ 3 & 4 \end{pmatrix}\mathbf{X}_p - \begin{pmatrix} 1 \\ 7 \end{pmatrix} e^t = \begin{pmatrix} 3e^t + te^t - e^t \\ 7e^t - te^t - 7e^t \end{pmatrix}$

$= \begin{pmatrix} 2e^t + te^t \\ -te^t \end{pmatrix} = \mathbf{X}_p'$

25. Let $\mathbf{X}_1 = \begin{pmatrix} 6 \\ -1 \\ -5 \end{pmatrix} e^{-t}$, $\mathbf{X}_2 = \begin{pmatrix} -3 \\ 1 \\ 1 \end{pmatrix} e^{-2t}$,

$\mathbf{X}_3 = \begin{pmatrix} 2 \\ 1 \\ 1 \end{pmatrix} e^{3t}$, and $\mathbf{A} = \begin{pmatrix} 0 & 6 & 0 \\ 1 & 0 & 1 \\ 1 & 1 & 0 \end{pmatrix}$. Then

$\mathbf{A}\mathbf{X}_1 = \begin{pmatrix} -6 \\ 1 \\ 5 \end{pmatrix} e^{-t} = \mathbf{X}_1'$,

$\mathbf{A}\mathbf{X}_2 = \begin{pmatrix} 6 \\ -2 \\ -2 \end{pmatrix} e^{-2t} = \mathbf{X}_2'$,

$\mathbf{A}\mathbf{X}_3 = \begin{pmatrix} 6 \\ 3 \\ 3 \end{pmatrix} e^{3t} = \mathbf{X}_3'$, and

$$W(\mathbf{X}_1, \mathbf{X}_2, \mathbf{X}_3) = \begin{vmatrix} 6e^{-t} & -3e^{-2t} & 2e^{3t} \\ -e^{-t} & e^{-2t} & e^{3t} \\ -5e^{-t} & e^{-2t} & e^{3t} \end{vmatrix} = 20 \neq 0.$$

Therefore, $\mathbf{X}_1, \mathbf{X}_2, \mathbf{X}_3$ form a fundamental set of solutions of $\mathbf{X}' = \mathbf{A}\mathbf{X}$ on $(-\infty, \infty)$. By definition

$$\mathbf{X} = c_1\mathbf{X}_1 + c_2\mathbf{X}_2 + c_3\mathbf{X}_3$$

is the general solution.

EXERCISES 4.2, Page 218

1. $\mathbf{X} = c_1\begin{pmatrix}1\\2\end{pmatrix}e^{5t} + c_2\begin{pmatrix}1\\-1\end{pmatrix}e^{-t}$

3. $\mathbf{X} = c_1\begin{pmatrix}2\\1\end{pmatrix}e^{-3t} + c_2\begin{pmatrix}2\\5\end{pmatrix}e^{t}$

5. $\mathbf{X} = c_1\begin{pmatrix}5\\2\end{pmatrix}e^{8t} + c_2\begin{pmatrix}1\\4\end{pmatrix}e^{-10t}$

7. $\mathbf{X} = c_1\begin{pmatrix}1\\0\\0\end{pmatrix}e^{t} + c_2\begin{pmatrix}2\\3\\1\end{pmatrix}e^{2t} + c_3\begin{pmatrix}1\\0\\2\end{pmatrix}e^{-t}$

9. $\mathbf{X} = c_1\begin{pmatrix}-1\\0\\1\end{pmatrix}e^{-t} + c_2\begin{pmatrix}1\\4\\3\end{pmatrix}e^{3t} + c_3\begin{pmatrix}1\\-1\\3\end{pmatrix}e^{-2t}$

11. $\mathbf{X} = c_1\begin{pmatrix}4\\0\\-1\end{pmatrix}e^{-t} + c_2\begin{pmatrix}-12\\6\\5\end{pmatrix}e^{-t/2} + c_3\begin{pmatrix}4\\2\\-1\end{pmatrix}e^{-3t/2}$

13. $\mathbf{X} = 3\begin{pmatrix}1\\1\end{pmatrix}e^{t/2} + 2\begin{pmatrix}0\\1\end{pmatrix}e^{-t/2}$

15. $\mathbf{X} = c_1\begin{pmatrix}1\\3\end{pmatrix} + c_2\left\{\begin{pmatrix}1\\3\end{pmatrix}t + \begin{pmatrix}\frac{1}{4}\\-\frac{1}{4}\end{pmatrix}\right\}$

17. $\mathbf{X} = c_1\begin{pmatrix}1\\1\end{pmatrix}e^{2t} + c_2\left\{\begin{pmatrix}1\\1\end{pmatrix}te^{2t} + \begin{pmatrix}-\frac{1}{3}\\0\end{pmatrix}e^{2t}\right\}$

19. $\mathbf{X} = c_1\begin{pmatrix}1\\1\\1\end{pmatrix}e^{t} + c_2\begin{pmatrix}1\\1\\0\end{pmatrix}e^{2t} + c_3\begin{pmatrix}1\\0\\1\end{pmatrix}e^{2t}$

21. $\mathbf{X} = c_1\begin{pmatrix}-4\\-5\\2\end{pmatrix} + c_2\begin{pmatrix}2\\0\\-1\end{pmatrix}e^{5t}$
$+ c_3\left\{\begin{pmatrix}2\\0\\-1\end{pmatrix}te^{5t} + \begin{pmatrix}-\frac{1}{2}\\-\frac{1}{2}\\-1\end{pmatrix}e^{5t}\right\}$

23. $\mathbf{X} = c_1\begin{pmatrix}0\\1\\1\end{pmatrix}e^{t} + c_2\left\{\begin{pmatrix}0\\1\\1\end{pmatrix}te^{t} + \begin{pmatrix}0\\1\\0\end{pmatrix}e^{t}\right\}$
$+ c_3\left\{\begin{pmatrix}0\\1\\1\end{pmatrix}\frac{t^2}{2}e^{t} + \begin{pmatrix}0\\1\\0\end{pmatrix}te^{t} + \begin{pmatrix}\frac{1}{2}\\0\\0\end{pmatrix}e^{t}\right\}$

25. $\mathbf{X} = -7\begin{pmatrix}2\\1\end{pmatrix}e^{4t} + 13\begin{pmatrix}2t+1\\t+1\end{pmatrix}e^{4t}$

27. $\mathbf{X} = c_1\begin{pmatrix}\cos t\\2\cos t + \sin t\end{pmatrix}e^{4t} + c_2\begin{pmatrix}\sin t\\2\sin t - \cos t\end{pmatrix}e^{4t}$

29. $\mathbf{X} = c_1\begin{pmatrix}\cos t\\-\cos t - \sin t\end{pmatrix}e^{4t} + c_2\begin{pmatrix}\sin t\\-\sin t + \cos t\end{pmatrix}e^{4t}$

31. $\mathbf{X} = c_1\begin{pmatrix}5\cos 3t\\4\cos 3t + 3\sin 3t\end{pmatrix} + c_2\begin{pmatrix}5\sin 3t\\4\sin 3t - 3\cos 3t\end{pmatrix}$

33. $\mathbf{X} = c_1\begin{pmatrix}1\\0\\0\end{pmatrix} + c_2\begin{pmatrix}-\cos t\\\cos t\\\sin t\end{pmatrix} + c_3\begin{pmatrix}\sin t\\-\sin t\\\cos t\end{pmatrix}$

35. $\mathbf{X} = c_1\begin{pmatrix}0\\2\\1\end{pmatrix}e^{t} + c_2\begin{pmatrix}\sin t\\\cos t\\\cos t\end{pmatrix}e^{t} + c_3\begin{pmatrix}\cos t\\-\sin t\\-\sin t\end{pmatrix}e^{t}$

37. $\mathbf{X} = \begin{pmatrix}28\\-5\\25\end{pmatrix}e^{2t} + c_2\begin{pmatrix}5\cos 3t\\-4\cos 3t - 3\sin 3t\\0\end{pmatrix}e^{-2t}$
$+ c_3\begin{pmatrix}5\sin 3t\\-4\sin 3t + 3\cos 3t\\0\end{pmatrix}e^{-2t}$

39. $\mathbf{X} = -\begin{pmatrix}25\\-7\\6\end{pmatrix}e^{t} - \begin{pmatrix}\cos 5t - 5\sin 5t\\\cos 5t\\\cos 5t\end{pmatrix}$
$+ 6\begin{pmatrix}5\cos 5t + \sin 5t\\\sin 5t\\\sin 5t\end{pmatrix}$

EXERCISES 4.3, Page 223

1. $\mathbf{X} = c_1\begin{pmatrix}1\\1\end{pmatrix} + c_2\begin{pmatrix}3\\2\end{pmatrix}e^{t} - \begin{pmatrix}11\\11\end{pmatrix}t - \begin{pmatrix}15\\10\end{pmatrix}$

3. $\mathbf{X} = c_1\begin{pmatrix}2\\1\end{pmatrix}e^{t/2} + c_2\begin{pmatrix}10\\3\end{pmatrix}e^{3t/2} - \begin{pmatrix}\frac{13}{2}\\\frac{13}{4}\end{pmatrix}te^{t/2} - \begin{pmatrix}\frac{15}{2}\\\frac{9}{4}\end{pmatrix}e^{t/2}$

5. $\mathbf{X} = c_1\begin{pmatrix}2\\1\end{pmatrix}e^{t} + c_2\begin{pmatrix}1\\1\end{pmatrix}e^{2t} + \begin{pmatrix}3\\3\end{pmatrix}e^{t} + \begin{pmatrix}4\\2\end{pmatrix}te^{t}$

7. $\mathbf{X} = c_1\begin{pmatrix}4\\1\end{pmatrix}e^{3t} + c_2\begin{pmatrix}-2\\1\end{pmatrix}e^{-3t} + \begin{pmatrix}-12\\0\end{pmatrix}t - \begin{pmatrix}\frac{4}{3}\\\frac{4}{3}\end{pmatrix}$

9. $\mathbf{X} = c_1\begin{pmatrix}1\\-1\end{pmatrix}e^{t} + c_2\begin{pmatrix}-t\\\frac{1}{2}-t\end{pmatrix}e^{t} + \begin{pmatrix}\frac{1}{2}\\-2\end{pmatrix}e^{-t}$

11. $\mathbf{X} = c_1\begin{pmatrix}\cos t\\\sin t\end{pmatrix} + c_2\begin{pmatrix}\sin t\\-\cos t\end{pmatrix} + \begin{pmatrix}\cos t\\\sin t\end{pmatrix}t + \begin{pmatrix}-\sin t\\\cos t\end{pmatrix}\ln|\cos t|$

13. $\mathbf{X} = c_1\begin{pmatrix}\cos t\\\sin t\end{pmatrix}e^{t} + c_2\begin{pmatrix}\sin t\\-\cos t\end{pmatrix}e^{t} + \begin{pmatrix}\cos t\\\sin t\end{pmatrix}te^{t}$

15. $\mathbf{X} = c_1\begin{pmatrix}\cos t\\-\sin t\end{pmatrix} + c_2\begin{pmatrix}\sin t\\\cos t\end{pmatrix} + \begin{pmatrix}\cos t\\-\sin t\end{pmatrix}t + \begin{pmatrix}-\sin t\\\sin t\tan t\end{pmatrix} - \begin{pmatrix}\sin t\\\cos t\end{pmatrix}\ln|\cos t|$

17. $\mathbf{X} = c_1\begin{pmatrix}2\sin t\\\cos t\end{pmatrix}e^{t} + c_2\begin{pmatrix}2\cos t\\-\sin t\end{pmatrix}e^{t} + \begin{pmatrix}3\sin t\\\frac{3}{2}\cos t\end{pmatrix}te^{t} + \begin{pmatrix}\cos t\\-\frac{1}{2}\sin t\end{pmatrix}e^{t}\ln|\sin t| + \begin{pmatrix}2\cos t\\-\sin t\end{pmatrix}e^{t}\ln|\cos t|$

19. $\mathbf{X} = c_1\begin{pmatrix}1\\-1\\0\end{pmatrix} + c_2\begin{pmatrix}1\\1\\0\end{pmatrix}e^{2t} + c_3\begin{pmatrix}0\\0\\1\end{pmatrix}e^{3t} + \begin{pmatrix}-\frac{1}{4}e^{2t} + \frac{1}{2}te^{2t}\\-e^{t} + \frac{1}{4}e^{2t} + \frac{1}{2}te^{2t}\\\frac{1}{2}t^2e^{3t}\end{pmatrix}$

21. $\mathbf{X} = \begin{pmatrix}2\\2\end{pmatrix}te^{2t} + \begin{pmatrix}-1\\1\end{pmatrix}e^{2t} + \begin{pmatrix}-2\\2\end{pmatrix}te^{4t} + \begin{pmatrix}2\\0\end{pmatrix}e^{4t}$

23. (b) $\begin{pmatrix}i_1\\i_2\end{pmatrix} = 2\begin{pmatrix}1\\3\end{pmatrix}e^{-2t} + \frac{6}{29}\begin{pmatrix}3\\-1\end{pmatrix}e^{-12t} + \begin{pmatrix}\frac{332}{29}\\\frac{276}{29}\end{pmatrix}\sin t - \begin{pmatrix}\frac{76}{29}\\\frac{168}{29}\end{pmatrix}\cos t$

EXERCISES 4.4, Page 241

1. (0, 0) is an unstable degenerate node.

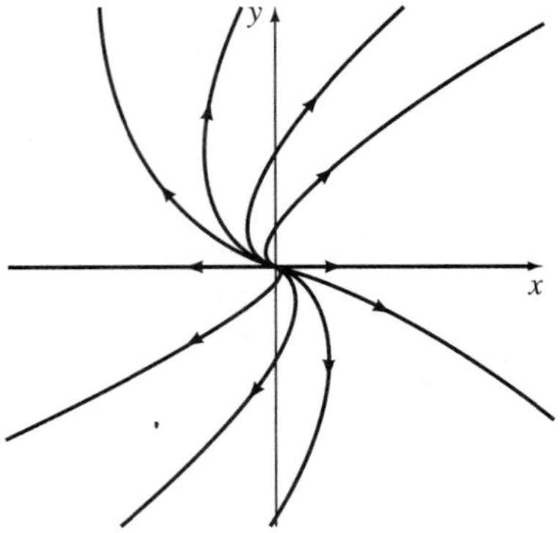

3. (0, 0) is an asymptotically stable node.

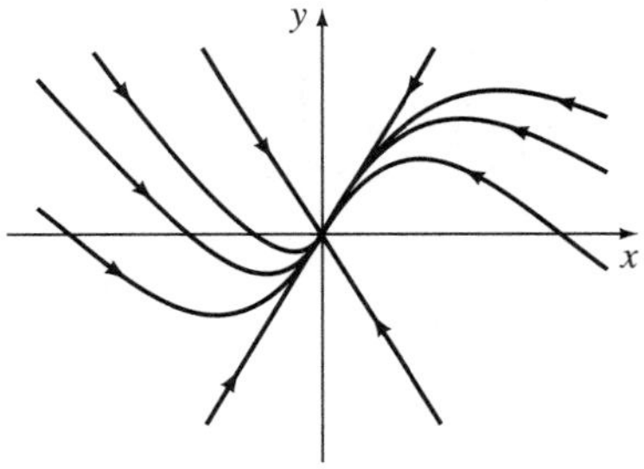

5. (0, 0) is a center.

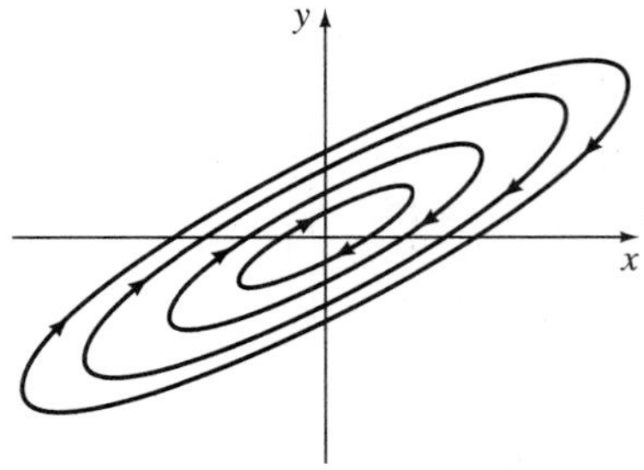

7. (0, 0) is an unstable spiral point.

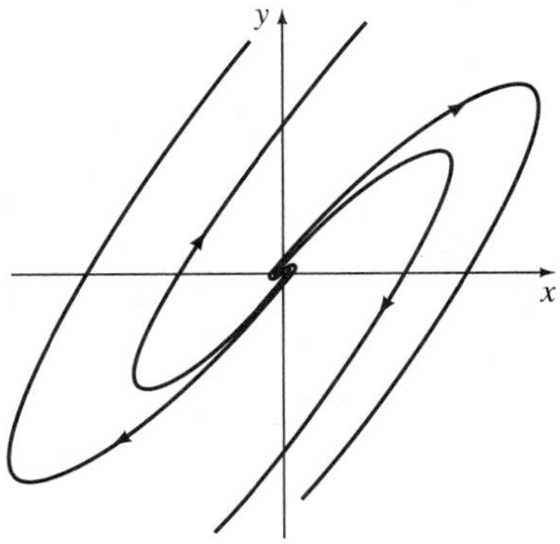

9. $(-3, 4)$ is a saddle point.

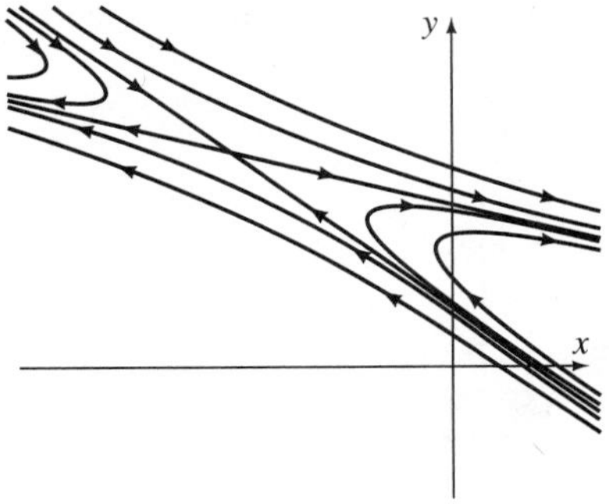

11. $(\frac{1}{2}, 2)$ is an unstable spiral point.

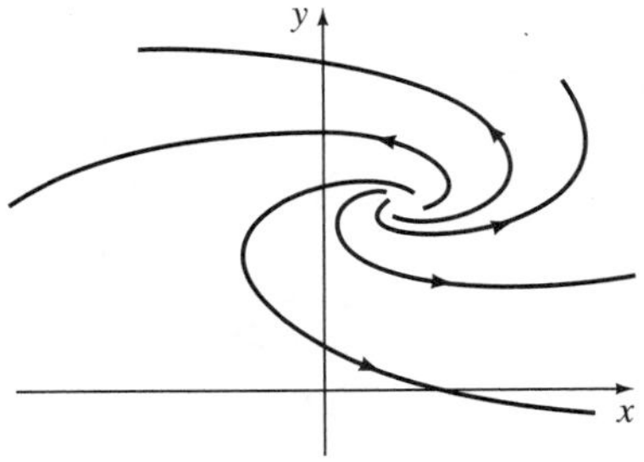

13. **(a)** $c < 0$ **(b)** $0 < c < 1$ **(c)** no values of c
(d) $c = 0$ **(e)** $c > 1$

15. $-3 < a < -1$

17. **(a)** Since λ_1 and λ_2 are roots of the characteristic equation, we have the identity

$$\begin{aligned}\lambda^2 - (a + d)\lambda + ad - bc &= (\lambda - \lambda_1)(\lambda - \lambda_2)\\ &= \lambda^2 - (\lambda_1 + \lambda_2)\lambda + \lambda_1\lambda_2\end{aligned}$$

and so $a + d = \lambda_1 + \lambda_2$, $ad - bc = \lambda_1\lambda_2$.

(b) $\det \mathbf{A} = ad - bc = \lambda_1\lambda_2$, so $\det \mathbf{A} < 0$ if and only if λ_1 and λ_2 have opposite signs. Therefore (0, 0) is a saddle point. If $\det \mathbf{A} > 0$ then (0, 0) could be a node, spiral point, or a center.

19. (0, 0) is an unstable node.

21. (0, 0) is an unstable spiral point.

23. (0, 0) is a saddle point.

25. (0, 0) is an unstable node.

27. (1, 1) is an unstable spiral point; $(-1, 1)$ is a saddle point.

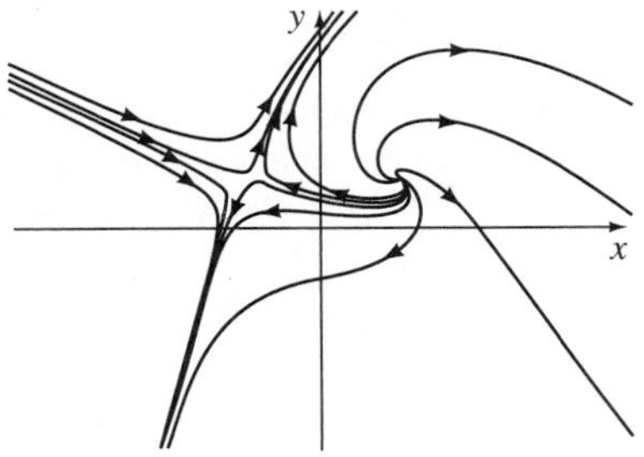

29. (1, 1) is a saddle point; (2, 1) is a saddle point; and $(\frac{3}{2}, \frac{3}{2})$ is an asymptotically stable spiral point.

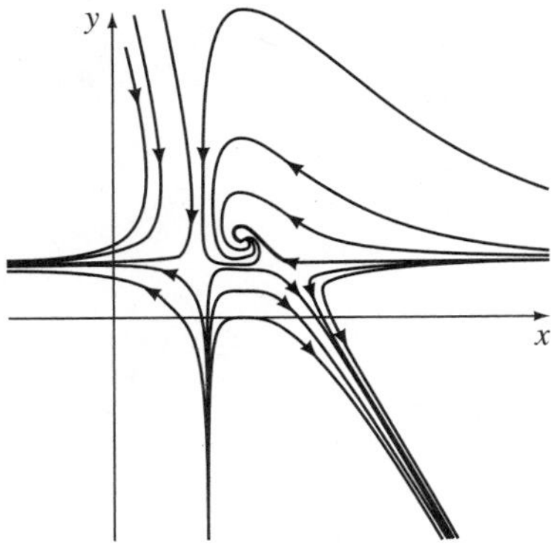

31. Critical points are $((2m+1)\pi/2, n\pi)$, $m = 0, \pm 1, \pm 2, \ldots$, $n = 0, \pm 1, \pm 2, \ldots$. On a vertical line $x = (2m+1)\pi/2$, the critical points alternate between centers and saddle points. For example, on $x = \pi/2$ $(m = 0)$, $(\pi/2, 2\pi)$ is a center, $(\pi/2, \pi)$ is a saddle point, $(\pi/2, 0)$ is a center, $(\pi/2, -\pi)$ is a saddle point, and so on.

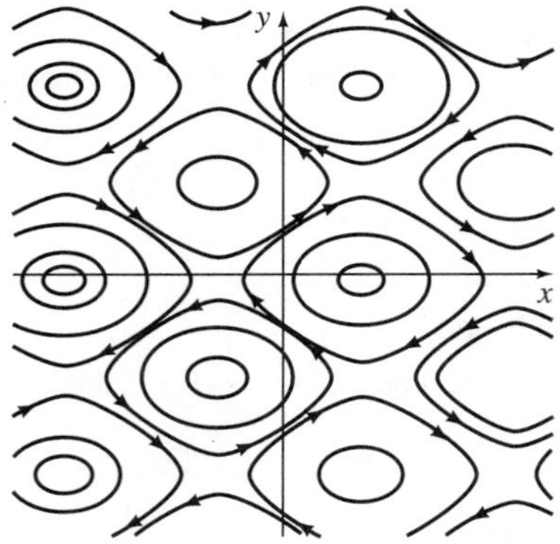

33. The only critical point is (0, 0). The phase portrait suggests that (0, 0) is stable but is not classifiable by type.

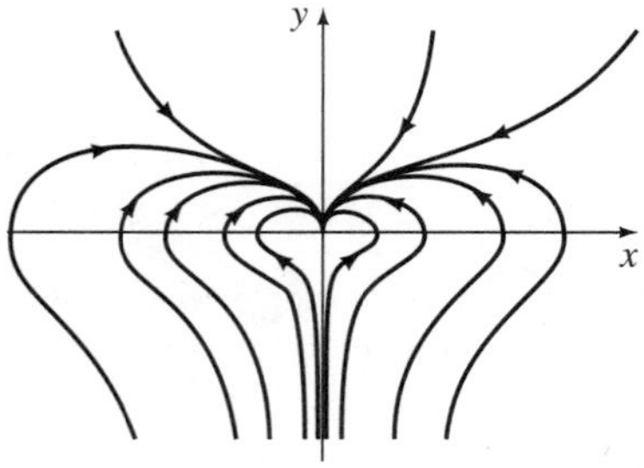

35. (a) The linearized system is

$$\frac{dx}{dt} = y$$

$$\frac{dy}{dt} = -x$$

(b) (0, 0) is a center.

(c) From the following phase portrait we see that (0, 0) is a center for the first system.

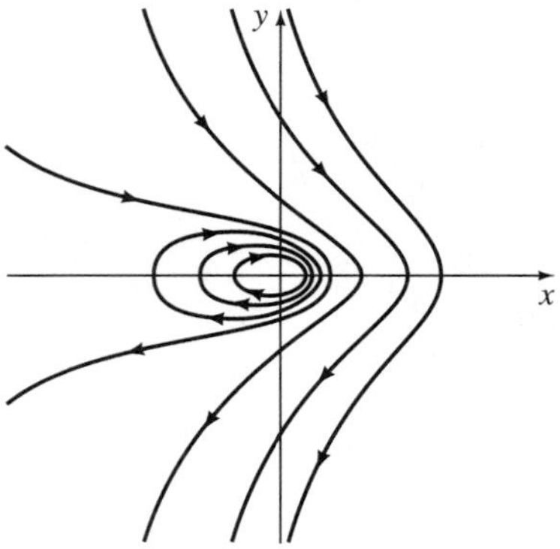

From the following phase portrait we see that (0, 0) is an asymptotically stable spiral point for the second system.

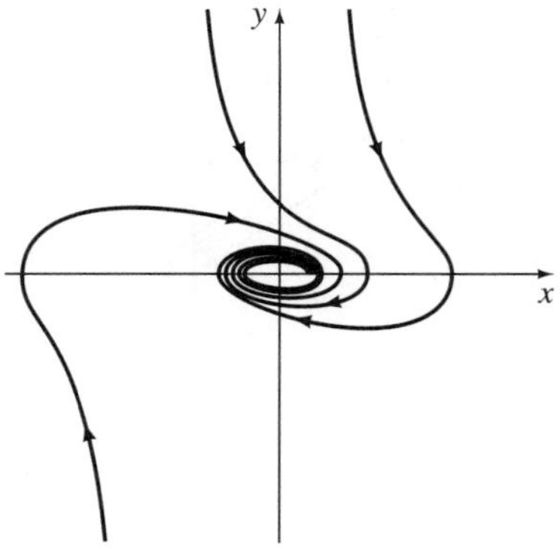

37. The following phase portrait suggests that the differential equation possesses no nonconstant periodic solutions.

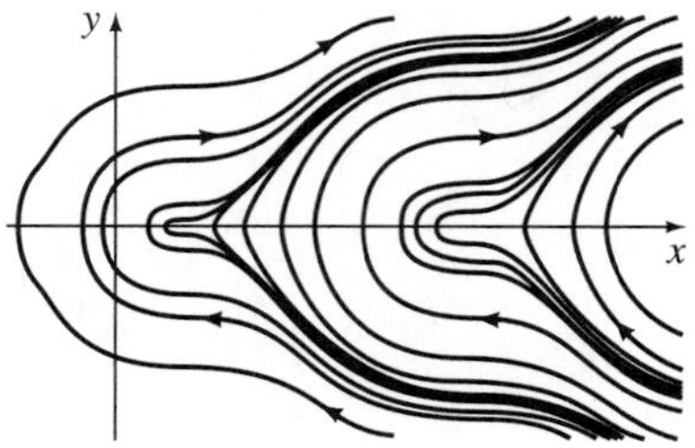

39. If a trajectory passes through (r_0, θ_0) when $t = 0$, then

$$r^2 = \frac{1}{\dfrac{1}{r_0^2} - 2t}, \quad \theta = t + \theta_0 \quad \text{or} \quad r^2 = \frac{1}{\dfrac{1}{r_0^2} - 2\theta + 2\theta_0}$$

describes a spiral curve. (0, 0) is an unstable spiral point.

43. (a) Critical point in the first quadrant is (5, 8).

(b)

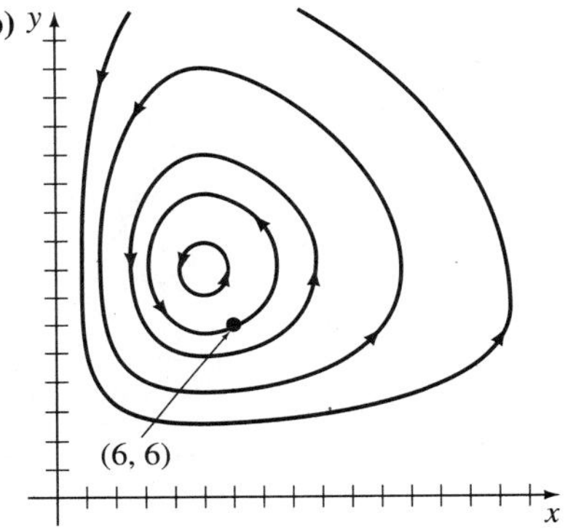

(c) Approximate period of $x(t)$, and $y(t)$ is $10\sqrt{2}\pi \approx 44.4$.

(d)

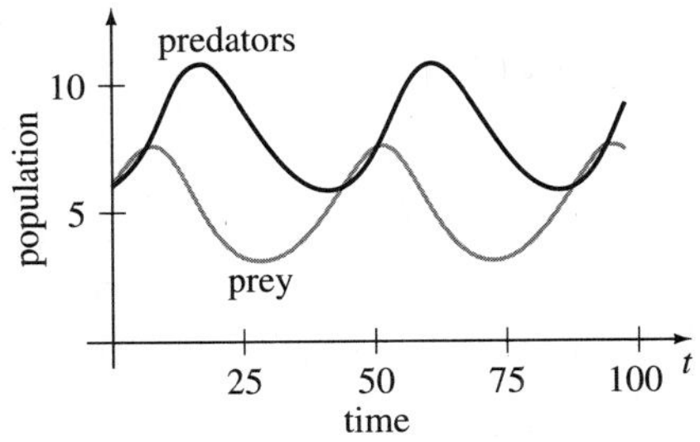

Populations are first equal at (approximately) $t = 6$. Approximate period is 45.

45. (a) Of the critical points (0, 0), (0, −2), (6, 0), and (4, 2) only (4, 2) is in the first quadrant. By linearization, verified by the phase portrait, we find that (4, 2) is an asymptotically stable spiral point. In other words, the populations $x(t)$ and $y(t)$ are not periodic in this model. After a long time both populations survive and approach equilibrium, that is, $x(t) \to 4$ and $y(t) \to 2$ as $t \to \infty$.

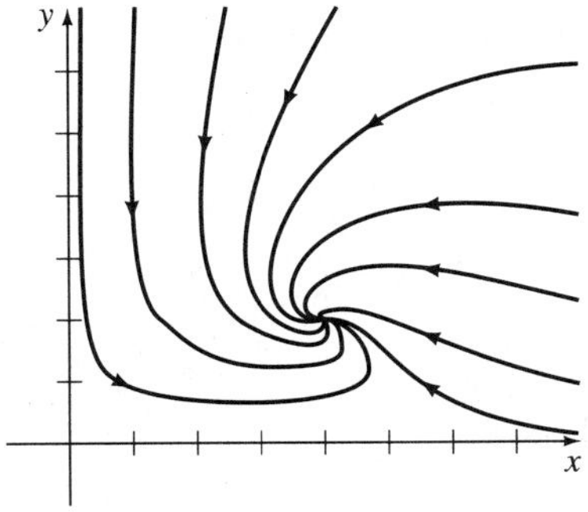

(b) None of the critical points (0, 0), (2, 0), and (0, −2) are in the first quadrant. Nothing is learned from linearization at (2, 0) but the phase portrait indicates that (2, 0) is an asymptotically stable node. Since $y(t) \to 0$ as one population (predators) becomes extinct, the other population (prey) survives because $x(t) \to 2$ as $t \to \infty$.

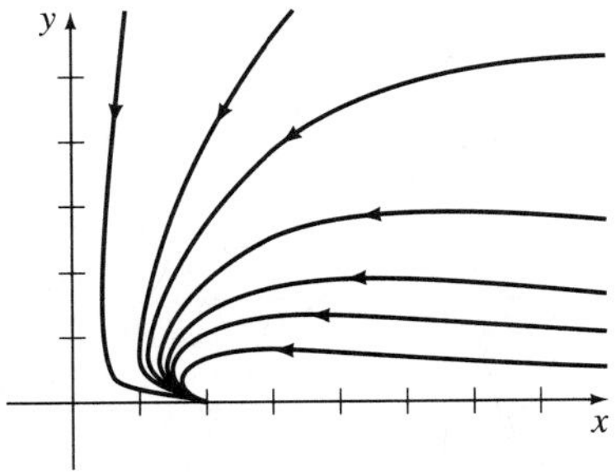

EXERCISES 5.1, Page 255

1. $\dfrac{2}{s}e^{-s} - \dfrac{1}{s}$ **3.** $\dfrac{1}{s^2} - \dfrac{1}{s^2}e^{-s}$ **5.** $\dfrac{1 + e^{-s\pi}}{s^2 + 1}$

7. $\dfrac{e^{-s}}{s} + \dfrac{e^{-s}}{s^2}$ **9.** $\dfrac{1}{s} - \dfrac{1}{s^2} + \dfrac{e^{-s}}{s^2}$ **11.** $\dfrac{e^7}{s - 1}$

13. $\dfrac{1}{(s-4)^2}$ **15.** $\dfrac{1}{s^2 + 2s + 2}$ **17.** $\dfrac{s^2 - 1}{(s^2 + 1)^2}$

19. $\dfrac{48}{s^5}$ **21.** $\dfrac{4}{s^2} - \dfrac{10}{s}$ **23.** $\dfrac{2}{s^3} + \dfrac{6}{s^2} - \dfrac{3}{s}$

25. $\dfrac{6}{s^4} + \dfrac{6}{s^3} + \dfrac{3}{s^2} + \dfrac{1}{s}$ **27.** $\dfrac{1}{s} + \dfrac{1}{s - 4}$

29. $\dfrac{1}{s} + \dfrac{2}{s - 2} + \dfrac{1}{s - 4}$ **31.** $\dfrac{8}{s^3} - \dfrac{15}{s^2 + 9}$

33. Use $\sinh kt = \dfrac{e^{kt} - e^{-kt}}{2}$ to show that

$$\mathscr{L}\{\sinh kt\} = \frac{k}{s^2 - k^2}.$$

35. $\dfrac{1}{2(s-2)} - \dfrac{1}{2s}$ **37.** $\dfrac{2}{s^2 + 16}$

EXERCISES 5.2, Page 261

1. $\frac{1}{2}t^2$ **3.** $t - 2t^4$ **5.** $1 + 3t + \frac{3}{2}t^2 + \frac{1}{6}t^3$
7. $t - 1 + e^{2t}$ **9.** $\frac{1}{4}e^{-t/4}$ **11.** $\frac{5}{7}\sin 7t$
13. $\cos(t/2)$ **15.** $\frac{1}{4}\sinh 4t$ **17.** $2\cos 3t - 2\sin 3t$
19. $\frac{1}{3} - \frac{1}{3}e^{-3t}$ **21.** $\frac{3}{4}e^{-3t} + \frac{1}{4}e^{t}$
23. $0.3e^{0.1t} + 0.6e^{-0.2t}$ **25.** $\frac{1}{2}e^{2t} - e^{3t} + \frac{1}{2}e^{6t}$
27. $-\frac{1}{3}e^{-t} + \frac{8}{15}e^{2t} - \frac{1}{5}e^{-3t}$ **29.** $\frac{1}{4}t - \frac{1}{8}\sin 2t$
31. $-\frac{1}{4}e^{-2t} + \frac{1}{4}\cos 2t + \frac{1}{4}\sin 2t$
33. $\frac{1}{3}\sin t - \frac{1}{6}\sin 2t$

EXERCISES 5.3, Page 270

1. $\dfrac{1}{(s-10)^2}$ **3.** $\dfrac{6}{(s+2)^4}$ **5.** $\dfrac{3}{(s-1)^2 + 9}$

7. $\dfrac{3}{(s-5)^2 - 9}$ **9.** $\dfrac{1}{(s-2)^2} + \dfrac{2}{(s-3)^2} + \dfrac{1}{(s-4)^2}$

11. $\dfrac{1}{2}\left[\dfrac{1}{s+1} - \dfrac{s+1}{(s+1)^2 + 4}\right]$ **13.** $\frac{1}{2}t^2e^{-2t}$

15. $e^{3t}\sin t$ **17.** $e^{-2t}\cos t - 2e^{-2t}\sin t$

19. $e^{-t} - te^{-t}$ **21.** $5 - t - 5e^{-t} - 4te^{-t} - \frac{3}{2}t^2e^{-t}$

23. $\dfrac{e^{-s}}{s^2}$ **25.** $\dfrac{e^{-2s}}{s^2} + 2\dfrac{e^{-2s}}{s}$ **27.** $\dfrac{s}{s^2+4}e^{-\pi s}$

29. $\dfrac{6e^{-s}}{(s-1)^4}$ **31.** $\frac{1}{2}(t-2)^2\mathscr{U}(t-2)$

33. $-\sin t\,\mathscr{U}(t-\pi)$ **35.** $\mathscr{U}(t-1) - e^{-(t-1)}\mathscr{U}(t-1)$

37. $\dfrac{s^2 - 4}{(s^2+4)^2}$ **39.** $\dfrac{6s^2 + 2}{(s^2-1)^3}$ **41.** $\dfrac{12s - 24}{[(s-2)^2 + 36]^2}$

43. $\frac{1}{2}t\sin t$ **45.** (c) **47.** (f) **49.** (a)

51. $f(t) = 2 - 4\mathscr{U}(t-3)$;

$$\mathscr{L}\{f(t)\} = \frac{2}{s} - \frac{4}{s}e^{-3s}$$

53. $f(t) = t^2\mathscr{U}(t-1)$

$= (t-1)^2\mathscr{U}(t-1) + 2(t-1)\mathscr{U}(t-1) + \mathscr{U}(t-1)$;

$$\mathscr{L}\{f(t)\} = 2\frac{e^{-s}}{s^3} + 2\frac{e^{-s}}{s^2} + \frac{e^{-s}}{s}$$

55. $f(t) = t - t\mathscr{U}(t-2)$

$= t - (t-2)\mathscr{U}(t-2) - 2\mathscr{U}(t-2)$;

$$\mathscr{L}\{f(t)\} = \frac{1}{s^2} - \frac{e^{-2s}}{s^2} - 2\frac{e^{-2s}}{s}$$

57. $f(t) = \mathscr{U}(t-a) - \mathscr{U}(t-b)$;

$$\mathscr{L}\{f(t)\} = \frac{e^{-as}}{s} - \frac{e^{-bs}}{s}$$

59.

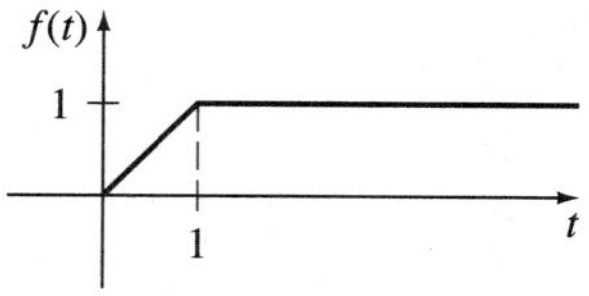

61. $\dfrac{e^{-t} - e^{3t}}{t}$

EXERCISES 5.4, Page 278

1. Since $f'(t) = e^t$, $f(0) = 1$, it follows from (1) that $\mathscr{L}\{e^t\} = s\mathscr{L}\{e^t\} - 1$. Solving gives $\mathscr{L}\{e^t\} = 1/(s-1)$.

3. $(s^2 + 3s)F(s) - s - 2$ **5.** $F(s) = \dfrac{2s - 1}{(s-1)^2}$

7. $\dfrac{1}{s(s-1)}$ **9.** $\dfrac{s+1}{s[(s+1)^2 + 1]}$ **11.** $\dfrac{1}{s^2(s-1)}$

13. $\dfrac{3s^2 + 1}{s^2(s^2+1)^2}$ **15.** $\dfrac{6}{s^5}$ **17.** $\dfrac{48}{s^8}$

19. $\dfrac{s-1}{(s+1)[(s-1)^2 + 1]}$ **21.** $\displaystyle\int_0^t f(\tau)e^{-5(t-\tau)}\,d\tau$

23. $1 - e^{-t}$ **25.** $-\frac{1}{3}e^{-t} + \frac{1}{3}e^{2t}$ **27.** $\frac{1}{4}t\sin 2t$

29. The result follows from letting $u = t - \tau$ in the first integral.

31. $\dfrac{(1 - e^{-as})^2}{s(1 - e^{-2as})} = \dfrac{1 - e^{-as}}{s(1 + e^{-as})}$ **33.** $\dfrac{a}{s}\left(\dfrac{1}{bs} - \dfrac{1}{e^{bs} - 1}\right)$

35. $\dfrac{\coth(\pi s/2)}{s^2 + 1}$ **37.** $\dfrac{1}{s^2 + 1}$

EXERCISES 5.5, Page 289

1. $y = -1 + e^t$

3. $y = te^{-4t} + 2e^{-4t}$

5. $y = \frac{4}{3}e^{-t} - \frac{1}{3}e^{-4t}$

7. $y = \frac{1}{9}t + \frac{2}{27} - \frac{2}{27}e^{3t} + \frac{10}{9}te^{3t}$

9. $y = \frac{1}{20}t^5e^{2t}$

11. $y = \cos t - \frac{1}{2}\sin t - \frac{1}{2}t\cos t$

13. $y = \frac{1}{2} - \frac{1}{2}e^t\cos t + \frac{1}{2}e^t\sin t$

15. $y = -\frac{8}{9}e^{-t/2} + \frac{1}{9}e^{-2t} + \frac{5}{18}e^t + \frac{1}{2}e^{-t}$

17. $y = \cos t$ **19.** $y = [5 - 5e^{-(t-1)}]\mathcal{U}(t-1)$

21. $y = -\frac{1}{4} + \frac{1}{2}t + \frac{1}{4}e^{-2t} - \frac{1}{4}\mathcal{U}(t-1)$
$-\frac{1}{2}(t-1)\mathcal{U}(t-1)$
$+\frac{1}{4}e^{-2(t-1)}\mathcal{U}(t-1)$

23. $y = \cos 2t - \frac{1}{6}\sin 2(t-2\pi)\mathcal{U}(t-2\pi)$
$+ \frac{1}{3}\sin(t-2\pi)\mathcal{U}(t-2\pi)$

25. $y = \sin t + [1 - \cos(t-\pi)]\mathcal{U}(t-\pi)$
$-[1 - \cos(t-2\pi)]\mathcal{U}(t-2\pi)$

27. $y = (e+1)te^{-t} + (e-1)e^{-t}$ **29.** $f(t) = \sin t$

31. $f(t) = -\frac{1}{8}e^{-t} + \frac{1}{8}e^t + \frac{3}{4}te^t + \frac{1}{4}t^2e^t$

33. $f(t) = e^{-t}$

35. $f(t) = \frac{3}{8}e^{2t} + \frac{1}{8}e^{-2t} + \frac{1}{2}\cos 2t + \frac{1}{4}\sin 2t$

37. $y = \sin t - \frac{1}{2}t\sin t$

39. $i(t) = 20{,}000[te^{-100t} - (t-1)e^{-100(t-1)}\mathcal{U}(t-1)]$

41. $q(t) = \dfrac{E_0C}{1 - kRC}(e^{-kt} - e^{-t/RC})$;

$$q(t) = \frac{E_0}{R}te^{-t/RC}$$

43. $q(t) = \frac{2}{5}\mathcal{U}(t-3) - \frac{2}{5}e^{-5(t-3)}\mathcal{U}(t-3)$

45. **(a)** $$i(t) = \frac{1}{101}e^{-10t} - \frac{1}{101}\cos t + \frac{10}{101}\sin t - \frac{10}{101}e^{-10(t-3\pi/2)}\mathcal{U}\left(t - \frac{3\pi}{2}\right) + \frac{10}{101}\cos\left(t - \frac{3\pi}{2}\right)\mathcal{U}\left(t - \frac{3\pi}{2}\right) + \frac{1}{101}\sin\left(t - \frac{3\pi}{2}\right)\mathcal{U}\left(t - \frac{3\pi}{2}\right)$$

(b) $i_{\max} \approx 0.1$ at $t \approx 1.6$,
$i_{\min} \approx -0.1$ at $t \approx 4.7$

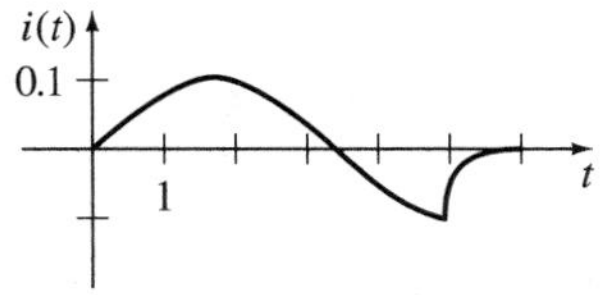

47. $$i(t) = \frac{t}{R} + \frac{L}{R^2}(e^{-Rt/L} - 1) + \frac{1}{R}\sum_{n=1}^{\infty}(e^{-R(t-n)/L} - 1)\mathcal{U}(t-n)$$

For $0 \le t < 2$,

$$i(t) = \begin{cases} \dfrac{t}{R} + \dfrac{L}{R^2}(e^{-Rt/L} - 1) & 0 \le t < 1 \\ \dfrac{t}{R} + \dfrac{L}{R^2}(e^{-Rt/L} - 1) + \dfrac{1}{R}(e^{-R(t-1)/L} - 1), & 1 \le t < 2 \end{cases}$$

49. $q(t) = \frac{3}{5}e^{-10t} + 6te^{-10t} - \frac{3}{5}\cos 10t$;
$i(t) = -60te^{-10t} + 6\sin 10t$; steady-state current is $6\sin 10t$

51. $$q(t) = \frac{E_0}{L\left(k^2 + \dfrac{1}{LC}\right)}[e^{-kt} - \cos(t/\sqrt{LC})] + \frac{kE_0\sqrt{C/L}}{k^2 + \dfrac{1}{LC}}\sin(t/\sqrt{LC})$$

53. $$x(t) = -\frac{3}{2}e^{-7t/2}\cos\frac{\sqrt{15}}{2}t - \frac{7\sqrt{15}}{10}e^{-7t/2}\sin\frac{\sqrt{15}}{2}t$$

55. $$y(x) = \frac{w_0L^2}{16EI}x^2 - \frac{w_0L}{12EI}x^3 + \frac{w_0}{24EI}x^4 - \frac{w_0}{24EI}\left(x - \frac{L}{2}\right)^4\mathcal{U}\left(x - \frac{L}{2}\right)$$

57. $$y(x) = \frac{w_0L^2}{48EI}x^2 - \frac{w_0L}{24EI}x^3 + \frac{w_0}{60EIL}\left[\frac{5L}{2}x^4 - x^5 + \left(x - \frac{L}{2}\right)^5\mathcal{U}\left(x - \frac{L}{2}\right)\right]$$

EXERCISES 5.6, Page 296

1. $y = e^{3(t-2)}\mathcal{U}(t-2)$ **3.** $y = \sin t + \sin t\,\mathcal{U}(t-2\pi)$

5. $y = -\cos t\,\mathcal{U}\left(t - \dfrac{\pi}{2}\right) + \cos t\,\mathcal{U}\left(t - \dfrac{3\pi}{2}\right)$

7. $y = \frac{1}{2} - \frac{1}{2}e^{-2t} + [\frac{1}{2} - \frac{1}{2}e^{-2(t-1)}]\mathcal{U}(t-1)$

9. $y = e^{-2(t-2\pi)}\sin t\,\mathcal{U}(t-2\pi)$

11. $y = e^{-2t}\cos 3t + \frac{2}{3}e^{-2t}\sin 3t + \frac{1}{3}e^{-2(t-\pi)}\sin 3(t-\pi)\mathcal{U}(t-\pi) + \frac{1}{3}e^{-2(t-3\pi)}\sin 3(t-3\pi)\mathcal{U}(t-3\pi)$

13. $$y(x) = \begin{cases} \dfrac{P_0}{EI}\left(\dfrac{L}{4}x^2 - \dfrac{1}{6}x^3\right), & 0 \le x < L/2 \\[2ex] \dfrac{P_0 L^2}{4EI}\left(\dfrac{1}{2}x - \dfrac{L}{12}\right), & L/2 \le x \le L \end{cases}$$

EXERCISES 5.7, Page 301

1. $x = -\frac{1}{3}e^{-2t} + \frac{1}{3}e^{t}$, $y = \frac{1}{3}e^{-2t} + \frac{2}{3}e^{t}$ **3.** $x = -\cos 3t - \frac{5}{3}\sin 3t$, $y = 2\cos 3t - \frac{7}{3}\sin 3t$

5. $x = -2e^{3t} + \frac{5}{2}e^{2t} - \frac{1}{2}$, $y = \frac{8}{3}e^{3t} - \frac{5}{2}e^{2t} - \frac{1}{6}$

7. $x = -\frac{1}{2}t - \frac{3}{4}\sqrt{2}\sin\sqrt{2}t$, $y = -\frac{1}{2}t + \frac{3}{4}\sqrt{2}\sin\sqrt{2}t$

9. $x = 8 + \dfrac{2}{3!}t^3 + \dfrac{1}{4!}t^4$, $y = -\dfrac{2}{3!}t^3 + \dfrac{1}{4!}t^4$

11. $x = \frac{1}{2}t^2 + t + 1 - e^{-t}$, $y = -\frac{1}{3} + \frac{1}{3}e^{-t} + \frac{1}{3}te^{-t}$

13. $x_1 = \dfrac{1}{5}\sin t + \dfrac{2\sqrt{6}}{15}\sin\sqrt{6}t + \dfrac{2}{5}\cos t - \dfrac{2}{5}\cos\sqrt{6}t$

$x_2 = \dfrac{2}{5}\sin t - \dfrac{\sqrt{6}}{15}\sin\sqrt{6}t + \dfrac{4}{5}\cos t + \dfrac{1}{5}\cos\sqrt{6}t$

15. **(b)** $i_2 = \frac{100}{9} - \frac{100}{9}e^{-900t}$, $i_3 = \frac{80}{9} - \frac{80}{9}e^{-900t}$
(c) $i_1 = 20 - 20e^{-900t}$

17. $i_2 = -\frac{20}{13}e^{-2t} + \frac{375}{1469}e^{-15t} + \frac{145}{113}\cos t + \frac{85}{113}\sin t$
$i_3 = \frac{30}{13}e^{-2t} + \frac{250}{1469}e^{-15t} - \frac{280}{113}\cos t + \frac{810}{113}\sin t$

19. $i_1 = \frac{6}{5} - \frac{6}{5}e^{-100t}\cos 100t$
$i_2 = \frac{6}{5} - \frac{6}{5}e^{-100t}\cos 100t - \frac{6}{5}e^{-100t}\sin 100t$

21. **(b)** $q = 50e^{-t}\sin(t-1)\mathcal{U}(t-1)$

EXERCISES 6.1, Page 314

1. $(-1, 1]$ **3.** $[-\frac{1}{2}, \frac{1}{2})$ **5.** $[2, 4]$

7. $(-5, 15)$ **9.** $\{0\}$

11. $x + x^2 + \frac{1}{3}x^3 - \frac{1}{30}x^5 + \cdots$

13. $x - \frac{2}{3}x^3 + \frac{2}{15}x^5 - \frac{4}{315}x^7 + \cdots$

15. $y = ce^{-x}$; $\quad y = c_0\displaystyle\sum_{n=0}^{\infty}\frac{(-1)^n}{n!}x^n$

17. $y = ce^{x^3/3}$; $\quad y = c_0\displaystyle\sum_{n=0}^{\infty}\frac{1}{n!}\left(\frac{x^3}{3}\right)^n$

19. $y = c/(1-x)$; $\quad y = c_0\displaystyle\sum_{n=0}^{\infty}x^n$

21. $y = C_1\cos x + C_2\sin x$;

$$y = c_0\sum_{n=0}^{\infty}\frac{(-1)^n}{(2n)!}x^{2n} + c_1\sum_{n=0}^{\infty}\frac{(-1)^n}{(2n+1)!}x^{2n+1}$$

23. $y = C_1 + C_2e^x$;

$$y = c_0 + c_1\sum_{n=1}^{\infty}\frac{x^n}{n!} = c_0 - c_1 + c_1\sum_{n=0}^{\infty}\frac{x^n}{n!}$$
$$= c_0 - c_1 + c_1e^x$$

25. $y(x) = \frac{1}{2} + \frac{1}{4}x - \frac{3}{8}x^2 + \frac{1}{16}x^3 + \cdots$

27. $y(x) = 2 - 3x - 5x^2 + \frac{13}{2}x^3 + \cdots$

EXERCISES 6.2, Page 323

1. $$y_1(x) = c_0\left[1 + \frac{1}{3\cdot 2}x^3 + \frac{1}{6\cdot 5\cdot 3\cdot 2}x^6 + \frac{1}{9\cdot 8\cdot 6\cdot 5\cdot 3\cdot 2}x^9 + \cdots\right]$$

$$y_2(x) = c_1\left[x + \frac{1}{4\cdot 3}x^4 + \frac{1}{7\cdot 6\cdot 4\cdot 3}x^7 + \frac{1}{10\cdot 9\cdot 7\cdot 6\cdot 4\cdot 3}x^{10} + \cdots\right]$$

3. $$y_1(x) = c_0\left[1 - \frac{1}{2!}x^2 - \frac{3}{4!}x^4 - \frac{21}{6!}x^6 - \cdots\right]$$

$$y_2(x) = c_1\left[x + \frac{1}{3!}x^3 + \frac{5}{5!}x^5 + \frac{45}{7!}x^7 + \cdots\right]$$

5. $y_1(x) = c_0\left[1 - \frac{1}{3!}x^3 + \frac{4^2}{6!}x^6 - \frac{7^2 \cdot 4^2}{9!}x^9 + \cdots\right]$

$y_2(x) = c_1\left[x - \frac{2^2}{4!}x^4 + \frac{5^2 \cdot 2^2}{7!}x^7 - \frac{8^2 \cdot 5^2 \cdot 2^2}{10!}x^{10} + \cdots\right]$

7. $y_1(x) = c_0;\quad y_2(x) = c_1 \sum_{n=1}^{\infty} \frac{1}{n}x^n$

9. $y_1(x) = c_0 \sum_{n=0}^{\infty} x^{2n};\quad y_2(x) = c_1 \sum_{n=0}^{\infty} x^{2n+1}$

11. $y_1(x) = c_0\left[1 + \frac{1}{4}x^2 - \frac{7}{4 \cdot 4!}x^4 + \frac{23 \cdot 7}{8 \cdot 6!}x^6 - \cdots\right]$

$y_2(x) = c_1\left[x - \frac{1}{6}x^3 + \frac{14}{2 \cdot 5!}x^5 - \frac{34 \cdot 14}{4 \cdot 7!}x^7 - \cdots\right]$

13. $y_1(x) = c_0[1 + \frac{1}{2}x^2 + \frac{1}{6}x^3 + \frac{1}{6}x^4 + \cdots]$
$y_2(x) = c_1[x + \frac{1}{2}x^2 + \frac{1}{2}x^3 + \frac{1}{4}x^4 + \cdots]$

15. $y(x) = -2\left[1 + \frac{1}{2!}x^2 + \frac{1}{3!}x^3 + \frac{1}{4!}x^4 + \cdots\right] + 6x$
$= 8x - 2e^x$

17. $y(x) = 3 - 12x^2 + 4x^4$

19. $y_1(x) = c_0[1 - \frac{1}{6}x^3 + \frac{1}{120}x^5 + \cdots]$
$y_2(x) = c_1[x - \frac{1}{12}x^4 + \frac{1}{180}x^6 + \cdots]$

21. $y_1(x) = c_0[1 - \frac{1}{2}x^2 + \frac{1}{6}x^3 - \frac{1}{40}x^5 + \cdots]$
$y_2(x) = c_1[x - \frac{1}{6}x^3 + \frac{1}{12}x^4 - \frac{1}{60}x^5 + \cdots]$

23. $y_1(x) = c_0\left[1 + \frac{1}{3!}x^3 + \frac{4}{6!}x^6 + \frac{7 \cdot 4}{9!}x^9 + \cdots\right]$

$+ c_1\left[x + \frac{2}{4!}x^4 + \frac{5 \cdot 2}{7!}x^7 + \frac{8 \cdot 5 \cdot 2}{10!}x^{10} + \cdots\right]$

$+ \frac{1}{2!}x^2 + \frac{3}{5!}x^5 + \frac{6 \cdot 3}{8!}x^8 + \frac{9 \cdot 6 \cdot 3}{11!}x^{11} + \cdots$

EXERCISES 6.3, Page 337

1. $x = 0$, irregular singular point

3. $x = -3$, regular singular point; $x = 3$, irregular singular point

5. $x = 0, 2i, -2i$, regular singular points

7. $x = -3, 2$, regular singular points

9. $x = 0$, irregular singular point; $x = -5, 5, 2$, regular singular points

11. $r_1 = \frac{3}{2},\ r_2 = 0;$

$y(x) = C_1x^{3/2}\left[1 - \frac{2}{5}x + \frac{2^2}{7 \cdot 5 \cdot 2}x^2 - \frac{2^3}{9 \cdot 7 \cdot 5 \cdot 3!}x^3 + \cdots\right]$

$+ C_2\left[1 + 2x - 2x^2 + \frac{2^3}{3 \cdot 3!}x^3 - \cdots\right]$

13. $r_1 = \frac{7}{8},\ r_2 = 0;$

$y(x) = C_1x^{7/8}\left[1 - \frac{2}{15}x + \frac{2^2}{23 \cdot 15 \cdot 2}x^2 - \frac{2^3}{31 \cdot 23 \cdot 15 \cdot 3!}x^3 + \cdots\right]$

$+ C_2\left[1 - 2x + \frac{2^2}{9 \cdot 2}x^2 - \frac{2^3}{17 \cdot 9 \cdot 3!}x^3 + \cdots\right]$

15. $r_1 = \frac{1}{3},\ r_2 = 0;$

$y(x) = C_1x^{1/3}\left[1 + \frac{1}{3}x + \frac{1}{3^2 \cdot 2}x^2 + \frac{1}{3^3 \cdot 3!}x^3 + \cdots\right]$

$+ C_2\left[1 + \frac{1}{2}x + \frac{1}{5 \cdot 2}x^2 + \frac{1}{8 \cdot 5 \cdot 2}x^3 + \cdots\right]$

17. $r_1 = \frac{5}{2},\ r_2 = 0;$

$y(x) = C_1x^{5/2}\left[1 + \frac{2 \cdot 2}{7}x + \frac{2^2 \cdot 3}{9 \cdot 7}x^2 + \frac{2^3 \cdot 4}{11 \cdot 9 \cdot 7}x^3 + \cdots\right]$

$+ C_2\left[1 + \frac{1}{3}x - \frac{1}{6}x^2 - \frac{1}{6}x^3 - \cdots\right]$

19. $r_1 = \frac{2}{3},\ r_2 = \frac{1}{3};$
$y(x) = C_1x^{2/3}[1 - \frac{1}{2}x + \frac{5}{28}x^2 - \frac{1}{21}x^3 + \cdots]$
$+ C_2x^{1/3}[1 - \frac{1}{2}x + \frac{1}{5}x^2 - \frac{7}{120}x^3 + \cdots]$

21. $r_1 = 1,\ r_2 = -\frac{1}{2};$

$y(x) = C_1x\left[1 + \frac{1}{5}x + \frac{1}{5 \cdot 7}x^2 + \frac{1}{5 \cdot 7 \cdot 9}x^3 + \cdots\right]$

$+ C_2x^{-1/2}\left[1 + \frac{1}{2}x + \frac{1}{2 \cdot 4}x^2 + \frac{1}{2 \cdot 4 \cdot 6}x^3 + \cdots\right]$

23. $r_1 = 0,\ r_2 = -1;$

$y(x) = C_1x^{-1} \sum_{n=0}^{\infty} \frac{1}{(2n)!}x^{2n} + C_2x^{-1} \sum_{n=0}^{\infty} \frac{1}{(2n+1)!}x^{2n+1}$

$= \frac{1}{x}[C_1 \cosh x + C_2 \sinh x]$

25. $r_1 = 4,\ r_2 = 0;$

$y(x) = C_1\left[1 + \frac{2}{3}x + \frac{1}{3}x^2\right] + C_2 \sum_{n=0}^{\infty} (n+1)x^{n+4}$

27. $r_1 = r_2 = 0$;

$$y(x) = C_1 y_1(x) + C_2 \left[y_1(x) \ln x + y_1(x) \times \left(-x + \frac{1}{4}x^2 - \frac{1}{3 \cdot 3!}x^3 + \frac{1}{4 \cdot 4!}x^4 - \cdots \right) \right],$$

where $y_1(x) = \sum_{n=0}^{\infty} \frac{1}{n!}x^n = e^x$

29. $r_1 = r_2 = 0$;
$$y(x) = C_1 y_1(x) + C_2[y_1(x) \ln x + y_1(x) \times (2x + \tfrac{5}{4}x^2 + \tfrac{23}{27}x^3 + \cdots)],$$

where $y_1(x) = \sum_{n=0}^{\infty} \frac{(-1)^n}{(n!)^{2n}} x^n$

EXERCISES 6.4, Page 344

1. $y = c_1 J_{1/3}(x) + c_2 J_{-1/3}(x)$

3. $y = c_1 J_{5/2}(x) + c_2 J_{-5/2}(x)$

5. $y = c_1 J_0(x) + c_2 Y_0(x)$

7. $y = c_1 J_2(3x) + c_2 Y_2(3x)$

9. After we use the change of variables, the differential equation becomes

$$x^2 w'' + xw' + (\lambda^2 x^2 - \tfrac{1}{4})w = 0.$$

Since the solution of the last equation is

$$w = c_1 J_{1/2}(\lambda x) + c_2 J_{-1/2}(\lambda x),$$

we find

$$y = c_1 x^{-1/2} J_{1/2}(\lambda x) + c_2 x^{-1/2} J_{-1/2}(\lambda x).$$

11. After substituting into the differential equation, we find

$$xy'' + (1 + 2n)y' + xy$$
$$= x^{-n-1}[x^2 J_n'' + xJ_n' + (x^2 - n^2)J_n]$$
$$= x^{-n-1} \cdot 0 = 0.$$

13. From Problem 10 with $n = \frac{1}{2}$ we find $y = x^{1/2} J_{1/2}(x)$; from Problem 11 with $n = -\frac{1}{2}$ we find $y = x^{1/2} J_{-1/2}(x)$.

15. From Problem 10 with $n = -1$ we find $y = x^{-1} J_{-1}(x)$; from Problem 11 with $n = 1$ we find $y = x^{-1} J_1(x)$ but since $J_{-1}(x) = -J_1(x)$, no new solution results.

17. From Problem 12 with $\lambda = 1$ and $v = \pm\frac{3}{2}$ we find $y = \sqrt{x} J_{3/2}(x)$ and $y = \sqrt{x} J_{-3/2}(x)$.

19. Using the hint, we can write

$$xJ_v'(x) = -v \sum_{n=0}^{\infty} \frac{(-1)^n}{n!\Gamma(1 + v + n)} \left(\frac{x}{2}\right)^{2n+v}$$
$$+ 2 \sum_{n=0}^{\infty} \frac{(-1)^n (n + v)}{n!(n + v)\Gamma(n + v)} \left(\frac{x}{2}\right)^{2n+v}$$
$$= -v \sum_{n=0}^{\infty} \frac{(-1)^n}{n!\Gamma(n + v + n)} \left(\frac{x}{2}\right)^{2n+v}$$
$$+ x \sum_{n=0}^{\infty} \frac{(-1)^n}{n!\Gamma(1 + v)} \left(\frac{x}{2}\right)^{2n+v-1}$$
$$= -vJ_v(x) + xJ_{v-1}(x).$$

21. Subtracting the equations

$$xJ_v'(x) = vJ_v(x) - xJ_{v+1}(x)$$
$$xJ_v'(x) = -vJ_v(x) + xJ_{v-1}(x)$$

gives $\quad 2vJ_v(x) = xJ_{v+1}(x) + xJ_{v-1}(x).$

23. **(b)** $y = c_1 x^{1/2} J_{1/3}(\frac{2}{3}\alpha x^{3/2}) + c_2 x^{1/2} J_{-1/3}(\frac{2}{3}\alpha x^{3/2})$

APPENDIX I EXERCISES, Page 364

1. **(a)** $\begin{pmatrix} 2 & 11 \\ 2 & -1 \end{pmatrix}$ **(b)** $\begin{pmatrix} -6 & 1 \\ 14 & -19 \end{pmatrix}$
(c) $\begin{pmatrix} 2 & 28 \\ 12 & -12 \end{pmatrix}$

3. **(a)** $\begin{pmatrix} -11 & 6 \\ 17 & -22 \end{pmatrix}$ **(b)** $\begin{pmatrix} -32 & 27 \\ -4 & -1 \end{pmatrix}$
(c) $\begin{pmatrix} 19 & -18 \\ -30 & 31 \end{pmatrix}$ **(d)** $\begin{pmatrix} 19 & 6 \\ 3 & 22 \end{pmatrix}$

5. **(a)** $\begin{pmatrix} 9 & 24 \\ 3 & 8 \end{pmatrix}$ **(b)** $\begin{pmatrix} 3 & 8 \\ -6 & -16 \end{pmatrix}$
(c) $\begin{pmatrix} 0 & 0 \\ 0 & 0 \end{pmatrix}$ **(d)** $\begin{pmatrix} -4 & -5 \\ 8 & 10 \end{pmatrix}$

7. **(a)** 180 **(b)** $\begin{pmatrix} 4 & 8 & 10 \\ 8 & 16 & 20 \\ 10 & 20 & 25 \end{pmatrix}$ **(c)** $\begin{pmatrix} 6 \\ 12 \\ -5 \end{pmatrix}$

9. **(a)** $\begin{pmatrix} 7 & 38 \\ 10 & 75 \end{pmatrix}$ **(b)** $\begin{pmatrix} 7 & 38 \\ 10 & 75 \end{pmatrix}$

11. $\begin{pmatrix} -14 \\ 1 \end{pmatrix}$ **13.** $\begin{pmatrix} -38 \\ -2 \end{pmatrix}$ **15.** singular

17. nonsingular; $\mathbf{A}^{-1} = \frac{1}{4}\begin{pmatrix} -5 & -8 \\ 3 & 4 \end{pmatrix}$

19. nonsingular; $\mathbf{A}^{-1} = \frac{1}{2}\begin{pmatrix} 0 & -1 & 1 \\ 2 & 2 & -2 \\ -4 & -3 & 5 \end{pmatrix}$

21. nonsingular; $\mathbf{A}^{-1} = -\frac{1}{9}\begin{pmatrix} -2 & -2 & -1 \\ -13 & 5 & 7 \\ 8 & -1 & -5 \end{pmatrix}$

23. $\det \mathbf{A}(t) = 2e^{3t} \neq 0$ for every value of t;

$$\mathbf{A}^{-1}(t) = \frac{1}{2e^{3t}}\begin{pmatrix} 3e^{4t} & -e^{4t} \\ -4e^{-t} & 2e^{-t} \end{pmatrix}$$

25. $\dfrac{d\mathbf{X}}{dt} = \begin{pmatrix} -5e^{-t} \\ -2e^{-t} \\ 7e^{-t} \end{pmatrix}$

27. $\dfrac{d\mathbf{X}}{dt} = 4\begin{pmatrix} 1 \\ -\frac{1}{4} & 1 \end{pmatrix} e^{2t} - 12\begin{pmatrix} 2 \\ 1 \end{pmatrix} e^{-3t}$

29. **(a)** $\begin{pmatrix} 4e^{4t} & -\pi \sin \pi t \\ 2 & 6t \end{pmatrix}$ **(b)** $\begin{pmatrix} \frac{1}{4}e^{8} - \frac{1}{4} & 0 \\ 4 & 6 \end{pmatrix}$

(c) $\begin{pmatrix} \frac{1}{4}e^{4t} - \frac{1}{4} & (1/\pi)\sin \pi t \\ t^2 & t^3 - t \end{pmatrix}$

31. $x = 3,\ y = 1,\ z = -5$

33. $x = 2 + 4t,\ y = -5 - t,\ z = t$

35. $x = -\frac{1}{2},\ y = \frac{3}{2},\ z = \frac{7}{2}$

37. $x_1 = 1,\ x_2 = 0,\ x_3 = 2,\ x_4 = 0$

41. $\lambda_1 = 6,\ \lambda_2 = 1,\ \mathbf{K}_1 = \begin{pmatrix} 2 \\ 7 \end{pmatrix},\ \mathbf{K}_2 = \begin{pmatrix} 1 \\ 1 \end{pmatrix}$

43. $\lambda_1 = \lambda_2 = -4,\ \mathbf{K}_1 = \begin{pmatrix} 1 \\ -4 \end{pmatrix}$

45. $\lambda_1 = 0,\ \lambda_2 = 4,\ \lambda_3 = -4,$

$$\mathbf{K}_1 = \begin{pmatrix} 9 \\ 45 \\ 25 \end{pmatrix},\ \mathbf{K}_2 = \begin{pmatrix} 1 \\ 1 \\ 1 \end{pmatrix},\ \mathbf{K}_3 = \begin{pmatrix} 1 \\ 9 \\ 1 \end{pmatrix}$$

47. $\lambda_1 = \lambda_2 = \lambda_3 = -2,$

$$\mathbf{K}_1 = \begin{pmatrix} 2 \\ -1 \\ 0 \end{pmatrix},\ \mathbf{K}_2 = \begin{pmatrix} 0 \\ 0 \\ 1 \end{pmatrix}$$

49. $\lambda_1 = 3i,\ \lambda_2 = -3i,$

$$\mathbf{K}_1 = \begin{pmatrix} 1 - 3i \\ 5 \end{pmatrix},\ \mathbf{K}_2 = \begin{pmatrix} 1 + 3i \\ 5 \end{pmatrix}$$

51. $\dfrac{d}{dt}\begin{pmatrix} a_{11}(t) & a_{12}(t) \\ a_{21}(t) & a_{22}(t) \end{pmatrix}\begin{pmatrix} x_1(t) \\ x_2(t) \end{pmatrix}$

$$= \frac{d}{dt}\begin{pmatrix} a_{11}(t)x_1(t) + a_{12}(t)x_2(t) \\ a_{21}(t)x_1(t) + a_{22}(t)x_2(t) \end{pmatrix}$$

$$= \begin{pmatrix} a_{11}(t)x_1'(t) + a_{11}'(t)x_1(t) + a_{12}(t)x_2'(t) + a_{12}'(t)x_2(t) \\ a_{21}(t)x_1'(t) + a_{21}'(t)x_1(t) + a_{22}(t)x_2'(t) + a_{22}'(t)x_2(t) \end{pmatrix}$$

$$= \begin{pmatrix} a_{11}(t)x_1'(t) + a_{12}(t)x_2'(t) + a_{11}'(t)x_1(t) + a_{12}'(t)x_2(t) \\ a_{21}(t)x_1'(t) + a_{22}(t)x_2'(t) + a_{21}'(t)x_1(t) + a_{22}'(t)x_2(t) \end{pmatrix}$$

$$= \begin{pmatrix} a_{11}(t) & a_{12}(t) \\ a_{21}(t) & a_{22}(t) \end{pmatrix}\begin{pmatrix} x_1'(t) \\ x_2'(t) \end{pmatrix} + \begin{pmatrix} a_{11}'(t) & a_{12}'(t) \\ a_{21}'(t) & a_{22}'(t) \end{pmatrix}\begin{pmatrix} x_1(t) \\ x_2(t) \end{pmatrix}$$

$$= \mathbf{A}(t)\mathbf{X}'(t) + \mathbf{A}'(t)\mathbf{X}(t)$$

53. Since $\mathbf{A}^{-1}$ exists, $\mathbf{AB} = \mathbf{AC}$ implies $\mathbf{A}^{-1}(\mathbf{AB}) = \mathbf{A}^{-1}(\mathbf{AC})$, $(\mathbf{A}^{-1}\mathbf{A})\mathbf{B} = (\mathbf{A}^{-1}\mathbf{A})\mathrm{C}$, $\mathbf{IB} = \mathbf{IC}$, or $\mathbf{B} = \mathbf{C}$.

55. No, since in general $\mathbf{AB} \neq \mathbf{BA}$.

Index

A page number in parentheses refers to the Computer Lab Manual

TABLE OF INTEGRALS

1. $\int u\,dv = uv - \int v\,du$

2. $\int u^n\,du = \frac{1}{n+1}u^{n+1} + C, n \neq -1$

3. $\int \frac{du}{u} = \ln|u| + C$

4. $\int e^u\,du = e^u + C$

5. $\int a^u\,du = \frac{1}{\ln a}a^u + C$

6. $\int \sin u\,du = -\cos u + C$

7. $\int \cos u\,du = \sin u + C$

8. $\int \sec^2 u\,du = \tan u + C$

9. $\int \csc^2 u\,du = -\cot u + C$

10. $\int \sec u \tan u\,du = \sec u + C$

11. $\int \csc u \cot u\,du = -\csc u + C$

12. $\int \tan u\,du = -\ln|\cos u| + C$

13. $\int \cot u\,du = \ln|\sin u| + C$

14. $\int \sec u\,du = \ln|\sec u + \tan u| + C$

15. $\int \csc u\,du = \ln|\csc u - \cot u| + C$

16. $\int \frac{du}{\sqrt{a^2 - u^2}} = \sin^{-1}\frac{u}{a} + C$

17. $\int \frac{du}{a^2 + u^2} = \frac{1}{a}\tan^{-1}\frac{u}{a} + C$

18. $\int \frac{du}{u\sqrt{u^2 - a^2}} = \frac{1}{a}\sec^{-1}\frac{u}{a} + C$

19. $\int \frac{du}{a^2 - u^2} = \frac{1}{2a}\ln\left|\frac{u+a}{u-a}\right| + C$

20. $\int \frac{du}{u^2 - a^2} = \frac{1}{2a}\ln\left|\frac{u-a}{u+a}\right| + C$

21. $\int \sin^2 u\,du = \frac{1}{2}u - \frac{1}{4}\sin 2u + C$

22. $\int \cos^2 u\,du = \frac{1}{2}u + \frac{1}{4}\sin 2u + C$

23. $\int \tan^2 u\,du = \tan u - u + C$

24. $\int \cot^2 u\,du = -\cot u - u + C$

25. $\int \sin^3 u\,du = -\frac{1}{3}(2 + \sin^2 u)\cos u + C$

26. $\int \cos^3 u\,du = \frac{1}{3}(2 + \cos^2 u)\sin u + C$

27. $\int \tan^3 u\,du = \frac{1}{2}\tan^2 u + \ln|\cos u| + C$

28. $\int \cot^3 u\,du = -\frac{1}{2}\cot^2 u - \ln|\sin u| + C$

29. $\int \sec^3 u\,du = \frac{1}{2}\sec u \tan u + \frac{1}{2}\ln|\sec u + \tan u| + C$

30. $\int \csc^3 u\,du = -\frac{1}{2}\csc u \cot u + \frac{1}{2}\ln|\csc u - \cot u| + C$

31. $\int \sin^n u\,du = -\frac{1}{n}\sin^{n-1}u\cos u + \frac{n-1}{n}\int \sin^{n-2}u\,du$

32. $\int \cos^n u\,du = \frac{1}{n}\cos^{n-1}u\sin u + \frac{n-1}{n}\int \cos^{n-2}u\,du$

33. $\int \tan^n u\,du = \frac{1}{n-1}\tan^{n-1}u - \int \tan^{n-2}u\,du$

34. $\int \cot^n u\,du = \frac{-1}{n-1}\cot^{n-1}u - \int \cot^{n-2}u\,du$

35. $\int \sec^n u\,du = \frac{1}{n-1}\tan u \sec^{n-2}u + \frac{n-2}{n-1}\int \sec^{n-2}u\,du$

36. $\int \csc^n u\,du = \frac{-1}{n-1}\cot u \csc^{n-2}u + \frac{n-2}{n-1}\int \csc^{n-2}u\,du$

37. $\int \sin au \sin bu\,du = \frac{\sin(a-b)u}{2(a-b)} - \frac{\sin(a+b)u}{2(a+b)} + C$

38. $\int \cos au \cos bu\,du = \frac{\sin(a-b)u}{2(a-b)} + \frac{\sin(a+b)u}{2(a+b)} + C$

39. $\int \sin au \cos bu\,du = -\frac{\cos(a-b)u}{2(a-b)} - \frac{\cos(a+b)u}{2(a+b)} + C$

40. $\int u \sin u\,du = \sin u - u\cos u + C$

41. $\int u \cos u\,du = \cos u + u\sin u + C$

42. $\int u^n \sin u\,du = -u^n\cos u + n\int u^{n-1}\cos u\,du$

43. $\int u^n \cos u\,du = u^n\sin u - n\int u^{n-1}\sin u\,du$

44. $\int \sin^n u \cos^m u\,du = -\frac{\sin^{n-1}u\cos^{m+1}u}{n+m} + \frac{n-1}{n+m}\int \sin^{n-2}u\cos^m u\,du = \frac{\sin^{n+1}u\cos^{m-1}u}{n+m} + \frac{m-1}{n+m}\int \sin^n u\cos^{m-2}u\,du$